D1203889

Statistical Cost Estimation

JOEL DEAN

Statistical Cost Estimation

INDIANA UNIVERSITY PRESS
Bloomington London

Published in Canada by Fitzhenry & Whiteside Limited, Don Mills, Ontario

Manufactured in the United States of America

Library of Congress Cataloging in Publication Data
Dean, Joel, 1906-
Statistical cost estimation.
Includes index.
1. Costs, Industrial–Statistical methods. 2. Cost–Statistical methods. I. Title
HD47.D4 658.1'552 74-378
ISBN 0-253-3544-5 1 2 3 4 5 80 79 78 77 76

Contents

Preface

The function of this book is to make accessible nine of my early statistical cost studies; seven of the studies have been out of print for decades and two are published here for the first time. To carry out this mission of documenting the prehistory of what is now called "Operations Research" and "Management Science," my published studies were reprinted without change, even though some repetition resulted. Previously unpublished studies were condensed and edited only enough to make the prose publishable.

Cost has many determinants. The two that have been most important in economic theory are rate of output and size of plant (or firm). Others have, of course, been recognized by economists, but they have usually been viewed as contaminants—forces to assume constant in order to examine the two critical cost determinants: output rate and plant size. To test empirically accepted economic theory about these two kinds of cost functions was the main purpose of the first six statistical cost studies. A second and implementary purpose was to measure the effects on cost behavior of those other determinants whose influence had to be held constant in order to isolate the relation of cost to output rate and to plant size.

The four studies in Part One concern the relation of cost to output in the short run (i.e., when adaptation to output rate is incomplete, being constrained by inability quickly to change plant equipment or managerial cadres). The orthodox cost–output hypothesis is a building block of accepted economic theory about how prices are set, how firms compete, and how society's resources are allocated. Three of the statistical cost studies test hypotheses about short-run relation of production cost to output: "Furniture Factory" (Study 1), "Hosiery Mill" (Study 2), and "Belt Shop" (Study 3).

Economic theory was built on hypotheses about the behavior of production costs. Selling costs were neglected, partly in blindness and partly in the belief that they were features of ephemeral monopoly that were inimical to consumer welfare and unimportant in a body of theory designed to explain the working of an enterprise system. The managerial importance of persuasion costs to a modern corporation is undeniable, and economists in the last 30 or 40 years have grudgingly grafted hypotheses about the behavior and function of selling costs onto the theory of the firm. "Department Store," the fourth study in Part One, concerns short-run selling costs mixed with pure production costs.

Removal of the constraints on complete adaptation of plant size to rate of output produces economic hypotheses about another kind of relation of cost to output;

these are tested by the studies in Part Two, "Cost and Plant Size." Economists have labeled this cost relation "long run," meaning after enough adjustment time to permit complete adaptation of all factors to rate of output. Two cost studies that test hypotheses about the relation of cost to size of plant are "Finance Chain" (Study 5) and "Shoe Chain" (Study 6). Like the department store study, these studies also examine selling costs mixed with production costs.

Part Three concerns the relation of cost to plant location. The geographic structure of wage rates within a nation is basic to the economic theory of regional trade and development. "Drug Chain" (Study 7) examines the topography of labor cost. Here the interest in discovering statistical "laws" concerning the geographic structure of wages was managerial: to make better decisions about pay and perquisites of the retail employees of a nationwide chain by economic enlightenment about the geographic structure of market prices of labor.

Profits, as reported by the accountant, are the primary measure of the success of a business firm. Yet during inflation the accountant's measurement of profit seriously overstates real profit. The professional manager, trying to maximize the real wealth of the shareholder and also his score in Wall Street as measured by accounted profits, has perplexing problems in developing forecasts of profits and costs that are pertinent to the economic choices he must make in pursuing these two conflicting goals. Inflation undermines efficiency by scrambling the signals of a competitive economy, frustrating its semiautomatic adjustments, and distorting the optimization decisions of managers of individual firms. The supply of capital for corporate investment, which is the basis for technical advance, is reduced by inflation because real tax rates of corporations and of individuals greatly exceed purported rates. Thus the power to save, as well as the incentives, are reduced for both individuals and corporations. Part Four, "Cost and Real Profits," contains two statistical studies of cost in relation to the measurement and prediction of a corporation's real profit, as contrasted with its accounted profit: "Electronics Producer" (Study 8) and "Machinery Manufacturer" (Study 9).

Each of the four parts is prefaced by an introductory essay in which I have sketched the theoretical background of the hypotheses tested by the research, alternative research methods, some of the problems encountered in the methods used, the salient findings, and their managerial usefulness. Partly because the findings of some of these statistical cost studies ran counter to accepted economic theory, they received considerable criticism. Although I tried to benefit from the criticism, I have not previously replied to it. In these introductory essays it now seems appropriate to do so.

J.D.

Hastings-on-Hudson
August 1975

Acknowledgments

I am indebted to many individuals and institutions for their help either in the original preparation of these studies or in connection with their publication in the present volume. I would like, first of all, to acknowledge the generous contributions to the original studies made by the following:

"Furniture Factory": Theodore O. Yntema, James B. Moffat, Garfield Cox, Henry Schultz, Jacob Viner, C. Lawrence Christenson, William Cleveland, Edward Duddy, Martin Freeman, Willard Graham, Samuel Nerlove, Benjamin Caplan, and Sune Carlsen gave guidance, advice, and encouragement; Mary Yellen, Emma Peterson, and Rondal Huffman contributed valuable technical assistance.

"Hosiery Mill": Edward Duddy, Roy W. Jastram, John H. Smith, Jacob Viner, Mary Hilton Wise, and Theodore O. Yntema read the manuscript and made helpful suggestions; Phyllis van Dyk, R. Warren James, Emma Manning, Mary Jo Lawley, Kathryn Withers, and Forrest Danson helped with the statistical calculations; the Cowles Commission for Research in Economics, the University of Chicago, and the National Bureau of Economic Research provided financial assistance.

"Belt Shop": Moses Abramovitz, Arthur F. Burns, Solomon Fabricant, Milton Friedman, Edward Mason, Frederic Mills, Leo Wollman, and Theodore O. Yntema read the manuscript and offered helpful suggestions; Phyllis van Dyk and the computational staff of the National Bureau of Economic Research helped in the preparation of the study.

"Department Store": Phyllis van Dyk assisted in carrying out the study; Mary Hilton Wise helped revise the manuscript for publication; Arthur F. Burns, R. Warren James, John H. Smith, Gerhard Tintner, and Theodore O. Yntema read the manuscript and made valuable suggestions; the Social Science Research Committee of the University of Chicago and the National Bureau of Economic Research provided funds for conducting the study.

"Finance Chain": Theodore O. Yntema supervised this research as a part of my Ph.D. dissertation, made arrangements with the finance chain to supply the cost data, and gave inspiration and counsel at all stages of the investigation.

"Shoe Chain:" Moses Abramovitz, George H. Brown, William Cooper, Edward Duddy, Solomon Fabricant, R. S. Howey, Leonid Hurwicz, Roy W. Jastram, John H. Smith, Phyllis van Dyk, Theodore O. Yntema, and Mary Hilton Wise read the manuscript and offered many helpful suggestions; Forrest Danson, Mary Jo Lawley, Emma Manning, and Kathryn Withers assisted with typing, computing, and graphics. The Cowles Commission for Research in Economics and the Social

Science Research Committee of the University of Chicago provided financial assistance. R. Warren James was coauthor.

"Drug Chain": Stanley Balmer, Justin Dart, and Marvin Bower read the original report and made helpful suggestions. Much of the research for this paper was conducted while I was associated with McKinsey & Company; the findings are presented here with their kind permission.

"Electronics Producer": Stephen Taylor helped in carrying out the study and preparing the report; the staff of the Treasurer of IBM provided the data and made computations; Albert Williams, Thomas J. Watson, Jr., James Lorie, and Winfield Smith read the draft and suggested revisions.

"Machinery Manufacturer": Stephen Taylor and Charles Dean assisted in conducting this study; James Bonbright, Carl Kaysen, Theodore Kiendl, Charles Stockton, and Charles Wyzansky read the manuscript and made encouraging suggestions. I am grateful to the company for help in conducting the research and for permission to publish the findings.

I should like to express particular thanks to the executives of the following companies for making available the data for study, for helping me to understand the company's operations, for permission to publish the research, initially without revealing the name of the company, and now, 20 to 40 years later, for their explicit permission to lift the veil of anonymity: "Furniture Factory," Showers Brothers Furniture Company, Bloomington, Indiana; "Hosiery Mill," Gotham Hosiery Mills, New York, N.Y.; "Belt Shop," Graton and Knight, Worcester, Mass.; "Department Store," Lord & Taylor, New York, N.Y.; "Finance Chain," Household Finance Corporation; "Shoe Chain," Melville Shoe Corporation; "Drug Chain," Rexall Drug Stores; "Electronics Producer," International Business Machines Corporation.

In assembling the present volume, I am indebted to Helen Kielkucki for conscientiously typing and retyping the introduction and the unpublished studies and for managing the venture, to Louise Marinis for editing the introductions and the previously unpublished studies, to Janet Rabinowitch for shepherding the manuscript through the Indiana University Press and for suggesting many improvements, to Herman B Wells, Chancellor, Indiana University, for his personal interest in this project because statistical cost curves were born in Bloomington, and to the Indiana University Foundation for administrative assistance in financing this publication. Without the help of all those mentioned, the studies could not have been made and the book could not have been published.

Finally, I wish to acknowledge my gratitude to the journals and publishers who graciously permitted reproduction of all or parts of the following articles and monographs:

"Furniture Factory," reprinted from *Statistical Determination of Cost, with Special Reference to Marginal Cost,* Studies in Business Administration, vol. 7, no. 1 (Chicago: University of Chicago Press, 1936), by permission of the publisher. Copyright 1936 by The University of Chicago.

"Appendix C" of "Furniture Factory," reprinted from "Correlation Analysis of Cost Variation," *Accounting Review*, vol. 12, no. 1 (March 1937) by permission of *Accounting Review*.

"Hosiery Mill," reprinted from *Statistical Cost Functions of a Hosiery Mill*, Studies in Business Administration, vol. 11, no. 4 (Chicago: University of Chicago Press, 1941), by permission of the publisher. Copyright 1941 by The University of Chicago.

"Belt Shop," reprinted from *The Relation of Cost to Output for a Leather Belt Shop*, by permission of the National Bureau of Economic Research. Copyright 1941 by National Bureau of Economic Research, Inc.

"Department Store," reprinted from "Department-Store Cost Functions," in *Studies in Mathematical Economics and Econometrics*, ed. Oskar Lange et al. (Chicago: University of Chicago Press, 1942), by permission of the publisher. Copyright 1942 by The University of Chicago Press.

"Shoe Chain," reprinted from Joel Dean and R. Warren James, *The Long-Run Behavior of Costs in a Chain of Shoe Stores: A Statistical Analysis*, Studies in Business Administration, vol. 12, no. 3 (Chicago: University of Chicago Press, 1941), by permission of the publisher. Copyright 1942 by The University of Chicago.

"Drug Chain," reprinted from "Geographical Salary Administration," in *Plant-Wide and Geographical Salary Administration*, American Management Association, Personnel Series, no. 114 (New York, 1947), by permission of American Management Association. Copyright 1947 by American Management Association.

"Machinery Manufacturer," reprinted from "Measurement of Real Economic Earnings of a Machinery Manufacturer," *Accounting Review*, vol. 29, no. 2 (April 1954), by permission of *Accounting Review*.

Portions of "Introduction to Part One" were adapted from Joel Dean, *Managerial Economics* © 1951. Reprinted by permission of Prentice-Hall, Inc., Englewood Cliffs, New Jersey.

PART ONE

Cost and Output Rate

Introduction to Part One

The statistical cost studies in Part One examine what happens to cost when the rate of utilization of fixed plant is changed. "Furniture Factory" (Study 1), "Hosiery Mill" (Study 2), and "Belt Shop" (Study 3) concern the short-run behavior of manufacturing costs. "Department Store" (Study 4) examines the short-run behavior of operating costs that are a composite of pure production costs and selling costs.

I. Hypothesis Tested

The purpose of these studies was to test the traditional cost–output hypothesis of economic theory, which shows the behavior of cost when plant size cannot be changed and when adaptation to rate of output is therefore incomplete. This hypothesis is diagramed in Chart One–1. In the right section of the chart, average

Chart One–1 Relation of Output Rate to Cost

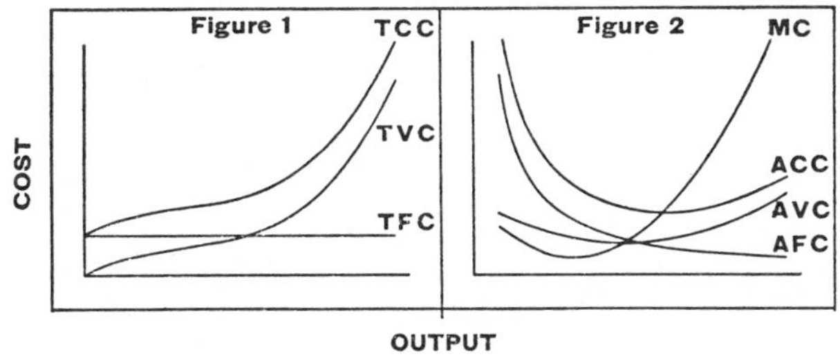

fixed cost per unit (*AFC*) declines in a rectangular hyperbola as the productive services of inputs that are fixed in total in the short run get utilized more fully and costs are spread thinner, with the increase in rate of output. Average variable cost per unit (*AVC*) is portrayed as first dropping at a declining rate and then rising at an increasing rate, with increases in rate of output. Average combined cost (*ACC*) is the sum of fixed and variable costs per unit, so it is U-shaped, with the declining phase caused mainly by spreading fixed overheads thinner and a rising phase caused by average variable cost going up at an accelerating rate, as output is increased. Marginal cost (*MC*) declines briefly and then rises continuously at an increasing rate over virtually the entire range of output.

By transforming the curves of cost per unit into corresponding curves of cost per month, we get the set of total cost curves diagramed in the left section of Chart One–1. Total fixed cost (*TFC*) is the same for all outputs and hence is horizontal. Total variable cost (*TVC*) rises (from zero) first at a declining rate, then at a constant rate, and then at an increasing rate. Summing the two, we get total combined cost (*TCC*), which rises from the intercept of *TFC* (at zero output), first briefly at a declining rate, then momentarily at a constant rate, then at a continuously increasing rate over almost the entire range of output—this concave section is of interest to economic theorists. This hypothesis is the set of short-run cost curves portrayed in economic textbooks. Rising short-run marginal (and average) cost of production is a building block of economic theory about how prices are set, how firms compete, and how society's resources are allocated.

Assumptions

What does the traditional hypothesis assume? It assumes, for simplicity, that output is a single homogeneous product, that a single input factor is variable and all other input factors are fixed, and that the law of diminishing product increments (diminishing returns) is operative.[1]

The hypothesis also assumes that the prices of input services are constant in the sense of being unaffected by their rate of use by the firm (elastic supply under conditions of perfect competition). This assumption rules out indirect increases in input prices in the form of degradation of their quality or premiums for overtime or for working on second or third shifts: thus rising marginal costs from these sources are excluded.[2]

Rationale

Economic theory purports to validate the accepted short-run cost curve model by revealing as its substructure the production function, which is stated with such mathematical elegance that it has been widely accepted as proof. This theoretical production function is a rigorously specified relationship between the factory's rate of output and the rates of input of productive services (machine hours, materials, labor hours, etc.), all measured in physical terms. It assumes that management is perfect, i.e., that executives will always achieve the maximum rate of output that is technologically possible with the specified rates of use of the productive services. It also assumes that those inputs which are not fixed in total can be varied continuously (i.e., by tiny increments) either individually or together, and that the resulting output will also increase smoothly by tiny increments. Finally, the model assumes the possibility of continuous substitution of one input service for another productive service without gaps or discontinuities in the input mix or the resultant output.

The production function that underlies the *short-run* cost curve specifies that the productive services of some of the inputs (e.g., buildings and machinery) cannot be changed. Thus the cost curve shows (under the perfect-management assumption) the best that can be done with incomplete adaptation of all the input services to the rate of output. This incompleteness has degrees and forms a spectrum. To speak of *the* short run is a conventional simplification. At one extreme is total adaptation to output rate, which we call the "long run."[3] All the rest of the spectrum of incomplete adaptation is labeled "short run."

Defects

The hypothesis of conventional theory has come under attack both on its own theoretical ground and by the discoveries of managerial economics. The theoretical difficulties are several. First, the assumptions that underlie the law of diminishing product increments are inappropriate for modern manufacturing technology and hence highly refutable. Second, the asserted behavior of short-run cost is not a relentless result of the assumptions: it is not really proven by going beneath the cost curves to the short-run production function. Third, the theory fails to define the fixed factor with adequate precision. Fourth, it neglects several important dimensions of output and generally refuses to cope with the complexity of multiple products, despite the fact that today homogeneous single-product output is a rarity.

The law of diminishing product increments is not (as is generally supposed) based upon an extensive body of validated technological data descriptive of modern industry and examined and authenticated by economists. Rather, it was derived mainly from casual observation of eighteenth-century English agriculture. Over the years, it has acquired the underpinning of a precisely stateable but nevertheless theoretical (empirically unproven) production function. The appropriateness today of the assumptions that underlie this model is highly refutable. Even if these assumptions are granted, the theoretical analysis does not establish the law of diminishing product increments as a necessary, logical conclusion from the assumptions.[4]

The rationale for the short-run cost behavior hypothesis of conventional theory is incompatible with the empirical discoveries of managerial economics in several important respects. The concept of "the fixed factor" must be redefined both more comprehensively and more adaptively than is implied by theory. Because of several kinds of plant segmentation, the presumed invariableness and indivisibility of the fixed factor are simply inappropriate for much of modern industrial technology. Moreover, the assumed substitutability of the fixed input for variable inputs in the process of changing the rate of output is not usually found today. Finally, in addition to neglecting multiproducts, the accepted theory ignores several other important dimensions of output, for example, variety of product, mix of products, size of manufacturing lot, and unfamiliarity of products (position on learning curve). Despite its frailties, this is the accepted hypothesis of economic theory and the one to test.[5]

Strictly, this hypothesis applies only to costs of production, not to selling costs. The first three studies of Part One ("Furniture Factory," "Hosiery Mill," and "Belt Shop") concern manufacturing costs only; for them the applicability of the hypothesis is clear. On the surface, its relevance to the fourth study (the hosiery, shoe, and coat departments of the Lord & Taylor Department Store) appears dubious because included here are selling costs as well as what are, from an economic viewpoint, production costs. On closer examination, it is clear that the department store study validly tests the traditional hypothesis. A high proportion of the activity of each of the three departments involves essentially production costs: moving products through the channels of distribution and supplying a sales service that is an ingredient in the customer-delivered "product." A brief consideration of the nature of selling costs supports this conclusion.

Selling Costs

Selling costs are all outlays designed to increase or maintain sales, i.e., to shift to the right the firm's price–quantity demand schedule for a product: selling costs augment demand; production costs serve demand. Thus costs incurred not only to fabricate the product but also to move it physically into the hands of the customer are not true selling costs except insofar as they additionally increase sales. Therefore the costs of a department store are a mix of pure selling and pure production costs. Separation is made more difficult because persuasion cost has many facets: the design, styling, and packaging of the product itself, advertising, display, demonstration, personal selling, and range of merchandise choice plus delivery, credit terms, and follow-up services. These complementary ingredients in the marketing mix are substitutes at the edges in augmenting sales, yet some of them also contain an element of pure production cost.

Most of the effects of department store advertising and other persuasion outlays are immediate, although some have delayed and cumulative impacts. Hence the hypothesis to be tested for selling costs is the short-run relationship between aggregate outlays on persuasion and the volume of sales. This relationship is diagramed in Chart One–2. For clarity, it is confined to a pure form of persuasion: direct-mail advertising of a book. The example assumes that advertising is the sole or main form of promotion, that the price remains the same, that seasonal and cyclical shifts of demand are absent, and that the incremental cost of manufacturing and physical distribution is constant per copy over the managerially relevant range. The curve indicates that the increase in sales attributable to persuasion becomes less and less as more and more advertising is applied. This conforms to common experience and is supported by considerable research and by a plausible rationale. In successive mailings, the initial advertising goes to the most productive mail lists, where it attracts the most susceptible customers. Subsequent mailings to inferior lists must be more and more intense to induce the less susceptible to become customers. The resulting incremental advertising cost curve is shown in Chart

Chart One–2 Effect of Advertising on Sales

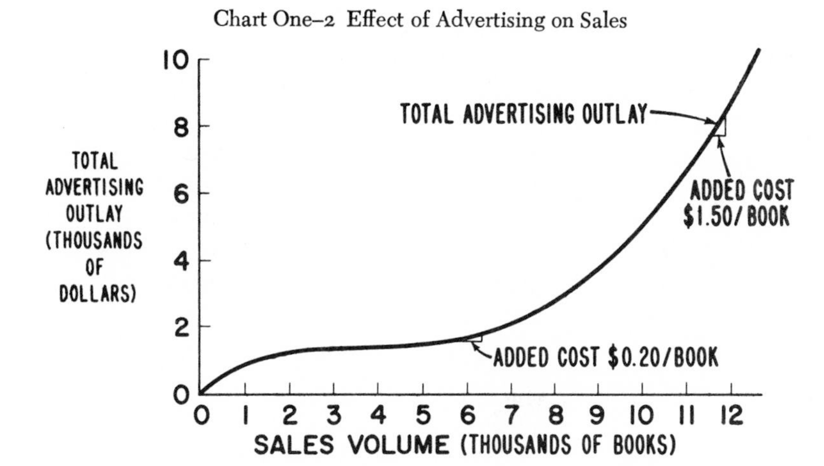

One–3, where it is compared with incremental pre-promotion profit to indicate how much to spend on advertising.[6]

This hypothesis for pure selling cost in the short run is similar in shape to the traditional hypothesis for pure production cost. That is fortunate, since separation of these two ingredients of a department store cost function is not possible.

This simplifying hypothesis abstracts from shifts in demand caused by outside forces. In a mature competitive market like that of the Lord & Taylor Depart-

Chart One–3 Effect of Advertising on Profits

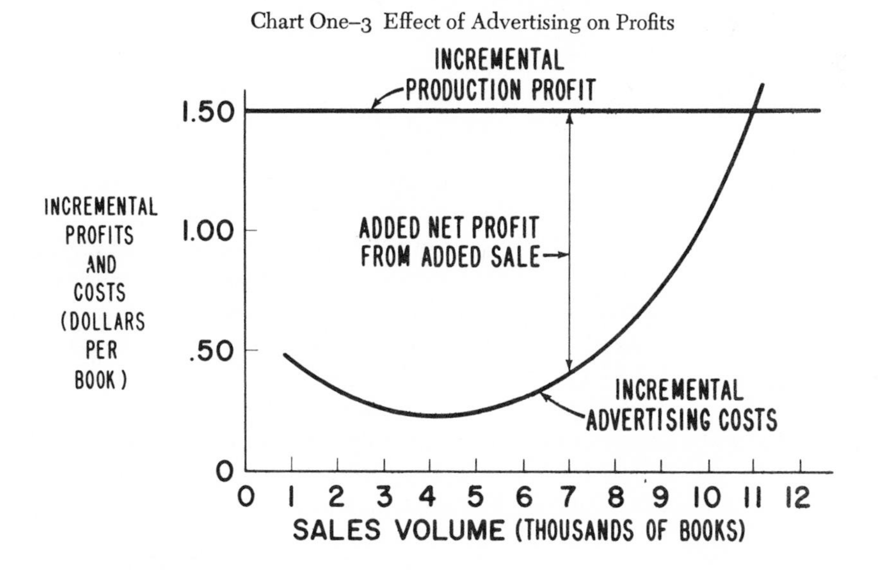

ment Store, a high proportion of selling expenditures have the effect of merely maintaining the firm's share of total industry sales, which fluctuate as total demand shifts seasonally and cyclically in response to exogenous changes in income, expectations, tastes, and custom. As a consequence, a spurious correlation between selling expenditures and sales could come about, and the statistical findings could not be interpreted as measuring the diminishing effectiveness of additional promotional outlays as prospective customers who are less susceptible or more inaccessible are tapped. Nor can our findings be interpreted as measuring the declining effectiveness of selling outlays as a firm expands its market share at the expense of rivals.

II. Methods of Measurement

Alternative Methods

There are three basic approaches to an estimate of the short-run cost function: (1) classification of accounts into fixed, variable, and semi-variable, on the basis of inspection and judgment; (2) estimation of the relationships of cost to output on the basis of engineering conjectures; and (3) determination of the functional relationship of cost to rate of output by statistical analysis of recorded cost, output, and other operating conditions. These approaches need not be mutually exclusive; indeed, two or more may supplement each other. While the statistical approach was the method used in these studies, data were derived from accounting records and the findings can be projected by means of engineering estimates. Hence each approach will be described briefly.

Accounting Approach

The accounting approach involves classification of expenses as (1) fixed, (2) variable, and (3) semi-variable, on the basis of inspection and experience. This approach is the simplest and least expensive of the three. Hence, it should normally be used whenever feasible as a supplement to the other methods if it is not used as the principal method. To be most successful the requisite conditions are: (1) experience with a wide range of fluctuation in output rate; (2) a detailed breakdown of accounts kept on the same basis over a period of years; and (3) relative constancy in wage rates, material prices, plant size, technology, and so forth.

Since the accounting approach provides no way to correct data explicitly for changes in cost prices or for changes in other conditions that affect cost behavior, a constancy of these cost conditions is essential if accurate results are to be obtained. The statistical method can tolerate more variation in underlying conditions because it possesses a means of dealing with these variations. Uniform coverage of the output range is not essential for success of the accounting method. A group of observations at each extreme of the range is sufficient. The accounting method isolates constant cost easily by inspection. It identifies variable cost easily, but determines less accurately the pattern of variation of these and of semi-variable costs. This approach needs to be supplemented by graphic statistical analysis to separate the variable and fixed components of semi-variable cost and to determine the linearity of output relationship for semi-variable and for variable costs.

Engineering Approach

In essence, the engineering method consists of systematic conjectures about what cost behavior ought to be in the future on the basis of what is known about the rated capacity of equipment, modified by experience with manpower requirements and efficiency factors, and with past cost behavior. Hence, it relies upon knowledge of physical relationships, supplemented by pooled judgments of practical operators. It should, and usually does, make use of whatever analyses of historical cost behavior are appropriate and available, as a means of making the judgment better. Typically, the engineering estimate is built up in terms of physical units, i.e., man hours, pounds of material, and so forth, and converted into dollars at current or prospective cost prices. The cost estimates are usually developed at a series of peg points that cover the contemplated or potential output range.

The engineering approach is the only feasible one when experience and records do not provide an adequate historical basis for measuring cost behavior. Engineering estimates are also a needed supplement to statistical or accounting analysis when it is desired to project cost behavior beyond the range of past output experience, or when it is necessary to estimate the effect of major changes of technology or plant size upon cost behavior over a familiar or unfamiliar output range.[7]

Statistical Approach

Statistical analysis of past behavior of costs deals explicitly with each major problem of determining cost relationships empirically. The accounting and engineering approaches encounter these problems but cope with them less consciously and therefore perhaps less successfully.

Statistical cost determination can be of two kinds: (1) simultaneous observation of costs of different (but similar) plants, operating at different rates of utilization, or (2) sequential observation of costs of the same plant over a period of time when it operates at different rates of output. To find a large number of plants that are sufficiently similar in equipment, management methods, and records (so as to proxy an identical plant) but that differ over a wide range in rate of use is not easy.[8] Nor is it easy to find a factory whose size, technology, and management methods have remained substantially constant over a period during which output rate has fluctuated widely. Cost data of the second kind were used in all the short-run cost studies in Part One.

This variant of the statistical approach, when conditions are favorable, is likely to achieve more reliable measurement of short-run cost relationships than the alternative methods. It is, however, more time-consuming and expensive. It uses multiple regression analysis to find a functional relationship between cost (preferably in terms of total outlays per month) and output rate by holding constant at their mean (as independent variables) other dimensions of output such as size of manufacturing lot and size and direction of change in output.

An important superiority of statistical analysis over the accounting and engineering methods lies in its ability to measure the fixed-cost ingredient of each component of cost and determine whether its marginal-cost ingredient is constant, rising, or falling, and to do this separately for those dimensions of output (and other cost determinants) that are found to affect cost importantly.

III. Method Used

The purpose of these statistical studies of short-run cost behavior was to determine the net effect of rate of output on cost when the influence of the remaining variables has been allowed for. If the statistical cost function can be determined in such a way as to eliminate the influence of all other cost determinants apart from output rate, the resultant cost function is the empirical counterpart of the theoretical cost function described above.

In general, three steps are needed to eliminate the influence of these irrelevant cost forces. The first is to select a plant and a period of observation in which dynamic elements, such as changes in the size of the plant, technical production methods, managerial efficiency, and so forth, were at a minimum. The second is to rectify the cost and output data recorded in the firm's accounts in order to remove the effect of remaining irrelevant factors, such as changes in wage rates, prices of materials, tax rates, special accounting allocations, lags caused by the production cycle, and so forth.

The third step in removing other cost influences is to hold their effect constant by means of multiple regression analysis. Cost may be affected by operating variables and other dimensions of output, e.g., size of production lot, change in output from the previous period, style variety, learning stage, and so forth. Hence it is necessary to take account of these additional independent determinants which reflect operating conditions suspected to exercise an important influence on short-period fluctuations in cost.

All three of these general methods of purification must usually be employed. Operationally, they break down into smaller steps. These may be conveniently discussed in terms of the following problems:

1. Selection of plant suitable for statistical analysis.
2. Measurement of output.
3. Determination of time unit of observation.

4. Choice of period of analysis.
5. Analysis of cost components.
6. Determination of form of cost observations.
7. Deflation of cost data.
8. Matching of costs with output.
9. Allowance for other dimensions of production.
10. Selection of form of function.

Each of these problems will be discussed in general terms in the light of experience in making statistical determinations of short-run cost behavior for the furniture factory, hosiery mill, belt shop, and department store.

Suitability of Plant

Some plants are appropriate for statistical determination of cost–output functions; some are not.[9] Six criteria are useful for judging the suitability of a plant for statistical analysis:

1. *Age of plant.* A large amount of accumulated data must be available for analysis. Usually, four or five years' operation is necessary before statistical analysis can be done with precision.

2. *Homogeneity of product.* The smaller the number of different products, the easier it is to measure output and to put cost estimates to practical use. Also, the less the products themselves have changed (in ways that affect cost), the easier and more durable is the study. For example, in this respect a ball-bearing plant is better than an automobile body plant.

3. *Homogeneity of equipment.* Similarity of machine units is desirable in order to eliminate cost variations due to use of more efficient machines for some levels of production than for others. In the hosiery industry, for example, shutting down the least efficient knitting machines during times of slack production is relatively unimportant, because union restrictions prohibit allotment of work in a way that discriminates against operators of inefficient equipment during periods of low or part-time employment.

4. *Technical changes.* If the production technology is altered through new developments during the analysis period, there is a change of cost function, and statistical results do not reflect a single type of cost behavior. However, if only minor changes have occurred, they can be allowed for in the analysis and a practical approximation to a single cost function can be made.[10]

5. *Length of production cycle.* The production cycle, which is the average length of time it takes the initial inputs in a process to emerge as finished products, should be short. This will minimize the reallocation of recorded costs and output necessary to place them in the accounting period which is appropriate, i.e., so that costs attributable to a certain output are actually matched with that output.

6. *Output variation.* The production volume should vary from month to month

so as to include a wide range of output rate and a fairly uniform coverage of this range.

Measurement of Output

Measurement of output is usually the hardest problem in statistical determination of cost. Theoretical cost functions assume that output consists of homogeneous units of a single product. Actually, however, almost all modern plants produce a number of varied products. This is true even when output appears at first blush to be homogeneous, as in the hosiery mill and the belt shop.

There are three ways to solve the problem of measuring heterogeneous output. The first, and best, solution is to determine the cost–output relationship separately for each product. This approach is available only when products are few, when processing is separable, when records of cost and output for each individual product are available, and when observations of output of each are spread fairly evenly over a wide enough range to permit fitting of cost regressions and computing of coefficients of variation. These conditions were not met by any of the plants studied here.[11]

The second way to measure variegated production is to introduce each significant aspect or dimension of multiproduct output as a separate independent variable in the multiple regression analysis. This solution is particularly useful when different dimensions of output have distinctive cost influences. It was used in the belt shop, where weight per square foot of belting was introduced as an independent variable. In each of the three units in the study of department store costs, two dimensions of output, namely number of transactions and average value of transaction, were used as separate independent variables.

The third solution for measurement of heterogeneous production is to develop an index of multiproduct output. Although this solution is inherently inadequate because no scheme of product weighting is fully satisfactory, nevertheless it is the only workable solution when the number of products is large, their mix fluctuating, and their individual costs nonseparable. It has no significant redeeming virtues, but is a resort of desperation. For statistical cost determination, the most suitable product weights are those based on identifiable ingredients of normalized variable inputs—for example, the standard cost at constant prices of the directly traceable, i.e., "direct," labor and materials of each individual product that was produced during the accounting period. This kind of index of output was computed for the furniture factory by weighting each article of furniture in proportion to its deflated standard cost. For the hosiery mill, the weights of the output index were based on the relative normal direct labor input of each product. For the belt shop, products were weighted on the basis of square feet of single-ply equivalent belting, primarily because the cost to manufacture was found to be more closely related to the area of a product than to its weight, dollar value, or standard costs, all of which were studied as alternative measures of output.

Time Unit of Observation

Should the unit of observation be a week, a month, or a year? The smaller the unit, the less chance there is that the effect of fluctuations in output rate during the observation unit will be missed. On the other hand, the smaller the observation unit, the greater is the problem of matching recorded output with the cost that caused it. The best unit of observation is the briefest period that will not cause serious gestation lags, arbitrary "distribution-lag" time allocations, or incompleteness of data. An observation period of a month was found to be satisfactory for most cost–output studies. For the furniture factory, a two-week time unit was used. For the hosiery mill, test studies were made for one-week and three-month (quarter) units as well.

Time Period for Data

The time period for collecting data should have the following characteristics: (1) wide range of output variability and uniform data coverage for the range; (2) constant size of plant; (3) little change in technology; (4) stable managerial methods; (5) uniform cost records covering changes in volume, cost, and other operating conditions; (6) number of observations large enough to permit generalization and yet small enough to be manageable in correlation analysis. The observation period, moreover, should be recent and, hopefully, relevant to future operations. In "Furniture Factory," for example, the study covered the years 1932–34, and 47 two-week accounting periods were used in the analysis.

Analysis of Cost Components

When, as is common, one plant of a multiplant firm is being studied, a decision must be made on what elements of cost should be included. Those that are arbitrarily allocated to the plant and that bear no apparent relationship to its operating conditions should be excluded. To find out whether the omission of overhead common to several plants caused understatement of the marginal cost of one plant, the omitted overhead should be correlated with the output of the plant (as was done for the hosiery mill and the belt shop).

Another problem is to decide whether the separable components of combined cost should be studied statistically. Analysis of individual elements of expense has several advantages. In the first place, individual accounts may require different corrective devices, corresponding to the varying influences which give rise to the need for rectification. Irrelevant influences may differ in kind for different expenses and may also operate with varying intensity on the various categories of cost. The same considerations apply also to the influence of independent variables. An independent variable may affect only certain components of cost. A flexible budget can be more detailed and more accurately adjusted to variations in operating conditions, and the estimates more precisely modified to accord with changed input prices, if the analysis is so made that separate correction and reflation of the components is possible. Again, if underlying conditions that affect only a few cost elements change, it is possible to readjust the costs affected.

Selection of Form of Cost Observations

Experimentation with alternative forms of cost observation in various cost studies has led me to believe that analysis in the form of total cost for the accounting period, rather than average unit cost, yields more reliable findings.[12] Cost in total form is easily converted to average and marginal form. Marginal cost, being the rate of increase in the total function, is the slope of the net regression line of total cost on output.

Deflation of Cost Data

To obtain empirical cost functions that are of lasting generality and are analogous to static theoretical functions, it is necessary to hold constant the prices of input factors. Two assumptions are made: (1) that substitution among the input factors does not take place as a result of changes in their relative prices; (2) that changes in the output rate of the enterprise exert no influence on the prices paid for materials, labor, and services. The short-run cost function which is sought here will not be descriptive of actual cost behavior over the course of a normal business cycle. Near-capacity rates of output of any one firm will generally occur in the rising phase of the business cycle when other firms will also be expanding operations. An individual firm may be in itself sufficiently small to have no influence on factor prices. Nevertheless, increases in its output will be correlated with a rise in the output of the industry, which will generally exert a significant influence on prices of some inputs. Consequently, if a firm increases its rate of output during the upsweep of a business cycle, the increase will be accompanied by a rise in factor prices with consequently rising marginal cost. Cyclical behavior of cost is a separate problem. The attempt here is to approximate a static model for which prices are assumed to remain unchanged no matter what the rate of output of the individual firm.[13]

Because factor prices (and other cost distortants) affect the individual elements of combined cost differently, elements must be rectified separately. Regretfully, the convenience of composite deflation must usually be passed up.[14]

Matching of Cost with Output

Rectification of the time lag between the recording of cost and the recording of the resulting output ordinarily requires two steps: (1) the determination of the proportion of cost recorded in a period different from the period in which the corresponding output is recorded; (2) the determination of the length of the recording time lag. Sometimes these magnitudes are measurable, but it is usually necessary to resort to estimates based on technical considerations and engineering opinions. These estimates can be supplemented by statistical analysis designed to test objectively the correctness of the engineering calculations.[15]

The recording of the cost of machinery repairs and depreciation involves timing problems that warrant special attention. Capricious fluctuations in repair expenditures tend to obscure the cost–output relation. Special studies of individual machines over a long period of time can sometimes determine what part of repairs are a function of operations and what part a function of the ravages of time. When there is wide discretion in the timing of maintenance, e.g., a railroad's right-of-way, a spurious relationship to output is sometimes produced to give the illusion of close "control" of expenses. Repair outlays are cumulative, as well as fortuitous, so that some kind of lag correction is usually required. In the belt shop, for example, machinery repairs were adjusted to be one-fifth of the current figure, and four-fifths of the outlay three months later. Fortunately, repair cost in all the studies was small enough so that its adjustment could not affect the form of the marginal cost function.

Depreciation presents a similar problem. Ideally, use-depreciation should be separated from time-depreciation, since it alone is relevant in determining the cost occasioned by different levels of operation. The shape of the marginal cost function depends upon whether use-depreciation is present and whether it is a linear, increasing, or decreasing function of intensity of utilization. This relation depends upon the effects of differing intensity of utilization upon the deterioration of equipment. Moreover, loss of value from more intense use must be greater than that caused by obsolescence to affect marginal cost. Depreciation caused by physical deterioration due solely to the passage of time and by losses in value as a result of technological progress or changes in product specification (obsolescence) affects merely the height of the intercept of the total cost function on the cost axis, not the shape of the function itself and not its marginal cost.

Unfortunately, depreciation is usually charged as if it were entirely time-depreciation, generally on a straight-line basis—i.e., as a linear function of time—so that time- and use-depreciation cannot be differentiated in these studies. Marginal cost was understated only to the extent that significant losses of value, in excess of time-depreciation, arose from use and exceeded that restored (or prevented) by outlays for maintenance.[16]

Allowance for Other Determinants

There are usually other determinants of cost whose influence has not been removed by selection of the sample, by rectification of the data, or by the devices for getting a comprehensive measure of output. The effect of some of these factors is averaged out in the course of the statistical analysis. When the cost determinant can be measured and is independent, multiple regression analysis can be used to determine explicitly the relation of these factors to cost.

In all the studies variables selected for testing were those that management thought had an important effect on cost. In addition we usually applied a statistical test that required the candidate variable to supply a significant additional systematic explanation of the behavior of cost. Each influence selected was examined separately in order to ascertain: (1) the reasons for its influence on cost; (2) the best statistical series available for its measurement; (3) its net correlation with cost. The list of candidate variables differed for each establishment. Those tested for the belt shop, for example, included change in output, variability in rate of output within accounting period, size of manufacturing lot, proportion of special orders, and rate of labor turnover. For the hosiery mill, time was used as an independent variable to reflect changes in efficiency, and other influences.[17]

Specification of Form of Function

Before the regression analysis can be made, it is necessary to specify the form of the functional relationship of cost to the independent variables. Most interest attaches to the output relationship, but other relationships present similar problems.

The choice of the form of the cost–output function that is to be fitted to the data is determined basically by the shape of the data (made visible by scatter diagrams). But the choice is instructed by theory about a plausible relationship and is constrained by the cost of fitting and by the feasibility of extrapolation of some functions. In these studies, the choice was explored and narrowed by graphic multiple regression analysis. Usually it was made finally by fitting alternative functions by least-squares regression analysis and then by using statistical tests of significance to determine whether a more complex function fitted the data better than a simpler one. The form of functional relationship between total cost and output that was usually selected was either a cubic function of the form

$$X_1 = b_1 + b_2X_2 + b_3X_2^2 + b_4X_2^3$$

or a linear function of the form

$$X_1 = b_1 + b_2X_2$$

In these equations X_1 is total cost, X_2 is output, and b_1 and b_2 are constants. To specify the shape of the total cost function, three investigations were made. First, the process of manufacturing was examined in terms of potential segmentation and the technological probability of variable proportions. Second, the statistical distribution of cost observations was examined by graphic multiple regression analysis. This usually included a study of the scatter of first differences of cost on output, as illustrated by Chart 2–11 in "Hosiery Mill." Here the linear relationship of first differences in the top panel, combined with the lack of any evidence of rising per unit cost at extreme output in the bottom panel, substantiated the hypothesis that the relationship of total cost was linear. A third step was to fit not only a straight line but also curves, usually parabolic and cubic regressions, for total cost and output. Then various statistical tests were applied to find which functional form fitted best.

In none of the three manufacturing plants studied did the first-difference tests or the tests of significance for the higher-degree functions indicate that a curvilinear total-cost curve fitted the data better than a straight line.

IV. Findings of Empirical Studies

For each of the three factories studied, total cost rose in a straight line as a function of the rate of output over the range of output observed. Hence marginal production cost was constant over the actual operating range of the period studied. Probably, however, there is some critical, capacity-straining output rate where marginal cost rises very steeply. These three factories did not, during the analysis period, push production beyond that critical level. Each of these studies was made during the 1930s, when low demand may have made unnecessary output rates that push the outer limits of plant capacity.

For the furniture factory, incremental cost was found to be constant regardless of the rate of output. The average unit cost curve declined at a diminishing rate as output increased, failing to rise even at the maximum of the range of output produced during the three years studied. In addition to rate of output, several other dimensions of output—size of production lot, position on learning curve (measured by number of new styles), and the size and direction of change in output rate—had independent impacts upon cost which were measured and held constant by multiple regression analysis.

For the belt shop, the total cost curve was found to be linear over the range of observed output rate. This yielded a hyperbolic average unit cost curve and a constant marginal cost curve. Variance analysis did not reject the linearity of the total cost function. As an additional test, the first differences in total cost, ratioed to corresponding first differences in output, were plotted against output rate. This resulted in a horizontal scatter of these incremental cost ratios. The mean ratio was very close to the value of constant marginal cost derived from the regression analysis. A cubic total cost function was also fitted. The coefficients of the higher terms were significantly different from zero at the 2% level, on application of the *t*-test. The cubic function was, however, rejected because of the high coefficient correlation in the linear total cost function, because of the constancy of marginal cost indicated by the independent analysis of incremental cost ratios, and because of the possibility that lower-quality inputs had to be used at high output rates, thereby raising their effective price per unit of service (which is a source of cost upturn excluded by the hypothesis tested).

For the hosiery mill, findings were based on monthly data for 1935 through 1939, when output varied over a wide range—from four to almost 48 on the output index. The total cost curve was linear and the marginal cost curve constant. Separate analysis of the relation to rate of output of month-to-month cost increments showed that these incremental cost ratios were not correlated with output rate. They had an average value very close to the constant marginal cost measured by the regression analysis. Parabolic and cubic total cost functions were also fitted. The coefficients of the higher terms were smaller than their standard errors. Thus the curvilinear total cost function that is called for by accepted economic theory was not supported by the empirical findings.

In addition to the three manufacturing plants, the costs of three departments of the Lord & Taylor Department Store were studied statistically on the basis of 60 monthly observations which covered the period 1931 to 1935. The three departments—shoes, coats, and hosiery—were picked because their output (essentially, volume of sales service) varied over a wide and evenly covered range and because there was little change in layout, operating methods, or managerial personnel. A linear total-cost function and constant marginal cost were found for the hosiery department and for the shoe department. Higher-order functions were fitted and subjected to critical ratio tests, which showed that the more complex total cost functions did not fit the data significantly better.

For the coat department, total cost rose at a diminishing rate as output (measured by number of transactions) increased, yielding a declining marginal cost curve. The decline may be explained by error in the index of output that was correlated with output rate. A transaction when sales personnel were frantically busy involved a smaller output of sales service, since customers had to wait longer or do more for themselves.

V. Usefulness of Findings

Knowledge of the behavior patterns of short-run total cost, average unit cost, and marginal cost of production has great practical value for management. It can be the basis for forecasts of various kinds of future short-run incremental costs which are needed to make decisions on output rate, pricing, product mix, distribution-channel mix, and the size of outlays on advertising and other forms of persuasion.

Findings about behavior of outlay cost of production have great practical use, for example, in forecasting future opportunity costs. The versatile capability of much of modern manufacturing and merchandising capacity, together with the prevalence of multiple products and sectored markets that differ widely in incremental profitability, give opportunity costs major importance for many kinds of short-run managerial decisions and make the prediction of opportunity costs an important application of knowledge of the behavior of outlay costs.

The estimates of short-run marginal cost for each of three departments of the Lord & Taylor store that were obtained from this study produced a measurement of short-run incremental profits that could be useful (1) for pricing, (2) for guiding investment in inventory, (3) for measuring the profit performance of the department manager, and (4) for rationing capital and selling space.

The most important and immediate practical use is for pricing. Estimates of short-run marginal cost supply a measurement of incremental profits and thus provide a tool for the manager to set prices (and promotional outlays) at that level which, taking into account the spectrum of his competitor's prices, will probably produce the largest total of incremental profits for his department. This price will be quite different from that which maximizes dollar profit *margin* on a particular transaction.

A second use is to guide the department manager and his supervisor in buying and inventory decisions that will optimize the amount of capital tied up. It would be wrong if the manager were to maximize dollars of contribution profits without regard to the amount of investment (in merchandise and in selling space) that is required to produce them. Treating capital as a free good would not maximize the wealth of stockholders, which is presumably the master goal of the department store. To accomplish this requires a measurement that correctly takes into account the amount and the cost of the capital which the department manager ties up and which is, over a broad range, controllable at his level. This measurement can be made quite simply by converting the short-run incremental profits of the department, or of any line of merchandise or item, into a rate of return on the associated controllable investment (mostly inventory). The resulting rate of incremental-profit return on investment that the department manager can control is the quantity he should maximize.

A third practical use is to develop a sophisticated but simple measure of managerial performance. Not only price and volume were controllable at his level, but also the amount invested in merchandise inventory. Maximizing incremental profits without regard to the amount of this capital tie-up would fail to maximize the welfare of the stockholders. This defect can be corrected quite simply by expressing the incremental profits measured by the study as a rate of return on the incremental investment in merchandise inventory that is controllable at the departmental level. This controllable incremental return on investment is what the department manager should maximize. It is also a measure of his profit performance which is compatible with and contributory to the master goal of maximizing stockholder wealth. This investment return is also compatible with the central concept of profit-center decentralization—namely, focusing the department manager's energies on those short-run incremental revenues, costs, and profits that are largely controllable at his level. His performance gauged by this measurement and stacked against a target return is the third practical use for the findings. The target return should, at minimum, be the firm's composite cost of capital expected on average for the long future. At maximum, it should be the return that could be earned if the space in this department were made available for the expansion of another department whose rate of controllable incremental return on investment is higher—in other words, the opportunity cost of the space now available to this department.

By this opportunity cost route, measurement of short-run marginal costs can be put to a fourth use, namely, to guide decisions on the continuous reallocation of merchandising floor space among rival departments. This fourth use steps over into store-wide capital-investment decisions. Here too the overriding goal is to maximize stockholder wealth, which is roughly to maximize return on investment. To achieve this goal departments should compete for limited space on the basis of prospective rate of incremental profit return on the total increment of investment involved in this reallocation of space. This total is comprised of the investment controllable at the department level (largely inventory) plus that avoidable investment controllable at higher levels.

VI. Criticisms and Limitations

Statistical determination of short-run cost functions has been, from the outset, vigorously criticized by some economic theorists, who have objected to its findings and its methods. Deficiencies are of two sorts: (1) defects of statistical cost curves as a test of economic theory, and (2) limitations of the curves as an aid to management decisions. Most criticisms are of the first type, centering on whether the empirical findings are a good test of the theory. The focus is on the linearity of statistical total cost functions, since constancy of marginal cost is unseemly to a conventional theorist. A number of the more interesting criticisms are examined below:[18]

1. The range of observations of rate of output in statistical cost studies is too narrow to test the behavior of short-run marginal costs at the extremes of crowding of capacity, when the law of diminishing returns would be expected to operate most forcefully.

This could be true. But if the observations cover the compass of actual operations, this inadequacy will have no practical significance; it will be of academic interest only. For example, in "Furniture Factory" the range of observed operations was wide and included a boom period as well as depressions. The observed peak output rate had not been exceeded in the previous decade. A rate that overloads capacity may cause marginal cost to rise so rapidly, so perilously, and so clearly that operating at that rate would be foolish. If the reason that there are no observations is the prohibitive cost penalties caused by overloading, then the behavior of costs at output rates at which no one operates should not be of ravenous interest to economic theorists.[19]

2. The measurement of multiproduct output is defective, causing spurious correlation with cost.

The index of multiproduct output is bound to be defective; all statistically workable solutions for this complex problem are intellectually unsatisfying.[20]

Two kinds of solutions were used in the studies. The first was to introduce as independent variables, in addition to rate of output, those aspects of the variegated product mix that had significant independent effect upon cost. This was done in the department store study where average value of transaction was used in addition to number of transactions to measure heterogeneous output. The second kind of solution was to create an index of multiproduct output by weighting products by their standard cost at constant prices of traceable inputs. In a belt shop, the index was square feet of single-ply equivalent belting. It was chosen because manufacturing cost was found to be more closely related to area than to the weight or dollar value of output. For the hosiery mill, the output index was weighted on the basis of each product's relative direct labor cost. For the furniture factory, each product was weighted in proportion to its deflated standard direct cost. Hans Staehle faulted this output index, arguing that it amounted to "determining output by costs, i.e., to introducing a spurious dependence where measurement of an independent relationship is really wanted."[21]

"Spurious dependence" was avoided (1) by using *standard* cost, which stayed the same for years, rather than fluctuating *actual* cost; (2) by confining the output index to direct costs, which for the furniture factory were *variable* inputs; and (3) by measuring these standard direct costs in *constant* factor prices. When products are numerous and when the product mix changes frequently and widely, as was the case for furniture, it is difficult as a practical matter to develop a better measure of output.

If output consisted of a single homogeneous product and if the proportion of input services were fixed at the same level for all output rates, then rate of output would be correctly measured by the deflated standard-cost index (which yields the same results as a physical count). Consequently, if the law of diminishing increments of product were in fact operating, it would show up in the cost function that was fitted statistically to the observations of output thus measured. If these conditions of technologically fixed proportions of input factors for the running costs (as opposed to set-up costs) of a particular product prevail for all products, a deflated standard-cost output index should have no more linearity distortion for multiproduct output than for a single-product output—provided, of course, that the possibility of continuous substitution between factors or products and the diminishing marginal rate of substitution of one factor for another, which is assumed in the multifactor, multiproduct model of Professor J. R. Hicks,[22] is *not* applicable to the modern plant.

3. Straight-line depreciation introduces a linear bias into the cost function that is solely attributable to the accounting technique.

Depreciation charges do not impart linearity to the statistical total cost function. This is true for double-declining balance, sum-of-the-digits, present-value, and all other time-curved depreciation functions, as well as for straight-line depreciation.

Linearity is not bestowed by depreciation because the bookkept cost is the same in monthly total regardless of the rate of output. It is a fixed cost, constant as to output rate. It changes, in the case of time-curved depreciation, not with the rate of use of the equipment, but only with its age. Thus straight-line depreciation is no more likely to cause a linearity bias than is time-curved depreciation because no kind of depreciation can cause it.

Marginal cost will be affected by depreciation only (1) if wear of the equipment from use reduces its market value faster than does obsolescence or size-inadequacy (whichever is faster) and (2) if this additional loss of market value due to wear is not restored by repair outlays.

Does the true cost of depreciation (as distinguished from its bookkept cost) cause marginal cost of manufacturing to be greater at high output rates than at low? Probably not. Use-depreciation is the only kind that enters marginal cost, and it does so only to the extent that it is not restored or prevented by repairs.

In modern manufacturing, use depreciation is unlikely to be significant. This is because most equipment becomes obsolete before it wears out. Moreover, maintenance captures in repair charges the small amount of use depreciation that may exist. Because machines run at the same speed for high output as for low and usually get about the same mechanical care, uncaptured use depreciation that is higher per unit at peak operations is unlikely. Use depreciation is therefore likely to be zero. If so, inaccurate prediction of either the economic life of equipment (which is unavoidable) or the time-shape of value loss will result only in an intercept of the total cost function that is too high or too low, not in any error in estimating its slope or its curvilinearity.

4. Rectification of the cost data for changes in wage rates and material prices biases the findings of statistical studies toward a linear total cost function. Price-motivated substitution of factors of production, which is an important ingredient in the operation of the law of diminishing returns, is deflated away in the rectification process.

Price-motivated substitution was not affected by the subsequent rectification. The high prices were not removed; they were actually paid and if they were going to cause substitution, they would have done so. Price rectification did not remove the stimulus; it only removed the distortion. Cost observations in all three short-run cost–output studies covered a period during which prices of materials and components and wages of labor fluctuated widely. Changes in input prices were, however, not caused by changes in the output rate of the firm; instead, they were caused mainly by changes in general business conditions. Input-price contamination had to be removed in order to lay bare the cost-output relationship. Recorded cost data were deflated by index numbers, in most cases specially created for this purpose. An alternative method (not used in the studies) is to recalculate costs by applying constant factor prices to actual physical inputs. However, rectification by this method will cause overstatement of the costs of every month except the one to which the selected set of factor prices relates.[23]

Hans Staehle saw in both kinds of rectification a potential linearity bias.[24] This criticism is based on an assumption that is often contrary to fact—that the proportions of input factors can be, and are, changed in response to changes in factor prices. Linearity bias from this source could have had no significant effect in the three studies because the proportions of input factor services were fixed technologically within narrow limits and could not, in the short run, have been changed in response to fluctuations in their relative prices.[25]

5. Marginal costs are bound to rise at high rates of output because the quality of input factors deteriorates; the need to fall back on obsolescent standby equipment and inexperienced temporary help and to pay high premiums for the night shift will, during cyclical and seasonal peak periods, inevitably cause marginal costs to rise.

Cost increases from this source, although they may be real and managerially important, are excluded from the hypothesis that is being tested. The law of diminishing increments of product assumes that the units of the variable input services are constant in price and homogeneous in quality. Diminishing returns are caused not by using less and less efficient machines or men but, instead, by using less efficiently men and machines of equal quality. Consequently, cost increases caused by the deterioration of the quality of input factors are in essence increases in the price of input factors. Hence they should be removed from the cost in order to test the theory of its relation to output.

To do this is very difficult, however, and it is doubtful that the distortion could be completely eliminated. A pattern of deterioration of the quality of inputs at cyclical and seasonal peaks is inherent in the confluent fluctuations of a competitive economy. If the fluctuations in the output of the plant are correlated with general economic activity, and if there is insufficient slack in input supply, then some rise of the marginal cost curve should be expected from this source.[26] To the extent that this distortion is imperfectly removed, the findings are biased toward *rising* marginal cost.

Thus inadequate rectification for cyclical changes in price and quality of input factors cannot cause marginal cost to be measured as constant. Instead, these inadequacies of rectification bias the statistical analysis in the direction that would create a cubic total cost function and hence a rise in marginal cost. With expanding industry demand, recourse may be had to input factors that are inferior in quality whereas when industry demand is low, these tend to be sloughed and superior workers and other factors retained. This is illustrated by the cyclical circumstances of those observations that lie above the fitted total cost line at highest output levels in the belt shop and of those observations that lie below it at lowest output levels.

6. Learning curves contaminate statistical cost–output findings, reducing costs at high outputs and thereby obscuring a rise of marginal cost.

Learning is costly and could have this effect under the conditions implicitly assumed by the criticism, namely, that learning costs are proportionately greater during months of high output rate than in low output months. It is true that the frequent changes in product specifications, product mix, and manufacturing processes that characterize modern multiproduct production, incur a large amount of learning cost. The incidence of this cost is hard to trace, which makes it difficult to determine whether it is proportionately greater in high output months. This uncertainty may make cost–output measurement for multiproduct manufacturers in dynamic, style-oriented industries of limited reliability for testing economic theory. Much reliance must be placed on (1) development of a good index of output, (2) presumed stability of processes, and (3) separation of learning curve costs. Even though not bookkept as such, capital investment in learning and in improved processes is, in a modern factory, continuous. So the statistical cost function is not a purely short-run cost behavior.

Learning (i.e., increased experience with the productive processes that are peculiar to the particular new product) affects the manufacturing costs of the individual new product. It reduces costs as a function of the cumulative total output of that product over its production lifetime. This output is different from and not necessarily correlated with the overall output rate of the plant. Consequently, a rising phase of short-run marginal manufacturing costs will be obscured by learning costs only to the extent that periods of high overall output are correlated with periods of maturity (low learning costs) in the manufacture of individual new products.

Distortion from this source can be removed at least partially by holding the cost effect of the learning curve constant at its mean. This was done in the furniture factory study, where learning costs were explicitly measured by introducing as an independent variable in the multiple regression analysis the number of new products (products whose costs were high on the learning curve).

7. Changes in management methods and technology were probable over the long observation period and these changes may have contaminated the purely short-run relationship of cost to output rate and obscured its curvilinearity.

It is likely that improvements did occur in management methods and possibly in technology. Perfection in holding constant these cost determinants is unattainable.[27] The result is a downward drift of the short-run cost–output function and cost observations scattered over its migration path.

Whether this inadequate purging biases the cost function toward linearity is, however, doubtful. Only if high output rates were concentrated at the end of the analysis period, when the fruits of improved management and technology were reaped, and if costs were thereby made lower than they would otherwise have been, could an understatement of the concavity of the total cost function in this upper range of output have resulted. In the studies under consideration, the time-distribution of output rates that would have been required for this explanation of cost linearity was absent.

What is measured as an increase in cost as a function of output is sometimes, instead, a cost-postponement path. For example, railroad track maintenance is usually timed cyclically to regularize reported profit by matching revenue and hence traffic, even though this is the most dangerous and expensive time to fix the track.[28] Another example is the notorious income sensitivity of research and development outlays. The consequence for statistical cost curves is to convert a cost that is inherently an investment, and is hence fixed in respect to volume, into one that appears to be variable.

8. Variation in unit variable cost will cause only small curvature in the total cost function, so that there is a danger that what is in reality a cubic function will, in statistical analysis, be taken for a straight line.[29]

That is possible. But whether or not there is a linearity bias will depend on how tests of statistical significance are used to reject a hypothesis because of low probability that the relationship being tested would have generated the behavior of the observations sampled. Tests of significance, together with direct analysis of the relation of total cost and of cost increments to output and regression analysis of average cost per unit, have tended to confirm a linear total cost function. Thus, generally speaking, probability tests have failed to reject the hypothesis of linear total cost and have indicated that no statistically significant improvement results from fitting a parabolic or cubic total cost function.

9. Constant marginal cost is at odds with observed output behavior. If short-run marginal costs were constant over a wide output range and then rose steeply near capacity output, the output of a competitive industry would, in the short run, change mainly by shutdowns, i.e., by changes in the number of plants or firms that operate. Output would not change significantly by means of fluctuations in the production rate of plants that stay open. But this is the opposite of observed behavior: marginally efficient firms and plants continue to remain in operation and there is wide disparity among them in rate of output.[30]

Since there are great differences in the level of marginal production cost, it would seem that shutdown of high-cost plants would be the main mechanism for reducing the output of an industry. Some plants do close, but most reduce output and/or step up selling efforts. They do this because some sales, being sheltered differentially from competition and hence priced discriminatorily, produce incremental profits. This makes continued operation, even at a low rate, preferable to shutdown. Moreover, anticipated future profits may warrant an investment of operating losses at low output rates. For the decision to close a plant, it is only the forecast of future sales and costs that matters. Thus statistical findings of constant marginal manufacturing costs are not contradicted by the observed facts of output behavior.

In practice a competitive industry changes its rate of production mainly by variations in output rate that differ among plants and firms rather than by plant shutdowns. This is the principal means of adjusting output to fluctuations in industry demand and is entirely consistent with constant short-run marginal manufacturing cost. It is consistent because most American industries operate under conditions of monopolistic, rather than atomistic, competition. Consequently, the output rate of any firm or plant is determined not by the rise of its marginal manufacturing cost but instead by the rise of its marginal selling costs as the firm vies for patronage at the edges of its limited locational or product-preference "territory." Ask a businessman why he doesn't manufacture more if his incremental manufacturing costs are constant and are low relative to unit combined cost. He will reply that he can't sell any more. He makes all the furniture or cars or cement that he can sell—meaning, of course, all that he estimates it will be profitable to sell in view of the steeply rising incremental selling, transport, and competitive-retaliation costs required to dispose of additional output. Economic theorists who say that the short-run marginal cost curve must turn up to achieve equilibrium are, in one sense, right. Rising costs are what determine the optimum output rate. However, these costs are not the costs of manufacturing more but, instead, the costs of selling more. Selling costs are in a different universe of discourse from the manufacturing costs whose behavior the theory purports to describe.

Selling costs are designed to augment demand, manufacturing costs to serve it. Selling costs are alternative and supplemental to reductions in price. They behave quite differently from manufacturing costs. Incremental selling costs rise continuously in the short run for any given level of industry demand and are the main determinant of the output rate of the plant and firm. Economic theorists sometimes portray rising selling costs as a downward sloping demand curve for the individual firm. Business executives view them as a rising cost of distribution or a shrinking of net-back (net receipts after persuasion outlays, price concessions, and transport cost). Either way, incremental selling costs make variation in the output rate of the individual plant or firm sensible, despite a horizontal marginal manufacturing cost and big differences in its level among plants.

10. The disparity between the time period of statistical observation and the idealized briefer time period of economic theory (in which it is assumed that the proportions between input services and rate of output are momentarily unchanged) causes a linear bias in statistical cost functions.

Hans Staehle suggests that by averaging the rate of output over the entire accounting period (e.g., a month) statistical analysis assumes a linear cost function for the range of output within that period. This assumption, he argues, biases the statistically determined cost function toward linearity because "the midpoint of a secant connecting any two points on a curve whose second derivative does not change sign lies closer to a straight line connecting the end points of the curve itself."[31] The degree of variation in the rate of output may be masked by the size of the observation unit. Fluctuations within the month are averaged away, and they cannot, because of the nature of cost records, be matched precisely with corresponding costs.

Professor Staehle's criticism is correct, but only for the range of output rate that occurs within each accounting period and is therefore averaged away. Even assuming that each such secant is cup-shaped, the worst that can happen is that the monthly average overstates the bottom of the cup. There is little likelihood that a linearity bias will result, i.e., that the overstatement will be greater for a high-output month than for a low-output month. In the hosiery mill study, for example, weekly, monthly, and quarterly observations produced the same linear total cost function.

As indicated earlier, in modern technology the proportions among factors of production (inputs) and the rate of output are quite inflexible in the short run and tend to remain unchanged during the production run of any one product. Changes are caused by shifts in product mix and by instability of output rate, not by the length of the accounting period compared with some theoretical and unobservable "unit time period" of unknowable brevity.[32]

11. The cost data, recorded outlays per month, imperfectly approximate the continuous rates of flow of input services that are specified in the cost curves of economic theory; hence the hypothesis is not precisely tested by the observed cost data.

That is true, but it is an unavoidable limitation. The hypothesis of theory is formulated in terms of continuous flows of input services, whereas the cost data of statistical studies are recorded in terms of outlays per month for input services, e.g., a man's pay, the monthly rental for a leased sewing machine, or the monthly depreciation (plus interest payments) for an owned machine. These outlay costs will not correspond precisely to the ideal of a continuous rate of flow of input services unless (1) payments for rented equipment are solely unit charges (e.g., per inch stitched or per output unit) and (2) depreciation charges are confined to that part of capital wastage which is caused by heavier use of the machine and which is not restored by repair outlays. Capital wastage that is caused by obsolescence has zero marginal cost. Yet in modern industry obsolescence (or inadequate capacity, which, for this point, is equivalent) causes the lion's share of capital wastage measured in market values.

Thus the outlay costs per month that were used in the statistical studies may be an imperfect proxy for the idealized flow of input services of the theoretical model. Nevertheless, it is the outlays, not the abstract rate of flow, that are alone measurable and that matter managerially. Perhaps in this respect the model should be brought into conformity with reality rather than vice versa. In any case, it is not probable that this unavoidable approximation to the mathematically convenient abstraction could cause a bias toward linearity in the statistical cost function.

12. The estimation of short-run marginal cost by the multiple regression analysis of cost behavior results in an understatement of incremental cost by reason of the nature and fixity of the accounted depreciation charge.

An understatement could result only if the loss of market value of equipment which resulted from greater output (more intensive use) exceeded the loss in value caused by obsolescence and not made up by maintenance. The only true incremental cost of more intense use of depreciable facilities is the excess in value-loss over that caused by obsolescence. This excess is small because routine maintenance (primarily for purposes of operating efficiency) usually repairs damage due to wear and tear mainly by piecemeal replacement of components. Higher output rates can cause greater repair outlays. If they do, and if their time-lag is correctly allowed for, then the resulting marginal cost is totally reflected in the regression analysis. Consequently no understatement of marginal costs comes from this source, even though accounted depreciation of the equipment is the same regardless of rate of output.

However, piecemeal replacement by routine maintenance does not *fully* restore equipment physically. So the question remains whether the hidden cost of this deficit, to the extent that it is caused by rate of use of equipment, is omitted from the regression analysis by reason of the constancy of the depreciation charge. This incomplete physical restoration can result in higher marginal cost only if it causes a loss in the market value of the equipment that exceeds the loss that occurs from obsolescence alone. A machine that is obsolete will sell as scrap at the same price regardless of its physical condition. Characteristically in modern plants, obsolescence wipes out the value of equipment much faster than wear and tear that is not made up by routine maintenance. As a consequence, capital consumption which is a function of use and in excess of that reflected in maintenance outlays (the only marginal-cost depreciation) is rarely important.

13. Multiple regression analysis of cost behavior is too complex to be believed by operating executives and too expensive to be warranted except for academic research.

As to the believability of multiple regression, many modern executives view it as less unrealistic than estimates that require gross simplification of reality, either by naive fixed-variable cost classification or by assuming costs to be a function of output alone. The regression technique permits probability statements about the reliability of cost estimates. An operating executive can combine these with his experienced judgment about known but unmeasured parameters in the sort of intuitive Bayesian probability analysis that a good factory superintendent uses daily.

High cost was a valid objection to multiple regression analysis in the past. But today with computers this limitation is no longer important for the regression analysis itself. However, it can still be a deterrent with respect to the costs of preparing the data, particularly when output is heterogeneous and cost determinants are numerous. Costs are significantly reduced by computerized accounting only when its cost classification and causal-factor measurement meet the requirements of multiple regression analysis.

> 14. Changes in the specifications of individual products and in the output mix, which occur almost continuously, impair the precision of cost functions established by multiple regression analysis and the reliability of cost forecasts based on this past behavior.

This criticism is correct for many plants and constitutes an important restriction on the practical usefulness of regression analysis of cost–output relationships for the same plant across time. As indicated earlier, avoidance of basic changes in product specifications, of extreme volatility of product mix, and of profound changes in the technology or the size of plant is crucial in selecting an establishment for statistical cost determination. However, total absence of change in the product or product-mix or production processes over a period long enough to permit statistical sampling is uncommon. So ways to cope with modest changes are needed and have been developed. Changes in the character of the individual product and in the product mix call for the development of a statistical index of output that expresses product characteristics in terms of their normal or probable cost consequences. This kind of output index was sought in each of the cost studies of Part One. To the extent that the index is correct, it permits measurement of reliable cost functions of the past and their projection into cost forecasts of the future, assuming adequate allowance for changes in production technology and plant size, or absence of such changes.

How well the index solves the problem of variegated output depends on a trade-off of fidelity against expense. The marginal cost of increasing its precision should be stacked against the resulting marginal revenue from losses avoided by not making mistakes such as passing up marginally profitable business or accepting orders whose marginal revenue does not beat marginal cost. Evaluation of an index-solution must also take account of the "abandonment-alternative" of settling for even poorer predictions of marginal cost obtainable by methods more primitive than multiple regression analysis.

VII. Alternative Theory

For economic theorists, the horizontal statistical marginal cost curve is what Charles Fort called one of the "damned facts—the facts that do not fit in." It cannot be handily reconciled with the construct used by economic theory to explain the behavior of the business firm and the short-run equilibrium of a competitive economy.

The inconvenient fact—one that does not fit the accepted theory—can be a productive one for advancing theory. However, most economic textbooks ignore the inconvenient fact of flat marginal manufacturing cost. They continue to show the short-run production cost curves as continuously rising, an assumption which is viewed as necessary for an elegantly consistent structure of concepts to explain the adjustments of the firm to change in its economic climate.

How can economic theory be rearranged to accommodate constant marginal short-run production cost? First, we need an expanded and more realistic theory of the role of various kinds of persuasion costs and other aspects of the product bundle in the determination of the output rate of the individual firm under conditions of monopolistic competition. Second, we need to formulate an alternative hypothesis for the behavior of short-run manufacturing costs.

Optimization under Monopolistic Competition

Each modern firm has some market power, however limited and ephemeral. The firm is sheltered from competition by locational advantage, by differentiation of its products in wanted ways (either physically or in the eye of the buyer), or by distribution services preferred by customers. The amount of market power differs widely among firms' products and among market sectors. Each firm strives to augment its own market power and to frustrate the efforts of rivals to do the same. In this process, the seller tries to differentiate his product bundle, which is the focal point of modern competition, so as to achieve greatest perceived value per dollar of price.

Several aspects of persuasion cost comprise this product bundle, and these complementary ingredients in the marketing mix are substitutes at their edges. The seller seeks that combination which will minimize his cost, in the sense of giving him the most persuasion per dollar. Using his imperfect approximation to this optimum mix, he increases outlays on all aspects of persuasion to the point where forecasted short-run selling cost just equals incremental profit after manufacturing costs but before persuasion outlays. Thus this short-run optimum output is determined by rising marginal selling costs and is completely compatible with horizontal marginal manufacturing costs.

New Theory of Production Cost

A new substitute theory of the behavior of short-run manufacturing costs has three traits that distinguish it from the conventional theory: (1) constancy of marginal cost over the managerially relevant range, (2) multiplicity of products and of variable inputs, (3) multidimensional output.

The first trait is diagramed in Chart One–4. Total combined cost (*TCC*), shown in the upper section of the chart, rises linearly over the range of output that matters. From a low rate of output near the intercept of the horizontal total fixed cost (*TFC*) it rises at a constant gradient parallel to total variable cost (*TVC*) until it reaches an output rate that crowds capacity so cruelly that it is rarely chosen. Bigger

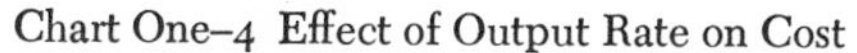
Chart One–4 Effect of Output Rate on Cost

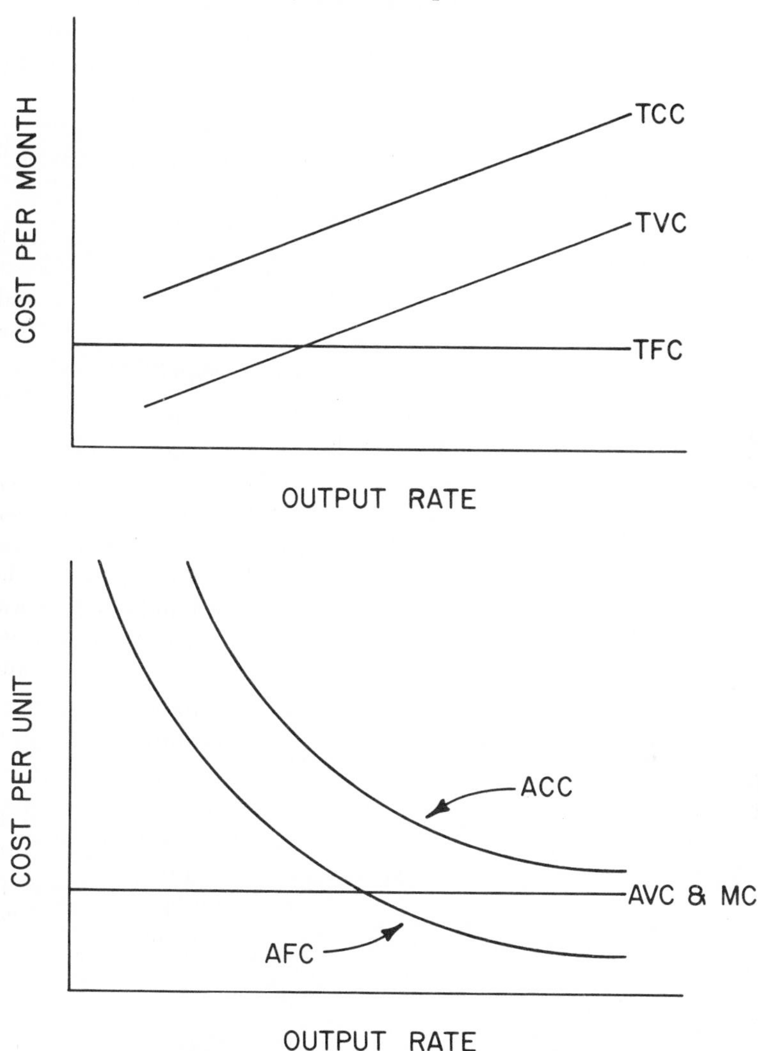

backlogs, even though they may risk loss of customers, are usually a preferable alternative to these marginal production cost penalties. In terms of unit cost, which is shown in the lower section of Chart One–4, marginal cost (*MC*) is constant and equal to average variable cost (*AVC*) over the managerially relevant range; then it probably rises steeply. Average unit fixed cost (*AFC*) declines in a rectangular hyperbola producing an average combined cost per unit (*ACC*) that declines continuously over the range of actual operations. Presumably, if the plant were overloaded sufficiently at this extreme of output, incremental cost would rise quite steeply. For example, an overloaded electric generator would burn out.

The second aspect of the alternative hypothesis is recognition that most plants make many products and have several variable input factors. Monistic production is a rarity today. Hence the simplifying assumption of a single homogeneous product and of a single variable input sacrifices too much reality.[33]

The third aspect of the alternative hypothesis is the recognition that output is multidimensional. For simplicity, economic theory has assumed not only that there is a single homogeneous product but also that rate of output has only one dimension. Both these simplifications are manifestly unrealistic. In a customer-responsive factory, output consists of many variegated products, and production has several dimensions, in addition to the simplistic "rate of output." Among them are magnitude and direction of change in output, proportion of individual products in the output mix, size of manufacturing lot, and proportion of new products for which learning costs are still important (position on learning curve). These dimensions, which can have an important effect on short-run cost behavior, are different and have different relative importance, depending on the products and processes of manufacture. Their independent impact on cost is hidden when they are lumped together in the simplified composite "rate of output" or are assumed away as "other things" that are "held constant." To average away or to assume away these other aspects of output neglects determinants of cost that have importance not only for management but for economic theory.

Rationale

What explains constant short-run marginal manufacturing cost? What are the forces that negate or offset the operation of the law of diminishing returns and make the total cost function linear until output presses the limits of plant capacity? Three technical conditions of modern manufacturing combine to cause constancy of marginal cost: (1) segmentation of equipment, (2) limited substitutability of variable inputs for the fixed factor, and (3) limited substitutability among variable inputs.

1. SEGMENTATION OF EQUIPMENT

The characteristics of a plant's fixed equipment[34] determine its potentiality for segmentation, which plays a major role in the shape of the short-run cost function. The critical characteristic is the plant's potentiality for varying the rate of flow of output without changing the proportions of variable inputs to fixed equipment in use (e.g., the ratio of man hours to printing press hours). Segmentation in this sense refers to the technical nature of the fixed equipment which permits a wide range of choice in the machine hours used per week.

The degree of segmentation differs widely among plants, ranging from none to complete and continuous segmentation. An example of zero segmentation is a single blast furnace, which must be operated continuously. An approximation of complete and continuous segmentation may occur in hydro-electric generation, e.g., at Grand Coulee Dam, where there are 18 generators, each of which can deliver power at any rate from almost zero to its full capacity. The position of a plant on this spectrum of potential segmentation depends on the technical nature of the equipment, the success of managerial efforts to segment, and the nature of the labor contract.

Three sources of segmentation can be distinguished. The first is physical divisibility, i.e., where fixed equipment consists of a large number of homogeneous units. An example of this is a hosiery mill where the knitting of stocking legs is done on 81 nearly identical knitting machines.[35] This may be called "unit segmentation." Under such circumstances the successive introduction or withdrawal of the machine units permits wide variation in the input services of the machinery despite its overall fixity. A second kind of segmentation is achieved by varying the number and hours of the shifts per period that the fixed equipment is employed. This we call "time segmentation." If the technical nature of the fixed plant is such that it can be used at varying speeds without varying the input proportions, a third kind, "speed segmentation," can be achieved by more intensive use of the equipment, i.e., by operating machines at faster or slower rates. The technical structure of production will determine the degree of segmentation that can be achieved, and the means of attaining segmentation will determine its continuousness. Unit segmentation is characterized by discreteness in the flow of services, as each successive unit of plant is brought into operation. If the units are small, however, the discontinuities are unimportant. Time segmentation may also have marked discontinuities, for example in plants that are inhibited by labor contracts or managerial prejudice from operating at other than full shifts (no overtime or part shifts). On the other hand, speed segmentation permits continuous variability in plant use, although the range within which machine speed-up, etc. is a practical method of achieving segmentation is narrow.

Thus, in sharp contrast to the indivisibility and fixity assumed in accepted theory, is the multidimensional divisibility of fixed factors achievable by segmentation. From this source marginal cost will have substantial constancy—with three provisos: (1) that the machines are reasonably uniform as to productivity, (2) that if they differ, management is deterred (e.g., by union rules) from favoring the best machines or best operators, and (3) that if the machines do differ in efficiency, there is enough first-line equipment to produce the required output (except for occasional emergencies). Characteristically, first-line equipment used over the full range of normal variation in output rate tends to be modern and homogeneous. (Obsolescent equipment occupies valuable floor space and requires maintenance and perishable operator know-how. Consequently, it is normally retained only as standby equipment in emergency reserve, or when, for some special products or small manufacturing lots, it has lower incremental costs than first-line equipment.)

Because of segmentation, what looks to the manufacturer like fixed cost and is properly measured as fixed can be a variable cost in the sense that the rate of use-up of the input service (e.g., machine hours) varies with volume, even though the payments for it do not. Thus segmentation causes marginal manufacturing costs to remain constant over a wide range of output rate.

2. LIMITED SUBSTITUTABILITY OF VARIABLE INPUTS FOR FIXED INPUTS

A second cause for cost constancy is the sharply restricted substitutability in the short run between fixed and variable input factors. In terms of flow of productive services and any particular finite segmentation of the fixed factor (e.g., a machine hour), variable input factors like labor and materials are not, as has been assumed in theory, freely substitutable today for fixed factors. Instead, their mix is technologically frozen. Variable inputs must be combined with the productive services of fixed factors (e.g., machine hours) in the proportions that are determined within a narrow range by the design of the equipment, the capabilities of its operators, and the social restraints upon them. In the furniture factory, for example, the typical woodworking machine had a rather inflexible operating complement of variable input services, electric power, light, lubricants, raw materials, and labor time, and ran at a speed that, with an experienced operator, would vary only narrowly. Because the range of equipment speed is narrow and the mix of other input services is invariable, marginal cost is constant (if other dimensions of output are held constant). Incremental cost is substantially the same whether the machine is run four, eight, sixteen, or twenty-four hours a day.[36]

Substitution of variable input factors for fixed factors has been assumed to occur and to cause incremental cost to rise. In modern industrial processes this is largely a fiction in the short run. This kind of substitution occurs only as a corrigible technological error, hopefully only during the learning period or as a result of a slowdown or other industrial sabotage.[37] Normally, the mix can be changed only (1) by shifting to a different existing process, or (2) by investment (intangible as well as tangible) in a new process, which, for short-run cost behavior, is ruled out. (A process change sometimes takes the form of a substitution of components: a die casting for a forging or a stamping. This pushes back to a supplier much of the investment required for substitution, although some change in the design and the processing of the finished product is usually involved.)

3. LIMITED SUBSTITUTABILITY AMONG VARIABLE INPUTS

A third explanation for horizontal marginal manufacturing cost is that substitutability of variable input factors for one another is, in a modern factory, narrowly restrained by technology, by standard shop practices, by union work rules, and by social mores. This is in sharp contrast to the unlimited substitutability assumed in accepted economic theory.

Unlike the models of multi-input factors and multiproducts developed by Professor Sune Carlson and by Sir John Hicks, the input factors in the three plants studied could not be flexibly substituted for one another. The amount of glue or wood or labor in a chair cannot in the short run be altered in response to changes in the relative prices of these inputs, or for any other reason. For each process, the mix of services of input factors was fixed by technology and by standard operating procedures. It could be changed only by changing the process. The number of existent alternative processes was quite limited, consisting mostly of a few pieces of obsolescent equipment held in emergency reserve. In the long run, of course, new equipment on the technological horizon could drastically change the input mix and its output transformation; but again, this would involve a new set of inflexible proportions, both among the so-called variable input factors (idealized in the model as freely substitutable) and in the ratio of the input service mix to the productive service rate of the equipment (e.g., knitting machine hour). This non-substitutability among inputs additionally contributes to the constancy of short-run marginal manufacturing cost.

Thus three factory conditions combine to cause marginal cost to remain constant over a wide range of output rate: (1) wide freedom to segment the plant and thus to vary the number of approximately homogeneous units of "fixed" factor inputs, e.g., machine hours, (2) a relatively constant, technologically determined mix of variable inputs with the input *services* of the fixed factor (e.g., machine hours), and (3) non-substitutability among variable inputs in the short run.

Each process for each product has its own production function—i.e., its own technologically fixed input mix and inflexible transformation rate—for uninterrupted production (running cost) once the machine has been adjusted for the particular product-process (set-up cost) and the processing routines have been mastered (learning cost). The proportion of these three costs is determined by the length of the run and the unfamiliarity of a product. Consequently, unit cost declines with learning and with lot size.

To make a different product requires a hop from one fixed-proportion process to another, taking the hurdles of set-up costs and learning costs. These are in a narrow sense short-lived investments, as are the costs of accelerating and decelerating the overall output by the other kinds of segmentation. The goal of production planning is to find the optimum trade-off of these operational investments against the cost of tying up capital in inventory.

The changes in input mix that result from variation in product mix, newness mix, lot size, and output rate instability are sometimes mistaken for variation in the proportion of factors in any one product process. Costs caused by output instability may also be correlated with output rate and mistakenly attributed to it. Low output need not result in smaller lot size, more new styles, or more changes in output rate. And even if it did, these cost influences would be irrelevant to the hypothesis being tested.

Thus if the productive services of input factors can be obtained by the firm in unlimited quantities at the market price, and if their prices are unaffected by the firm's output rate, then their unit costs in a fixed-input-mix process will be constant regardless of the level of output.

Study 1

FURNITURE FACTORY

Contents

Chapter I

Summary and Conclusions

Objectives

The principal objectives of this study are, first, to investigate the potentialities and limitations of various statistical methods for analyzing cost behavior, with a

Reprinted from Joel Dean, *Statistical Determination of Cost, with Special Reference to Marginal Cost,* Studies in Business Administration, vol. 7, no. 1 (Chicago: University of Chicago Press, 1936), by permission of the pub-

view to developing practical procedures for determining the short-period behavior of average and marginal cost; and, second, to examine by means of these methods the behavior of average and marginal cost for a particular manufacturing establishment.

By average cost is meant the cost per unit of output, i.e., the total cost for the entire period divided by the number of units produced during that period. And by marginal cost (also called differential and incremental cost) is meant that addition to the total cost caused by the production of an additional unit of product, i.e., that particular increase in total cost which is associated with a unit increase in output. The short period may be defined as a time span too brief to permit alterations in the scale or character of the fixed equipment. For the purposes of this study accounting periods of two weeks' duration were used.

These objectives were deemed worthwhile because, in the first place, a knowledge of the short-period behavior of average and marginal cost is of considerable utility in the management of any business and, in the second place, few systematic procedures for securing this knowledge have been developed.

An understanding of the behavior of average cost may be useful to the business executive on several accounts: (1) it will aid in establishing flexible cost standards which can be used for controlling cost,[1] (2) it will enable the management to institute programs of cost reduction based upon more accurate knowledge of the causes underlying cost behavior; and (3) it will assist in computing estimates of cost and of total profit under predicted conditions of future operation.

The superiority of adjustable cost standards as control devices results from the fact that, if standards have been fitted to operating conditions, executives are less likely to be held responsible for cost variation which is beyond their control, and are more likely to be charged with responsibility for results which are within their control.[2] Standards of this type can be developed only after a systematic study has been made of the actual effects upon cost behavior of various operating conditions, such as rate of output, size of manufacturing order, rate of labor turnover, etc.

The second use to which the knowledge of the behavior of per-unit cost can be put is in the reduction of cost itself. Systematic analysis of the effects upon cost of various changes in operating conditions is likely to extend the executive's understanding of the causes of cost behavior and bring to light possibilities of reduction of costs.

The third use which business men can derive from this knowledge of per-unit cost behavior is in forecasting costs and profits under anticipated operating conditions. Since profit represents the difference between total cost and total income, any improvement in the precision with which costs can be estimated is likely to

---

lisher. Appendix C, not published as a part of this monograph, is reprinted from Joel Dean, "Correlation Analysis of Cost Variation," *Accounting Review*, vol. 12, no. 1 (March 1937), pp. 55–60, by permission of *Accounting Review*.

result in a corresponding increase in the accuracy with which the profit-and-loss statement can be projected into the future.

Even more important use may be made by business firms of knowledge of marginal cost behavior. Most short-period managerial decisions should, if maximization of profit were the sole test of correctness, be made by balancing the additional cost caused by the suggested policy against the consequent additional revenue. As a general rule, this applies to any type of decision—the introduction of a new product, a new manufacturing method, or a new sales campaign. But most frequently utilization of cost data in this manner is essential if optimum decisions are to be made with respect to the price and the output policy of the firm and, more specifically still, when the acceptance of an additional order is being considered. The increment in revenue which the order brings should be balanced against the addition to total cost which it causes. This additional revenue produced by a unit increase in output is called "marginal revenue," just as the increase in total cost caused by a unit increase in output is called "marginal cost." Thus it appears that a knowledge of marginal cost (and of marginal revenue) is of great importance in forming the rational basis for many of a firm's most important short-period decisions.[3]

More and more, of late, the attention of business men has been directed to the importance of these marginal costs. Notwithstanding the fact that the behavior of cost, as output is altered, has been the object of considerable speculation, the problem, until recently, has elicited comparatively little statistical investigation. Nevertheless, it is becoming increasingly recognized in some quarters that cost compiled by routine traditional accounting procedures (with their arbitrary appraisals, allocations, and imputations) has little value as a guide to short-run business decisions and therefore has small significance in determining price.[4]

It is important for management to know the short-period behavior of its average and marginal costs, yet such information does not usually exist. Orthodox accounting costs tell little about the effect of altered operating conditions upon average costs and nothing about the behavior of short-run marginal cost. Furthermore, there seem to have been surprisingly few efforts to study cost behavior systematically. A survey of published studies indicates a paucity of methodical procedures for explaining cost behavior, a widespread reliance upon rule-of-thumb guesses, and, consequently, a generally recognized inexactness of cost forecasts.[5] Information required for balancing the expected additional revenue to be derived from a given course of action against the expected additional cost is not supplied to the management of business firms through the regular channels of accounting routine. It is derived primarily from guesses or estimates made by the enterpriser himself, or by engineers, or, more rarely still, by accountants or statisticians. Such a state of affairs invites investigation aimed at developing practical methods for studying average and marginal cost behavior, which should systematically provide management with the estimates of short-run average and incremental costs required for intelligent business decisions.

Short-period cost behavior, rather than long-period cost behavior, was made the subject of this investigation because the majority of executive decisions are probably made for the short period. Moreover, accounting costs are often moderately satisfactory approximations to long-period marginal costs.[6]

Results

METHODS DEVELOPED

The results of this investigation may be summarized under two general headings: (1) methods and (2) findings concerning cost behavior. As has been previously indicated, the first objective of this study was to develop practical systematic methods for determining the short-period behavior of average and marginal cost. With this in mind, an examination of published studies was made in order to evaluate procedures previously employed. Few studies were found which had a direct bearing upon the problem at hand. Other studies, however, with different objectives, suggested techniques which proved useful. Procedures thus suggested were empirically tested by making a statistical study of the cost behavior of a certain furniture factory.

The particular plant chosen provided suitable data for studying the relationship between short-period changes in output (and other factors) and short-run changes in cost. Its cost data were reported at two-week intervals, and were subject to large fluctuations. Moreover, costs were available in terms of total expenditures associated with a particular total output, rather than being assigned, by means of a specific order cost system, to a particular unit of product. This total cost, furthermore, was broken down into its components (i.e., components of combined cost), each of which was recorded as a lump sum for the reporting period.

The general method which was developed as a result of the analysis of the cost data of this factory constitutes, on the whole, the most useful result of this study. In developing this method a common denominator for output was first found which would measure the great variety of models and styles produced by the factory under investigation. The term used for this measure, "old warehouse value," is one which the management of this factory used, and it has been adhered to throughout this discussion. This unit of measure is the deflated standard cost of each item, computed upon the basis of a standard quality of material (red gum), standard labor and machine efficiency, and a normal level of output. Furthermore, since it is deflated to a base period, it does not reflect fluctuations in the market price of either the finished product or the input factors. It consequently represents an approximation to a weighted physical measure of output.

The second step in the development of this general method consisted in reducing the cost data to usable form. Since the dollar was selected as the unit of measurement in preference to various possible physical measures, it was necessary to correct the cost data for changes in the prices of materials and in the wages of workers. Cost data were then adjusted for discrepancies occurring between the time at which costs were recorded and the time at which output was reported. This adjustment was made for each of the several cost components on the basis of an individually estimated time lag. After the cost data had been thus corrected, they were, in the case of most items of cost, reduced to averages per unit of output.

The third step was to measure and study the effect of several other operating factors which influence cost behavior—such as the rate of output, size of production order, etc. This was accomplished by introducing these cost influences as independent variables in a graphic multiple correlation analysis. By this procedure the effect of each of these factors upon average cost or upon total cost was recorded in the form of a curve showing estimated costs for various levels of output, for various sizes of production order, and the like. These curves were then available for utilization in computing flexible cost standards, and for forecasting and controlling cost behavior.

The last step was concerned with the analysis of marginal cost behavior. To accomplish this the net relationship between rate of output and total cost was first obtained. This was accomplished either by direct correlation analysis of the data in the form of totals or by obtaining the average cost function and converting it into a total cost relationship. Next, this total cost function was differentiated, with the purpose of ascertaining the additions to total cost which corresponded to unit increases in output. The resulting series of additions, when smoothed, were taken as the marginal cost function.

The general methodology just outlined is, of course, an oversimplification of the procedure actually developed. As a matter of fact, alternative methods for achieving each of the four steps listed above were carefully considered and evaluated in the light of their apparent appropriateness to the particular task at hand, and the more promising of these techniques were used in the analysis of some of the cost data. These tests were chiefly applied to:

1. Alternative degrees of rectification of the cost data.
2. Alternative independent variables to be used in the multiple correlation analysis.
3. Alternative procedures for deriving marginal cost estimates.
4. Alternative breakdowns of combined cost.
5. Alternative dimensions or aspects of marginal cost.

For the purpose of determining the amount of rectification which cost data would ordinarily require in order to yield satisfactory results the entire procedure outlined above was repeated with cost figures representing several stages of correction. The conclusion reached on the basis of this methodological experiment was that all data would have to be fully corrected for changes in prices, output lag, and all other operating conditions which had important effects upon cost, in order to secure workable estimates of cost behavior.

Experiments to determine which types of operating conditions were important led to the conclusion that a small number of factors could account for a large proportion of cost variance but that the importance of the several operating conditions varied for different cost components. Also, the proportion of total variation in cost accounted for by all the pertinent independent factors employed in the analysis was different for each cost item. The operating conditions which were found to be the most important cost factors were:

1. Output (X_2)
2. Size of production order (X_3)
3. Number of new styles (X_4)
4. Changes in output from previous period (X_5)
5. Labor turnover (X_7)
6. Quality of material (X_8)

Experiments to determine the best methods for deriving marginal costs were made in two directions. The first attempted to discover whether better results could be secured when cost data were kept in terms of totals than when they were reduced to averages per unit. On the basis of experimentation with cost data it was concluded that more sensible and reliable marginal cost estimates were obtained from data expressed as totals. Marginal cost curves derived from average cost data were found to have large and probably spurious fluctuations attributable to slight alterations in the curvature of the function fitted to average costs. It is preferable, therefore, to secure marginal cost from a function which is fitted to total cost residuals rather than to average cost residuals.

Testing of alternative methods of fitting and differentiating an average or a total cost series was the other direction of this experimentation. These alternatives may be classified under three heads: (1) graphic methods—those which transform visually fitted average or total cost curves into tabular series of total cost and which then smooth the first differences secured from the total cost series to obtain the marginal cost function; (2) mathematical methods—those which fit mathematical functions to the cost data by least-squares procedures and differentiate the resulting total cost function by means of differential calculus; and (3) hybrid methods—those which combine parts of these two methods. Conclusions reached on the basis of this methodological experimentation are as follows:

1. The graphic-tabular method appears preferable in deriving marginal cost from average cost data. It is economical and easy to understand. Moreover, its greater flexibility enables it to fit complex functions and transcend accidental aberrations in the data. There is danger, however, that complex curves, fitted by either graphic or least-squares methods to crude average cost data, will result in spurious and meaningless fluctuations in marginal cost curves.

2. The derivative method may have some superiority in dealing with total cost residuals, especially when simple functions are employed. An accuracy and objectivity which are not altogether spurious can be achieved by fitting a straight line to total cost residuals. A more complex function should not be employed unless both empirical and a priori evidence indicate statistically significant curvilinearity. More reliable tests of significant departures from linearity can be applied to functions fitted by least squares.[7]

3. A hybrid of these two techniques was developed which appears to enjoy some of the advantages of each of its parents. It demonstrated several treacherous imperfections, however, in actual use with data in the form of averages which make it untrustworthy unless employed with great caution. More satisfactory results were obtained when it was used with data in the form of totals.

Another phase of the general methodological experimentation consisted in studying several component elements of combined cost to determine their average, total, and marginal cost behavior. It was hoped that the experimentation with modifications of the basic technique which were required by peculiarities in the behavior of the several cost components would prove valuable and that the resulting knowledge of the behavior of cost components would be helpful in controlling and forecasting cost in detail and, at the same time, would serve as a check upon the accuracy of the combined cost findings.

A full statement of the conclusions reached for each of these components cannot be made here. In general, the analysis indicated (at least to the author) that detailed study of minor components did not yield results commensurate with the effort expended. Modifications in technique proved unimportant. While the behavior of the average cost of components would be of considerable managerial usefulness, their isolated, individual cost behavior is difficult to ascertain with accuracy in any real situation because of the impossibility of holding constant the inputs of the remaining cost components. The practical usefulness of the imperfect approximations obtained in this study may not be very great. The use of the findings of these subsidiary analyses as a check on the results of combined cost studies was prevented by the absence of suitable data necessary for the analysis of certain of the components.

A fifth type of exploration of methods grew out of the recognition that marginal cost has many dimensions in addition to the more generally emphasized output dimension. Since total cost is a function of many variables, cost may be regarded as marginal with respect to each of these variables. Increasing the number of new styles, for example, has a marginal cost analogous to that resulting from increasing the number of units of output. Rough estimates of some of the more important of these other margins of cost were computed, more with the hope of indicating the nature and importance of this problem and of suggesting a tentative technique than with the conviction that any very reliable cost functions had been determined.

It must be borne in mind that the conclusions regarding general procedures which have been set forth above are subject to certain limitations. For one thing, the marked superiority of one particular technique over some other may have been due, in part, to peculiarities of the special problem under investigation. Thus the methods selected may not necessarily be the best procedures for use in other investigations of cost behavior. In the second place, the methodological impediments encountered in this study may have been fewer and probably were (in large part) different from those to be met with if a similar study were made of another firm.

The procedures used in this analysis cannot, therefore, be borrowed *in toto*. However, in view of the fundamental similarity of many of the methodological problems, it is hoped that the methods developed in this investigation will be found useful in subsequent studies of average and marginal cost behavior.

It should also be emphasized that no new statistical techniques were developed in this study. The procedures proposed and illustrated in this investigation are simply new combinations of well-known methods. The only originality which can be claimed is that these procedures constitute (as far as the investigator can learn) a new approach to an important and neglected problem.

COST-BEHAVIOR FINDINGS

Although the methodological results of this study are of central importance, its findings concerning actual cost behavior are also of interest, first, as illustrative of the type of result which can be secured and, second, because of the clearer understanding thereby obtained of cost behavior in the short period.

In examining the behavior of average cost per unit the relationship between this per-unit cost and several cost factors (such as rate of output, size of lot, number of new styles, and the like) was determined by graphic multiple correlation analysis. The results were expressed in the form of net regression curves which showed the effect upon average cost of changes in each of the several operating conditions. These curves themselves provide the best type of summary for the average cost findings. The various items of cost for which such a set of curves was worked out are listed below.[8]

Corrected combined cost (Chart 1–2)
Uncorrected combined cost
Direct labor cost (Chart 1–9)
Direct materials cost
Indirect combined cost
Indirect labor cost (Chart 1–10)
Inspection labor cost (Chart 1–11)
Yard labor cost (Chart 1–12)
Power and watch labor cost
Indirect materials cost (Chart 1–13)
Power cost (Chart 1–14)
Maintenance cost (Chart 1–15)

From an examination of these figures it is apparent that the effect of the several operating conditions varies both quantitatively and qualitatively with the type of cost. In general, the effect of altering each operating factor was about what was expected. (Exceptions probably represent errors in data or method.) Hence the contribution here made consists chiefly in a quantitative statement of expected relationships. These quantitative statements are, however, subject to serious limitations. In some instances curves have probably been fitted too closely to the observed data, thus producing an illusion of accurate knowledge when, in reality, only the general tendency of behavior is significantly indicated. In general, the scatter of observations about the curves was such that a wide band of standard error must be drawn about the curve indicating that only its general level and direction are statistically reliable.[9]

The average cost findings of combined cost are presented in Charts 1–2 and 1–3, which are discussed in chapter vi. The observed average cost behavior of several components of combined cost is presented in Charts 1–9 through 1–15, which are dealt with in Appendix A. These findings provide the raw material for building up flexible cost standards, for obtaining a clearer understanding of the causes for cost behavior, and for forecasting expenses more accurately. This raw material, however, must be adjusted and modified in different ways for different purposes. Problems encountered in making these adjustments are discussed in chapter viii.

Not only the average cost behavior but also that of marginal cost was determined for each of the foregoing expense items. The magnitude of marginal cost was computed at each of a series of output levels and was expressed either as an equation, a curve, or a series of marginal cost magnitudes, depending upon the technique of derivation employed. Marginal cost estimates for combined cost were obtained from data representing several levels of correction. Marginal estimates were also computed by several alternate procedures from fully corrected data in order to evaluate the various methods. These findings for combined cost are presented in Charts 1–4 through 1–8 and in Tables 1–4 and 1–5. Since they are fully discussed in chapter vii, it is sufficient at this point to mention only a few of the general characteristics of the findings.

In the first place, marginal combined cost appears to be constant throughout most of the usual range of output,[10] but appears to rise as the output falls toward zero. In the second place, estimates of marginal combined cost obtained from total cost data representing several levels of correction varied considerably and improved in reliability with each successive stage of data rectification. From this we can assume that complete refinement of the data was necessary in order to obtain satisfactory results for this firm during this particular period. In the third place, the curves of marginal combined cost which were obtained by the several alternative methods differed considerably in their fluctuations (we may consider these fluctuations as obviously spurious), but held to the same general level and exhibited considerable uniformity with respect to an upturn for small rates of output.

The marginal cost estimates for components of combined cost are shown in the form of equations and tables. The undulations of most of these marginal cost curves are probably not significant but are attributable, in part at least, to the use of data in the form of averages and to errors in sampling, curve-fitting, and accounting classification. These findings are presented in Tables 1–13 through 1–22 in Appendix B.

In closing, the limitations and inaccuracies of these cost findings should be emphasized. The treacherous nature of social-science data, the questionable reliability of the sample, and dubious appropriateness of many of the procedures employed in analyzing the data have often led to grave misgivings in the mind of the investigator as to the statistical significance of many of the findings of this study. But despite the fact that the estimates of average and marginal cost behavior for this furniture factory are not completely reliable, they are nevertheless considerably more trustworthy than the crude guesses which preceded them. Moreover, a method has been developed which, it is hoped, will, with repeated study of similar data, yield increasingly reliable results.

CHAPTER II

Survey of Previous Studies of Cost Behavior

This summary of a survey of the available literature dealing with cost behavior is restricted to those studies which are concerned primarily with the effect of output upon cost. Output is usually one of the most important factors causing variation in total cost and in average cost[11] and is the significant factor in deriving marginal costs.[12] Comparatively few systematic studies of the effect of other factors upon cost were found. These are cited in connection with the net regression curves for the respective cost influences.

As opposed to the numerous studies of average and total cost, there was a surprising dearth of studies of marginal cost. This is especially noteworthy in view of the importance of marginal cost to general price theory and its even greater importance to internal management of business firms. Studies of this type have, however, in all probability been made by private firms, but have not found their way into print because of the confidential character of the data.

Two basic types of methodologies may be distinguished in these previous studies of cost behavior: (1) that which relies heavily upon rule-of-thumb estimates for arriving at the cost at various output levels; and (2) that which statistically analyzes actual cost data at various levels of ouput. The demarcation between these two methods is, of course, somewhat arbitrary.

Method of Estimates

The first methodology, the method of estimates, is based primarily upon a reclassification of cost accounts according to their degree of variability in response to changes in the level of output. Two variants of this method can be discerned: (1) a twofold classification of all cost items as either absolutely fixed or proportionately variable; and (2) a tripartite classification which includes, in addition to these two categories, an intermediary one composed of semivariable or composite items.

Although the twofold concept of constant and variable cost is convenient, it is oversimplified. By defining constant cost as that part of cost which continues when the plant is idle,[13] and by assuming that all other expenses vary proportionately with output, an important intermediate type of cost has been ignored. It is true that many overhead costs do fall neatly into one class or the other. Nevertheless, there are many expenses which are intermittently variable and which fluctuate in discrete steps at certain critical resistance points.

Both J. O. McKinsey[14] (in 1923) and John H. Williams[15] (in 1922) early recognized the need for introducing a third, intermediate category of semivariable expense. This recognition of the complexity of cost behavior led to the development of more elaborate methods of compiling and portraying cost estimates. To this development the advocates of the break-even chart added much. The major contributions of this group are: (1) acknowledgment of the need for an intermediary category of semivariable costs; (2) more explicit definition and isolation of fixed costs; (3) recognition of the controllable character of many fixed costs; (4) stimulation, by the development of the flexible budget, of continuous improvement in the accuracy of estimates of the effect of varying volumes upon cost; and (5) development of a graphic method for portraying the relationship between costs, profits, and output.

Of all these contributions of the break-even chart advocates perhaps the most significant is the clear, vivid presentation of the way in which costs, sales, and profits vary with output. In a break-even chart the abscissa records rate of output, and the ordinate, the dollars of total sales revenue or total cost. Both sales revenue and total cost are assumed to rise at a constant rate and are hence drawn as upward sloping straight lines which intersect at a break-even point, beyond which lie profits.[16]

Certain limitations of this break-even chart technique should, however, be pointed out. Its basic weakness lies in the methods used to arrive at the cost estimates which it so ably portrays and manipulates. Unrealistic assumptions of a sharp dichotomy of perfectly fixed and proportionately variable components, of the complete independence of cost components from one another, and of a monistic, functional relationship between cost and output are usually tacitly and unquestioningly made. The assumption of a cost component which is proportionately variable at a constant rate, regardless of the time span of output variation, sets aside the principle of diminishing returns and ignores the effect of varying lengths of time required for the adjustment of different cost items to altered output. Moreover, cost is a function of many variables, of which output is but one. By ignoring these other factors this method tacitly assumes that they do not exist or, at least, that they do not significantly affect the output relationship.

Little effort appears to have been expended in systematic analyses to improve the accuracy of estimates of cost behavior, and little recognition appears to have been given to the limited accuracy of estimates made without the benefits of such analyses. These weaknesses in the methods of estimating cost behavior have been corrected in some measure by important refinements in the procedures of analyzing the hitherto unstudied semivariable items. It is these composite items which give trouble since a large proportion of the cost items fall in this intermediate class and since the perfectly fixed and proportionately variable costs are not difficult to forecast. In general, an attempt is made to divide the composite item into a perfectly fixed portion and a proportionately variable addition. Five different methods of accomplishing this separation may be mentioned:

1. By means of rule-of-thumb estimates.
2. By means of estimated cost amounts at maximum and minimum output levels.
3. By means of the actual cost amounts at typical high and low output levels.
4. By means of a series of separate estimates of the amount of each individual composite item at each of a series of production levels.
5. By means of an analysis of the time span of variability of each composite item.

Improvment in accuracy of rule-of-thumb estimates as a result of experience in budgeting was noted by Joseph Geschelin.[17] G. E. F. Smith[18] and J. H. Williams[19] sponsor the method of estimating cost amounts at maximum and minimum output levels, and then interpolating between these extremes. A third method of breaking composite items into fixed and variable portions is advocated by H. P. Dutton. He employs the same cost formula in his analysis that Williams used, but, instead of utilizing estimated costs, he takes the actually recorded cost figures of the composite item for two operating periods which differ materially in rate of output.[20]

Still another method of segregating the variable components of a composite cost item could be developed by the use of hyperbolic cost paper. R. M. Barnes describes the advantages of this ingeniously ruled paper in an article which, however, does not mention this particular use.[21] The scale of this paper is so designed that the rectangular hyperbola of average fixed cost per unit assumes the form of a straight line sloping downward to the left. The per-unit cost curve of a directly variable cost appears as a horizontal line. Floyd F. Hovey makes the suggestion that the per-unit cost curve of a semivariable cost will be a sloping straight line whose ordinate at infinity will show the per-unit value of its variable portion.[22]

A system of handling composite items which is very largely free from the questionable assumption that the fluctuating residue of a composite cost is proportionately variable over infinitesimal steps (an assumption implicit in the technique just noted and also in the Dutton method) has been suggested by J. M. Clark,[23] Walter Rautenstrauch,[24] G. H. Bates,[25] and others. Clark, for example, suggests that an estimate be made of each subdivision of cost for each of a series of outputs on the basis of a survey of actual cost experience, and also on the basis of an appraisal of current conditions. The total cost series thus built up makes no a priori assumptions regarding the nature of cost behavior. Since the result is frankly a forecast of how costs will behave, rather than a chronicle of how they actually did vary, it is perhaps free from some of the defects of a mechanical extrapolation of past experience. Such an estimate, of course, like any other, can attain only approximate accuracy.

G. H. Bates shows a clear comprehension of the value of understanding how cost varies with output. He has made some distinct contributions to the methods of analysis by constructing a variable budget for the Staten Island Shipbuilding Company, which was designed primarily to give the following information: "a standard of expense to cover any volume of business, ... a determination of overhead in percentages for differing volumes of business, ... [and] a disclosure of additional expenses occasioned by increasing the volume of business."[26] The contributions made to the methods of estimating cost variation consist of: (1) a detailed schedule of estimated expenditures at frequent output intervals; (2) a variable burden rate for each department; (3) an allocation of fluctuating, service department expenses to the direct departments on a variable basis; and (4) a clear recognition of the principle of marginal costs.

A few studies employing the method of estimates have been carried out with the clear objective of determining marginal costs. J. M. Clark's schedule of estimated cost at a series of output levels was designed specifically to illustrate a method of computing these incremental costs. The difference between the total cost at one level and that at the next higher level was taken as the increase in expense, or the differential cost.[27] G. E. F. Smith advocates the use of his flexible budget for somewhat similar purposes. A sales order, the price of which covered the variable direct costs plus the fluctuating overhead, could be profitably taken despite the failure of its price to cover average cost. That is to say, per-unit variable cost may be used as a rough equivalent of marginal cost.[28] G. H. Bates also saw the application of his flexible budget to the problem of deciding upon the acceptance of low-priced business. To facilitate this application, Bates used his budget figures to work out a fairly accurate schedule of estimated differential costs for particular departments at various levels of output.[29] T. H. Saunders likewise suggested the construction of an elastic budget as a means of arriving at an estimate of marginal costs.[30]

The pertinence of differential (or marginal) cost in deciding the question of the acceptance of additional business at a given price has been recognized by some engineers. Estimates of this cost increase have been developed, partly on the basis of known physical input and output relationships consequent upon machine production and partly on the basis of judgment and experience. The accuracy of many of these estimates of marginal cost is questionable. In all probability they have frequently been too low because of the tendency to treat immediate out-of-pocket outlays as though they represented the entire incremental costs. This neglect of variable and postponable elements of overhead cost has often resulted in an underestimation of marginal costs.[31]

Method of Statistical Analysis

A systematic statistical analysis of the actual cost data for past periods is one very fruitful way of developing standards for estimating future cost behavior. Hence it is surprising to note the paucity of studies employing this statistical approach. Those few statistical studies which were uncovered fall under two general heads: (1) those primarily concerned with average or total cost behavior and (2) those whose principal objective is the examination of marginal cost behavior. Each of these two general headings can be still further subdivided into: (*a*) studies which recognize only simple relationships and (*b*) studies which make specific use of multiple relationships.

Data used in these four kinds of studies are of two types: (1) simultaneous observations of a number of similar plants (i.e., all observations were taken at exactly the same time) and (2) successive observations of the same firm spread out over a period of time. These observations may produce either long- or short-run results, depending upon whether it is the scale of the plant or the intensity of its use which has changed between successive observations. Difficulty is usually encountered in isolating variation which is purely short-period or purely long-period.

STUDIES OF AVERAGE AND TOTAL COST

An example of simple relationships determined from data of the second type (i.e., successive observations of the costs of the same firm) is seen in an analysis made by D. D. Kennedy of the expenses of a machine shop over a period of years. By computing the coefficient of correlation between each expense item and direct labor hours (a supposedly variable item of cost) a classification was developed of the individual items of overhead according to their variability with production.[32] The classification thus built up could have been used (though this application was not contemplated by Kennedy) to formulate rough estimates of average and marginal cost at a series of output levels. The procedure followed by Kennedy is inadequate for our purposes in that the regression is simple rather than partial and is not related to output but to an item of variable input. It is consequently of limited usefulness in determining the pure relationship between average or total cost and the rate of output. Thus it is really of little direct value in computing marginal cost unless it can be assumed that direct labor hours vary proportionately with, and therefore accurately measure, output.

Instead of leaning heavily on estimates of the supposed behavior of overhead cost D. S. Cole used actual overhead at a series of output levels in developing a linear relationship. After skimming off fixed overhead items, he plotted the average monthly totals of variable overhead against the average monthly productive labor hours (his measure of the rate of production). He then drew a freehand line which represented the average relationship between overhead cost and a supposedly proportionately variable input factor.[33] The limitations of this simple but crude method are obvious. The curve was roughly drawn upon the basis of a few yearly averages which concealed within the year much significant variability in cost. Moreover, it represented a gross rather than a net relationship since no allowance was made for the influence of cost factors other than the rate of production.

The foregoing statistical studies treated cost behavior as though it were a simple function of output. A more realistic view of cost fluctuations can be taken if the effect upon cost of influences other than output is clearly recognized.[34] Such factors as size of the production order, level of wages and prices, efficiency of the labor force, and number of new styles manufactured are all constantly affecting cost variation and thereby distorting the pure relationship between cost and output. A complete grasp of the changing cost situation can be obtained only through an understanding of the influence of changes in the variables upon the resulting cost.

Two general methods are available for allowing for the distorting effect of these other influences. One is to remove the effect of these distorting factors upon the cost data before attempting to analyze the relation of cost to the rate of production. A second method is to measure such distorting influences at the same time that the analysis of the major relationship of cost and output is made. This can be done by means of multiple classification, multiple correlation, or similar techniques. In actual practice, however, these two methods can best be combined. Data which have been corrected for certain dynamic variables, such as price changes and long-term trends, are then subjected to a multiple correlation analysis in order to stabilize the effect of still other factors not conveniently removed from the data.

STUDIES OF MARGINAL COST

In additon to these analyses of average and total cost some studies of marginal cost were found. Even though not directly focused upon the determination of incremental costs, quantitative farm management studies have, nevertheless, made several significant contributions to the problem under investigation. These contributions may be classified under three general heads:

1. The techniques of multiple correlation, joint correlation, and multiple classification have been refined and developed through much use. (This is, perhaps, the most important contribution of all the farm cost studies.)

2. Increments in physical products associated with physical increments in input factors have been computed in an effort to aid the farmer in balancing marginal product with marginal input.

3. Regressions and part correlations have been established for the relationships between labor income and size of farm (or of component enterprise). The direct usefulness of these regression lines in determining marginal cost is relatively limited, however, because of the heterogeneity of the cost data, the difficulty of measuring size, and the irrelevance of the dependent variable commonly used, namely, labor income.

In addition to these agricultural studies a few industrial analyses were found which determined marginal cost by statistical methods. The most outstanding of these contributions which came to the attention of the investigator were made by W. L. Crum, R. T. Livingston, and T. O. Yntema. By determining multiple regression equations for the effect of passenger and freight traffic upon the operating costs of railroads Crum secured a measurement of the increment in cost which was occasioned by a ton-mile of freight traffic.[35] This marginal cost was constant with respect to output since only linear regressions were determined. But the irregularity of the data made this first approximation appear sufficient. Only two independent factors (both of them phases of output) were employed in the analysis on the ground that other influences were probably randomly distributed with respect to the relationship under investigation.

R. T. Livingston fitted a straight line, by the method of least squares, to data representing the total operating expenses of a steam-generating plant at various levels of output. As an indication of the degree of success attained by this simple linear correlation he computed the standard error of estimate and the coefficient of correlation. As one aspect of the economic significance of this linear equation of total cost he computed marginal cost, which he found to be constant for all levels of output, since the total cost regression was linear.[36]

This remarkable statistical determination of marginal cost employed a simple and objective technique which doubtless will yield useful practical results under certain conditions. There are, however, certain limitations to its general applicability. First, no allowance was made for the effect of changes in prices and in wage rates upon the historical total costs used as data.[37] Second, no provision was made for allowing for the time which often elapses between the recording of a cost and the recording of output. Third, no effort was made to hold constant various other distorting factors, by means of multiple correlation or multiple tabulation. And, last, no attempt was made to investigate the realism of the assumption that the total cost function is a straight line. This assumption of constant marginal cost appeared to be reasonable from Livingston's data; however, the output range was limited. The a priori expectancy was that the operation of the law of diminishing returns would cause the marginal cost to rise as the plant was pushed to capacity. The linearity of the total cost function should therefore be tested, not merely assumed. And even if it is established, the limited output range over which it is likely to hold should be specifically recognized.

Some of these limitations of the simple linear correlation method were not found in a multiple curvilinear correlation procedure used by T. O. Yntema[38] and W. W. Peterson.[39] Allowance for the distorting effect of various cost influences other than output was made by means of graphic curvilinear multiple correlation analysis. The flexible net regression curve of average cost upon output, which was secured by this procedure, made no presuppositions regarding the shape of the marginal cost curve. This net regression curve for average cost was then converted into a total cost function. The first differences of this total cost function were then computed and smoothed to secure a marginal cost curve.

The technique developed by Yntema for investigating long-period cost behavior was made the basis for the methods used in this study of short-period costs. However, since the problems are, in many respects, different, the procedure was amplified and modified in several respects. First, the cost data were laboriously rectified before commencing the multiple correlation analysis. This was found to be necessary because the data used in the present investigation appeared to be markedly affected by variations in prices and wages, and by lag in recording output. (Errors of this type were less serious in the data used by Yntema since short-period fluctuations tended to wash out in the annual averages which he used as data.) Second, several alternative procedures were tested and compared. Alternative degrees of rectification of the cost data, alternative methods of curve fitting, and alternative procedures for deriving marginal cost were tried out in order to note the differences produced in the results. Third, in addition to the analysis of the average and marginal behavior of combined cost, several of the more important components of combined cost were also analyzed. And, fourth, other margins of cost, in addition to the traditional output margin, were computed.

CHAPTER III

Methods of Preparing Cost Data for Analysis

An examination of previous investigations of cost behavior indicated that these studies failed to emphasize two of the most difficult problems in short-period cost analyses, namely, (1) the selection of the cost data and (2) the correction of defects in these data. The methods followed in this study in the collecting and sifting of the cost data, and then in correcting these data for the effects of irrelevant influences, are discussed under the following headings:

1. Collection of cost data
   A. Selection of establishment whose costs were studied
   B. Selection of cost items for which data should be collected
   C. Selection of unit for measuring output
   D. Selection of unit for measuring cost

2. Correction of defects in cost data
   A. Distortion prevented by holding certain disturbing factors constant
   B. Distortion caused by some of the irrelevant variables removed from the cost data
   C. Distortion caused by certain factors allowed for by introducing them as independent variables in a multiple correlation analysis

Collection of Cost Data

SELECTION OF ESTABLISHMENT WHOSE COSTS WERE STUDIED

In studying the short-period cost behavior of a particular business firm it was essential that an establishment be found which met with all the requirements for such a study. A factory in a small city, manufacturing medium-grade furniture, was decided upon. It made available to the investigator complete, detailed records, covering several years, of the actual costs for each two-week accounting period. Moreover, these years included frequent and drastic changes in the level of output and in the magnitude of the several cost items, without any accompanying changes in the scale of the plant.

Of the several plants operated by this company two were selected for investigation. They presented more difficult methodological problems than did the others because they were the largest, manufactured a greater variety of products, and experienced wider fluctuations in output and in size of production order. These two plants were treated as one unit because they were located side by side; had products which were almost identical; were substantially alike in degree of mechanization and in methods of oganization; were fed by the same veneer mill and dimension mill; and, lastly, because they were really operated together (since one method of running part time was to shut down one plant completely).

SELECTION OF COST ITEMS FOR WHICH DATA SHOULD BE COLLECTED

Only factory cost data were collected for analysis since information concerning selling and general administrative expenses was not made available to this investigator. These data were collected in as detailed an expense breakdown as the records would permit.[40] This was done because separate correction of the individual components seemed to improve the accuracy of these adjustments, and because separate correction was a necessary preliminary to a separate study of the cost behavior of these components, which study promised to broaden the methodological experimentation and enhance the usefulness of the entire study.[41]

SELECTION OF UNIT FOR MEASURING OUTPUT

The problem of selecting the proper unit for measuring output is an important one because output is the factor whose relation to cost is of primary concern in this study. This problem is a difficult one because of the great variety of products which constitute output and the numerous dimensions according to which output could be measured.[42] No single one of the many dimensions of output will be precisely suitable for all elements of cost. However, since the determination of the combined marginal cost of increases in the quantity of output is one of the principal objectives of this study, it is necessary to discover some one measure of output which will serve as a common denominator for the various types of finished product.

Output may be reflected by some input item which is proportional to output[43] or it may be measured more directly by some index of output. The various alternative measures which were systematically studied may be divided into two main types of units—physical units and value units. These may be further subdivided as follows:

Physical Units	*Value Units*
Measures of input	Measures of input
Board feet of lumber	Value of direct materials
Direct labor hours	Dollars of direct wages
Measures of output	Measures of output
Number of board feet	Wholesale price
Number of cases (pieces)	Standard cost
Number of each type of case	Old warehouse value

Previous studies of cost behavior throw some light upon this problem of choosing among these various measures of production activity. In general, it may be said that in agricultural studies physical units prevail. They are also used quite commonly in industrial studies as measures of input and, occasionally, as measures of output for firms free from product diversity. Value units, however, are used more frequently in industrial studies, particularly as measures of output.

After examining each of the foregoing alternative units for the purpose of determining a specific measure of production activity, no one of the physical input units listed for the furniture factory was considered quite suitable. Board feet of lumber was unsatisfactory because it is not a dominant input item and does not accurately reflect the relative cost-causing importance of the various articles produced. Direct labor hours were also rejected, partly because of the disparity of wage rates among departments, partly because of the relatively small importance of this input factor, and partly because it was not proportional to output. In a like manner physical measures of output were also found wanting. Number of board feet of output had the same limitations as board feet of input. Number of cases was insensitive to variation in the type of piece produced. An attempt to correct this defect by classifying cases by product types and thereby developing a weighted physical unit was unsuccessful because of the great variation in value which was found to exist within the type.

Value units appeared to be more promising. Value of direct materials is a more inclusive measure than a physical measure of input, and, what is more, it reflects variations in the complexity of product more accurately. However, this measure also reflects changes in the price of materials—a type of fluctuation which is irrelevant to the relationship under consideration but which could be removed by deflation. Perhaps the most satisfactory value measure of the input type is dollars of direct wages. However, it too is affected by rate fluctuations and fails to reflect output accurately.

Among the value measures of the output type, wholesale price and standard costs appear to have serious limitations. Wholesale price data are exceedingly laborious to collect and free from the numerous trade and credit discounts. Moreover, these data reflect shifts in the demand schedules for the various finished products and in the purchasing power of money. Standard cost, on the other hand, is free from some of these defects (e.g., it does not reflect shifts in the demand for finished products). But then, again, it is sensitive to changes in the prices of various input factors (since it is revised frequently to account for such changes). Consequently, standard cost has many of the deficiencies of a market-value measure of output.

Old warehouse value, as a value measure of output, is definitely superior to the other two value units just mentioned. It is a deflated standard cost computed for each separate model of furniture by summing the three following normal cost components:

1. Standard labor cost, deflated to the 1925-base wage level.
2. Standard material cost, figured on the basis of standard quality red gum and similarly deflated for price change.
3. Machine overhead, allocated upon the basis of a normal level of output.

It is important to notice that, once a warehouse value has been established for any given item, it remains unaltered regardless of changes in the rate of output, size of lot, and the like, and is changed only by major alterations in production procedures. This output unit is also unaffected by changes in the prices either of raw materials or of finished products.

Thus we can see that old warehouse value is superior to standard costs in that it is wholly unaffected by changes in the wages of labor, in the prices of materials, and in the quality of materials. And, unlike wholesale prices, it is impervious to shifts in the demand schedules of finished products, to changes in the purchasing power of the dollar, and to alterations in the discounts deducted consequent upon shifts in the channels of distribution. From one point of view total old warehouse value is an index of physical inputs under constant price and operating conditions. From the point of view pertinent to this study, however, total old warehouse value is a weighted sum of the various items of output and, consequently, is an index of physical volume of output.

SELECTION OF UNIT FOR MEASURING COST

Selection of a unit for measuring cost was not difficult. The choice lay between physical and monetary units. Physical units had certain significant advantages:

(1) no allowance would need to be made for changes in price level; (2) utilization of these results in making business decisions regarding future periods would be facilitated because the difficulty engendered by both absolute and relative price shifts would be obviated; (3) permanence of results would be increased because relatively changeless physical units would be employed rather than any unstable monetary unit.

Difficulties, however, barred the way to the use of physical units in this case. Complete cost data were not available in physical terms, and the difficulty of combining the cost elements thus measured was unsurmountable owing to the fact that there were thousands of types and qualities of cost constituents. Obviously, each of these constituents could not have been separately analyzed in terms of its particular physical unit. Hence, these diverse physical units would have had to be combined in proportion to their relative values at some base period. The resulting physical unit would have little superiority over the monetary unit in coping with absolute and relative price changes.

It was therefore decided to use the deflated dollar as the common denominator of cost. The dollar was the unit in which the data were most completely available, and also the unit which was familiar to executives and truly significant for business decisions.

Correction of Defects in Cost Data

After the cost data had been selected and the units of measurement chosen, the most obvious defects in the data were removed before the systematic analysis was begun. The methods used to free the cost-output relationship from the distorting effect of irrelevant variables may be grouped under three main heads:

A. Distortion was prevented by holding certain disturbing factors constant
   1. Factors which would have obscured the cost-output relationship were held constant by careful selection of the period for which the data were studied
   2. Cost items which could not be corrected for the serious effects of irrelevant influences were omitted

B. Distortion caused by some of the irrelevant variables was removed from the cost data
   1. Effects of certain disturbing variables were "averaged out"
   2. Certain cost items were stabilized at their three-year average
   3. Many cost elements were adjusted by means of index numbers of prices and wage rates
   4. Portions of certain costs were reallocated to the accounting period in which the output they contributed to was recorded

C. Distortion caused by certain factors was allowed for by introducing these factors as independent variables in a multiple correlation analysis

The first two of these groups of methods will be discussed in this chapter, and the third, which may more properly be regarded as a method of analyzing cost behavior, in the next chapter.

DISTORTION PREVENTED BY HOLDING CERTAIN DISTURBING FACTORS CONSTANT

By judicious selection of the period for analysis certain disturbing variables, whose effects are very difficult to remove, can be reduced approximately to constants. Three years were chosen during which technological progress in these plants was almost at a standstill and in which no growth of "fixed plant" occurred. Continuance of the same group of mature executives, moreover, kept managerial skill at a fairly constant level. Finally, accounting routines were not altered in any fundamental respect. What changes did occur in overhead allocation were allowed for.

Certain items of overhead cost were dropped from the analysis because their fluctuations were relatively great as well as irrelevant, and also were difficult to correct by other means. Most of these items depended upon arbitrary decisions of the accounting department and hence fluctuated erratically and irrelevantly. Items handled in this way were: (1) prorations of plant superintendence cost to other minor plants; (2) central administrative cost prorated to plants studied; (3) teaming and trucking among plants; and (4) traveling expenses of buyers and superintendents. Certain accounting periods were also dropped from the sample in order to remove the effect of a particular disturbing variable. Such action is dangerous, and is only justifiable when the distortion thus removed more than compensates for the loss in size and representativeness of the sample.

DISTORTION CAUSED BY SOME OF THE IRRELEVANT VARIABLES REMOVED FROM THE COST DATA

The two procedures discussed above attempted to prevent distortion of costs by holding constant the disturbing variables themselves. The second group of methods attempted to remove distortions from costs by means of direct modification of the data. This procedure was employed when the consequences of various disturbing variables could not be prevented and when the approximate effect of these irrelevant factors upon the cost data could be rather definitely determined.

A satisfactory sample being given, the effects of certain disturbing factors tend to "average out." This holds true, however, only when the errors are small and evenly distributed with respect to the relationship under investigation.[44] If the errors caused by disturbing influences affect only the values of the dependent variable and are not correlated with its true values, their presence does not tend to change the slope of the regression lines from their true slope. It does, however, tend to lower the correlation and increase the standard error of estimate.

A cost item was stabilized at its three-year average when it became clear that the causes for its fluctuation were not related to output. This was the case for three items of overhead cost: (1) depreciation, (2) property taxes, and (3) fire insurance. The variation in depreciation rates was not due to differences in intensity of use but rather to the firm's practice of ceasing to charge depreciation when a machine was fully depreciated. The cost-output relationship was in no way affected when a machine's useful life exceeded its estimated life. Variation in property taxes was a function of the judgment of appraisers and of the city tax policy. It was not a function of output. Fire-insurance rates moved up and down with the success of the town's fire department, in a manner wholly unrelated to changes in production activity.

Since there were violent fluctuations in material prices and in wage rates during the period chosen for analysis, it was necessary to correct the cost data for changes in the prices of these input factors. Construction of an index number for direct materials[45] involved problems of collecting data, sampling, weighting, and averaging. In collecting the data price quotations were secured from the purchase records. Whenever there were purchases at more than one price during any given two-week accounting period, a weighted average of these prices was taken.

A completely accurate index of material prices would include every one of the hundreds of items of material used in the manufacture of furniture. Since this was out of the question, a representative sample was selected. Commodities for which price-making forces were similar were grouped together, and samples were selected to represent each of these groups. The criteria used in selecting these group representatives were: (1) continuity of usage, (2) constancy of the percentage of total material costs, (3) typicalness of price fluctuations, (4) relative importance, and (5) frequency of price quotations.

The prices of the samples which had been thus selected as representatives of groups were then converted into relatives with January, 1932, as the base. In combining these relatives a weighted arithmetic mean was used and the price relatives of sample commodities were weighted according to the 1932 value consumption of the commodity group which they represented. This average was considered sufficiently accurate for the purpose of this investigation.

Construction of index numbers for indirect materials involved problems similar to those dealt with in computing an index number for direct materials. But two differences in the methods of computation are significant. In the first place, no single index number was developed to represent all the indirect materials; and, in the second place, some of the indirect material price indexes were much cruder than the direct material price index.

The index numbers thus constructed possess several limitations. In the first place, they do not correct for variations in price alone, but tend also to adjust partially for some of the differences in the quality of the materials. Furthermore, weights which reflect the relative importance of usage of a commodity in a base period may not measure usage at some later period. Under these conditions it is impossible to construct an index number which is unequivocal. Finally, even if this study be successful in removing most of the distorting effect of changes in relative prices during the years under investigation, it has only succeeded in artificially freezing prices into the relationship they had to each other for the base period (January, 1932). The cost-relationship formulas derived by this study are predicated upon this particular status of relative prices and/or relative quantities.

In addition to removing from the data the effect of changes in material prices it was also found necessary to correct the cost data for fluctuations in wage rates. Great changes occurred in wage rates during this three-year period and the effect of these changes upon cost was striking. Wage cuts, late in coming to the firm, hurried fast on one another during 1932. The average hourly rate for direct labor was forced down by April, 1933, to 48.6 per cent of its January, 1932, level. Wages for indirect labor followed a similar course. However, almost immediately afterward, wage rates were stepped up sharply, and by July, 1933, had almost reached the January, 1932, level (see Chart 1–1).

Chart 1–1 Index Numbers for Wages of Direct and Indirect Labor

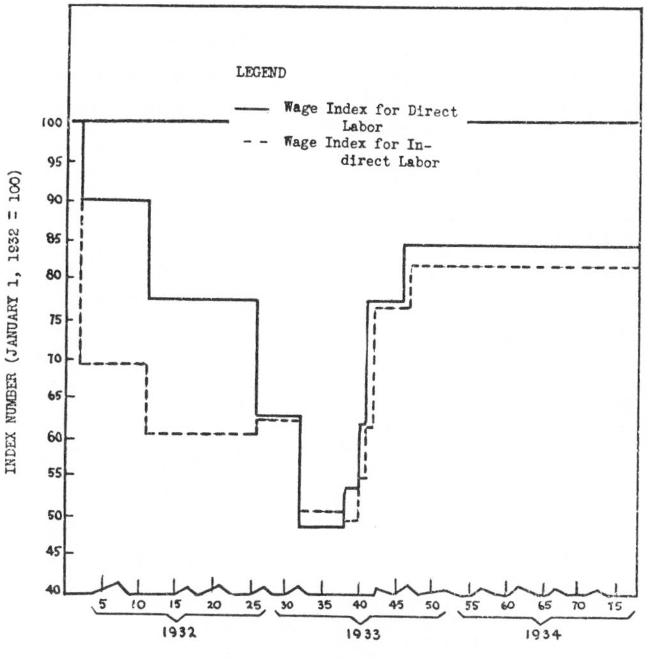

The correction for the effect of these sudden changes was accomplished by computing a separate index number for each direct and indirect labor department. It was thought advisable to compute these individual index numbers because: (1) there was some diversity in wage-rate fluctuations between departments; (2) the time lag adjustments, which were required in order to match costs with output, were different for various departments; and (3) many of the labor departments were to be made the subject of a special study, and this study would be more accurate if a specific, rather than an average, index were used for correction.

The data used for computing these indexes were the hourly wage rates found on pay-roll cards. A simple arithmetic average rate was computed for each direct and indirect labor department, and expressed as a relative with the rate of January, 1932, as the base. In addition to these separate departmental wage indexes a general index for direct and for indirect labor was computed by the same method. These index numbers are plotted in Chart 1–1 to show the extent of wage-rate fluctuations.

In order to correct the data for time lag (i.e., for a varying lapse of time between the recording of cost and of output) a reallocation of expenses to accounting periods was found necessary. The process-cost system used by this firm recorded output in the accounting period during which the product passed from the cabinet room to the finishing room. The majority of costs, on the other hand, were recorded in the accounting period during which they were applied to goods in production. Since goods usually remained in process at least two weeks, costs were frequently recorded in a different period from the one in which the output to which they contributed was recorded. Thus, comparison of costs as recorded with output as recorded would fail to associate output with the costs actually incurred in its production. Throughout the production cycle (which averaged about three weeks) costs of the departments which worked on the product prior to its arrival in the finishing room were certain to be recorded at an earlier date, and might be recorded in an earlier period than was the output to which they contributed.

In view of the seriousness of this time discrepancy it was clear that a mere comparison of recorded costs with recorded output would have failed to associate output with the costs actually incurred in bringing it into existence. Either costs or output would have to be shifted to their proper period, in order that the two could be matched up. But owing to the lack of departmental production records output could not be reassigned to previous records. Consequently, costs had to be corrected for this discrepancy. In making this cost correction two kinds of problems presented themselves. The first concerned methods of determining the amount of lag or lead for each cost item. The second centered about methods of adjusting cost for the lead thus determined. The treatment accorded direct departments was somewhat different from that given indirect departments; hence they are discussed separately.

As an a priori method of determining the extent of time lag for costs suggestions were elicited from production and accounting executives and were supplemented by an examination of accounting routines. Two kinds of discrepancy were ferreted out: (1) the lapse of time between the recording of the cost and the performance of the service and (2) the lapse of time between the performance of the cost service and the recording of output.

These a priori hypotheses regarding cost lag were then subjected to three kinds of empirical tests: production schedule analysis, time series comparison, and multiple correlation analysis. Production schedule analysis was the only method used for direct labor departments; time series analysis was used primarily for indirect labor departments; and multiple correlation analysis was applied only to tools and to finishing materials.

In the first of these empirical tests an informal analysis of production schedules by the planning department determined the average number of days an order remained in each department. The cumulative figure represented the number of days by which costs in that department anticipated the recording of output; then these figures were transformed into percentages of the ten-day accounting period. This percentage indicated the proportion of the costs of this department which was to be shifted forward or back one period. This procedure, however, was open to several objections: (1) the averages, upon which the whole scheme was based, covered up a great deal of variability; (2) differences in the type and quality of product produced during a particular period caused considerable variation in the operating schedule which was not reflected in the schedule of adjustments; and (3) complexity introduced at this point in the analysis would have to be cleared up when results were applied to forecasting costs. Nevertheless, this rough method was used to test suggested lags and leads of direct labor departments.

Time series analysis (the second of the empirical tests) was carried out usually by means of time charts and scatter-diagrams. For constructing time charts the cost elements were first corrected for a price change and reduced to a per-unit basis. Then these figures were plotted as a time series and compared over a light box with output (also plotted as a time series). In some cases a scatter-diagram was drawn relating output with a cost item which had been corrected for a suggested lag or lead. This was then compared with a scatter-diagram relating output with the uncorrected values of the cost element. By these informal statistical devices various deductively suggested lags and leads were tested for optimum fit.

On the basis of this test it was possible to group overhead costs into five categories according to their indicated approximate lag or lead-over-output reportings: (1) no indication; (2) no lag or lead; (3) lag of less than one period; (4) lead of less than one period; (5) lead of one period or more. For numerous overhead items no adjustment for lag or lead seemed advisable. Only when empirical and a priori considerations clearly indicated the need for a lag or lead adjustment was one made.

In closing it should be reiterated that this chapter is but a summary description of the methods actually used[46] to free the cost–output relationship from the distorting effects of irrelevant variables. Moreover, the correction devices used in this study are, to some extent, peculiar to this case. Nevertheless, the problem of correcting the data for various types of distortion will be encountered in any short-period cost study.

DISTORTION CAUSED BY CERTAIN FACTORS ALLOWED FOR BY MULTIPLE CORRELATION ANALYSIS

Finally, when the consequences of irrelevant factors could not be prevented and when their effects could not be measured independently, then, in order to remove them from the data, a third general method for preventing cost distortion was employed. These recalcitrant factors (if important and susceptible to statistical treatment) were introduced as independent variables in a multiple correlation analysis. Thus their effects were removed by studying the relationship of cost to one independent factor (output) while holding the other independent variables constant at their means. This method of allowing for the effects of cost influences is treated in the next chapter, where quantitative analysis of average cost behavior is discussed.

CHAPTER IV

Methods of Analyzing Cost Data

Up to this point the difficulties encountered in preparing the cost data for analysis have been discussed. In the present chapter the problems involved in analyzing these rectified cost data are considered. These problems center upon the determination of the relationship between cost and the various factors which affect its behavior. For the purpose of the present discussion they may be grouped under the following major headings:

A. Selection of technique for analyzing cost behavior.
B. Selection of cost constituents to be analyzed.
C. Selection of accounting periods for analysis.
D. Choice of factors affecting cost behavior.
E. Methods of deriving marginal cost.

Obviously, all these problems are actually too closely interrelated to be considered separately. Decisions as to the kind of technique to be used depend upon the independent variables available, the nature of the sample, etc. For clarity in presentation, however, these problems have been separately considered as listed above.[47]

Selection of Technique for Analyzing Cost Behavior

In order to secure usable knowledge of average and marginal cost behavior the following definite kinds of information were required:

1. Some definite measure of the net effect of several important cost influences upon cost behavior. Such a measure should take the form of a curve or a schedule of per-unit or of total cost at a series of values for each important factor affecting cost. In such form it is directly usable for setting standards and making forecasts of average and of total cost, and also for determining marginal cost.

2. Some indication of the combined importance of the effect of these causal factors upon cost. Such a measure is highly serviceable in determining the limits of usefulness of cost standards, cost predictions, and of marginal costs computed on the basis of the foregoing relationships.

3. Some measure of the reliance which can be placed upon the curves or schedules as representations of cost behavior. This kind of measure is of particular importance in the case of the curve showing the relationship between output and cost, since this curve is the basis for the determination of the marginal cost curve.

Several alternative techniques for analyzing cost behavior were considered in order to ascertain which would give the required information most completely and most economically. The most important of these methods are:

1. Estimates of cost behavior on the basis of a classification of costs according to their variability as indicated by a priori notions.
2. Multiple tabulation of actual cost data, sorted on the basis of the values of the various factors which influence cost.
3. Joint correlation analysis of the effects of factors which operate together.
4. Mathematical multiple correlation analysis of the effects of influential cost factors.
5. Graphic multiple correlation analysis of the effects of important cost factors.

From the foregoing methods graphic multiple correlation analysis was chosen as the most suitable for this project. This choice was made despite the fact that the graphic technique is inferior to the others in certain respects: It is more laborious than the method of estimates; it is less satisfactory than joint correlation if joint relation with the dependent variable exists; and it is inferior to mathematical multiple correlation both in objectivity and in the definiteness of its error formulas. Nevertheless, it was preferred because of its economy (as compared with all except the estimates method) and flexibility, and because the regression curves and error formulas which are used as implements of analysis also serve admirably as methods of presenting the average and total cost findings and lead conveniently to the derivation of marginal cost curves. The well-known Bean-Ezekiel technique of graphic multiple correlation analysis was employed. It is fully described by Ezekiel, in *Methods of Correlation Analysis* (New York: John Wiley & Sons, Inc., 1930), chapter xvi.

Selection of Cost Constituents to Be Analyzed

In deciding which cost constituents should be thus analyzed the choice lay between limiting the study to an analysis of corrected combined cost alone, or broadening the scope of the investigation to include an examination of the various constituents of combined cost.

The decision depended upon whether or not the objectives of this investigation could be realized short of the more detailed study. These objectives were: (1) exploration of methodology; (2) development of a practical procedure; (3) investigation of average cost behavior for the purpose of (*a*) determining flexible cost standards, (*b*) improving cost forecasts; and (4) determination of marginal costs. Despite the fact that combined cost is of dominant importance, both for managerial control and for marginal cost determination, it was clear that nothing short of detailed multiple correlation analysis of cost elements would bring an approximate realization of the objectives mentioned above.

The following definite contributions to methodology resulted from the analysis of the cost constituents: (1) Detailed correction of the data for cost elements was made necessary, thereby developing some refinements in correction technique. (2) Additional independent variables were introduced which were of peculiar importance to a particular cost element, but of relatively little significance for the analysis of combined cost. (3) Methods of allocating and combining cost elements were developed in an effort to apply the results of these detailed studies to improving cost standards and cost forecasts.[48] (4) A rough test of the value of correcting data for rate changes and for time lag was developed by means of the analysis of uncorrected combined cost.

As a result of these additional detailed analyses flexible cost standards were determined more accurately and were made more useful to management. Accuracy was increased by the employment of independent variables which were thought to be of peculiar significance to the specific cost being analyzed. The usefulness of cost standards was enhanced because costs are more easily controllable in detail than in total and causes for departure can be determined more accurately and responsibility for adherence to standards assigned more definitely.

The accuracy and usefulness of cost forecasts were also increased by the addition of detailed analyses. Accuracy was increased, partly because prediction was based upon a somewhat better understanding of the causes of cost behavior and a quantitative measure of the past effect of various factors upon cost constituents, and partly because it was safer and easier to make a priori allowances for the effect of changed conditions upon individual cost constituents than upon their total. The usefulness of cost forecasts was extended because they were now available in more controllable detail and also in a form enabling allocation of estimated cost more specifically to particular articles or output lots.

Estimates of marginal cost presumably could be assigned more definitely to a particular article or a particular order[49] by using the marginal curves of cost elements, instead of assigning combined marginal cost to the article on the basis of its old warehouse value alone. This procedure, however, assumes the separability of these cost constituents and presupposes considerable accuracy in the estimates of their marginal cost.

For each of the following elements of cost a separable multiple correlation analysis and marginal cost derivation were made:

1. Corrected combined cost
2. Uncorrected combined cost
3. Direct material cost
4. Direct labor cost
5. Combined indirect cost
6. Combined indirect labor cost
7. Inspection labor cost
8. Yard labor cost
9. Power and watch labor cost
10. Indirect materials
11. Power
12. Maintenance

Selection of Accounting Periods for Analysis

Data for all the available accounting periods could not be used in the analysis, partly because of defects in the records, and partly because of the limits placed by the graphic method itself upon the number of observations which could be efficiently analyzed. It was decided that a three-year period would provide a sufficient number of observations. The years 1932–34, inclusive, which were chosen, had certain advantages: (1) the short-run rather than long-run character of cost adjustments during this period; (2) the pronounced variability in all operating conditions; and (3) the recency of these three years.

Of the seventy-eight two-week accounting periods in these three years, thirty-one had to be discarded because of various types of errors and omissions in the data. The remaining forty-seven accounting periods were employed in the analysis of the following major items of cost:

1. Corrected combined cost per unit*[50]
2. Uncorrected combined cost per unit*
3. Direct labor cost per unit*
4. Direct material cost per unit
5. All indirect cost per unit
6. Indirect labor cost per unit*

For the analysis of subsidiary items of overhead cost a smaller sample of twenty-five picked accounting periods was used. The principal criteria used in the selection of this sample were: (1) that satisfactory data be available and (2) that the values of the independent variables be fairly evenly distributed over their range. The sample thus chosen was employed in the following subsidiary analyses:

1. Inspection labor cost (as a total)*
2. Power and watch labor cost (as a total)
3. Yard labor cost (as a total)*
4. Indirect material cost (per unit)*
5. Power cost (per unit)*
6. Maintenance cost (per unit)*

Choice of Factors Affecting Cost Behavior

The decision being made to use graphic multiple correlation techniques for analyzing the foregoing cost items, it was necessary, first, to choose as independent variables those factors which were most important as influences upon cost behavior and, second, to select the most suitable statistical series with which to represent these factors.

The choice of cost factors was made on the basis of the following criteria: (1) the factor should have a significant influence upon cost; (2) the factor should represent a cost influence which is distinct, that is, which is not to any important degree already included in some other independent variable; (3) the factor should be susceptible to statistical measurement.

Not all the factors which were used satisfied all the foregoing conditions, but it would have been difficult to find factors which would. The factors selected on the basis of these criteria were:

X_2—Output
X_3—Size of production order
X_4—Unfamiliarity of patterns
X_5—Change in output from previous period
X_7—Instability of labor force
X_8—Quality of material
X_9—Receipts of lumber
X_{10}—Fuel outlay

The second step in selecting the independent variables was to choose statistical series for representing the factors listed. Output was measured by old warehouse value; size of production order, by the average number of pieces per manufacturing lot; unfamiliarity of patterns, by the number of new styles produced for the first time in the accounting period in question. Change in output from the previous period was represented by two alternative statistical series. The series used in most of the analyses were increases or decreases in output from the previous period measured in dollars of old warehouse value. In analyzing direct labor and direct materials, however, percentage increases and decreases were used. Instability of the labor force was measured by the ratio of the number of additions to the force to the total number of men on the pay-roll; quality of materials, by the ratio of the standard cost of output to the old warehouse value of output;[51] receipts of lumber, by the number of board feet unloaded; and fuel outlay, by the dollars of expenditure for coal.

Some of these independent variables were found to be of greater significance for certain cost items than for others. Table 1–2 shows the set of independent factors employed in the analysis of each cost item.

Methods of Deriving Marginal Cost

These multiple correlation analyses of average and total cost provided rectified data for the determination of marginal cost. By holding constant the effect of other variables the pure relationship between cost and each of the independent variables was brought out. Thus the determination of the increment in cost which was associated with a change in one of these cost conditions was made easier. Although the incremental cost of each of the several independent factors may be of practical significance, the marginal cost of increases in the quantity of output is easily of greatest importance.[52]

In deriving marginal cost curves from these data one of two alternative general procedures could be followed. The data could be studied either in the form of totals or in the form of averages. In the first procedure a function was fitted to the total cost residuals secured from the net regression chart for output. This total cost function was then differentiated, and its first derivative taken as the expression for estimated marginal cost.

Three techniques were employed in fitting a function to these total cost residuals: (1) visual, (2) selected points, and (3) least squares. The visual method was used as a part of the graphic correlation analysis in the case of those items whose data had been kept in terms of totals (yard, inspection, and power labor costs). The method of selected points provided a short-cut technique for converting these flexible visually fitted curves into mathematical expressions which could be conveniently differentiated to secure an equation for marginal cost. For example, for yard labor cost the resulting equations, together with a series of substituted values, are shown in Table 1–9. The method of least squares was used for combined cost. By this third technique both a straight line and a parabola were fitted to the total cost residuals of combined cost. The results of this experiment are shown in Chart 1–4 and Table 1–4.

As an alternative to this total cost procedure, a second general technique was developed which derived marginal cost by manipulating a function fitted directly to the individual average cost residuals. The total cost function was secured in this case by transforming the fitted average cost function rather than by transforming the individual residuals to total cost residuals and then fitting a function to these.

The basic procedure was very simple. First, the net relationship between output and average cost per unit was determined. This relationship was recorded either as a curve which had been visually fitted to regularly spaced group averages found in the final regression chart, or as an equation which had been mathematically fitted to these group averages (or to the individual observations). Second, the resulting average cost function (curve or equation) was converted into a function which expressed these cost values as totals, rather than as averages per unit of output. Third, the first differences of this total cost series were taken. If the relationship was expressed as an equation, then the first derivative of the total cost equation gave the equation for the marginal cost curve. If, on the other hand, total cost was represented as a curve or a table of values, then its first difference series had to be smoothed by means of a moving average and by inspection to secure the required marginal cost curve.

Four variants of this fundamental average cost procedure may be distinguished primarily on the basis of the methods of curve fitting. They may be called for ready reference:

1. The "first differences method," in which no curve was fitted to the average cost data.
2. The "graphic method," in which curves were fitted visually and then differentiated by tabular methods.
3. The "derivative method," in which the least-squares method was used for curve fitting and the calculus for differentiating.
4. The "graphic-derivative method," in which the visually fitted curves were transformed into equations by the method of selected points and differentiated by the calculus.

In the method of first differences the relationship between average cost and output was recorded and manipulated in the form of a series of regularly spaced group averages of per-unit cost. Next, the series of interval averages of per-unit cost was converted into a series of total cost values by multiplying each item of the per-unit cost series by its corresponding output value. And, finally, the first differences of the resulting total cost series were then computed and smoothed to obtain the marginal cost curve. Chart 1–5 shows the results obtained by this method for combined cost.

The graphic procedure is somewhat simpler than the method of first differences. Here the visually fitted net regression curve of average cost on output represents the relationship between per-unit cost and the rate of production. The per-unit cost series (regularly spaced readings from the curve) was converted into a total cost series by multiplying each member by the output at which this curve reading was taken. The first differences of the resulting total cost series were then smoothed to yield the marginal cost curve. The resulting curve is shown in Chart 1–5.

The method of derivatives differed from those just discussed in that the relationship between output and average cost per unit was expressed in the form of an equation rather than as a series or a curve. This equation was obtained by the method of least squares, from the per-unit cost observations found in the final net regression chart of average cost and output. When this average cost equation had been converted into the total cost equation, its first derivative was taken as the equation for the marginal cost curve. Results secured by this method are shown in Chart 1–5.

The hybrid method combines the first steps of the graphic procedure with the last steps of the derivative technique. This method took as the average cost curve the graphically fitted final net regression curve of per-unit cost on output. To this curve an equation was fitted by the method of selected points. The average cost equation thus secured was then transformed into an equation for total cost, whose first derivative was the mathematical expression for the marginal cost curve. This hybrid method was used as one procedure for deriving the marginal cost of most of the components of combined cost.

An evaluation of each of the foregoing procedures, based upon a comparison of the results secured by each method, is found in the chapter dealing with the marginal curve for combined cost (chap. vii). A complete comparative description of these procedures is also found in the author's dissertation previously cited.[53]

CHAPTER V

Reliability of Findings

As a preface to the presentation of the specific findings of this study, certain general limitations upon the reliance which can be placed in these results should be stated. These limitations apply to both average and marginal cost behavior. They also apply both to the accuracy with which the cost formulas represent the actual cost behavior of the period analyzed and, more especially, to the dependability of the formulas as a forecast of future cost behavior. The accuracy and stability of these cost relationships are of considerable interest since their reliability determines, to a great extent, the managerial usefulness of the findings of this study. Hence it was thought advisable to summarize and evaluate in one chapter the various limitations so that they could be readily kept in mind in considering the later chapters of this study, which present the findings of this investigation into cost behavior.

These limitations will be discussed under the following four heads:

1. General assumptions regarding cost behavior.
2. Defects in the data.
3. Imperfections in the technique of analysis.
4. Limited applicability of the error formulas developed.

General Assumptions Regarding Cost Behavior

Certain general assumptions regarding cost behavior have been tacitly made in the course of this project. In the first place, the cost function of the firm studied is assumed to be sufficiently separable from the demand function to permit its independent analysis. This assumption appears to be substantially valid, for the functional relationship between manufacturing cost and its causal factors does not seem to be affected to a quantitatively significant degree by the shape and position of the demand schedules for the products made by this firm.[54]

A second assumption, namely, that the cost curves for this firm are sufficiently reversible and sufficiently continuous to be portrayed as two-direction curves or bands, is more questionable. Cost adjustments required to increase output from x to $(x + 1)$ output are not the exact reverse of those required to reduce output from $(x + 1)$ back to x. A rise in output which is followed by a drop to its former level is unlikely to return cost immediately to its previous figure. An attempt has been made to allow for this distortion by introducing, as an independent correlation variable, the amount and direction of change in output from that of the previous period. This correction, however, is too simple to allow for the complex patterns of change. Some account should also be taken of the number and amount of successive decreases or increases in output which preceded the current adjustment and, if possible, of the length of time at which it was expected that production would remain at the new level.

Doubt also attaches to the concept of the continuity of the cost functions. A smoothed curve is a statistical approach to a reality which could probably be more correctly represented by a staircase of discrete steps of cost adjustment. However, no great violence appears to have been done by this conventional assumption of continuity.

In the third place, it is assumed that the proportion in which the cost constituents are combined is affected by changes in the rate of production, size of lot, new styles, etc., but that it is not affected by changes in the relative prices of constituents. Examination of the regression curves of the components shown in Charts 1–9 through 1–15 of Appendix A indicates that the proportion of cost elements does vary with the rate of output and with the other independent variables. The second part of the assumption would appear to be more questionable. But, for the breakdown of cost used in this analysis, the proportion actually seems not to be affected by changes in the relative prices of the factors, within their observed range.

The technology of production is fairly rigid in the short run in a firm so highly mechanized as the one studied. Hence the resources are not very adequate substitutes for one another, since significant substitutions among the main categories of cost usually involve considerable rearrangement of the techniques of production. For the period studied in this investigation the technical coefficients of production at any given level of output, order size, etc., have probably remained fairly stable regardless of the considerable shifts which took place in relative prices. Substitution caused by changed relative prices has probably been confined mainly to shifts within the classifications of cost used in this study, e.g., within the general category of indirect materials, and has not extended to shifts among these classes of cost. With continued stability of technology it appears unlikely that the findings for individual cost components will be invalidated for future use by an increase in substitution, since shifts in relative prices greater than those which occurred during the period of study seem improbable. Thus it appears that, although some substitution will to a certain degree limit the significance of the separate analysis of cost components, it will not seriously damage its usefulness.

The fourth assumption is that the underlying causes for the relationships between cost and the several independent variables have remained essentially the same during the lapse of time covered by this study. This assumption appears to be fairly correct for the particular period analyzed, since a series of years was selected during which technology, production routines, and management personnel remained about the same. The general stability of the underlying conditions in these three years is in striking contrast to the marked variability in the intensity of use of the fixed factors and in the prices of certain of the variable input ingredients. This stability of plant scale and other fundamental conditions makes the observed behavior of cost in response to short-period changes in output fair approximations to the theoretical neoclassical cost functions developed from the "timeless" assumptions of static economics.

The cost formulas determined on the basis of past cost behavior in this period of exceptional stability in production techniques, accompanied by exceptional instability in output, size of lot, and the like, and in input prices, cannot be expected adequately to portray the cost relationships of future periods. If underlying conditions change rapidly in a manner which cannot be allowed for, then the results of this study will be of little direct value to management in setting standards, making forecasts, and deciding upon price and output policy; but it will still be useful as indicating a procedure for dealing with the changed circumstances.

The fact that these relationships were determined for data representing a depression period throws considerable doubt upon the value of these findings in making forecasts for normal periods. Cost behavior for depression years may be considerably different for several reasons. First, the altered morale of the labor force doubtless affects cost. On the one hand, fear of unemployment tends to drive men to attempt greater accomplishment, while the psychological accompaniments of a depression may unfavorably affect the productive capacity of the more sensitive workers. The realization of the imminence of one of the periodic shutdowns may cause some loafing at the tail end of a run. Both these tendencies would alter cost relationships. Second, selection of top-grade workmen through selective layoff should result in varying reductions of labor cost covering all departments.[55] Third, reduced schedules narrow the range of actual output levels, and thus fail to provide information as to cost relationships for high rates of production activity. Fourth, managerial skimping on postponable expenses may have altered temporarily the proportion of such items of cost.

The strong probability that changes in the cost relationships will occur in the future for the firm studied indicates the need for devices for adjusting the formulas for some changes and for recomputing them when this adjustment is no longer reliable. In applying the cost relationships developed in this study it should be recognized (1) that their dependability is limited by some inevitable changes in cost conditions during the period covered by the analysis and (2) that the unreliability of these results for prediction purposes increases with the magnitude of such changes in the future and with the length of time which elapses between the analysis and the forecast.

A closely associated assumption is that changes in managerial control of costs have not distorted the cost relationships under analysis. It is unreasonable to assume that management is omniscient and errorless; but it is not unreasonable to assume that its failings remain fairly constant for the period under investigation. The truth most likely is that the character and degree of managerial control over cost probably have not remained entirely constant. Hence this assumption of constant managerial efficiency is not entirely accurate.

The error in the foregoing assumption affects the reliability of predictions more than it does the accuracy of the statement of cost relationships during the period studied. Managerial changes far more drastic than those which occurred during this period may transpire in the future to curtail substantially the usefulness of these particular results as standards and forecasts.

It is also assumed that old warehouse value is a satisfactory measure of output. To be satisfactory this measure should have the following characteristics: (1) It should not reflect changes in physical inputs per unit of output caused by shifts in the rate of production. In other words, it should measure output rather than input if the two do not vary together. (2) It should not reflect changes in the value of physical output units caused by shifts in desirability or by changes in the price level.

As far as the investigator could ascertain, old warehouse value seems to be fairly satisfactory in each of these two respects. One other condition must be satisfied, however, if this measure is to be entirely adequate. The proportions of the models making up output must not seriously affect the combination of cost components. It must be true that the proportions between different models remain constant, that changes in the proportions average out, or that the same cost components are used in approximately the same way regardless of the character of the particular piece of output being made. The partial realization of all three of the foregoing conditions probably eliminates most of the error which could come from this source.

Defects in the Data

A second general type of limitation upon the reliability of the findings of this study consists of defects in the data. These imperfections have already been discussed in connection with attempts to correct them. It is therefore sufficient, at this point, to recall them to the attention of the reader in order that he may be on his guard against the residuum of error remaining after their incomplete correction. Limitations of this type are not peculiar to this investigation alone, but will be found in any study of a similar nature. Serious though these limitations are, they prove upon examination less damaging than their mere mention would indicate. Many of them may be randomly distributed with respect to the cost functions and hence may not prove distorting. The principal defects in the data may be classified under the following general headings:

1. Measurement of the phenomena studied was often indirect and frequently inexact. Although the units finally selected for measuring cost and the various independent variables seemed to be the best available, they still were not all that could be desired. The various adjustments of the reported cost figures removed most of the error associated with accounting routines, but here, too, correction was not perfect.

2. The manner of recording cost and output may have covered up significant variations. Daily fluctuations in cost, output, etc. were lost by totaling or averaging for the two-week accounting periods. Also the combining of varied types of product by using old warehouse value as a common denominator may have hidden cost variations arising from the diversity of output. This imperfection, however, is probably not as serious as it appears since the assortment of output is actually fairly constant and the processes are about the same for all units of output.

3. Correction of the cost data for irrelevant cost influences was not complete. These corrections were, however, as refined as will prove practical in commercial investigations and were probably sufficiently adequate for determining the general character of the cost functions.

4. The representativeness of the sample may be questioned: It was small and covered three depression years. The sample, however, was as large as was manageable with the technique chosen, and the advantages of the particular three-year period (as to recency, stability, and multiplicity of methodological problems) appeared to outweigh its cyclical disadvantage.

Imperfections in the Technique of Analysis

A third group of general limitations grows out of the methods of analysis. These limitations have already been discussed; consequently, they will only be listed here as a reminder of the extent to which they impair the reliability of the results of this study.

1. The partial correlation measures of cost behavior often embrace more territory than they purport to. One of the properties of the net regression curve is that it ascribes to any specific independent variable, such as output, not merely the behavior of the cost due to this variable alone, but also the variation which is due to such other independent variables as are correlated with this variable and have not been separately considered in the study.[56]

2. There were some important cost factors which were not separately included in this study. It is possible, however, that the omitted factors were randomly distributed with respect to the main cost relationships; and it is improbable that their inclusion would have had a pronounced effect upon the cost-output relationship.

3. The curve fitted by the graphic method may not represent accurately the observations upon which it is based. Visual fitting lacks, to some extent, a desirable

objectivity since so much is left to the personal judgment of the investigator. Moreover, the ease with which complex curves can be fitted is a temptation for their too generous use.

4. Marginal cost curves derived from flexible, visually fitted average cost curves frequently exhibit spurious and illogical fluctuations. This is due to the fact that a slight flexion of the average cost curve causes a magnified fluctuation in estimated marginal cost. It is often the case that only the general level and direction of marginal cost curves derived by this method are significant. On the other hand, those derived from total cost observations are more reliable since marginal cost is a simple function of the total cost curve (its slope) and hence unwarranted fluctuations in marginal cost can be more easily avoided.

Limited Applicability of the Error Formulas Developed

A fourth type of general limitation concerns the confidence which can be placed in the formulas for standard error. These formulas purport to state in precise terms the degree of reliability of the results obtained. The use of these formulas, however, is based upon certain assumptions not completely realized in this case.

1. A random sample from an unchanging universe is assumed.[57] It is clear that our sample is not a random one. The major sample of forty-seven accounting periods represents a fairly complete coverage of a presumably typical three-year period. If the cost behavior was typical during this period, then the sample is representative. The minor sample of twenty-five cases is a "purposive" sample[58] selected to cover the range of the independent variables. Not only is the sample not random but our universe is clearly not altogether unchanging.

Although the sample is not random, it may, nevertheless, be fairly representative of cost behavior during a period remarkably free from changes in the underlying cost conditions. Whether or not the universe is unchanging depends upon the meaning given the term "universe." In a time series study such as this, the universe could be conceived of as an imaginary one in which operating conditions, prices, etc. remained roughly within the same range as during the three years studied. With this view of the matter the error formulas would indicate whether or not successive samples of this type taken from this assumed universe would show that the independent factors did affect cost approximately as specified in the past period which was analyzed.[59] It appears therefore that, although the assumption of random sampling from an unchanging universe is not satisfied by the conditions of this study, the error formulas may nevertheless be useful, even if not completely satisfactory in judging reliability.

2. Normal frequency distribution of the data is assumed by the least-squares error formulas. The validity of this assumption may be tested by the examination of the plotted observations. Such inspection indicates that normal frequency distribution was approximated in a large part of the data used for this study.[60]

3. Error formulas computed by graphic methods are assumed to have a signifi-

cance parallel to that of those computed by classical least-squares methods. The graphically determined standard error of estimate is fairly well established. But the graphic methods suggested by Ezekiel for computing the standard error of a curve are still provisional. The general principles upon which these formulas are based, however, appear reasonable, and the results secured have stood up well under extensive experimental testing. Moreover, it should also be pointed out that these graphic error formulas were supplemented in this study by other criteria of significance computed by orthodox methods.

In general, it appears that the assumptions underlying these error formulas are not as far from the actual conditions of the investigation as might at first appear. A representative sample of a practically static and fairly typical period approximates the ideal of a random sample of a changeless universe. Almost normal distribution was found in the case of many of the variables. And graphic error formulas, although provisional and not exactly parallel in significance to least-squares formulas, appeared useful in indicating the significance of observed cost relationships, especially when supplemented by other criteria of significance.

In conclusion, a survey of the various limitations seems to indicate that: (1) a fair amount of faith can be placed in the general shape and location of most of the average and total cost functions; (2) less confidence can be placed in the marginal cost curves which were derived from average cost curves than in those computed from total cost functions; and (3) the dependability of the cost relationships for use in the future diminishes with the amount of change in cost conditions which takes place in the time which elapses.

Chapter VI

Behavior of Average Combined Cost

Knowledge of the behavior of average cost is useful to management for several practical purposes:

1. Cost control can be made fairer and more effective. Cost standards are made flexible so that they allow automatically for operating conditions over which production management has no control and for which it should therefore not be held responsible. Standards thus set are alibi-proof, for, if current average cost departs from these adjusted standards, the factory management alone can be held accountable.

2. Budgeting of expenses and profits can be made more accurate and flexible. Forecasts of costs can be made with greater precision and can be adjusted more gradually and more certainly for changed operating conditions when they are based upon knowledge of the actual effect of important cost influences. Flexible budgets of this character are more useful as control devices and more reliable as a basis for profit predictions than are fixed budgets. With this basic knowledge of average cost behavior the budget can be made alive and adaptable; it can be tailored to fit the actual operating conditions of the period and speedily remodeled as these conditions change.

3. Evaluation of cost-reduction policies is made more objective and accurate. A proposal to take steps to diminish the number of new styles, for instance, could be better appraised if the average net effect of the introduction of a new model had been measured.

Attainment of these practical objectives requires, first of all, a careful measurement of the effects of various operating conditions upon average costs.[61] The first section of this chapter presents a statement of the observed character of the relationship of corrected combined cost to each of several important cost influences, together with an appraisal of the degree of reliability of these functions and a measurement of the degree of closeness of the relationships.[62] In the second section of this chapter the behavior of uncorrected combined cost is presented with a view to determining whether or not the elaborate correction of the data was worth while as judged by its effect upon average cost results.

In describing the various findings for average cost behavior several general types of results of the graphic multiple correlation analyses may be noted.

1. Statements of the relationship between cost and the several independent factors
   a) Net regression curves showing cost at various values of the several independent factors
   b) Tables of estimated cost under various operating conditions, presenting in a more understandable form the information portrayed by the regression curves
2. Measures showing the probable reliability of results
   a) Standard error of estimate of cost
   b) Standard error of regression curve
3. Indexes showing the closeness of the relationship between cost and those independent factors affecting cost[63]
   a) Index of multiple correlation
   b) Index of multiple determination

Corrected Combined Cost

STATEMENTS OF THE RELATIONSHIP BETWEEN COST AND THE SEVERAL INDEPENDENT FACTORS

A net regression curve states the pure relationship between cost and one independent factor (such as output) while the remaining independent variables are, in effect, held constant at their means. The relationship of cost to output is stated as the average of observed cost associated with particular levels of output; readings from this curve show the expected cost at the corresponding level of output. The relationship of cost to any other independent factor is stated as the average deviations of cost associated with various values of such factor. Curve readings for these other factors are added to or subtracted from (as the case may be) the output curve reading in order to obtain the expected cost for a particular combination of output and other factors.

It is desirable to compare the observed cost behavior indicated by each of these regression curves with the behavior which might theoretically be expected. Such a comparison, by testing the reasonableness of results and attempting to account for illogical departures, enhances the confidence with which practical application of these results can be made.

The regression curve of average combined cost per unit on output would be expected to be asymptotic with the *Y*-axis, then decline to the right for a time, and finally rise again. In the short run (defined as a period short enough to permit no changes in the scale of plant, but only variation in the intensity of its use) the average cost curve would theoretically be expected to be asymptotic to the *Y*-axis because of the irreducible minimum of expense which must be spread over an infinitely small output. As the rate of production is increased, the curve declines to the right because of the spreading of this irreducible constant cost over a larger and larger output. Eventually the decline of average fixed cost is counterbalanced by a rise of the other component of combined cost, namely, average variable cost. This increase in variable cost per unit will eventually come about through the operation of the law of diminishing returns. As more and more of a variable input factor is applied to a constant factor, the output tends to increase at a diminishing rate. As a result the variable costs per unit rise sufficiently to counteract the decline of fixed cost and result eventually in an upturn of the combined cost curve.[64]

Thus we see that the average combined cost per unit would be expected, in the short-run period, to have high values for small output, to fall at a diminishing rate as output is increased, and, finally, to rise at an increasing rate as output is still further expanded.

From Chart 1–2 it can be seen that the observed output regression curve corresponds rather closely to expected behavior over most of its course. It has high values for small output and falls at a diminishing rate as output is increased. However, the curve fails to rise at the maximum of the range of output found for the three years studied.[65] Investigation of records for a period of years showed that output was rarely pushed beyond the limits included in this investigation. This suggests to the investigator that under conditions of monopolistic competition, with sharply rising selling costs, a firm may not normally push production to the point at which short-run average manufacturing costs actually rise.

In examining the regression curve for size of production order the normal expectation is that an increase in the number of cases per production order will cause

Chart 1–2 Partial Regression Curves for Corrected Average Combined Cost with Six Independent Factors: Output (X_2), Size of Production Order (X_3), Number of New Styles (X_4), Increase in Production (X_5), Labor Turnover (X_7), and Quality of Material (X_8)

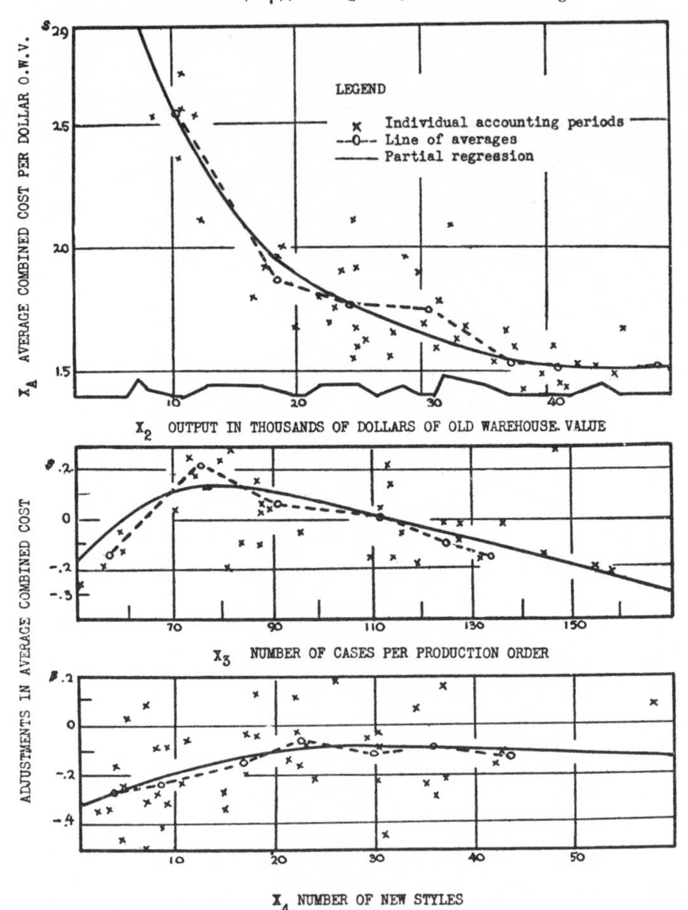

some reduction in the per-unit cost. This is because a constant setup time would be spread over a larger number of items with a consequent reduction in the per-unit cost of machine operations. In addition, increased familiarity with a particular assembly would tend to diminish the cost of hand operations. These tendencies would not be offset in this case by larger inventory cost, for carrying costs have not been included in the cost data.

The observed effect of increasing the size of the production order corresponds in general with the expected effect. According to the net regression curve for number of cases per production order in Chart 1–2, cost declines over most of the range as the size of the lot increases.[66]

The regression curve for the number of new styles produced per period (X_4)

Chart 1–2 *Cont.*

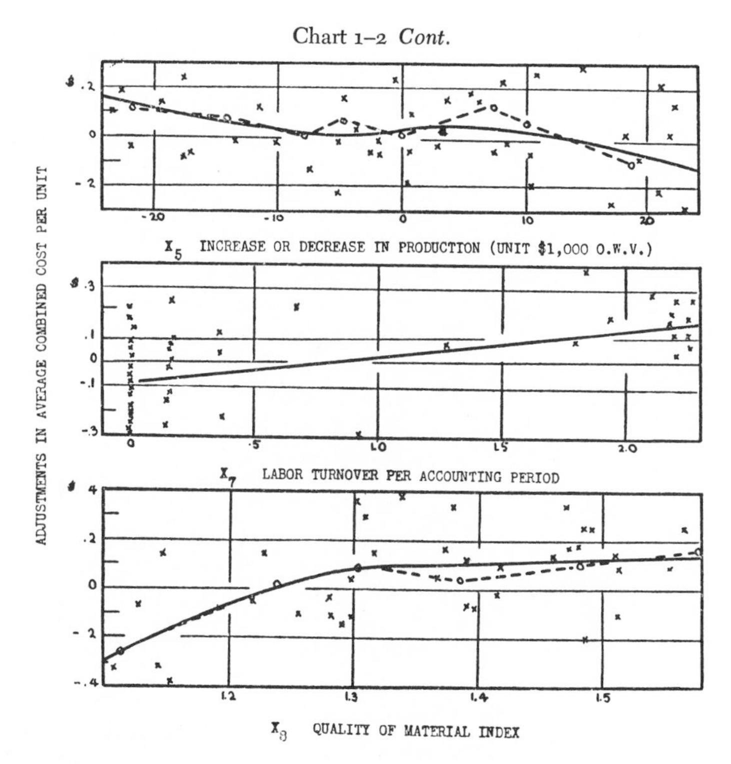

would be expected to rise continuously. Usually, the introduction of a new style affects costs unfavorably in both a direct and an indirect manner. Direct increases in cost are caused by new tools and new cutting patterns, which are charged immediately to the period in which the style is first made, rather than spread over the life of the tool or pattern. However, indirect increases in cost are perhaps more important. They arise from lack of familiarity on the part of foremen and workers with the cutting, assembly, and finishing processes required for the new product.

In the main, the actual course of the combined regression curve for new styles corresponds to its expected behavior: its general direction is upward as the number of new styles increases (Chart 1–2). Beyond thirty new styles, however, the curve flattens out, and beyond forty it actually declines very slightly. It may, nevertheless, be true that this phase of the curve is not statistically significant. This condition is suggested by the wide scatter of observations about the regression curve (Chart 1–2).

The a priori expectation of the regression curve which was designed to measure the effect of sudden changes in the rate of output from that of the previous period is not entirely clear. Increases and decreases in the rate of production would be expected to produce definite effects upon cost since change itself is expensive and causes readjustments in routines, redivision of labor, and general disorganization. It is not the amount of the change alone which must be considered, however, but also its direction. Decreases in output would be expected to raise cost per unit, whereas increases might well lower them, for it is more difficult to contract most expenses than it is to expand them.[67] The curve of cost adjustments for increases and decreases in output would then be expected to have high values for large decreases in output, lower values for smaller decreases, and possibly still lower values for increases in output.

The observed curve (Chart 1–2) follows in general this expected behavior. It declines from a high of $—0.20 for $24,000 decreases in output to a low of $—0.12 for $24,000 increases. However, the decline was not at a constant rate, but was interrupted by a plateau in the vicinity of small increases. No very satisfactory a priori explanation could be found for this interruption. An effort to trace it to some particular element of cost also failed. Consequently, it was deemed probable that this slight aberration from the expected behavior was not significant, but was due to incomplete data rectification, to sampling error, to the use of a cost function which was too flexible, or to the failure to allow adequately by this simple variable for the complicated pattern of change.[68]

The regression curve for labor turnover rate would be expected to rise as the values of this independent variable increase. Additions to the labor force cause direct expenses of hiring and training, and indirect expenses of low production, spoilage, and general confusion. These expenses are reflected in the index which was chosen to measure turnover rate, since this index is the ratio of the number of new men hired to the average number of men on the force. The regression curve obtained for labor turnover behaved in the manner anticipated (Chart 1–2): it rose continuously as the value of the turnover ratio increased.

The regression curve for quality of material would be expected to rise. Clearly, improvement in the quality of materials would logically be associated with higher material, labor, and supervisory costs. Whether or not this increase in cost should occur at a constant rate is not entirely clear. It may be that, beyond a certain point, further increases in the expensiveness of materials do not bring proportional refinements of workmanship. In that event cost would be expected to increase at a declining rather than at a constant rate. The regression curve obtained for quality of materials (X_8) actually does rise at a declining rate in a gentle convex curve (Chart 1–2).

In summary, then, the behavior of the net regression curves for corrected combined average cost is largely in harmony with theoretical expectations.

Another method of stating the relationship of average cost to the several independent factors, which is portrayed by the foregoing regression curves, is by means of a tabular exhibit such as Table 1–1. This table shows estimates of cost for various combinations of magnitudes of the independent factors. Such a table may prove to be easier to understand and to use than the regression curves, since estimates of the expected average cost under any given set of operating conditions can easily be made by entering the table at the proper points. However, since the table was prepared from regularly spaced readings from the regression curves, it is merely a more convenient, but slightly less accurate way of presenting these average cost findings.

MEASURES SHOWING THE PROBABLE RELIABILITY OF RESULTS

A second general type of findings resulting from the correlation analysis of average cost behavior consists of measures which show the probable reliability of the average cost curves. Two kinds of reliability measures were computed:[69] (*a*) the standard error of the estimate of cost made on the basis of all the regression curves determined for that particular cost item; (*b*) the standard error of the departures of the output regression curve from the curve reading at the output mean.

If the combined cost per unit for any individual period should be estimated on the basis of all six of the regression curves just described, the estimate would not be exactly correct, even for costs during the three-year period under investigation. A measure of the probable extent of this error is found in the standard error of estimate, which is the standard deviation of differences between actual and estimated cost. If these differences are normally distributed, this standard error of estimate is the range, plus or minus the estimated cost figure, within which the chances are two out of three that the actual cost figure will fall. It should be emphasized that this range is applicable only to accounting periods having fundamental operating conditions similar to those analyzed in this study. The value of the standard error of estimate was found to be $0.192 (Table 1–2). This is considerably less than the standard deviation of corrected cost ($0.447), which indicates that considerable improvement in cost estimating resulted from the use of the regression curves.

The second measure of unreliability applies only to the output regression curve. As a provisional indication of the statistical significance of this function, the standard error of departures of the curve from a horizontal line running through the curve value at the output mean ($1.603) was computed from experimental formulas suggested by Ezekiel.[70] Table 1–3 shows these departures and their standard errors at nine regularly spaced curve points. On the basis of this measure of significance the following conclusions were reached: (1) The deviations of the regression curve from the horizontal are of limited statistical significance in the vicinity of $45,000 and $50,000 output. (2) Considerable significance can be attached to these curve departures between $10,000 and $40,000 output.

INDEXES SHOWING THE CLOSENESS OF THE RELATIONSHIP BETWEEN COST AND THOSE INDEPENDENT FACTORS AFFECTING COST

The third general phase of the description of the behavior of average cost is the measurement of the closeness of the relationship between cost and the indepen-

Table 1-1

ESTIMATES OF AVERAGE COMBINED COST UNDER VARIOUS OPERATING CONDITIONS

OUTPUT (UNIT $1,000 O.W.V.)	INCREASE OR DECREASE OVER PREVIOUS PERIOD	AVERAGE NUMBER OF PIECES PER PRODUCTION ORDER														
		60			80			100			120			140		
		Number of New Styles			Number of New Styles			Number of New Styles			Number of New Styles			Number of New Styles		
		10	25	40	10	25	40	10	25	40	10	25	40	10	25	40
5.....	−$15,000	$2.864	$2.969	$2.959	$2.971	$3.076	$3.066	$2.904	$3.009	$2.999	$2.793	$2.898	$2.888	$2.687	$2.792	$2.782
	0	2.806	2.911	2.901	2.913	3.018	3.008	2.846	2.951	2.941	2.735	2.840	2.830	2.629	2.734	2.724
	+ 15,000	2.765	2.870	2.860	2.872	2.977	2.967	2.805	2.910	2.900	2.694	2.799	2.789	2.588	2.693	2.683
10.....	− 15,000	2.394	2.499	2.489	2.401	2.506	2.596	2.334	2.439	2.529	2.223	2.328	2.418	2.117	2.222	2.312
	0	2.336	2.441	2.431	2.443	2.658	2.538	2.376	2.591	2.471	2.265	2.480	2.360	2.159	2.374	2.254
	+ 15,000	2.295	2.400	2.390	2.302	2.507	2.497	2.235	2.440	2.430	2.124	2.329	2.319	2.018	2.223	2.213
15.....	− 15,000	1.954	2.059	2.049	2.061	2.166	2.156	1.994	2.099	2.089	1.883	1.988	1.978	1.777	1.882	1.872
	0	1.896	2.001	1.991	2.003	2.108	2.098	1.936	2.041	2.031	1.825	1.930	1.920	1.719	1.825	1.814
	+ 15,000	1.856	1.960	1.950	1.963	2.067	2.057	1.896	2.000	1.990	1.785	1.889	1.879	1.679	1.783	1.773
20.....	− 15,000	1.708	1.813	1.803	1.815	1.920	1.910	1.748	1.853	1.843	1.637	1.742	1.732	1.531	1.636	1.626
	0	1.650	1.755	1.745	1.757	1.862	1.852	1.690	1.795	1.785	1.579	1.684	1.674	1.573	1.578	1.568
	+ 15,000	1.609	1.714	1.704	1.716	1.821	1.811	1.649	1.754	1.744	1.538	1.643	1.633	1.532	1.537	1.527
25.....	− 15,000	1.564	1.669	1.659	1.681	1.776	1.766	1.614	1.709	1.699	1.503	1.608	1.588	1.397	1.502	1.482
	0	1.506	1.611	1.601	1.613	1.718	1.708	1.546	1.651	1.641	1.435	1.540	1.530	1.329	1.434	1.424
	+ 15,000	1.465	1.570	1.560	1.572	1.677	1.667	1.505	1.610	1.600	1.394	1.499	1.489	1.388	1.393	1.383

* As an illustration of a method of presenting results, this fragment of a complete table is shown.

Table 1-1—*Continued*

Output (Unit $1,000 O.W.V.)	Increase or Decrease over Previous Period	Average Number of Pieces per Production Order														
		60			80			100			120			140		
		Number of New Styles			Number of New Styles			Number of New Styles			Number of New Styles			Number of New Styles		
		10	25	40	10	25	40	10	25	40	10	25	40	10	25	40
30.....	−$15,000	$1.460	$1.565	$1.555	$1.567	$1.672	$1.662	$1.500	$1.605	$1.595	$1.389	$1.494	$1.484	$1.383	$1.388	$1.378
	0	1.518	1.623	1.613	1.625	1.730	1.720	1.558	1.763	1.753	1.447	1.652	1.642	1.341	1.546	1.536
	+ 15,000	1.477	1.582	1.572	1.584	1.689	1.679	1.517	2.122	1.612	1.406	2.011	1.501	1.300	1.905	1.396
35.....	− 15,000	1.374	1.479	1.469	1.481	1.586	1.576	1.414	1.519	1.509	1.303	1.408	1.398	1.197	1.302	1.292
	0	1.316	1.421	1.411	1.423	1.528	1.518	1.356	1.461	1.451	1.245	1.350	1.340	1.139	1.244	1.234
	+ 15,000	1.275	1.380	1.370	1.382	1.487	1.477	1.315	1.420	1.410	1.204	1.309	1.299	1.098	1.203	1.293
40.....	− 15,000	1.324	1.429	1.419	1.431	1.536	1.526	1.364	1.469	1.459	1.253	1.358	1.348	1.147	1.252	1.242
	0	1.266	1.371	1.361	1.373	1.488	1.468	1.306	1.421	1.401	1.185	1.310	1.290	1.079	1.204	1.184
	+ 15,000	1.225	1.330	1.320	1.332	1.437	1.427	1.365	1.370	1.360	1.254	1.259	1.249	1.148	1.153	1.143
45.....	− 15,000	1.310	1.415	1.405	1.417	1.522	1.512	1.350	1.455	1.445	1.239	1.344	1.334	1.133	1.238	1.228
	0	1.252	1.357	1.347	1.359	1.464	1.454	1.292	1.397	1.387	1.181	1.286	1.276	1.075	1.075	1.170
	+ 15,000	1.211	1.316	1.306	1.318	1.423	1.413	1.251	1.356	1.346	1.302	1.245	1.235	1.196	1.034	1.129
50.....	− 15,000	1.308	1.413	1.403	1.415	1.520	1.510	1.248	1.453	1.443	1.399	1.342	1.332	1.293	1.031	1.226
	0	1.250	1.355	1.345	1.357	1.462	1.452	1.290	1.395	1.385	1.341	1.284	1.274	1.235	1.073	1.168
	+ 15,000	1.209	1.314	1.304	1.316	1.421	1.411	1.249	1.354	1.344	1.300	1.243	1.233	1.194	1.032	1.127

Table 1–2

SUMMARY OF RESULTS OF MULTIPLE CORRELATION ANALYSES

(Comparison of several types of cost with respect to their standard deviation, standard error of estimate, index of multiple correlation, and index of multiple determination)

Dependent Variable	Figure Showing Regression Curves	Independent Variables‡ (Code Symbols)	Standard Deviation of Dependent Variable	Standard Error of Estimate		Index of Multiple Correlation		Index of Multiple Determination	
				Adjusted for Degrees of Freedom	Unadjusted	Adjusted for Degrees of Freedom	Unadjusted	Adjusted for Degrees of Freedom	Unadjusted
Average combined cost*	2	$X_2, X_3, X_4, X_5, X_7, X_8$	$ 0.447	$ 0.192	$ 0.163	$.906	$.931	$.821	$.867
Average combined cost (uncorrected)	3	$X_2, X_3, X_4, X_5, X_7, X_8$	0.295	0.230	0.196	.639	.750	.408	.562
Average direct labor cost	9	X_2, X_3, X_4, X_5, X_7	0.2283	0.108	0.095	.867	.878	.752	.771
Average combined indirect cost	....	X_2, X_3	0.0748	0.040	0.038	.986	.987	.972	.974
Average indirect labor cost	10	X_2, X_4, X_5	0.0693	0.022	0.020	.949	.958	.901	.918
Average indirect materials cost	13	X_2, X_3, X_4, X_5, X_8	0.0072	0.0068	0.0047	.344	.643	.118	.413
Total inspection labor cost	11	X_2, X_3, X_4, X_8	87.72	79.59	65.65	.456	.663	.208	.439
Total yard labor cost	12	X_2, X_3, X_5, X_9	530.87	392.46	323.82	.689	.793	.475	.628
Total power and watch labor cost	....	X_2, X_5, X_9	138.19	119.82	92.82	.527	.590	.278	.348
Average power cost	14	X_2, X_3, X_4	0.079	0.062	0.051	.637	.761	.406	.579
Average maintenance cost	15	X_2, X_3, X_5	0.076	0.0296	0.0258	.924	.940	.853	.884
Output†	....	X_3, X_4, X_5, X_7	11.62	11.71	9.08	.135	.622	.018	.387

* All costs not otherwise specified have been corrected for price changes and cost-output lags.

† Output was correlated with the other independent variables in order to determine the amount of interdependence among them. The unit is $1,000 old warehouse value.

‡ X_2 = output; X_3 = size of production order; X_4 = number of new styles; X_5 = change in output; X_7 = labor turnover; X_8 = quality of materials; X_9 = lumber receipts; X_{10} = fuel outlays.

Table 1–3

MEASURE OF THE STATISTICAL SIGNIFICANCE OF AVERAGE CORRECTED COMBINED COST FUNCTION

(Standard error of difference between partial regression curve at specified value of output and partial regression curve at average value of output)

OUTPUT (UNIT $1,000 IN O.W.V.)	ESTIMATED COST		STANDARD ERROR OF DEVIATIONS IN COL. 3‡
	Reading from Curve	Deviation from Reading at Mean Output†	
(1)	(2)	(3)	(4)
10	$2.469	$.866	$.201
15	2.038	.405	.246
20	1.794	.191	.154
25	1.651	.048	.043
30	1.548	− .055	.019
35	1.458	− .145	.096
40	1.409	− .194	.126
45	1.397	− .206	.216
50	1.397	− .206	.353

* The method of computing this standard error is described in Ezekiel, *Methods of Correlation Analysis*, p. 384. The formulas are provisional and give only a rough approximation.

† Departures of curve readings from the curve value $1.603, which corresponds to the mean of output ($29,720).

‡ Standard error refers to departures described in the preceding footnote.

dent factors. Two kinds of measures of the multiple relationship of all the independent factors to cost were computed for each cost item: (*a*) the index of multiple correlation and (*b*) the index of multiple determination.

The index of multiple correlation measures the closeness of the relationship between cost and the independent factors. It shows the degree of simple correlation between the actual cost magnitudes and the cost values which would be estimated on the basis of the net regression curves. This correlation was found to be high for corrected combined cost. The adjusted value of the index of multiple correlation was computed as .906 (Table 1–2).

The index of multiple determination is the square of the index of multiple correlation. It states the proportion of variance (squared variation) in cost which is "accounted for" by the independent factors. It thus measures the combined importance of the several independent variables as a means of explaining the difference in cost between accounting periods.[71] About 82.1 per cent of the variance of combined cost had been accounted for according to the index of multiple determination. The high values of these two correlation constants indicate that most of the influences upon cost have been included in this study.

Uncorrected Combined Cost

In a methodological study it is of interest to know whether the various laborious refinements in procedure make any difference in the results. Did correcting the cost data and allowing for independent factors by means of multiple correlation result in different and more sensible average cost findings than simpler methods would have given? To answer this question two special analyses of uncorrected combined cost were made: (1) a simple visual regression analysis of the gross relationship of uncorrected combined costs with output and (2) a complete graphic multiple

Chart 1–3 Partial Regression Curves for Corrected Average Combined Cost Compared with Partial Regressions for Cost Not Corrected for Price Change and Output Lag and with Gross Regression of Wholly Uncorrected Cost—Partial Regressions for X_2, X_3, X_4, X_5, X_7, X_8; Gross Regression for X_2

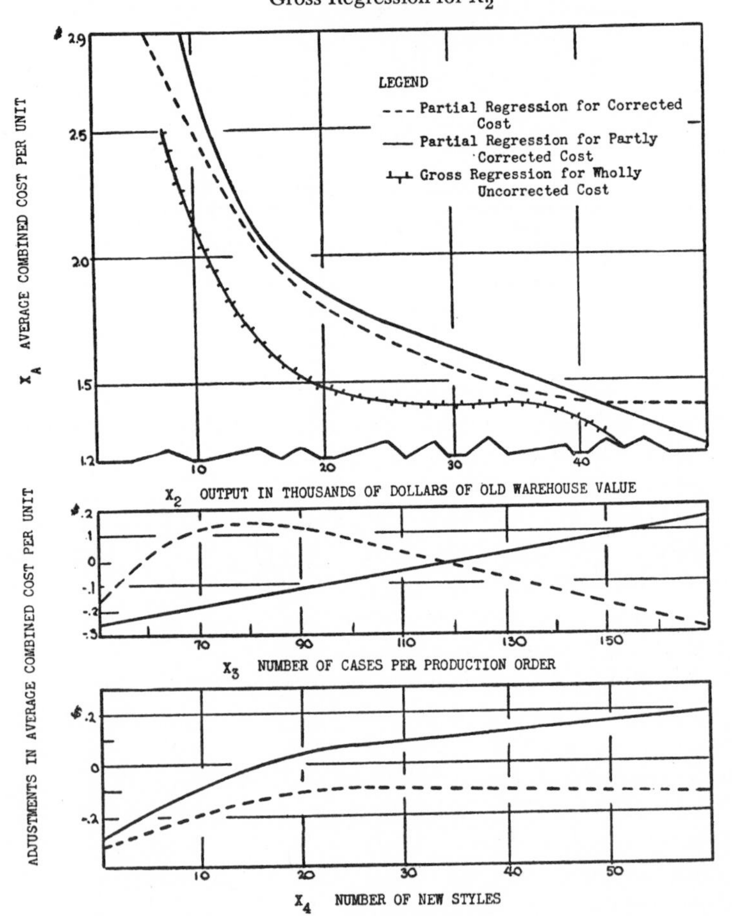

Chart 1–3 *Cont.*

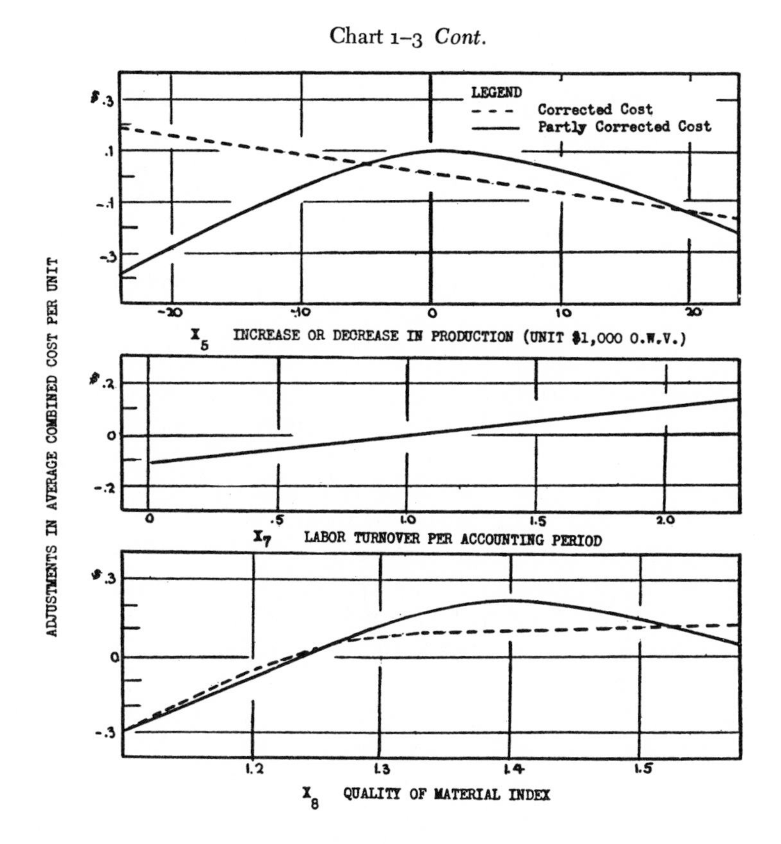

correlation analysis of uncorrected cost with the same independent variables which were employed for corrected combined cost.

The first of these special analyses obtained an output regression curve for average cost by visually fitting a gross rather than a partial regression curve to the uncorrected cost data. A comparison of this average cost curve with the partial regression curve for corrected cost (Chart 1–3) reveals several important differences. The former curve falls more steeply, flattens earlier, and then renews its fall more precipitously when large output levels are reached. Since these departures do not appear reasonable, they indicate that either data correction or multiple correlation analysis, or both, are essential steps in obtaining satisfactory average cost results.

The second subsidiary analysis of uncorrected cost was designed to find out whether both these steps in the preparatory analysis were necessary. A complete multiple correlation analysis identical to that made for corrected cost was made with data which were not corrected for rate changes or for output lag. The partial regression curves obtained by this analysis are contrasted with those for corrected cost in Chart 1–3. Differences in the results of these two analyses were obviously ascribable solely to data corrections. Differences between the net output regression curve for uncorrected cost and the gross regression for uncorrected cost, shown in Chart 1–3, on the other hand, are clearly attributable to the effects of multiple correlation analysis.

A careful comparison of these three sets of curves and a comparison of the corresponding correlation constants (Table 1–2) led to the following tentative conclusions regarding the justification for the foregoing two steps in data rectification, as indicated by their effect upon the behavior of average cost in the particular firm studied:[72]

1. Multiple correlation analysis appeared to be necessary in order to stabilize the effects of other cost influences and thus secure a sensible regression curve for average cost and rate of output.

2. The correction of cost data for price change and for output lag seemed to be a necessary preliminary step for the determination of satisfactory curves for several of the independent factors. Rectification of the cost data apparently brought the curves more nearly into accord, in several important respects, with reasonable a priori expectations. In the first place, the output regression curve (X_2) for corrected data, instead of declining continuously, flattened toward the horizontal in the region of high production activity. Second, the slope of the regression for size of the production order (X_3) was reversed and thus made reasonable. Third, the unreasonable slope of that portion of the regression curve for changes in output (X_5), which applies to decreases, was reversed in the corrected curve; thus the entire curve was made to accord more nearly with expectations. Fourth, the illogical downturn of the uncorrected regression for quality of material (X_8) was removed, leaving a corrected curve which rose reasonably over its entire course.

3. The accuracy with which costs can be estimated on the basis of the values of the independent factors was increased by rectifying the cost data. This is shown by the reduction of the adjusted standard error of estimate from $0.230 for uncorrected cost to $0.192 for corrected cost.

4. The degree of success with which cost behavior was accounted for by the independent variable was enhanced by rectifying the cost data. The adjusted index of multiple correlation rose from .639 for uncorrected cost to .906 for corrected cost, and the adjusted index of multiple determination rose from .408 to .826.

Apparently, clearing away the effects of irrelevant influences, such as price change and time lag, brought out more definitely the true relationship between cost and the various independent factors. The practical importance of a reasonably accurate determination of these crucial relationships is clear. If these curves are to be the basis for determining adjustable standards for cost control, it is important that these adjustments be correct. Otherwise the standards will not command respect. Similarly, if the curves are to be used as aids in budgeting expenses (and profits), rectification of data which will greatly improve the accuracy of estimates is highly desirable.

CHAPTER VII

Behavior of Marginal Combined Cost

The present chapter deals with the findings for marginal combined cost.[73] These findings are of interest to management because estimated future marginal cost, considered in connection with anticipated marginal revenue, is necessary in determining the selling price and the output level which will yield the maximum profit. These findings are also of interest to economists since the behavior of the marginal cost function has been the subject of much speculation on their part but has, notwithstanding, elicited little statistical investigation. And, finally, these results may interest the statistician because they throw some light upon various alternative procedures for obtaining estimates of marginal cost.

Summary of Conclusions

Extensive experiments with several methods of deriving marginal cost from corrected data in the form of averages and in the form of totals lead to the following tentative conclusions.

1. More satisfactory estimates of marginal cost can be obtained from data in the form of totals than from data in the form of averages.

2. In working with total cost data least-squares curve fitting gives satisfactory results when the proper function is chosen. Visual fitting is more economical, however, and apparently almost as reliable particularly when, as in this case, the total cost function is linear.

3. The most reliable estimate of marginal cost which can be made from these data is that it is constant at $1.12 per-dollar increase in old warehouse value of output regardless of the rate of output. Curvilinear functions were also fitted which yielded varying marginal costs, but they are not particularly reasonable and they squeeze more out of the data than their limited reliability justifies.

4. In dealing with cost data in the form of averages none of the methods employed gives very reliable marginal cost estimates. Of the procedures tried, the graphic method proved most satisfactory and the method of first differences least

trustworthy. Although both the derivative (least-squares) and the graphic-derivative procedures demonstrated important advantages, they proved to have weaknesses which were particularly dangerous in dealing with variable data in average cost form.

5. The adequacy and reliability of the best estimate of marginal cost ($1.12) are limited by several considerations: (*a*) this estimate represents marginal manufacturing costs only; (*b*) correction of the data is not complete; (*c*) the sample may not be representative of subsequent periods; and (*d*) measures of the reliability of average and total cost functions which are based upon the dispersion of the data and the size of the sample exaggerate somewhat the degree of confidence which can be placed in these functions and also the reliability of the marginal cost estimates derived from them.

Experimentation with data representing several stages of correction led to the general conclusion that full rectification of the data was necessary in order to obtain reliable estimates of marginal cost. This conclusion was based upon the following findings:

1. Estimates of constant marginal cost made from incompletely corrected data were different from those obtained from corrected data.

2. These estimates were also less reliable since the dispersion of observations was greater.

3. Correction of the data brought significant changes in the shape of the marginal cost function.

4. Partial correction of the data for price change and time lag alone did not suffice when, as in this case, there were other important operating conditions which affected cost behavior.

5. Partial correction by multiple correlation analyses alone did not give reliable marginal cost estimates, at least not during periods of pronounced price and volume variability.

6. A short-cut procedure yielding quick, rough estimates of constant marginal cost by visually fitting a straight line to the raw total cost data might prove usable under conditions of greater price and volume stability and when other cost factors did not cause important distortion of the cost-output relationship.

Marginal Cost Estimated from Corrected Data by Alternative Methods

The first phase of the experiment involved a test of several alternative procedures for deriving marginal cost from fully corrected combined cost data. These procedures may be classified in two groups according to the form of cost data used. The first group employed data in the form of the adjusted total cost for each accounting period, whereas the second used data in the form of the adjusted average cost per unit of output for the period. To the first type of data two kinds of function were fitted both by the method of least squares and by visual methods: (*a*) a straight line and (*b*) a second-degree parabola. To the second kind of data, curves of various types were fitted, and marginal cost was derived by several fairly distinct procedures. It will be noted that these procedures differ not so much according to the type of function fitted, as according to the method of fitting the function and of deriving marginal cost from it. The methods used in studying these average cost data may be called:

1. The graphic method (visual curve fitting).
2. The first-differences method (no curve fitted).
3. The derivative method (least-squares curve fitting).
4. The graphic-derivative method (selected-points curve fitting).

The relative merits of the total as against the average cost methods should be noted at this point. Total cost data proved more reliable because marginal cost is a simple function of the total cost curve (i.e., its slope). Hence, in fitting the total cost function with this in mind, its slope can be prevented from having unwarranted fluctuations. On the other hand, marginal cost is a complicated function of the average cost curve. Hence, in fitting this curve, spurious fluctuations of its marginal cost function cannot be easily prevented. If, however, comparable functions had been fitted both to totals and to averages, for example:

$$X_T = a + bX_2 \text{ (for totals)}$$

$$X_A = \frac{X_T}{X_2} = \frac{a}{X_2} + b \text{ (for averages)},$$

then results of about equal reliability could have been obtained from the average cost data.

Since parallel functions of this type were not employed (they would have been difficult to hit upon by graphic methods), observations of total cost were found to be superior to observations of average cost as a basis for deriving estimates of marginal cost. Two chief advantages resulted from using data in this form. First, marginal cost curves appeared to be somewhat more reliable and more reasonable since they were freed from the spurious fluctuations given them by amplification of faint erroneous changes of slope in the average cost curves. Second, marginal cost curves could be obtained more economically since, when the entire analysis was carried out in terms of totals, two steps were obviated: (1) conversion of individual values into per-unit observations and (2) subsequent transformation of the average cost function into a total cost function.[74]

METHODS EMPLOYED FOR TOTAL COST DATA

The best estimate of marginal cost which could be made from the available data was that it is constant at about $1.12 per dollar of old warehouse value (see Table

Table 1–4
COMPARISON OF MARGINAL COMBINED COST OBTAINED BY VARIOUS METHODS FROM FULLY CORRECTED COST DATA IN THE FORM OF AVERAGES AND IN THE FORM OF TOTALS

OUTPUT (UNIT $1,000 O.W.V.)	DERIVED FROM AVERAGE COST DATA†			DERIVED FROM TOTAL COST DATA‡	
	First-Differences Method§	Derivative Method‖	Graphic Method¶	Least-Squares Straight Line Function	Least-Squares Parabolic Function
10	1.52	1.52	1.62	1.12	1.54
15	1.02	1.06	1.17	1.12	1.43
20	1.01	0.98	1.18	1.12	1.31
25	1.14	1.06	1.17	1.12	1.20
30	1.50	1.21	1.07	1.12	1.08
35	0.10	1.30	1.04	1.12	0.96
40	1.25	1.15	1.27	1.12	0.85
45	1.43	0.70	1.41	1.12	0.73

* Data used in each of these procedures were adjusted by price index and time lag correction and by multiple correlation analysis.

† Derived by fitting functions to average cost residuals, then transforming these functions into terms of total cost and taking the first derivative of the latter function.

‡ Derived by first transforming individual average cost residuals into total cost residuals then fitting functions to these observations.

§ Tabular-graphic derivation from unsmoothed group averages of per-unit cost residuals (Fig. 5).

‖ Mathematical derivation from cubic equation fitted to average cost residuals (Fig. 5).

¶ Tabular-graphic derivation from smoothed curve fitted visually to average cost residuals (Fig. 5).

1–4). This figure was arrived at by fitting a straight line by the method of least squares to fully rectified cost data in the form of totals and taking the first derivative of the resulting total cost equation.

The least-squares method was chosen for fitting a curve to these preferred total cost data because it is more objective than visual fitting and because marginal cost cannot only be derived precisely from the resulting total cost equation but also, can be expressed concisely in the form of an equation. But a line fitted by visual methods would have given approximately the same results in a fraction of the time. Graphic methods were used satisfactorily in deriving marginal cost from uncorrected combined total cost data and from the total cost observations of its components.

A straight line was chosen as the preferred function to be fitted by least squares to these total cost data. In the first place, the distribution of these corrected data appeared to be approximately linear (Chart 1–4). In the second place, the data, even in this refined form, were thought to be too imperfect to justify representation by complex functions. A second-degree parabola,[75] which was fitted to these data by least-squares procedures, did not represent the data well at large outputs, as can be seen from Chart 1–4. From Table 1–4, where the first derivative of this

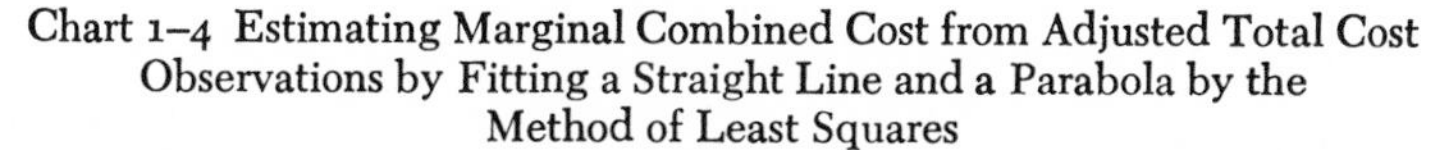

Chart 1–4 Estimating Marginal Combined Cost from Adjusted Total Cost Observations by Fitting a Straight Line and a Parabola by the Method of Least Squares

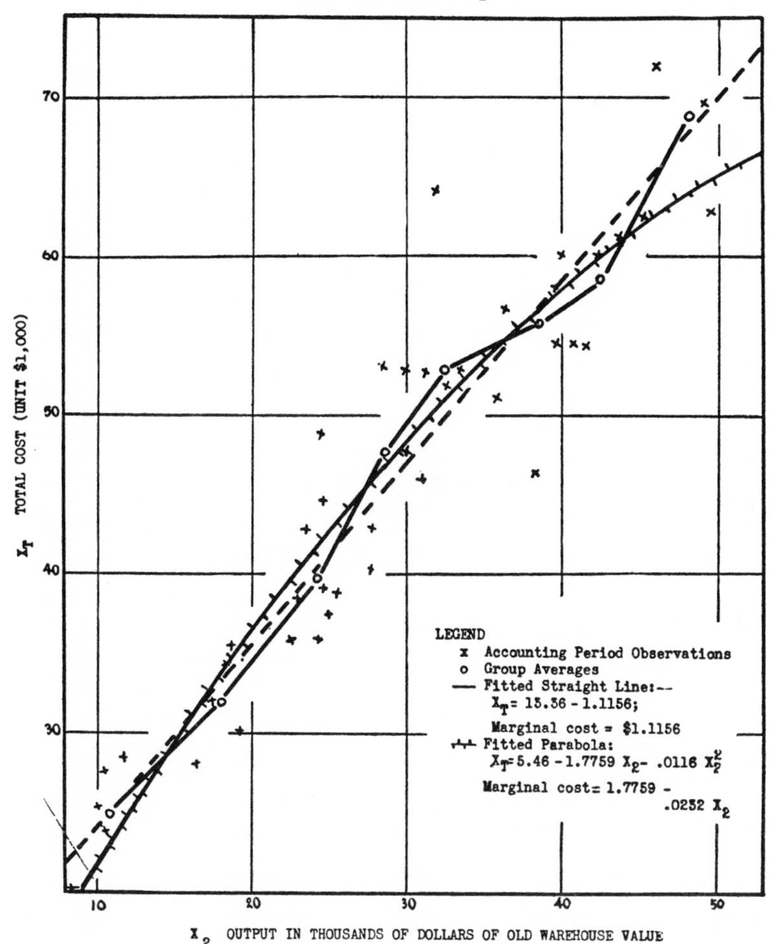

parabolic function is evaluated, it is clear that this estimate indicates that marginal cost declines continuously as output increases.

In the short period both constant marginal cost and declining marginal cost go counter to ordinary theoretical expectations since such behavior is generally not considered compatible with individual equilibrium for firms operating under conditions of effective competition.[76] Under conditions of monopolistic competition, on the other hand, with the consequent tendency toward continuous overcapacity, with a sharply rising selling cost curve and a resulting declining net individual demand schedule, and with considerable friction, constant or declining marginal manufacturing costs over the range of practical operation are compatible with equilibrium for the individual firm.[77]

METHODS EMPLOYED FOR AVERAGE COST DATA

The second group of methods used data in the form of average cost per dollar of old warehouse value for each accounting period. Only by trying data of this second type and comparing the results with those obtained from total cost data could the superiority of the total cost procedures be established. Such a comparison of results is facilitated by Table 1–4 which contrasts the estimates of marginal cost derived from total cost data with those obtained by four methods from average cost data. Chart 1–5 shows the same information in graphic form. Of the various procedures for deriving marginal cost from average cost data the graphic method proved to be most satisfactory.[78] The incremental cost curve obtained by this method is shown in Chart 1–5.

From the fact that the two striking peculiarities of this curve—its brief initial declining phase and its upturn at the end for high rates of production—appear reasonable under assumptions of effective competition, it might seem that the graphic procedure is to be preferred over other methods of deriving marginal cost from average cost functions. However, it must be borne in mind that the assumptions of atomistic competition probably do not hold for the present firm's manufacturing cost. It has been noted that, under conditions of imperfect competition, with rising marginal selling costs and declining individual demand schedules, marginal manufacturing cost may not rise at all over the relevant range of output. Hence, the shape of the marginal curve obtained in this case cannot be advanced as an argument for the use of the graphic method in other investigations, particularly in view of the fact that the fluctuations of this curve do not appear to be statistically significant (judging from the linearity of the distribution of the total cost residuals in Chart 1–4).

It is for sounder reasons that the graphic method is preferred for commercial derivations of marginal cost from average cost data. (These advantages apply equally to total cost data.) First, visual curve fitting is flexible since choice of the proper function and evaluation of its parameters is simplified; second, it is able to abstract from non-typical variations in the data; third, the method is economical, particularly when graphic multiple correlation analysis is a step in the rectification of the cost data; and, fourth, the procedure and its results are easily understood.

The method of first differences gave less satisfactory results than the graphic procedure. The marginal cost curve secured from fully rectified data by this method is depicted in Chart 1–5. Fluctuations of this curve are so violent and so irrational that they are of no managerial interest. Marginal cost estimates obtained by its use were unreliable, were stated with little precision, and were determined at no saving in labor.

When a cubic of the progressive series was fitted to the average cost data by the method of least squares and the resulting equation converted into a total cost function and differentiated, a different sort of marginal cost curve was obtained. This curve is compared in Table 1–4 and in Chart 1–5 with estimates computed by other methods. The most important departures of this curve from the estimate of constant marginal cost are: (1) it has high values for small outputs and declines sharply as output increases in this early range; (2) it rises gently in the middle range of outputs; and (3) it falls continuously for large outputs (Chart 1–5).

Chart 1–5 Comparison of Estimates of Marginal Combined Cost Obtained from Corrected Data by Several Alternative Methods Using Data in the Form Both of Averages and of Totals

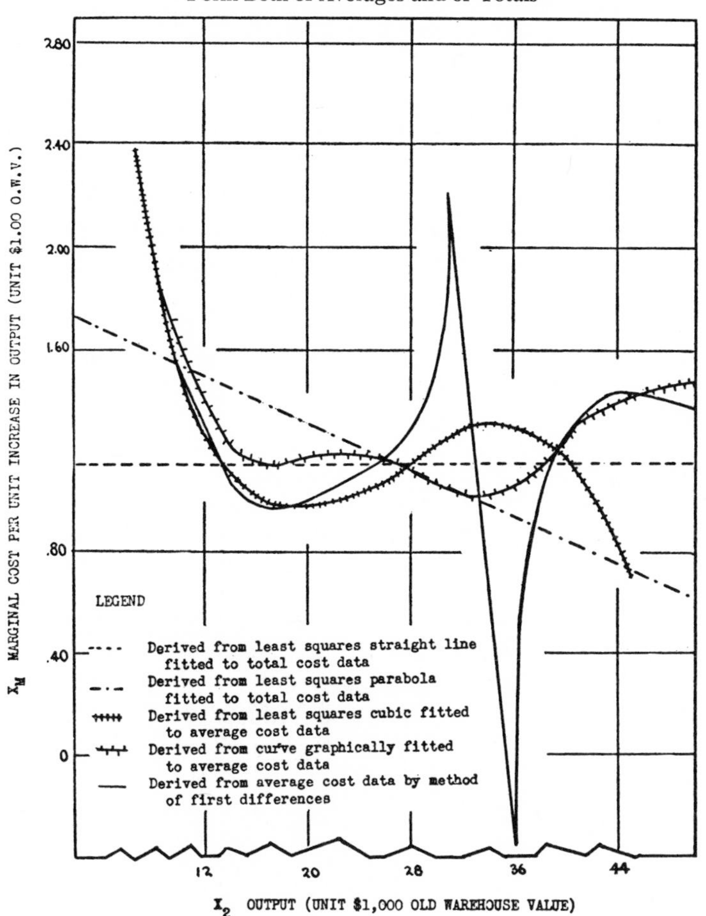

The peculiarities in the behavior of this curve are largely attributable (*a*) to the use of data in the form of averages rather than in the form of totals, (*b*) to the type of function chosen, and (*c*) to the method employed in fitting it to the data. Because the data were of inferior form, because the function chosen was not the proper type,[79] and because the method of least squares does not abstract from what appear to be non-typical data, the marginal curve cannot be considered reliable. It is merely another example of what not to do.[80]

The graphic-derivative method is a hybrid technique which attempts to preserve some of the advantages of the two procedures which it combines. From the graphic method it possesses advantages in curve fitting in the form of economy, flexibility, and freedom from sampling errors; and from the derivative method, precision and conciseness in differentiating and expressing functions. However, these advantages are offset in part by three defects which are, in some cases, important: (1) location of the curve by visual methods is somewhat subjective and arbitrary; (2) there is a possibility that the average cost equation will not accurately represent the average cost curve; and (3) it is certain that added labor will be required for the two-way transformation of graphic results into terms of mathematical symbols, then back again into terms of tables or graphs. Estimates of partially corrected combined marginal cost obtained by this method are shown in Table 1–5. Estimates of components of combined cost, derived by this same procedure, are found in Appendix B. This method is not considered trustworthy when average cost data are used.

In closing, it should be emphasized that it is not recommended that marginal cost be derived, by any of these methods, from data in the form of averages.

RELIABILITY OF MARGINAL COST FINDINGS

The best estimate of marginal cost which could be made from the fully corrected data was a constant, $1.12. This estimate, although more trustworthy than estimates computed by other methods and those made from less refined data, is still subject to several limitations which restrict its reliability. (1) It represents only marginal manufacturing cost. To this must be added an estimate of marginal selling and administrative cost before a figure can be arrived at which will be serviceable for price and output policy.[81] (2) There are some doubts about the completeness of the rectification of the data. (3) There is some question about the representative character of the sample. Eagerness to select a period full of challenging methodological problems may have resulted in findings which are less typical of subsequent periods than those which could have been secured from data of a more stable period than the depression years of 1932, 1933, and 1934.[82] And (4) the degree of reliability of this estimate, which is measured on the basis of the size of the sample and the variability of the data, cannot be very satisfactorily determined.[83]

The considerations mentioned above must be kept in mind in using this estimate

($1.12). They should not, however, lead us away from the important truth that this computed marginal cost is reasonably reliable, especially for the middle portion of the output range (fundamental cost conditions remaining the same). The data are essentially accurate, though incomplete, and the rectification was sufficiently satisfactory for obtaining serviceable marginal cost estimates.

Alternative Degrees of Data Refinement

In addition to testing alternative procedures for deriving cost from corrected data a second general experiment in methodology was performed in order to determine whether or not the correction of these data was necessary. This is a very practical question since the expense of rectifying data by means of specially computed index numbers, lag allowances, and multiple correlation analyses is great. If equally satisfactory marginal cost estimates can be obtained without this modification of the cost data, the saving in research costs will be great.

Some light has been thrown upon this problem in the course of the planning of the correction procedures, but not enough, however, to constitute convincing evidence that the correction procedure was necessary. Errors which were removed by correction may have been equally distributed with respect to the relationships under study. In that event their distorting effect would have been negligible.[84] The significant test of the necessity for the rectification procedure lies not in its effects upon the data, but in its effects upon the marginal cost results. Previous investigation of the effect of correction upon the observed curves of average cost indicated that nothing short of complete rectification of the data would give satisfactory knowledge of average cost behavior. It might be held that this is sufficient evidence of its effect upon marginal cost since, if the average cost curves are not reliable, marginal cost functions derived from them will be even more distorted. It will be of interest, however, to perform a parallel experiment by computing marginal cost estimates from total as well as from average cost data representing successive stages in the process of rectification in order to compare these results with those obtained from fully corrected data.

Three steps or levels of data correction were represented in this experiment. On the lowest level stand wholly uncorrected cost observations.[85] On the second level of refinement stand data which have been corrected for changes in prices and for discrepancies in the time at which price and output were recorded, but which have not been adjusted (by means of a multiple correlation analysis) for other influences which affect production cost and distort the relationship between cost and output. The third stage is represented by data which were not corrected for price fluctuations and output lag, but which were rectified by means of a multiple correlation analysis.

Data representing these various stages were studied in the form of totals and then in the form of averages in order to check the conclusions regarding this important practical question. Comparison of the marginal cost functions derived from the incompletely corrected data in both the average and the total form, with the best results obtained from corresponding corrected data, led to the general conclusion that full rectification of the data was necessary in order to secure reliable marginal cost results.

EXPERIMENTS WITH UNCORRECTED DATA IN THE FORM OF TOTALS

On the basis of experiments with data in the form of totals several conclusions regarding data correction may be drawn. First, estimates of constant marginal cost made by fitting a straight line to these uncorrected total cost observations were

Chart 1–6 Marginal Combined Cost Estimated from Wholly Uncorrected Total Cost Observations

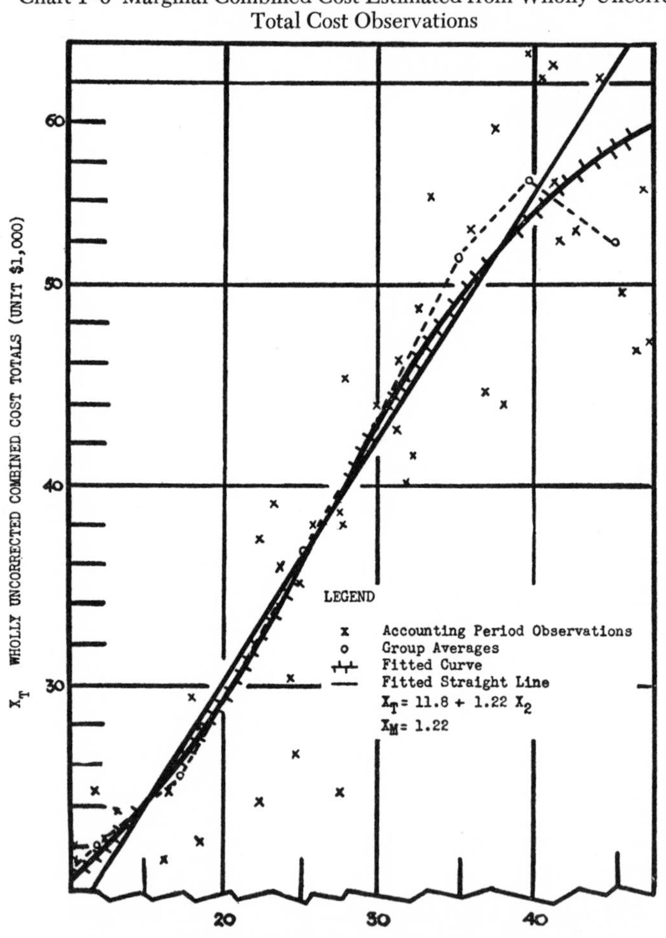

different from those obtained from corrected data. For wholly uncorrected data the estimate was \$1.22 per dollar of old warehouse value; for price and lag corrected data it was \$1.09; for data corrected by multiple correlation analysis alone marginal cost was \$0.79; and for fully corrected data it was \$1.12. This disparity throws doubt upon estimates based upon other than fully corrected data. Second, the dispersion of observations was materially reduced by rectification, as can readily be seen from a comparison of Charts 1–4, 1–6, and 1–8. This indicates that the confidence which can be placed in the estimates of marginal cost was considerably increased by correcting the data.

Third, in addition to reducing the dispersion, correction of the data brought the form of the total cost distribution nearer to linearity. Examination of Charts 1–4,

Chart 1–7 Marginal Combined Cost Estimated from Total Cost Observations Corrected for Price Change and Output Lag Only

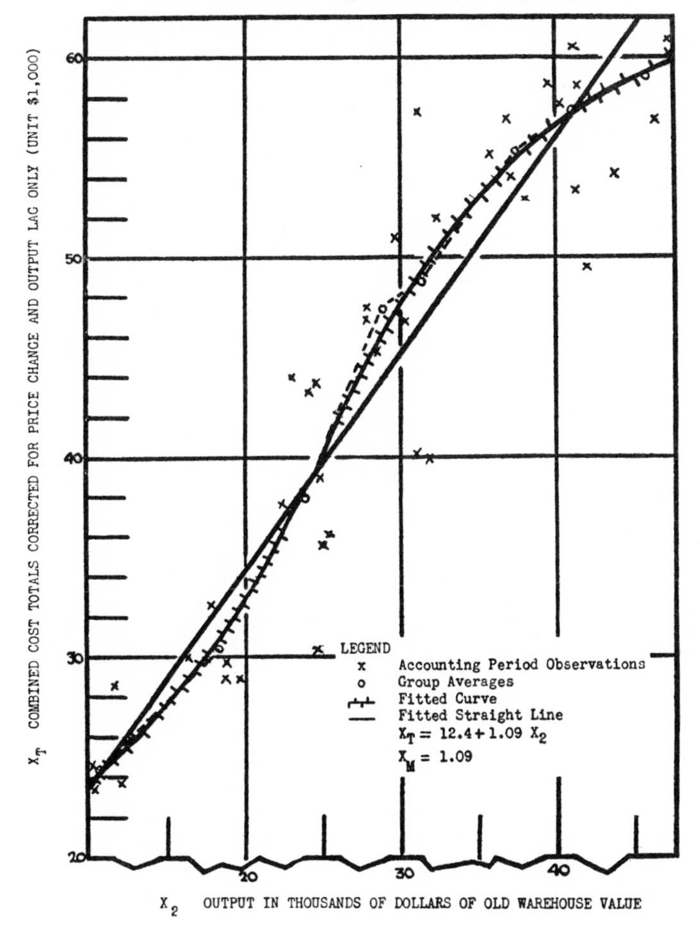

1–6, and 1–8 shows that each of these distributions could be better represented by a curve than by a straight line; on the other hand, the distribution of corrected cost in Chart 1–4 is approximately linear. This effect of rectification buttresses the case for constant marginal cost. From Chart 1–6 it can be seen that wholly uncorrected cost data would yield estimates of declining rather than of constant marginal cost and that this erroneous estimate would be poorly supported by the widely scattered data. Chart 1–7 shows that, even after correcting the data for price change and output lag, the distribution of total cost observations remained definitely curvilinear and indicated a declining, rather than a constant, marginal cost.[86] From Chart 1–8 it is clear that data corrected by multiple correlation analysis alone yielded unsatisfactory marginal cost estimates. A convex parabola is clearly called for by the dis-

Chart 1–8 Marginal Combined Cost Estimated from Total Cost Observations Corrected by Multiple Correlation Analysis Only

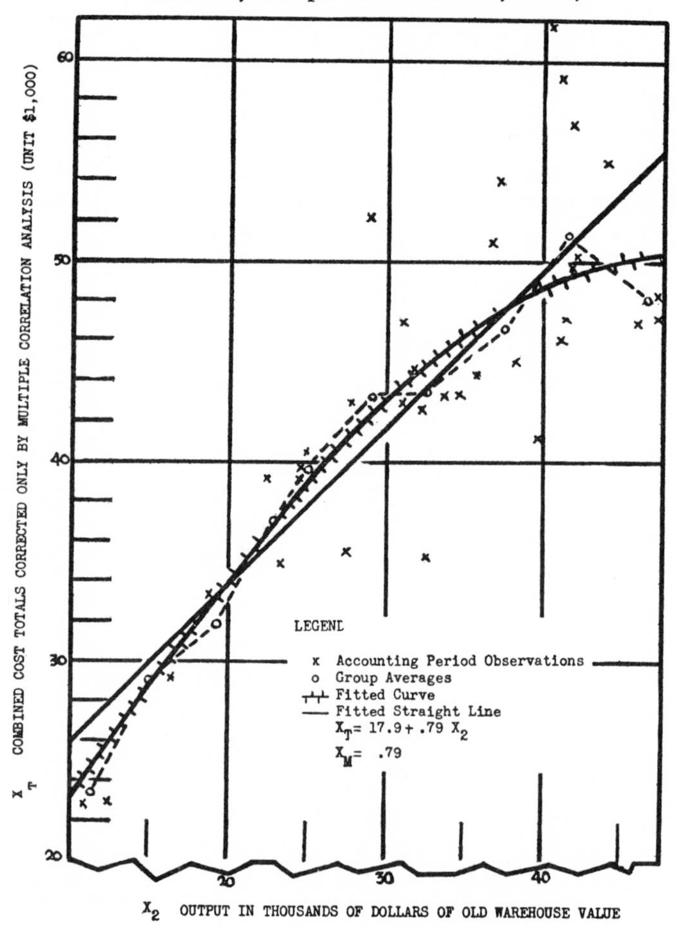

tribution of total cost observations. This type of total cost function will yield a continuously declining straight line as the estimate of marginal cost.

It is clear, from what has been said, that partial rectification of the data by means of price and time-lag correction alone, or by means of multiple correlation analysis alone, will not suffice during periods of great price and volume variability. Correction for changes in prices and wage rates may not, however, be necessary in studies covering periods of greater price stability. Correction for the lag of output behind recorded cost, on the other hand, although particularly needed in this period of output instability, would probably have to be retained even during a more stable period for those firms in which this type of disparity existed. Correction by means of multiple correlation analysis could be dispensed with in similar studies only if the effect of other variables upon cost was either unimportant or randomly distributed. The distribution of Chart 1–7 holds out considerable hope for this sort of short cut.[87]

EXPERIMENTS WITH UNCORRECTED DATA IN THE FORM OF AVERAGES

The general conclusions drawn from experiments with data in the form of totals were substantiated when the effects of the correction process were tested with data in the form of averages. The principal additional contribution of this latter experiment was to show explicitly the changes in the shape of the marginal cost function brought about by correction. These alterations are shown in Table 1–5, where mar-

Table 1–5
COMPARISON OF ESTIMATED MARGINAL COSTS FROM CRUDE, FROM PARTIALLY CORRECTED, AND FROM REFINED AVERAGE COST DATA

OUTPUT (UNIT $1,000 IN O.W.V.)	MARGINAL COST ESTIMATES		
	From Crude Data*	From Partially Corrected Data†	From Refined Data‡
10	$0.94	$1.50	$1.50
15	0.72	1.14	1.07
20	1.02	0.97	1.09
25	1.27	0.91	1.07
30	1.39	0.86	0.98
35	1.30	0.75	0.93
40	0.55	0.48	1.07
45	−0.20	−0.03	1.33
50	....	−0.87	1.38

* Data were not corrected in any way; and marginal cost was derived by the graphic method.

† Data were adjusted by multiple correlation analysis but were not corrected for price change or for output lag; marginal cost was derived by the graphic-derivative method.

‡ Data were adjusted by correlation analysis and by rate and time lag correction; marginal cost was derived by the graphic method.

ginal cost curves derived from wholly uncorrected average cost data and from data corrected by multiple correlation analysis alone (called partially corrected) are contrasted with the curve derived by the same average cost method from completely corrected data. Examination of Table 1–5 shows that agreement among these curves is confined to their initial falling phase; beyond $15,000 output the disparity is pronounced. Evidently, correction of the data substantially altered the shape of the marginal cost function.

Other Margins of Combined Cost

IMPORTANCE

Up to this point in the analysis of marginal cost we have devoted attention exclusively to the output margin of cost. But from our multiple correlation analyses we know that cost is really a function of many operating conditions of which the rate of output is but one. Consequently, cost has other margins beside the output margin. For example, increasing the number of new styles has an incremental cost which is analogous to that resulting from increase in output. Moreover, each of these independent variables may itself have several dimensions. Output, for example, often has an important quality margin as well as quantity margin.[88]

These other margins of cost may be exceedingly important for certain types of policies since the rational basis for any business decision may be viewed as a comparison of the anticipated margins. Marginal cost should be balanced against the marginal revenue attributable to a given order or a given course of action. The marginal cost of one method is also to be compared with the incremental cost of performing the service in a different way, and much business thinking goes on at this level without reference to the cost-revenue marginal equality. In either case, estimates of the differential costs of alternatives are needed, and these alternatives frequently involve the computation of margins of cost for other operating conditions in addition to output.[89]

Frequently, several of these margins must be considered in combination for a single decision. If the decision in question involves an increase of both output and the number of new styles, for instance, then the increment in total cost due to new styles must be added to the increment caused by additional output. To put it more generally, since cost is a function of several variables:

$$X_T = f(X_2\, X_3\, X_4 \,.\,.\,.\,.\, X_n)\,,$$

the addition to cost which is pertinent to a decision involving changes in several of these factors is the sum of the increments attributable to these various factors, i.e.,

Price theory usually ignores all these cost increments except that for quantity of output.[90] This simplification is correct when other margins are unimportant or when decisions can be formed in such a way as to involve changes only in one operating condition at a time.[91] For many practical pricing decisions, however, this approach is too simple since simultaneous change in several operating conditions is involved and summation of their respective cost increments is therefore required. For example, the particular firm studied has important orders from large mail-order houses. The acceptance of one of these orders often involves, in addition to an increase in the quantity of output produced, the design and production of new styles, increases in the average size of production order, sharp changes in the level of output from that of the previous period, and some change in the rate of labor turnover. Changes in each of these important conditions of operation involves some increment (or decrement) in cost. Consequently, to ignore all except the output increment might involve serious error.[92]

COMPUTATION OF OTHER MARGINS OF COMBINED COST

Several conditions must be satisfied before it is possible to compute the marginal cost of varying other operating factors. First, a numerical measure of the factor must be available; otherwise it cannot be treated mathematically. Second, the factor in question must be an independent variable; that is, the proportion in which this cost factor is combined with other operating conditions must be in some degree variable. Without this flexibility it is not possible to measure the partial derivative, and, even if this derivative were measured, it would not be possible for management to profit by this knowledge.

Rough, provisional estimates were made of margins of combined cost for four operating conditions: size of production order (X_3), new styles (X_4), changes in rate of output (X_5), and rate of labor turnover (X_7). These computations were made more to indicate the nature of the problem than to provide a final solution for it.

The procedure followed is similar to that used for calculating the output margin. The average slope of the net regression curve for size of production order and cost per unit of output (Chart 1–2) was obtained graphically by fitting a straight line freehand to the net residuals of the multiple correlation analysis. This slope (—.004) was taken as the approximate partial derivative of the average cost function.[93] It was transformed into the partial derivative of the total cost function by using the following relationship:

$$X_A = f(X_3) \qquad \text{(net regression line for average cost)}$$

$$X_T = X_2 \cdot f(X_3) \qquad \text{(net regression line for total cost)}$$

$$\frac{\partial X_T}{\partial X_3} = \partial \frac{[X_2 \cdot f(X_3)]}{dX_3}$$

Assuming that the derivative of output with respect to the size of production order is zero, then

$$\frac{\partial X_T}{\partial X_3} = X_2 \cdot \frac{df(X_3)}{dX_3}$$

$$= \text{(Rate of output)} \times \text{(Slope of net regression for production order)} .$$

By similar procedures the following estimates of other increments of cost were obtained:

Size of production order (X_3):[94]

$$\frac{\partial X_T}{\partial X_3} = X_2 \cdot (-\$0.004)$$ (for an increase of one case in average size of production order)

Number of new styles (X_4):

$$\frac{\partial X_T}{\partial X_4} = X_2 \cdot (+\$0.0055)$$ (for an increase of one new style per period)

Change in rate of output (X_5):

$$\frac{\partial X_T}{\partial X_5} = X_2 \cdot (-\$0.0044)$$ (for an increase in output above the previous period of \$1,000 O.W.V.)

Labor turnover (X_7):

$$\frac{\partial X_T}{\partial X_7} = X_2 \cdot (+\$0.11)$$ (for an increase of 1 per cent in labor turnover for accounting period)

Estimates of curvilinear marginal cost were also made for two independent variables: size of production order (X_3) and number of new styles (X_4). These functions are shown in chapter viii of the author's dissertation previously cited.

The following example is given of methods of combining the various constant margins of cost into a single estimate of the addition to cost consequent upon acceptance of an order which changes three conditions of operation—X_2, X_3, and X_4.

Since

$$X_T = f(X_2 X_3 X_4) ,$$

then

$$dX_T = \frac{\partial f}{\partial X_2} dX_2 + \frac{\partial f}{\partial X_3} dX_3 + \frac{\partial f}{\partial X_4} dX_4 .$$

If the partial derivatives are constant at the values given below, if the contemplated increment of each variable is small, if other factors than X_2, X_3, and X_4 are negligible, and if

$dX_2 = 1$ (dollar old warehouse value)
$dX_3 = 1$ (case production order)
$dX_4 = 1$ (new style per accounting period)

and

$$\frac{\partial f}{\partial X_2} = \$1.116$$

$$\frac{\partial f}{\partial X_3} = -.004$$

$$\frac{\partial f}{\partial X_4} = +.0055$$

then

$$dX_T = (\$1.116)\cdot(1) + (\$-0.004)\cdot(1) + (\$0.0055)\cdot(1) = \$1.1775$$
$$= \text{additional cost for this decision.}$$

If, on the other hand, the incremental functions are curves, rather than constants, the problem is not greatly altered since it is then simply necessary to evaluate the incremental functions for the appropriate value of each independent variable. If the total cost is correlated jointly with the various independent variables, the determination of the incremental costs is, of course, much more complex.

CHAPTER VIII

Practical Applications of Findings

The purpose of this chapter is to indicate the practical uses to which the findings of this study can be put and to show how to convert the average and the marginal cost estimates discussed in chapters vi and vii and in Appendixes A and B into a form which will serve these practical ends. The discussion is divided into two parts, the first of which deals with average cost findings and the second with marginal cost findings.

Average Cost Findings

PRACTICAL USES FOR AVERAGE COST FINDINGS

The most important practical use to which the average cost findings of this study can be put is in the determination of flexible standards for cost control. Control of expenses is often inadequate when the customary fixed standards are relied upon because these expense standards make no allowance for the effect upon cost of changes in operating conditions. Flexible cost standards which are free from these defects can be developed from the net regression curves of the multiple correlation analyses of combined cost found in chapter vi and in Appendix A.[95] By reading from these curves standard costs can be determined which will reflect the actual operating conditions of the period (e.g., rate of output, size of production order, number of new styles, etc.). Cost criteria thus developed are superior as control devices because a departure from the flexible standard is more strictly due to variation in efficiency than to any other factor.[96]

This type of control device deserves more general use than it now enjoys. Cost is commonly a function of other less uncontrollable factors as well as of managerial efficiency. Hence these other relationships must be recognized if expense standards which will deal justly with executives are to be set up.

In addition to this important application of average cost findings there are certain minor purposes for which these cost estimates may be used. One of these is to evaluate proposals for reducing cost. Another is to aid in forecasting future expenditures and profits under anticipated operating conditions.[97]

The average cost findings of this study have not been presented in a form which will best serve the practical ends discussed above. To achieve these purposes more effectively certain adjustments must be made in the regression curve magnitudes. The cost estimates must be placed on a parity with current (or future) cost with respect to (1) the level of input prices and (2) the lag and lead correction.

The adjustment of these estimates to correspond to the current level of prices is necessary because the cost data from which these curves were derived were reduced to the price level of January, 1932. Price adjustments of this type can be readily made, either upon the basis of a composite weighted index number of the prices of the input factors, or upon the basis of individual price indexes applied separately to each of the components of combined cost. The former method is simpler but may be less accurate, partly because the weighted average of the composite index becomes less representative as the lapse of time between the corrected period and the base period increases. Here separate correction of the components will probably be more satisfactory since estimates of component elements must be corrected individually if they are to be used for purposes of control or forecast. Individual correction will not be costly since indexes for the components have already been worked out and need only be kept up to date.

In addition to adjusting the cost estimates so as to make them correspond with the current level of prices these estimates must also be made comparable with current costs with respect to lag and lead correction. Cost data used in developing the estimates were adjusted so as to correspond to the output to which the actual cost contributed. Since, on the other hand, most cost entries anticipate output, the recorded cost of a current month has not been thus adjusted and hence does not correspond exactly to the recorded output of that period. Therefore, either the curve estimate or the current figure must be adjusted if the two are to be strictly comparable. If the cost estimate is to be used to predict future profits, the lag correction should not be removed, for the cost estimate, in this instance, should remain associated with the actual output to which it contributed. If, on the other hand, the objective is to compare present recorded cost with a flexible standard developed from the regression curves, then the recorded cost with which the estimates are being contrasted should be subjected to the same lag correction as the estimated standard.

In making the adjustments for output lag two methods are available. (1) Combined cost for the current period can be corrected as a whole by a weighted average of the time lag adjustments employed for its separate component elements. (2) Each constituent of current cost can be adjusted separately by the lag correction which was developed for rectifying the original data. The former procedure appears to be simpler, but the latter will be preferred if the cost components are to be used separately for purposes of control or prediction.

DEVICES WHICH FACILITATE THE USE OF AVERAGE COST FINDINGS

To maximize the practical usefulness of the average cost findings of such a study as this not only is it desirable to convert the cost estimates into comparable form (with respect to price level and output lag), but it is also important to call attention to devices for recording and manipulating these cost estimates—devices which will facilitate the use of these findings and make it convenient for management to apply them. These devices may be grouped as follows:

1. Net regression curves.
2. Tables of estimated cost under various operating conditions.
3. Charts relating costs to income at various levels of output.
4. Flexible budgets.

The net regression curves of the multiple correlation analyses are themselves a convenient and flexible record of results. Algebraic summation of readings from a set of these curves for any combination of current or anticipated operating conditions gives a rough estimate of average cost. These regression curves can also be converted into a second device for presenting results, namely, classified tables which show estimated cost under various operating conditions. Such a table was constructed for combined cost as a convenience to the management of the particular furniture factory studied (Table 1–1).

A third type of device is the break-even chart which compares cost and income at various levels of output. To construct such a chart the expected total magnitude of combined cost and the expected total net sales revenue are plotted as the ordinate, and the rate of output as the abscissa. The distance between the total cost line and the income line represents profits or losses.[98]

Visual administration of flexible cost standards by calling attention to departures of current cost from the net or gross regression curve which is used as the performance standard is a promising development. Two-dimensional charting, however, makes budgets flexible with respect to only one operating condition at a time. Such simplification is warranted only when this one variable is of dominant importance. Budgets which are flexible in a fuller sense of the term must be adjusted simultaneously for the effect of several operating conditions and must be revamped as the expectations with regard to these conditions change. The labor of constructing and revising such a budget can be reduced by tabulating on permanent data sheets the cost adjustments for a series of values of each operating condition read from the net regression curves.

Marginal Cost Findings

The practical application of marginal cost findings plays a subordinate rôle in a study whose principal concern is with methods of determining marginal cost rather than with ways of using them. The discussion which follows is limited, therefore, to the three types of considerations which affect the practical usefulness of marginal cost estimates in an individual firm:

1. Uses to which marginal cost findings can be put.
2. Problems of transforming the marginal cost estimates of this study into a form which is readily usable for these purposes.
3. Limitations upon the utilization of these transformed marginal cost estimates for practical business decisions.

PRACTICAL USES FOR MARGINAL COST ESTIMATES

The principal use to which marginal cost estimates can be put is to aid in determining the most profitable price and output policy. The selling price which yields the highest profit can be determined, within limits to be discussed later, from the estimated future marginal cost curve when this curve is considered in conjunction with the future marginal revenue curve. This maximum profit price is the price which corresponds to the output at which marginal cost and marginal revenue are equated.[99]

Even when the demand schedule and marginal revenue curve of the firm in question are not known precisely, marginal cost may be of some service in determining a rational price policy. If the market is sectionalized and a series of class prices prevails, then the marginal cost sets a limit below which the firm cannot, except under unusual circumstances, afford to allow its lowest class price to drop. For example, if chain stores constitute the minimum price sector of the market, then the lowest price at which it will be profitable to sell to that sector will be the incremental cost of these sales (sales at higher prices, to other sectors of the market, are assumed to be unaffected).[100] Since such buyers often bid on the basis of assumed knowledge of the differential cost of manufacture, it is well for the producer to know what his incremental cost actually is.

Another closely associated use for marginal cost estimates is to help determine the wisdom of introducing a new product or of varying the proportion of the divers articles manufactured. When the marginal costs of various ways of modifying a product are known, and when the probable increment in revenue resulting therefrom can be estimated with any degree of accuracy, then the equating of these two estimates provides a rational guide to the most profitable type of alteration of the product. In like manner marginal cost is the rational criterion for deciding upon the proportion of the various articles which should be manufactured. This guide should, however, be supplemented by other information since estimates of marginal cost are not allocated to particular products with great accuracy, and since increments in revenue attributable to the various articles and combinations of articles can be estimated only within wide margin of error. Nevertheless, both these margins can be estimated with sufficient precision to make such studies highly useful for this important purpose.

The marginal cost findings of this study are not usable as a guide to the proper proportioning of the factors of production, for the components of marginal combined cost do not represent partial derivatives of the total cost function and, therefore, are not measurements of the marginal productivity of the factors.

ADJUSTING MARGINAL COST ESTIMATES

To make the marginal cost estimates of this study usable for the objectives discussed above they must be adjusted in several ways. First, they must be corrected for those changes in the prices of input factors which have occurred between the base period of the analysis (January, 1932) and the time at which these estimates are being applied. This adjustment can be made by methods analogous to those suggested for correcting estimates of average cost.

A second type of adjustment consists of recognizing the effect upon marginal cost of operating conditions other than the rate of output.[101] Since total cost is a function of many variables it has many margins—one for each factor with which it varies. Frequently several of these margins, in addition to the commonly recognized output margin, are of quantitative significance. In the particular firm studied, as has been noted in chapter vii, the cost increments associated with increasing the size of lot, the number of new styles, and the rate of labor turnover have been roughly measured. When acceptance of a given order involves change in these other operating conditions, then their respective increments of cost must be added to the marginal cost of increasing the output in order to obtain the actual increment in cost caused by the order in question.

The allocation of marginal cost to particular products is a third type of adjustment. This disposition is especially important in getting the cost estimates into usable form because price and output decisions are commonly concerned with particular types and articles of product, rather than with output in general.

Typical practical questions which arise are: whether or not to accept an order for a given style at a specific price; whether or not to alter an article in a particular fashion in order to make it more salable; whether or not to place more stress on a particular line of products in the coming sales campaign; and the like. In dealing with these problems the incremental costs of particular articles of product are of dominant interest to the management. Some method, therefore, must be devised for attaching the marginal cost of general output to specific articles of product. At first sight the task of assigning the marginal cost of general output to specific articles and lots of produce appears to be hopeless. Theoretically, the only accurate solution for the problem is to vary the proportions of the several products freely and extensively, and then study the effects upon the total cost resulting from increasing the amount of product A while holding constant, in various proportions, products B, C, and D. Such a procedure is, of course, quite out of the question when there are, as in the present case, hundreds of products. Therefore, some rough, practical approximation to this theoretically correct solution must be devised.

The simplest approximation is to allocate marginal combined cost to the article in question on the basis of the index of output. Suppose that it is necessary to determine the minimum price at which a particular order for one hundred units of a specific model of furniture can be accepted. First the total old warehouse value (deflated standard cost) of this order will be computed, say at $1,000. Second, this value will be multiplied by the estimated marginal cost per dollar of old warehouse value at the expected level of output, perhaps $1.12, giving a marginal cost of $1,120 for the order. This estimate, in turn, can then be adjusted if necessary for the forecast price level and for the effect of other cost influences. Although it requires some rather questionable assumptions,[102] this method's simplicity and economy recommend it highly when dealing with marginal cost estimates of limited accuracy.

Another method of estimating the marginal cost of a particular article is to build up the combined marginal cost for this item from its various components by combining the cost constituents in the proportion peculiar to this particular article of product. This method is described in considerable detail in Appendix B. There is some question, however, whether this additive method will give more accurate marginal cost estimates than will the simpler procedure of allocating combined marginal cost according to the old warehouse value of the item.[103]

Even when marginal cost estimates have been correctly converted into usable form, their value is further restricted by other limitations. First, the expense of securing information of this type may put it beyond the reach of most firms. Second, shifts in cost relationships caused by technological progress, reorganization of production routines, etc., may quickly render expensively computed marginal cost estimates obsolete. Third, lack of accurate estimates of marginal revenue may seriously limit the usefulness of estimates of marginal cost.

Finally, there are certain imponderables such as the effect upon trade connections, attitude of customers, morale of the working force, etc., which, although they cannot be measured, nevertheless do influence rational price policy decisions.

Hence, even with accurate information regarding incremental revenue and incremental cost, there remain wide margins of uncertainty within which habit, business philosophy, and social pressure will sway decisions.

In conclusion, it should be pointed out that in the practical use of marginal cost estimates much thought should be given to the determination of the time period which is really appropriate for a given decision. Many policies which actually involve long-period commitments and ramifications are incorrectly formulated in the light of short-period considerations (and vice versa). For numerous types of decisions an intermediate-period rather than a short-period or a long-period marginal cost is required, since the psychological, social, and economic repercussions of the decision stretch forward some distance into the future. Rough estimates of the required intermediate-period marginal cost can be made on the basis of the findings of this study by taking a figure somewhere between the future short-period marginal cost[104] and the extrapolated average cost (which often constitutes a moderately good approximation to long-period marginal cost).

Summary

In order to facilitate the practical application of these *average cost* findings certain commercial uses have been pointed out. The most important managerial benefits are: (1) the determination of flexible cost standards which will make expense control more effective; (2) a saner evaluation of proposals for reducing cost—the result of more complete knowledge of the forces which affect cost behavior; and (3) a more accurate forecasting of expenditures and profits.

To put the cost findings into a form which will serve these managerial ends these estimates must be made comparable with current (or future) costs with respect both to the level of input prices and to the correction for lag and lead. Both of these adjustments can be made either by means of a composite correction index (for price level and for output lag) or by means of a separate adjustment of the individual components of combined cost. The latter method seems to be preferable for both types of cost adjustments since the components are to be used separately for control and for prediction purposes.

To facilitate the use of these adjusted cost estimates attention has been called to certain devices which make the findings available in convenient form. The most useful of these are the net regression curves of the correlation analysis, tabular summaries of these curves, break-even charts compiled from adjusted estimates, and permanent data sheets for flexible budgets.

The major practical use for *marginal cost* estimates is to determine the most profitable price and output policy. But this use has wide ramifications since a large proportion of business decisions can and should be put in terms of balancing the added cost of a certain policy against the added revenue resulting from this step. The findings of this study are not sufficiently refined, however, for use in many of these types of decisions.

Three types of adjustments must be made in the marginal cost estimates before they are usable for the practical purposes just mentioned. The adjustments are:

1. Marginal cost estimates should be adjusted to correspond to the price level of the period in which the estimates are to be used.

2. These corrected estimates should take into consideration the cost margins of other factors, such as size of production order, number of new styles, etc., as well as the marginal cost of increasing output.

3. The adjusted marginal cost estimates must be allocated to particular articles of product. This allocation can be accomplished in a rough way by assigning marginal cost to a particular article in proportion to that item's old warehouse value. A method which is more laborious, but may not be more accurate, is to build up the combined incremental cost of an article from the components of combined marginal cost.

Even if these adjustments are properly made, the usefulness of these marginal cost estimates is restricted by certain additional limitations: the expense of securing incremental cost figures and the limited life of these estimates owing to changes in fundamental cost conditions, the lack of information concerning marginal revenue, and the presence of certain imponderable considerations which must also enter into rational price decisions.

But these qualifications must not obscure the fundamental need of business men for better estimates of marginal costs and the importance of improving methods for making such estimates.

Appendix A

Behavior of Components of Average Combined Cost

The behavior of the average and total cost curves of constituent elements of combined cost is, for several reasons, of considerable interest to management. The most important service rendered by such knowledge is that it aids in the effective control of cost by providing the basis for setting up flexible cost standards in sufficient detail to be managerially usable. A second benefit derived from this detailed knowledge of the behavior patterns of cost elements is that such information makes possible more specific and exact allowances for dynamic changes such as shifts in particular prices or wage rates and minor alterations of production technology. Such allowances tend to increase the accuracy of predictions of combined cost. A third reason for managerial interest in the findings of this section is the light which they throw upon proposals for cost reduction and for alteration of production routines. A more exact appraisal of the effect of such policies can be made if the knowledge of combined cost is supplemented by an analysis of the behavior of its component elements. Finally, the analysis of the average or total cost behavior of these constituents by means of multiple correlation is a highly desirable preliminary step in the reliable estimation of the components of marginal combined cost—a type of information which enhances the flexibility and accuracy of marginal cost estimates.

In using these multiple correlation results for the foregoing purposes the method and the underlying assumption of the separate analysis of these components must be kept in mind. The method followed was to take the actual recorded expenditure for each component without attempting to hold constant the expenditures for other components. All components were allowed to vary together in the manner which the management had regarded as the most advisable combination in the light of the operating conditions of the period. The recorded expenditure for each component was separately corrected for price changes and for output lag and was then subjected to a graphic multiple correlation analysis. The basic assumption which underlay this procedure was that negligible substitution between cost components took place as a result of changes in their relative prices (pronounced though these price changes were during this period). However, substitution as a result of changes in the volume of output and in other operating conditions was fully reflected and, consequently, must be regarded as the major cause for the observed cost behavior. Thus, we have a dynamic record of the proportion in which management saw fit to combine resources in response to short-period changes in output, size of production order, number of new styles, and the like.[105]

Direct Labor Cost

The net regression curve for output (Chart 1–9) indicates by its continuous decline that direct labor cost per unit diminishes as the rate of production activity increases. Such cost behavior appears to be in accord with expectations.[106] The second regression of Chart 1–9 indicates that direct labor cost per unit also diminishes when the product is manufactured in larger lots. Such behavior is to be expected in view of the high ratio of setup time to running time in the machine operations of this factory, for the spreading of a fixed setup cost over a large number of pieces results in a lowering of the cost per piece. In accord with reasonable expectations the middle curve of Chart 1–9 shows that average labor cost rises at a uniform rate as the number of new styles is increased; new models usually involve special instruction of workers and are likely to cause confusion in production routines. The fourth curve shows that a striking increase in labor cost per unit accompanies a reduction in the level of output from that of the previous accounting period. This behavior may be partly caused by failure to reduce and reorganize the working force for the new low level of production. The bottom curve of Chart 1–9 shows, as would be expected, that higher labor turnover is associated with high labor cost.

Chart 1–9 Partial Regression Curves of Average Direct Labor Cost Per Unit with Five Independent Factors: Output (X_2), Size of Production Order (X_3), New Styles (X_4), Production Increases (X_5), and Labor Turnover (X_7)

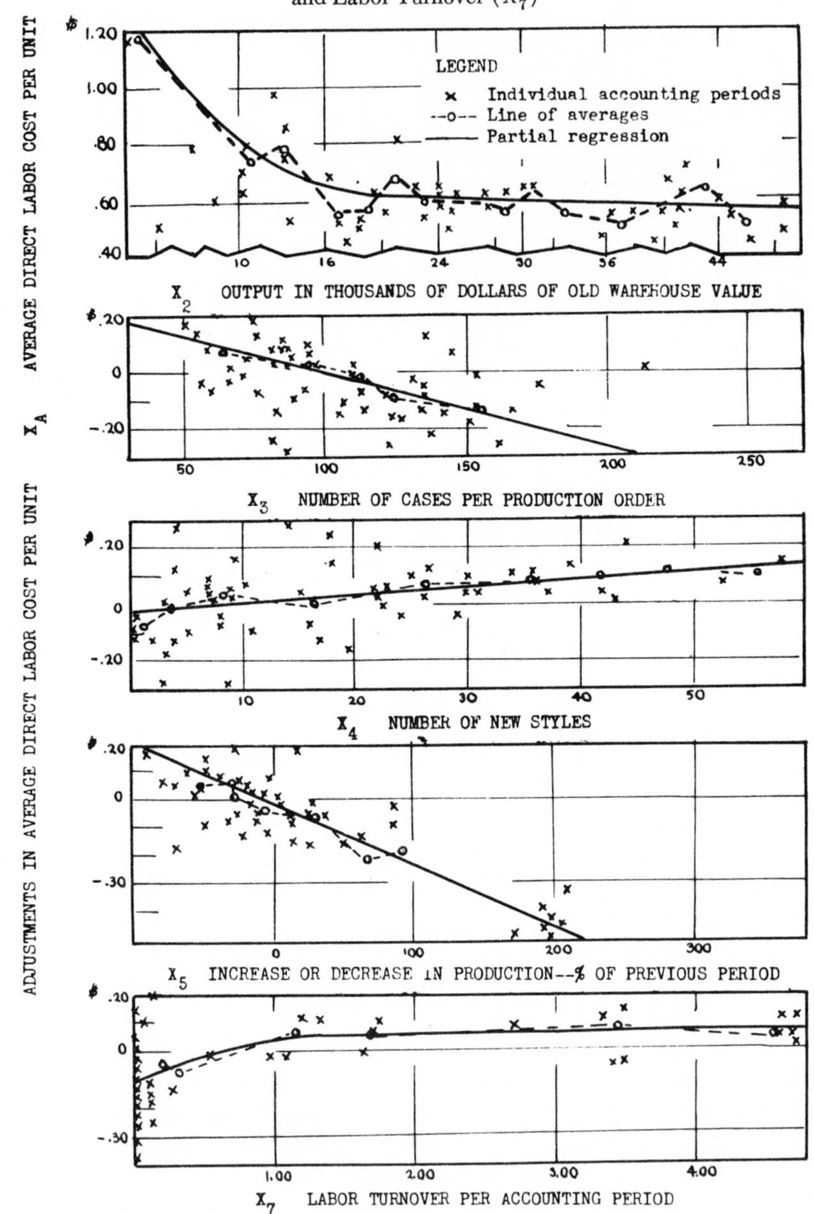

Table 1–6

MEASURE OF THE UNRELIABILITY OF AVERAGE DIRECT LABOR COST FUNCTION*

(Standard error of difference between partial regression curve at specified values of output and partial regression curve at average value of output)

OUTPUT (UNIT $1,000 IN O.W.V.)	ESTIMATED COST		STANDARD ERROR OF DEVIATIONS IN COL. 3‡
	Reading from Curve	Deviation from Reading at Mean Output†	
(1)	(2)	(3)	(4)
5	$1.070	$.483	$.077
10	0.814	.227	.056
15	0.672	.085	.039
20	0.613	.026	.023
25	0.597	.010	.020
30	0.588	.001	.005
35	0.580	— .007	.025
40	0.572	— .015	.031
45	0.563	— .024	.060
50	0.550	— .037	.098

*The method of computing this standard error is described in Ezekiel, *Methods of Correlation Analysis*, p. 384. The formulas are provisional and give only a rough approximation.

†Departures of curve readings from the curve value $0.587, which corresponds to the mean of output ($29,720).

‡Standard error refers to departures described in the preceding footnote.

On the basis of the standard error of departures of the output regression curve from its value at the mean level of output (shown in Table 1–6) the following conclusions were reached: (1) For the outputs greater than $25,000, O.W.V., little statistical significance can be attached to departures of this curve from its reading at the mean of output. (2) Only a moderate degree of confidence can be placed in the curve departures in the vicinity of $20,000 output. (3) Considerable significance can be attached to the curve below $20,000 output. It must be remembered, nevertheless, that the reliability of this curve as a basis for estimating cost behavior is not measured by these standard errors.

Other conditions revealed by the correlation analysis indicate that the net regression curves may be used with considerable confidence in setting standards and in forecasting results, provided, of course, that the underlying conditions remain the same as during the period of study. Each of the net regressions indicated cost behavior which was sensible. Used together, these curves accounted for a very large proportion of cost variance (75.2 per cent). Moreover, cost estimates made from them had a narrow band of standard error ($0.108) and were highly correlated with actual labor cost (multiple correlation index of .867); see Table 1–2.

Rough measurements were also made of the relative importance of the influences

Table 1–7

RELATIVE IMPORTANCE OF THE SEVERAL INDEPENDENT FACTORS INFLUENCING LABOR COST AS SHOWN BY THEIR COEFFICIENTS OF NET REGRESSION, COEFFICIENTS OF PART CORRELATION, AND BETA COEFFICIENTS

Independent Factor	Coefficient of Net Regression ($b_{12.3457}$, etc.)	Coefficients of Part Correlation ($_{12}r_{3457}$, etc.)	Beta Coefficients ($B_{12.3457}$, etc.)
X_2 (output).........	Curve	.905	Curve
X_3 (order size).......	.00266	.54	.317
X_4 (new styles)......	.002408	.31	.163
X_5 (change in output)	.00226	.67	.445
X_7 (labor turnover)...	Curve	.18	Curve

of the several independent factors upon average direct labor cost. These indexes are shown in Table 1–7. As indicated by the somewhat unreliable coefficients of part correlation, output (X_2) was the most important factor (.905); change in production (X_5) was second (.67); size of lot (X_3) third (.54); new styles (X_4) fourth (.31); and labor turnover (X_7) had least importance. This hierarchy of relative importance is substantiated by the Beta coefficients and is, moreover, in general accord with reasonable expectations. The importance of changes in output (X_5) is somewhat surprising, and the relatively small effect of the rate of labor turnover (X_7) upon direct labor cost will be of interest to personnel managers. It is dangerous, however, to generalize from the observed cost behavior of one firm.

Indirect Labor Cost

The behavior of average indirect labor cost which is indicated by the net regression curves of Chart 1–10 is, on the whole, in accord with a priori expectations. This partially fixed cost falls continuously, at a diminishing rate, as output increases. The number of new styles appeared to have no clear influence upon indirect labor cost. However, changes in output from that of the previous period had the expected effect: Decreases in output caused larger per-unit magnitudes of this relatively inflexible cost whereas small increases were associated with smaller cost since this inflexibility then worked to the firm's advantage.

These curvilinear paths of cost behavior accounted for about 90 per cent of the total variance of this cost item, according to the index of multiple determination. Estimates of cost made on the basis of these curves were highly correlated with the actual cost values as indicated by the coefficient of multiple correlation, .949 (see Table 1–2). The standard error of estimate for indirect labor cost was $0.022, which compares favorably with the standard deviation, $0.069. From these several measures it appears that a good deal of confidence can be placed in these regression results in accounting for the behavior of average indirect labor cost.

Chart 1–10 Partial Regression Curves of Average Indirect Labor Cost Per Unit for Three Independent Factors: Output (X_2), Number of New Styles (X_4), and Increases in Production (X_5)

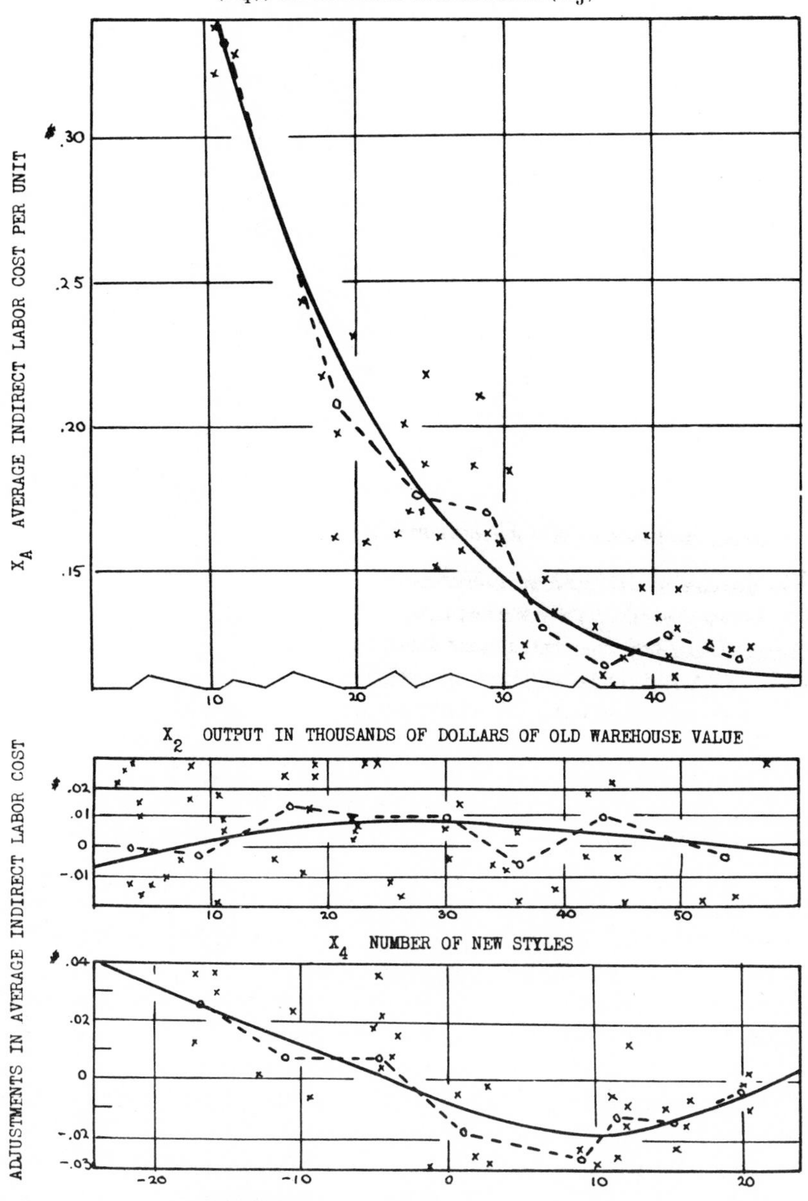

The standard error of departures of the output regression curve from the curve value at the output mean are shown in Table 1–8.

Table 1–8

MEASURE OF THE UNRELIABILITY OF THE AVERAGE INDIRECT LABOR COST FUNCTION*

(Standard error of difference between partial regression curve at specified values of output and partial regression curve at average value of output)

OUTPUT (UNIT $1,000 IN O.W.V.)	ESTIMATED COST		STANDARD ERROR OF DEVIATIONS IN COL. 3‡
	Reading from Curve	Deviation from Reading at Mean Output†	
(1)	(2)	(3)	(4)
10	$.3590	$.2113	$.0362
15	.2665	.1188	.0442
20	.2104	.0627	.0207
25	.1735	.0258	.0107
30	.1466	—.0011	.0029
35	.1288	—.0189	.0142
40	.1198	—.0279	.0174
45	.1151	—.0326	.0285
50	.1140	—.0337	.0734

*The method of computing this standard error is described in Ezekiel, *Methods of Correlation Analysis*, p. 384. The formulas are provisional and give only a rough approximation.

†Departures of curve readings from the curve value $0.1477, which corresponds to the mean of output ($29,720).

‡Standard error refers to departures described in the preceding footnote.

Inspection Labor Cost

The surprising characteristic of total inspection labor cost which is brought out by its output regression curve of Chart 1–11 is its flexibility. This flexibility was apparently brought about by transfer or partial layoff of inspectors. In the second chart of this figure it is indicated that large manufacturing lots, reasonably enough, bring a reduction in the total inspection cost. But increases in the number of new styles are not associated with the expected increase in this cost but, instead, with a strange decrease. Perhaps the inspection task for newly introduced styles is really performed by foremen, engineers, and workmen, thus leaving less rather than more for formal inspection. The curve for quality of materials also shows a somewhat illogical cost behavior. Cost should rise continuously, instead of first rising and then falling off for high qualities. Perhaps the illogical shape of this regression is explained by the poor curve-fitting; a rising straight line could have been more reasonably employed.

A study of the reliability of the regression curve for output (Table 1–9) indicates that a straight line was as flexible a function as the data justified. The undu-

Chart 1–11 Partial Regression Curves of Inspection Labor Total Cost for Four Independent Factors: Output (X_2), Size of Production Order (X_3), Number of New Styles (X_4), and Quality of Material (X_8)

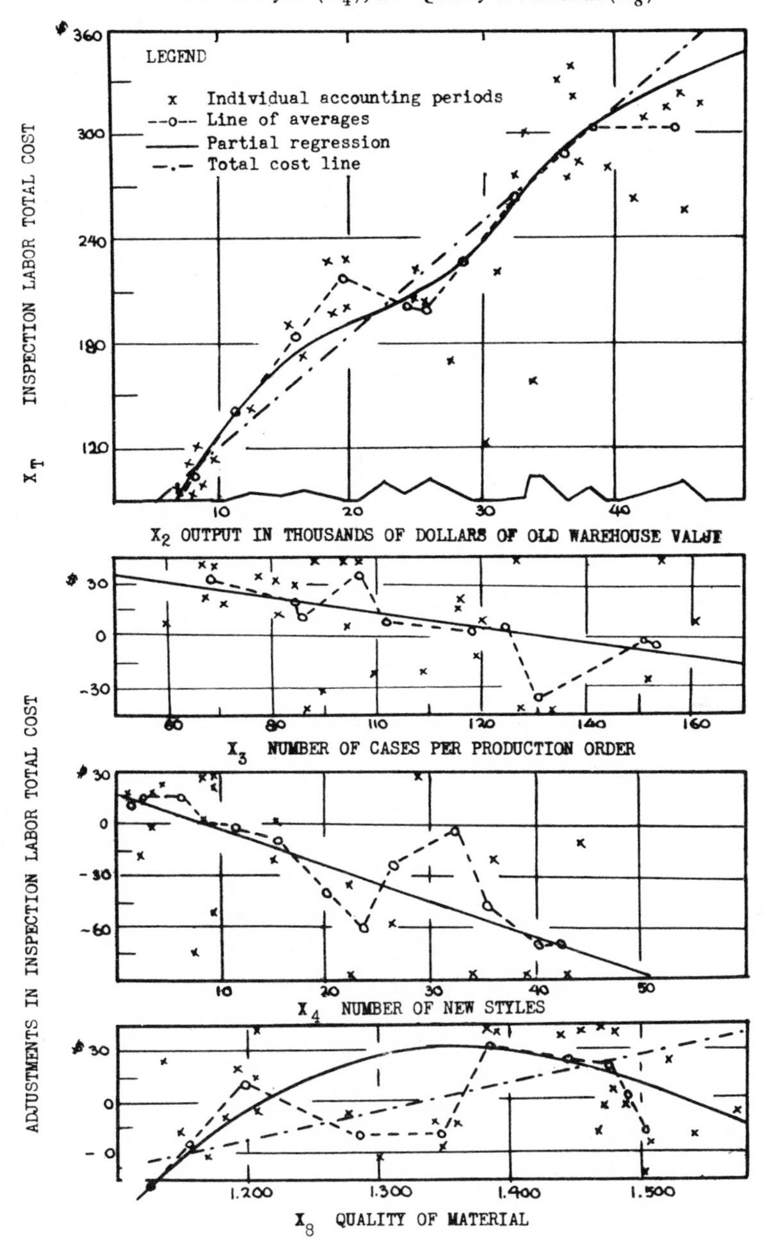

lations of the central section of the curve fitted to these data were clearly not statistically significant.

Table 1-9

MEASURE OF THE UNRELIABILITY OF THE TOTAL INSPECTION LABOR COST FUNCTION*

(Standard error of difference between partial regression curve at specified values of output and partial regression curve at average value of output)

OUTPUT (UNIT $1,000 IN O.W.V.) (1)	ESTIMATED COST		STANDARD ERROR OF DEVIATIONS IN COL. 3‡ (4)
	Reading from Curve (2)	Deviations from Reading at Mean Output† (3)	
5	$ 80	$—168	$64.41
10	140	—108	70.45
15	182	— 66	86.08
20	202	— 45	49.46
25	219	— 29	24.37
30	249	1	5.93
35	296	48	25.77
40	323	75	41.55
45	329	81	62.02
50	359	111	.....

*The method of computing this standard error is described in Ezekiel, *Methods of Correlation Analysis*, p. 384. The formulas are provisional and give only a rough approximation.

†Departures of curve readings from the curve value $248, which corresponds to the mean of output ($29,720).

‡Standard error refers to departures described in the preceding footnote.

Measurements of the degree of success attained in this analysis of total inspection cost behavior are not encouraging. The standard error of estimate was found to be $79.59, as compared with a standard deviation of $87.72, while the index of multiple correlation stood at .456 and the index of multiple determination at .208. Evidently estimates made from these curves must be supplemented by allowance for other causal facts not included in this analysis.

Yard Labor Cost

The upper chart of Chart 1-12 shows that the output regression curve of total yard labor cost rose continuously from an apparently irreducible minimum of about $600 to an asymptotic top of about $1,900. This behavior accorded well with expectations except that the flexibility of yard labor in adjusting itself to short-period changes in the level of production was rather surprising. It can be seen from the second chart of Chart 1-12 that increasing the size of the production order had an uncertain effect upon yard labor cost. The net regression curve at first rose, then

Chart 1–12 Partial Regression Curves of Yard Labor Total Cost for Four Independent Factors: Output (X_2), Size of Production Order (X_3), Increase in Production (X_5), and Lumber Receipts (X_9)

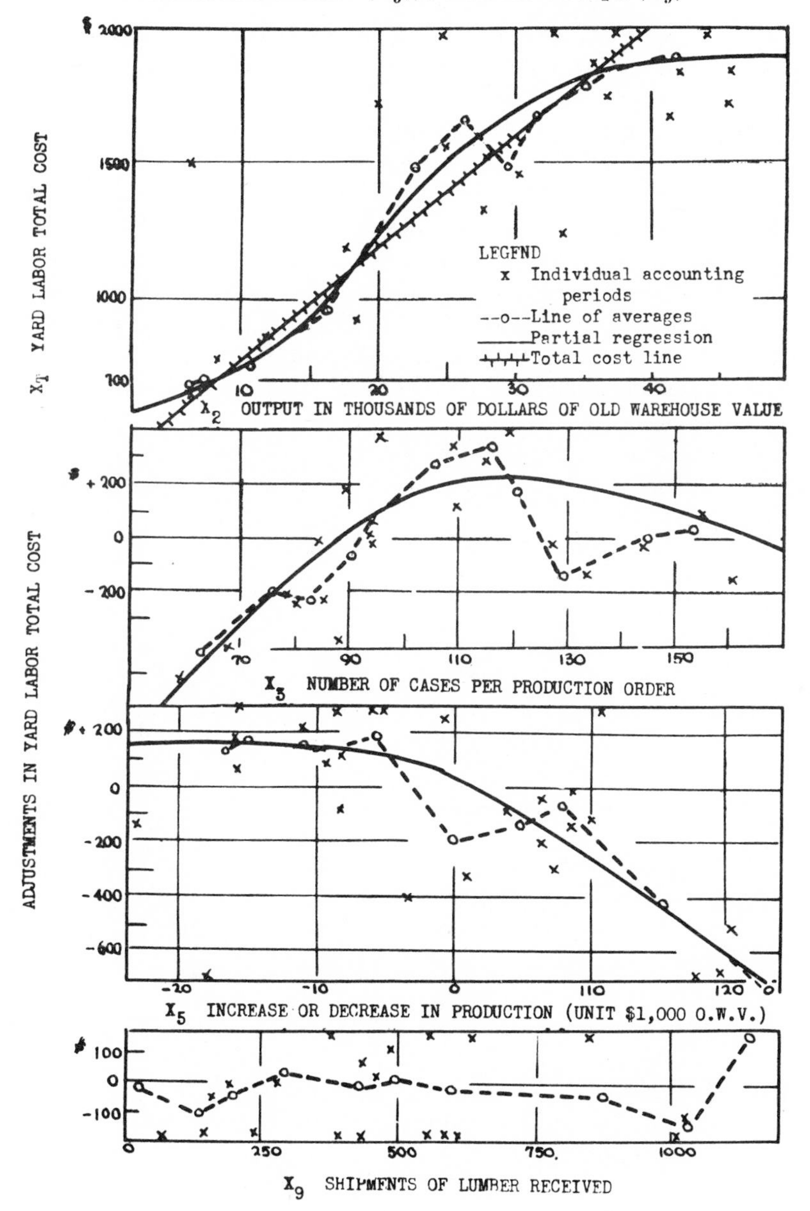

declined, with wide see-sawing of the line of averages. The regression curve for the third chart shows that a striking increase in the total of yard labor cost accompanies a reduction in the level of output from that of the previous period. And the bottom chart of Chart 1–12 shows that no relationship between yard labor cost and lumber receipts was discovered.

The adjusted standard error of the estimates of total yard labor cost made on the basis of these net regression curves was $392.46. This indicates that the chances are only one in three that the estimate of this cost will miss the actual cost by more than about $400. The index of multiple correlation (.689) showed that estimates of cost based upon the regression curves correlated only fairly well with the actual cost figures. The index of total determination was found to be .475, indicating that only a relatively small proportion of the variance of yard labor cost was accounted for by the independent factors used in this study. Estimates made on the basis of these curves should, therefore, be used cautiously in making managerial decisions and should be supplemented, where possible, with estimates of the effects of factors not included in this study.

The computed standard error of the departures of the output curve from its value at the mean level of output are shown in Table 1–10. Examination of this

Table 1–10

MEASURE OF THE UNRELIABILITY OF TOTAL YARD LABOR COST FUNCTION*

(Standard error of difference between partial regression curve at specified values of output and partial regression curve at average value of output)

OUTPUT (UNIT $1,000 IN O.W.V.) (1)	ESTIMATED COST		STANDARD ERROR OF DEVIATIONS IN COL. 3‡ (4)
	Reading from Curve (2)	Deviations from Reading at Mean Output† (3)	
5	$ 610	$—1,072	$390.09
10	766	— 916	348.41
15	964	— 718	368.68
20	1,236	— 446	244.62
25	1,473	— 209	147.74
30	1,649	— 33	32.95
35	1,788	106	139.65
40	1,779	197	251.58
45	1,892	210	306.69
50	1,899	217	.....

*The method of computing this standard error is described in Ezekiel, *Methods of Correlation Analysis*, p. 384. The formulas are provisional and give only a rough approximation.

†Departures of curve readings from the curve value $1,682, which corresponds to the mean of output ($29,720).

‡Standard error refers to departures described in the preceding footnote.

table led to the conclusion that between $5,000 and $30,000 output the curve is significantly different from its mean value, but that little significance can be attached to it beyond $35,000 output.

Indirect Material Cost

From Chart 1–13 it appears that indirect material cost per unit first declines and then rises as the rate of output is increased. This behavior may be caused by the composite character of this cost, which is made up of sandpaper, tools, and finishing materials. Per-unit tool cost should reasonably decline since it is comparatively fixed in total. Sandpaper cost, on the other hand, probably rises with heavy production because of pecuniary diseconomies, substitution of wasteful (i.e., wasteful of sandpaper) machine processes for hand sanding, and operation of these machines at uneconomical speeds. Finishing material cost probably behaves similarly because of the wastes of machine processes and of high-speed production.

Certain savings, particularly in finishing materials, appear to accompany large manufacturing lots. This expectation is substantiated by the second part of Chart 1–13. Increases in the number of new styles are associated first with an increase in indirect material cost, and later with a decrease, which decline is poorly supported

Table 1–11

MEASURE OF THE UNRELIABILITY OF THE AVERAGE INDIRECT MATERIALS COST FUNCTION*

(Standard error of difference between partial regression curve at specified values of output and partial regression curve at average value of output)

OUTPUT (UNIT $1,000 IN O.W.V.)	ESTIMATED COST		STANDARD ERROR OF DEVIATIONS IN COL. 3‡
	Reading from Curve	Deviations from Reading at Mean Output†	
(1)	(2)	(3)	(4)
5	$.03937	$.01095	$.00539
10	.03561	.00719	.00482
15	.03224	.00382	.00510
20	.02972	.00140	.00338
25	.02820	.00022	.00204
30	.02844	.00002	.00049
35	.02958	.00116	.00216
40	.03083	.00241	.00348
45	.03224	.00382	.00424
50	.03381	.00539	.00847

*The method of computing this standard error is described in Ezekiel, *Methods of Correlation Analysis*, p. 384. The formulas are provisional and give only a rough approximation.

†Departures of curve readings from the curve value $0.02842, which corresponds to the mean of output ($29,720).

‡Standard error refers to departures described in the preceding footnote.

Chart 1–13 Partial Regression Curves for Average Indirect Material Cost Per Unit with Five Independent Factors: Output (X_2), Size of Production Order (X_3), New Styles (X_4), Production Increases (X_5), and Quality of Materials (X_8)

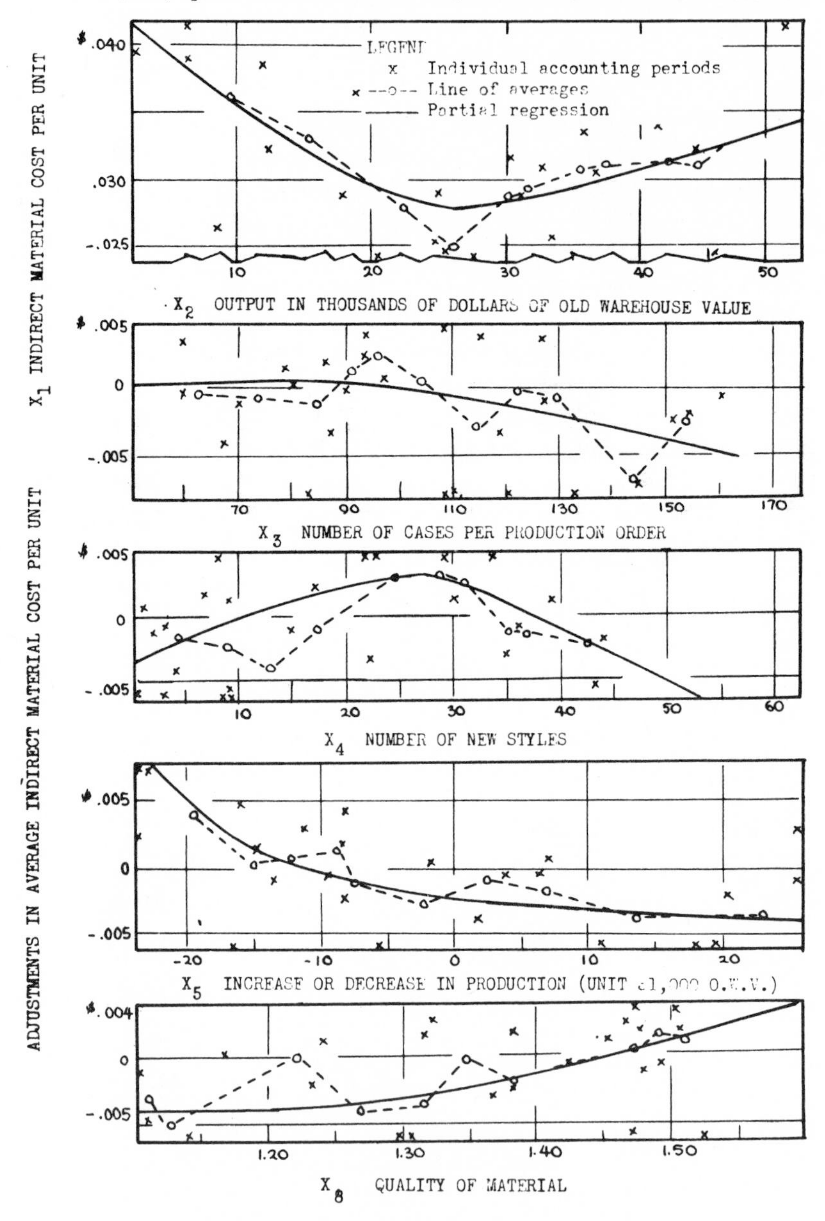

both rationally and empirically. The effect of sharp increases and decreases which is indicated by the fourth curve is entirely reasonable in view of the inflexibility of certain elements of indirect material cost. The rise of the regression curve for quality of materials is also in full accord with expectations, since more expensive finishing materials and larger proportions of sanded surfaces usually accompany a higher quality of product.

The standard error of departures of the output regression curve from its value at the mean level of output indicates that little confidence can be placed in the major portion of this curve (see Table 1–11). Moreover, the adjusted index of multiple correlation was found to have a value of only .343, which does not indicate sufficiently high correlation between estimated indirect material cost and actual cost to justify great confidence in the results of the analysis. Apparently some factors which have an important influence upon indirect material cost have been omitted. Hence, only limited reliance can be placed in these regression curves despite their apparent reasonableness.

Power Cost

Chart 1–14 shows the regression curves for average purchased power cost. From the top section of this figure it can be seen that this cost falls as output increases. This behavior is altogether reasonable since the public utility rate structure penalizes the firm for idleness, and since efficiency in the use of power is increased and transmission wastes are reduced by full operation. The middle chart of Chart 1–14 shows an entirely logical decline of power cost as the size of manufacturing lot is increased. The lower chart is also sensible, for it indicates that the number of new styles has no significant effect upon power cost per unit of output.

From Table 1–12 we see that small reliance in the departures of the regression curve for output from its value at the mean level of output is indicated by the standard error of this curve. The three regression curves taken together, however, account for a fair proportion of the variance of average power cost. The index of multiple correlation was found to be .637, and the standard error of estimate $0.062 (see Table 1–2).

Maintenance Cost

Average maintenance cost declines as output increases, as can be seen from the top chart of Chart 1–15. This behavior seems to be partially due to the fact that some of the wear and tear which causes maintenance is a function of time rather than of output.[107] But it is probably largely caused by the policy of this firm of using postponable maintenance as a sort of unemployment benefit.[108] In the middle chart of Chart 1–15 a slight decline of maintenance cost as a result of increasing the size of the production order is indicated. The third chart shows that quick changes in the rate of production have no significant effect upon maintenance cost.

Chart 1–14 Partial Regression Curves for Average Purchased Power Cost Per Unit with Three Independent Factors: Output (X_2), Size of Production Order (X_3), and New Styles (X_4)

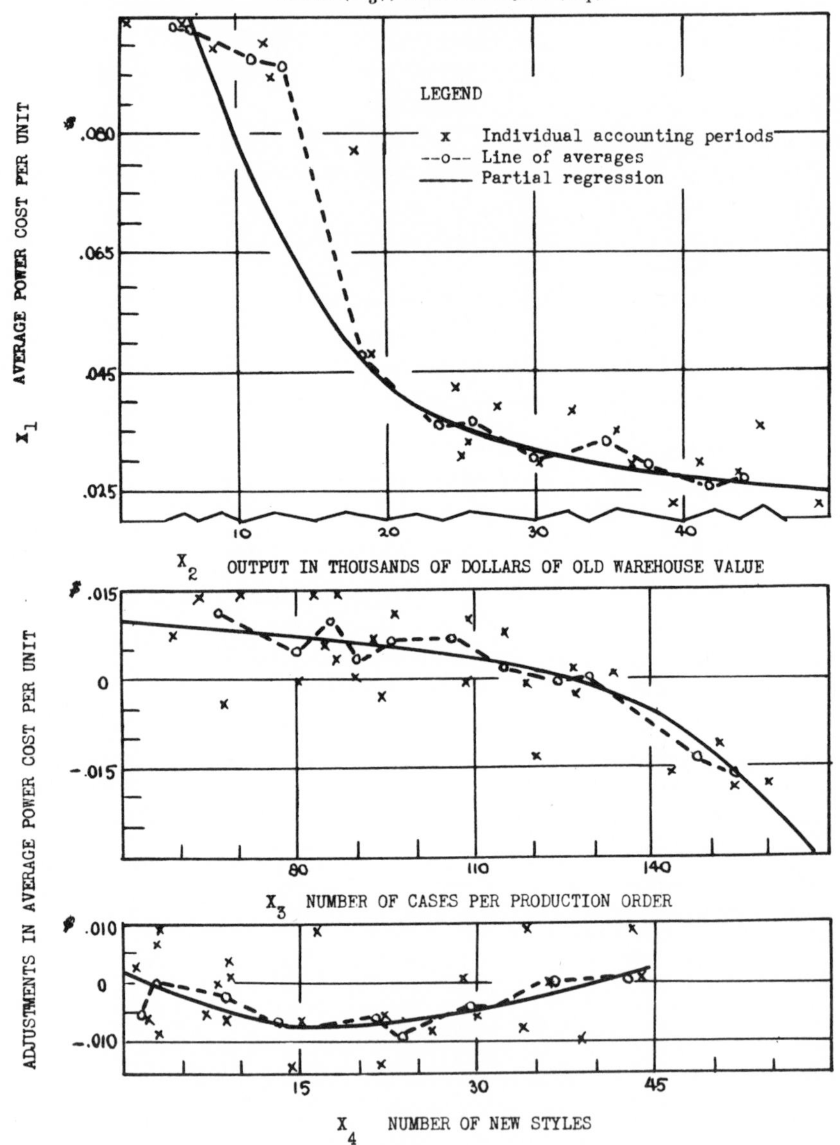

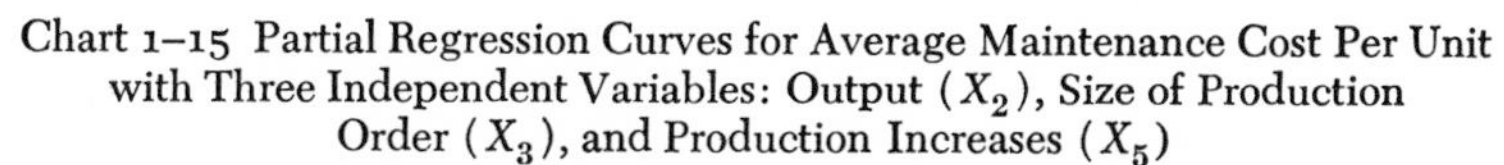
Chart 1–15 Partial Regression Curves for Average Maintenance Cost Per Unit with Three Independent Variables: Output (X_2), Size of Production Order (X_3), and Production Increases (X_5)

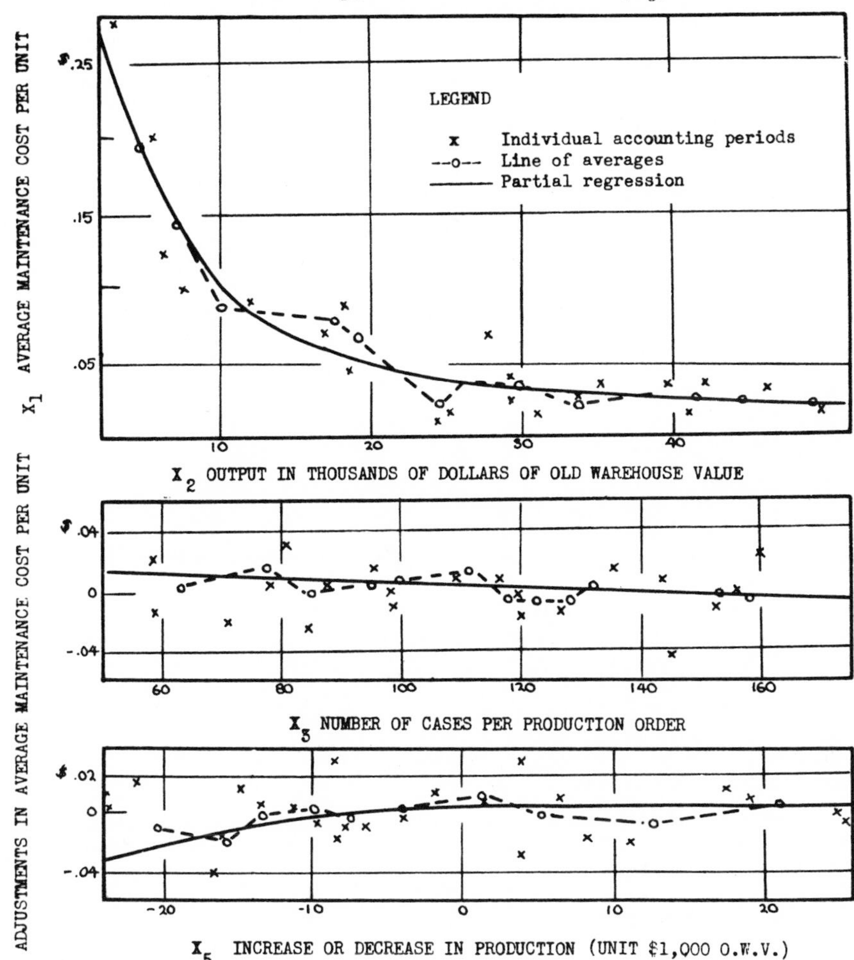

By means of these three regression curves average maintenance cost could be estimated with a high degree of accuracy, as is shown by the small size of the standard error of estimate, $0.0296, as compared with a standard deviation of $0.076. The correlation between maintenance cost and the three independent variables was high. The adjusted index of multiple correlation was computed as .924. This cost item ranks high among the average cost items with respect to the size of its adjusted correlations index (see Table 1–2).[109] Hence, these regression curves appear to represent a very satisfactory description of the behavior of this cost item.

Three cost items—combined indirect cost, direct material cost, and power and

Table 1–12

MEASURE OF THE UNRELIABILITY OF THE AVERAGE POWER COST FUNCTION*

(Standard error of difference between partial regression curve at specified values of output and partial regression curve at average value of output)

OUTPUT (UNIT $1,000 IN O.W.V.) (1)	ESTIMATED COST: Reading from Curve (2)	ESTIMATED COST: Deviations from Reading at Mean Output† (3)	STANDARD ERROR OF DEVIATIONS IN COL. 3‡ (4)
10	$.0843	$.0527	$.0517
15	.0562	.0246	.0547
20	.0426	.0110	.0363
25	.0354	.0038	.0219
30	.0313	—.0003	.0053
35	.0287	—.0029	.0232
40	.0271	—.0045	.0373
45	.0257	—.0059	.0455
50	.0244	—.0072	.0908

*The method of computing this standard error is described in Ezekiel, *Methods of Correlation Analysis*, p. 384. The formulas are provisional and give only a rough approximation.

†Departures of curve readings from the curve value $0.0316, which corresponds to the mean of output ($29,720).

‡Standard error refers to departures described in the preceding footnote.

watch labor cost—were also analyzed. Because of limited space the results are not presented in this abstract but may be found in the author's dissertation,[110] where those cost items which are presented in this abstract are also discussed more fully. Several components of the long-period cost of a finance company were also studied. These, too, could not be discussed here, but are found in Appendix E of the author's dissertation.

APPENDIX B

Behavior of Components of Marginal Combined Cost

Introduction

Before undertaking a discussion of the components of marginal combined cost it is important to define the meaning of this phrase. The direct labor component, for example, is not the partial increment of cost associated with increasing the output by varying labor cost alone. Instead, it is the increment in direct labor cost associated with an increase in output of one unit ($1.00 O.W.V.) when such an increment of output is obtained in the customary way, i.e., with the other variable cost elements increasing as they typically do. As was indicated in an earlier chapter, such a study is of use only when, for given operating conditions, the combination of cost constituents is fixed within rather narrow limits.

Since this condition appears to be approximated in the firm studied, some practical justification for a study such as this may be found in the expectation that a knowledge of the behavior of the components of marginal cost may increase the accuracy of forecasting and allocating combined marginal cost. The accuracy of forecasts may be enhanced by such a study partly because specific allowance can then be made for changes in conditions which affect a particular cost component. For example, an expected increase of 20 per cent in the wage rates of yard labor could be reflected fairly accurately by a 20 per cent increase of the estimate of the yard labor component of marginal cost, but could not be allowed for at all in an estimate of marginal combined cost without a knowledge of the relative size of the yard labor cost constituent.

A second way in which study of the behavior of these constituent elements may enhance accuracy is by providing a rough check upon the estimate of combined marginal cost. This can be accomplished by building up the combined marginal cost function by summation of its components. The estimate of each component of marginal cost is arrived at independently from the recorded totals of this particular type of expense. If the sum of these constituents equals the marginal cost obtained by differentiating total combined cost, faith in the accuracy of derivation is buttressed.[111]

In addition to the hope that a knowledge of the individual components may increase the accuracy of the estimate of marginal combined cost there is also a possibility that such information may improve the precision with which this estimated cost is allocated to particular items of product. Without such knowledge the average of combined marginal cost of all products is allocated to a particular product on the basis of our common denominator of output (old warehouse value). This procedure assumes, first, that old warehouse value is a satisfactory measure of output; second, that the components of marginal cost behave in approximately the same fashion regardless of the product being manufactured; and, third, that the proportion in which these various components are combined remains approximately constant for all articles of product.

It is with this third assumption that we are particularly concerned here, for the first two form the basis of the proposed refined method of allocation as well as for the average method. The assumption that articles do not differ in the proportion in which their several cost components are combined is patently incorrect. Certain types of bedroom furniture, for example, employ an inordinately large proportion of indirect materials and indirect labor.

A suggested alternative procedure for allocation is to build up an estimate of the combined marginal cost of an article as the sum of the marginal components. This could be accomplished as follows:

1. The standard average cost of an article is broken down into its components.
2. The normal operating conditions for which this standard was set are specified.
3. The deflated average cost of each component is determined from the regression curves for these specified normal conditions.
4. The deflated marginal cost of each component is ascertained from the regression curves and from estimated marginal cost functions for the operating conditions under which the article is expected to be produced.
5. The ratio of the marginal cost determined in (4) to the deflated average cost found in (3) is obtained.
6. These ratios are used to weight the respective components of standard average cost in (1), which are then summed to give the estimated marginal combined cost.

There are several important assumptions underlying such a procedure:

1. Standard average cost and its components are satisfactory.
2. Conditions for which such standards are computed can be specified.
3. The ratios of marginal cost under any given conditions to average cost under fixed normal or standard conditions are approximately the same for all articles. If the marginal cost of components is quite variable with output, this assumption would probably necessitate that the combination of cost elements at any given output be fairly stable regardless of the articles composing that output. Even if this were true, it is conceivable that these ratios might vary considerably for different articles.
4. For any given size of output composed of specified articles the proportion of cost constituents is fixed within rather narrow limits.
5. Old warehouse value measures output accurately.
6. Curve-fitting errors of the estimates of the marginal cost of components are somewhat compensating, so that accuracy is not destroyed by cumulative errors.

In determining which components of marginal cost were to be studied an effort was made to choose, first, cost items which would be of greatest interest to management; and, second, costs whose behavior would be difficult to estimate short of a systematic analysis—combined indirect labor, inspection labor, yard labor, power and watch labor, indirect materials, power, maintenance, direct materials, direct labor, and combined indirect cost. The methods employed in studying these costs were similar to those used for combined cost. Corrected data for most components were studied both in the form of totals and in the form of averages.[112]

A quick grasp of the findings of this phase of the study can be had by examining Tables 1–13 through 1–22. The results obtained by fitting straight lines to data in the form of totals are exhibited in Table 1–13, where the estimate of constant marginal cost together with the total cost equation from which it was derived are presented. The findings obtained by fitting curvilinear functions to average and to total cost data are shown separately for each cost item in Tables 1–15 through 1–22. Each of these tables shows estimates of the average, total, and marginal cost of the component at regularly spaced output levels. Below the table are presented the equations from which the estimates were obtained.[113] All estimates of marginal cost are expressed as the increment in deflated dollar cost (as of January 1, 1932) which accompanies an increment in output of one dollar of old warehouse value.

Before considering the findings it will be well to call attention to certain factors which affect their reliability. In general, the trustworthiness of the estimates of a particular component of marginal cost depends upon:

1. The form in which the data were studied. Totals appear to give more reliable results than averages since there is less danger of fitting spuriously complex functions to total cost observations.
2. The degree of faithfulness with which the graphically fitted average or total cost curve represents the data. Often the scatter is great and the function too flexible.
3. The exactness with which the graphic curve is described by the average or total cost equation. For straight lines this source of error is not important. But when the function is a complex curve, then an equation obtained by the method of selected points does not always fit the curve closely at all points.

Table 1–13

ROUGH CONSTANT ESTIMATE OF MARGINAL COST FOR SEVERAL ITEMS OF COST

(These components of combined cost have been adjusted for price changes, output lag, and for the effects of independent variables of the correlation analysis)

Cost Item	Total Cost Equation (X = \$1,000 O.W.V.)	Marginal Cost Estimate (Per Dollar O.W.V.)
Components:		
Indirect labor	$Y = \$\ 3{,}700 + 24\ X$	\$0.024
Inspection labor	$Y = 68 + 6.4X$	0.0064
Yard labor	$Y = 390 + 40\ X$	0.04
Power and watch labor	$Y = 640 + 13\ X$	0.013
Power cost	$Y = 548 + 13.8X$	0.0138
Maintenance	$Y = 964 + 2.4X$	0.0024
Direct labor	$Y = 2{,}700 + 505\ X$	0.505
All indirect cost	$Y = 7{,}040 + 71\ X$	0.071
Combined cost:		
Wholly uncorrected	$Y = 11{,}800 + 1{,}220\ X$	1.22
Corrected for price and lag alone	$Y = 12{,}400 + 1{,}090\ X$	1.09
Corrected by multiple correlation analyses only	$Y = 17{,}900 + 790\ X$	0.79
Fully corrected	$Y = 13{,}360 + 1{,}115.6X$	1.116

Direct Labor Cost

The best estimate of the direct labor component of marginal cost which could be made from these data is that it declines at first as output is increased up to about

25 per cent capacity, then remains approximately constant over the remaining observed range of output. The estimated constant cost, obtained by fitting a straight line to the total cost data, is $0.505. This estimate was supplemented by a curvilinear incremental cost function obtained by the graphic method from data in the form of averages. The shape and position of this marginal cost curve substantiate, in general, the estimate of constant incremental cost, for the curve departs significantly from this horizontal line only for small outputs. In this latter region the curve seems to indicate, as did the distribution of the total cost data, that marginal cost declines as output is increased from very low levels.

Except for this initial falling phase of the marginal cost curve only its general level appears to be significant. The minor undulations seem to be due to the fact that slight changes in the slope of the average cost curve bring pronounced and, according to the distribution of the total cost function, unwarranted fluctuations in the marginal cost function.

Table 1–14

AVERAGE AND MARGINAL COST OF DIRECT LABOR

(Selected items from determination by the graphic method)

Output* (Unit $1,000 O.W.V.)	Average Cost†	Moving Average of Marginal Cost‡
5	$1.0700	$.8288
10	0.8140	.4480
15	0.6720	.3885
20	0.6130	.4796
25	0.5967	.5510
30	0.5883	.5402
35	0.5800	.5236
40	0.5717	.5061
45	0.5633	.4903

*Items selected at intervals of $5,000 O.W.V. from a table computed at intervals of $1,000 O.W.V.

†Direct labor cost per dollar of old warehouse value, read from net regression curve of average labor cost on output.

‡Five-item moving average of first differences of the total cost series obtained by transformation of reading, from the average cost curve, cost increment corresponding to an output increment of $1.00 O.W.V.

Direct Materials Cost

Direct materials include all supplies used in the manufacture of furniture which can be assigned directly to particular lots or product. The most important components of direct materials are lumber, veneer, and glue. The average, total, and marginal cost of direct materials for various levels of output, as determined from average cost observations by the graphic-derivative method, are summarized in Table 1–15. From this table it appears that marginal cost first declines slightly, then rises continuously as output is increased, exceeding average cost at all levels

Table 1–15

AVERAGE, TOTAL, AND MARGINAL COST OF DIRECT MATERIALS

(Determined by the graphic-derivative method at regular intervals of output)

Output (Unit $1,000 O.W.V.)	Average Cost (Per Dollar of O.W.V.)*	Total Cost†	Marginal Cost‡
5..........	$.657	$ 3,285	$0.641
10..........	.646	6,465	0.629
15..........	.641	9,615	0.639
20..........	.643	12,860	0.671
25..........	.654	16,350	0.726
30..........	.673	20,190	0.805
35..........	.698	24,430	0.905
40..........	.732	29,280	1.028
45..........	.773	34,785	1.174
50..........	.821	41,050	1.343
10.31.......	....	......	Minimum

*Average cost equation: $X_A = .677 - .00467X + .000151X^2$.
†Total cost equation: $X_T = .677X - .00467X^2 + .000151X^3$.
‡Marginal cost equation: $X_M = .677 - .00934X + .000453X^2$.

beyond $20,000 output and ranging between about $0.60 and $1.35. However, the degree of reliance which can be placed in this schedule of marginal cost estimates probably is not great, for the average cost curve from which it was derived is of limited trustworthiness.[114] An estimate of about $0.65 for constant marginal cost is probably safer than these curve estimates.

Combined Indirect Cost

Combined indirect cost included all overhead expenses investigated in this study, together with a few minor items not analyzed in detail. Probably the most serviceable estimate of marginal indirect cost is that it is constant at about $0.071 (Table 1–13). Besides this constant value a curvilinear marginal cost function was also computed. The equation for this curve and for the average and total cost curves from which it was derived are shown and evaluated in Table 1–16. Examination of the marginal cost magnitudes in this table shows that marginal cost is great for small outputs, then falls sharply to the right as output is increased, and finally rises again for large outputs.

The reliance which can be placed in the curve for marginal cost is not great because the data were analyzed in the form of averages rather than in the form of totals, and because the average cost equation did not fit its curve well for outputs in excess of $30,000. Possibly the initial falling phase is significant, for the average cost curve is well established and faithfully reflected by its equation in this range. Moreover, this behavior is a reasonable accompaniment of the disorganization

Table 1–16

AVERAGE, TOTAL, AND MARGINAL ESTIMATES OF COMBINED INDIRECT COST

(Determined by the graphic-derivative method at regular intervals of output)

Output (Unit $1,000 O.W.V.)	Average Cost (Per Dollar of O.W.V.)*	Total Cost†	Marginal Cost‡
5..........	$1.260	$ 6,300	$0.743
10..........	0.769	7,690	0.234
15..........	0.554	8,310	−0.115
20..........	0.440	8,800	−0.306
25..........	0.371	9,275	−0.337
30..........	0.323	9,690	−0.210
35..........	0.292	10,220	0.077
40..........	0.282	11,280	0.522
45..........	0.300	13,500	1.127
23.64........	.....	......	Minimum

*Average cost equation: $X_A = 1.410 - .0747X_2 + .00106X_2^2$.
†Total cost equation: $X_T = 1.410X - .0747X_2^2 + .00106X_2^3$.
‡Marginal cost equation: $X_M = 1.410 - .1494X_2 + .00318X_2^2$.

costs of operating considerably below the designed scale. Thus, the best estimate which can be made from this study is that differential indirect cost is constant at about $0.071 for outputs in excess of 30 per cent capacity, and that it probably falls from a much higher level for small outputs before settling to this constant level at about 30 per cent capacity ($15,000 O.W.V.).

Combined Indirect Labor Costs

Combined indirect labor cost is composed of the wage costs of the following indirect departments: inspection, power and watch, stockrooms, yard, and supervison. Two estimates were made of the magnitude of this important component. The first, which was arrived at by fitting a straight line to total cost of observations, is a constant—$0.024 (Table 1–13). This estimate is probably most nearly correct for the middle range of output. The second estimate is a parabolic marginal cost function derived from a curve fitted to average cost observations. Its equation is given and evaluated in Table 1–17. From this table it appears that marginal cost increases at both extremes of the output range—a behavior pattern which appears reasonable in the light of the disorganization costs which are likely to accompany any substantial departure from the operating level for which the plant was designed. Limited confidence, however, can be placed in the undulations of this marginal cost curve.[115] Probably all that can be reliably stated about this component is that its magnitude declines from high values for small outputs to a level of about $0.024 for production in excess of about 50 per cent capacity.

Table 1–17

AVERAGE, TOTAL, AND MARGINAL COST OF INDIRECT LABOR

(Determined by the graphic-derivative method at regular intervals of output)

Output (Unit $1,000 O.W.V.)	Average Cost (Per Dollar of O.W.V.)*	Total Cost†	Marginal Cost‡
10	$.3602	$3,590	$.1919
15	.2665	3,998	.1050
20	.2104	4,208	.0458
25	.1735	4,338	.0144
30	.1472	4,398	.0107
35	.1291	4,508	.0348
40	.1198	4,792	.0866
45	.1202	5,409	.1662
50	.1240	6,200	.2735
28.162	.....	.....	Minimum

*Average cost equation: $X_A = .4490 - .01563X_2 + .000185X_2^2$.
†Total cost equation: $X_T = .4490X_2 - .01563X_2^2 + .000185X_2^3$.
‡Marginal cost equation: $X_M = .4490 - .03126X_2 + .000555X_2^2$.

Inspection Labor Cost

As an experiment in methodology, inspection labor cost and two other components—yard labor and power and watch labor—were kept in terms of totals rather than in terms of averages per unit of output throughout the multiple correlation analysis. This experiment substantiated the conclusions reached from the analysis of combined cost, namely, that marginal cost could be derived more economically and more reliably when data were analyzed in the form of totals than when studied in the form of averages.

Two types of total cost functions were fitted visually to these adjusted total cost observation residuals—a straight line and a curve. From the straight line an estimate of constant marginal cost of $0.0064 was derived (Table 1–13). From the curve a parabolic marginal cost function was computed. Its equation and the total cost equation from which it was derived are given and evaluated in Table 1–18. At most, only the general level of this marginal cost curve is significant,[116] and probably the best estimate to be made from the data is that marginal cost is constant at about $0.0064 regardless of the rate of output.

Yard Labor Cost

Another important type of indirect labor in a furniture factory is yard labor. In this cost are summarized the wages of the men whose functions are (1) to unload incoming lumber, (2) to stack it for storage, and (3) to transport it to the kilns

Table 1–18

TOTAL, AVERAGE, AND MARGINAL COST OF INSPECTION LABOR

(Determined by the graphic-derivative method at regular intervals of output)

Output (Unit $1,000 O.W.V.)	Total Cost*	Average Cost (Per Dollar of O.W.V.)	Marginal Cost†
5..........	$ 80	$.01600	$−.01770
10..........	140	.01400	−.00817
15..........	182	.01213	−.00110
20..........	202	.01013	.00405
25..........	219	.00876	.00710
30..........	261	.00870	.00805
35..........	296	.00846	.00690
40..........	323	.00808	.00365
45..........	345	.00766	−.00170
50..........	359	.00718	−.00915
29.76........	...	......	Maximum

*Total cost equation:
$X_T = 385.25 - 29.15X_2 + 1.25X_2^2 - .014X_2^3$.
†Marginal cost equation: $X_M = 29.15 + 2.5X_2 - .042X_2^2$.

and to the factory. Like inspection labor, yard labor cost was kept in the form of totals throughout the multiple correlation analysis. To the adjusted observations in the output regression chart of this analysis two types of functions were fitted—a straight line and a cubic parabola. From the straight line marginal cost was estimated to be constant at $0.04 per dollar of old warehouse value (Table 1–13). However, there may be grounds for believing that the cost rises from about $0.03 for small amounts to about $0.05 for medium outputs, and then declines somewhat for high rates of production, for from the cubic parabola of total cost a curvilinear marginal cost function of this general form was derived. Its equation is given and evaluated in Table 1–19. Greater reliance, however, can be placed in the estimate of constant marginal cost than in the curve.

Indirect Materials

Indirect materials consist of sandpaper, tools, and finishing supplies. Although they are properly called indirect, the expenditures for these items fluctuate in rough proportion to output. Table 1–20 shows the estimated marginal cost function derived from a parabolic curve fitted to the adjusted average cost observations. It is clear from this table that marginal cost is approximately constant, although it rises slightly for large rates of output. Moreover, it is significant to note that it does not depart greatly from average cost at any output level although it may exceed average cost at large outputs. It probably lies between $0.025 and $0.05, with its most likely

Table 1–19

TOTAL, AVERAGE, AND MARGINAL COST OF YARD LABOR

(Determined by the graphic-derivative method at regular intervals of output)

Output (Unit $1,000 O.W.V.)	Total Cost*	Average Cost (Per Dollar of O.W.V.)	Marginal Cost†
5	$ 610	$.125	$ .03105
10	766	.076	.04232
15	964	.064	.04794
20	1,236	.062	.04792
25	1,473	.059	.04224
30	1,649	.055	.03092
35	1,718	.051	.01394
40	1,779	.044	−.00868
45	1,755	.039	−.03696
50	1,700	.034	−.10784
17.48	.....	....	Maximum

*Total cost equation:
$X_T = 464.98 + 14.119X_2 + 1.975X_2^2 - .03767X_2^3$.
†Marginal cost equation: $X_M = 14.119 + 3.95X_2 - .11301X_2^2$.

Table 1–20

AVERAGE, TOTAL, AND MARGINAL COST OF INDIRECT MATERIALS

(Determined by the graphic-derivative method at regular intervals of output)

Output (Unit $1,000 O.W.V.)	Average Cost (Per Dollar of O.W.V.)*	Total Cost†	Marginal Cost‡
5	$.0458	$ 229	$.0360
10	.0374	374	.0241
15	.03224	483	.0200
20	.03030	606	.0215
25	.02820	705	.0266
30	.02850	855	.0331
35	.02958	1,035	.0391
40	.03100	1,240	.0424
45	.03224	1,451	.0459
50	.03560	1,780	.0546

*Average cost equation:
$X_A = .0575188 - .00271275X_2 + .00007885X_2^2$.
†Total cost equation:
$X_T = .0575188X_2 - .00271275X_2^2 + .00007885X_2^3 - .00000069X_2^4$.
‡Marginal cost equation:
$X_M = .0575188 - .0054255X_2 + .00023655X_2^2 - .00000276X_2^3$.

constant value at about $0.03. The reliability of the curvilinear marginal cost function is probably not great, despite the fact that the average cost equation from which it was derived fits its graphically determined curve well.[117]

Power Cost

Power cost includes electric current purchased from the public utility to run machines and to light the buildings. The best estimate of its marginal cost behavior which could be made from the data at hand was that incremental cost falls sharply as output is increased from very low levels, then settles to a constant of about $0.014 for outputs in excess of about $20,000 O.W.V. In addition to this estimate of constant differential cost, which was based upon a straight line fitted to total cost data (Table 1–13), a curve of marginal cost having considerably less reliability was derived from average cost data by means of the graphic-derivative method.[118] Regularly spaced values along this curve, together with those for average and total cost, are shown in Table 1–21.

Table 1–21

AVERAGE AND MARGINAL COST OF PURCHASED POWER

(Determined by the graphic-derivative method at regular intervals of output)

Output (Unit $1,000 O.W.V.)	Average Cost (Per Dollar of O.W.V.)*	Marginal Cost†
5	$.1097	$.0821
10	.0843	.0138
15	.0635	.0077
20	.0471	−.0565
25	.0354	−.0127
30	.0279	.0027
35	.0251	.0209
40	.0271	.0579
45	.0329	.1085
50	.0435	.1725
23.78	.....	Minimum

*Average cost equation: $X_A = .1395 - .00642X + .000090X^2$.
†Marginal cost equation: $X_M = .1395 - .01284X + .00027X^2$.

Maintenance Cost

Maintenance cost represents a consolidation of three accounts: (1) maintenance labor, (2) materials for the repair of buildings, and (3) materials for the repair of equipment. Each of these components was corrected separately for price change and time lag before being combined. The marginal cost of maintenance appears to be negligible. Fitting a straight line to data in the form of totals yielded an estimate of constant maintenance cost of about $0.0024 (see Table 1–13 for equation). But

Table 1–22

AVERAGE AND MARGINAL COST OF MAINTENANCE

(Determined by the graphic-derivative method at regular intervals of output)

Output (Unit $1,000 O.W.V.)	Average Cost (Per Dollar of O.W.V.)*	Marginal Cost†
5	$.1865	$.1335
10	.1376	.0482
15	.0969	−.0129
20	.0645	−.0493
25	.0401	−.0615
30	.0238	−.0491
35	.0157	−.0124
40	.0158	.0488
45	.0240	.1344
50	.0403	.2445
24.98	.....	Minimum

*Average cost equation: $X_A = .2434 - .012205X_2 + .000162875X_2^2$.
†Marginal cost equation: $X_M = .2434 - .024410X_2 + .00048863X_2^2$.

the rate of increase of total cost appears to fluctuate somewhat; hence, this estimate is not entirely satisfactory. Actually, marginal cost of maintenance appears to become negative in portions of the middle output range. A schedule of fluctuating marginal cost was therefore also computed. This schedule is shown in Table 1–22, together with the equation from which these tabulated values were obtained.

A fair amount of faith may be placed in the curvilinear marginal cost estimates up to about $35,000 output. Beyond this point, however, the estimates are invalidated by the erroneous departure of the average cost equations from the curve to which they were fitted. Since the form of the data did not appear to be strictly linear, only limited confidence can be placed in the estimate of constant marginal cost.

APPENDIX C

Correlation Analysis of Variations from Standard Cost

To maximize the managerial usefulness of a standard-cost system, it is necessary to analyze the variance of actual cost from standard cost. Only by studying the causes for this variation is it possible to separate the variance which can be con-

Appendix C was not a part of the furniture factory study. It is quite a separate study, being a multiple regression analysis of the departures of actual cost from standard cost. Its purpose was to find the main causes of these

trolled[119] by the production department from that which can not. Such a separation will make it easier to assign that portion of the controllable variance to the particular individual responsible for the variation. This paper suggests a method of isolating a portion of variance which is for the most part noncontrollable,[120] and, thus, of calling attention to that part of variance for which management can legitimately be held accountable. By means of multiple correlation analysis of past variance, the portion which can be related statistically to cost factors which are beyond the control of the production executives is isolated and subjected to numerical measurement. By this statistical segregation of presumably noncontrollable variance the portion of cost disparity which is most clearly a function of managerial efficiency[121] is dramatically brought to the attention of the executive so that it can be made the subject of a special investigation.

A. *Methods*

The firm whose cost data was studied manufactures a wide variety of medium-grade furniture. For each article of finished product a standard cost is computed; this figure is revised frequently to take into account changes in production methods and changes in the prices of material and labor. The firm reports cost and output at two-week intervals. In gathering data for this study, the total standard cost of the output of each of 47 typical accounting periods, selected from the years 1932, 1933, and 1934, was computed. Then the actual cost of each of these accounting periods was compiled, and the difference between actual cost total and standard cost total for each period was obtained and was expressed as percentages of standard cost. These percentage departures were employed as the dependent variable in the graphic multiple correlation analysis. Three statistically measurable causes for these standard cost departures were used as independent variables in the analysis: output, X_2 (expressed as a percentage of the rated capacity);[122] average size of production order, X_3 (expressed as the average number of the units of finished product per manufacturing lot for the accounting period); and number of new styles produced for the first time during the accounting period, X_4.[123] A net regression curve was established for each of these independent factors.[124]

B. *Findings*

Chart 1–16 shows these regression curves, together with the observations from which they were obtained. Each of these curves represents the average or typical

departures and determine how much of the variation can be ascribed to each. With this knowledge, forecasts of actual costs under predicted future operating conditions could be built up, based on the standard cost of the individual product, modified as indicated by the regression analysis. It is included here as an appendix because cost data are from the same factory and because it built on techniques developed in the furniture factory study.

Chart 1–16 Multiple Correlation Analysis of Variations from Standard Cost Expressed as Percentages of Standard Cost—Net Regression Curves for Two Independent Factors: Output (X_2), Size of Production Order (X_3)

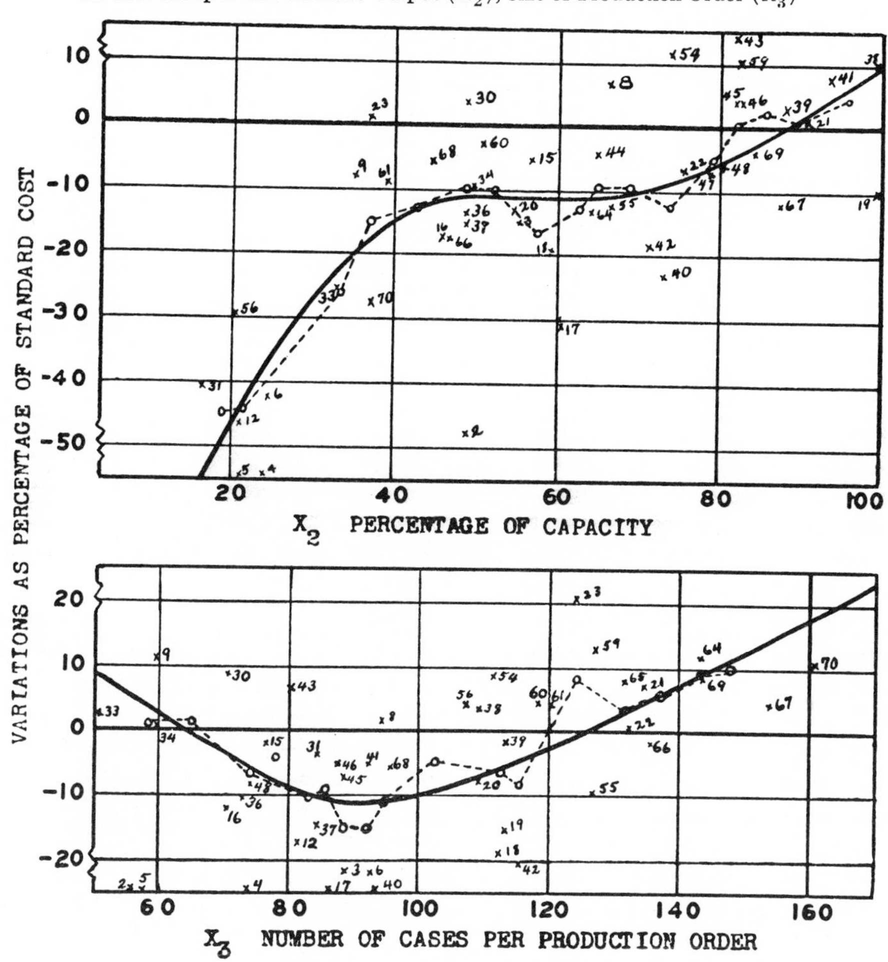

relationship between one independent factor and the discrepancy between actual cost and standard cost, in other words it indicates the standard cost variance which is attributable to one of the three independent variables. Signs of cost variance have been reversed on Chart 1–16 so that the curves indicate directly the modification of the fixed standard cost which is required to bring it into line with the actual operating conditions of the period. Thus in the top section, since actual cost is typically about 50% above standard cost when the plant is running at 20% capacity, the standard should be reduced 50% to fit this operating condition.[125] For cost control purposes, the significant information in Chart 1–16 is the departures of individual accounting periods from the norms set by the regression curve. These de-

Table 1–23

CORRECTION OF STANDARD COST VARIANCE BY ALLOWING FOR OPERATING CONDITIONS

Output[c]	Size of Production Order[a]														
	50			75			100			125			150		
	New Styles[b]			New Styles			New Styles			New Styles			New Styles		
	5	25	45	5	25	45	5	25	45	5	25	45	5	25	45
10%	−58.7	−64.4	−67.2	−73.3	−79.0	−81.8	−77.3	−83.0	−85.8	−67.3	−73.0	−75.8	−54.7	−60.4	−63.2
20%	−35.2	−40.9	−43.7	−49.8	−55.5	−53.8	−53.8	−59.5	−62.3	−43.8	−49.5	−52.3	−31.2	−36.9	−39.7
30%	−21.7	−27.4	−30.2	−36.3	−42.0	−44.8	−40.3	−46.0	−48.8	−30.3	−36.0	−38.8	−17.7	−23.4	−26.2
40%	− 2.8	− 8.5	−11.3	−17.4	−23.1	−25.9	−21.4	−27.1	−29.9	−11.4	−17.1	−19.9	+ 1.2	− 4.5	− 7.3
50%	+ 1.5	− 4.2	− 7.0	−13.1	−18.8	−21.6	−17.1	−22.8	−25.6	− 7.1	−12.8	−15.6	+ 5.5	−00.2	− 3.0
60%	+ 1.3	− 4.4	− 7.2	−13.3	−19.0	−21.8	−17.3	−23.0	−25.8	− 7.3	−13.0	−15.8	+ 5.3	−00.4	− 3.2
70%	+ 2.2	− 3.5	− 6.3	−12.4	−18.1	−20.9	−16.4	−22.1	−24.9	− 6.4	−12.1	−14.9	+ 6.2	+00.5	− 2.3
80%	+ 6.5	+ 0.8	− 2.0	− 8.1	−13.8	−16.6	−12.1	−17.8	−20.6	− 2.1	− 7.8	−10.6	+10.5	+ 4.8	+ 2.0
90%	+12.8	+ 7.1	+ 4.3	− 1.8	− 7.5	−10.3	− 5.8	−11.5	−14.3	+ 4.2	− 1.5	− 4.3	+16.8	+11.1	+ 8.3
100%	+18.1	+12.4	+ 9.6	+ 3.5	− 2.2	− 5.0	− 0.5	− 6.2	− 9.0	+ 9.5	+ 3.8	+ 1.0	+22.1	+16.4	+13.6

[a] Average number of cases per production order for the accounting period.
[b] Number of new styles manufactured for accounting period.
[c] Percentage of plant capacity.

partures are of great managerial concern for they indicate cost variance which is not due to the statistically treated operating conditions or to changes in price structure (since both these types of variance have been allowed for);[126] but are rather due to fluctuations in efficiency. Any accounting period which, when plotted against these regression curves, shows marked departure should, therefore, be made the subject of special investigation.[127]

To facilitate this type of cost control Table 1–23 was constructed from regularly spaced readings taken from these regression curves. This table shows for any given combination of operating conditions such as size of output, size of production order, and new styles, the amount of standard cost variance which is to be expected or countenanced. By entering this table at the particular combination found for the current accounting period, the chief executive may determine in advance the allowable departure of actual cost from standard cost. Any variance in excess of this should be closely scrutinized.

As an additional executive aid, Table 1–24 has been constructed. This table

Table 1–24

COMPARISON OF EXPECTED STANDARD COST VARIATION WITH ACTUAL VARIATION

Accounting Period	Output Allowance	Size of Lot Allowance	New Styles Allowance	Expected Standard Variance	Actual Cost Variation	Variation in Excess of Expected Variation
2	−11.3	+ 5.0	−4.4	−10.7	−33.3	−22.6
3	−11.5	−11.0	−2.9	−25.4	−23.5	+ 1.9
4	−38.6	− 5.5	+1.0	−43.1	−96.8	−53.7
5	−45.0	+ 4.5	+1.0	−39.5	−98.5	−59.0
6	−37.3	−11.0	+4.5	−43.8	−45.5	− 1.7
8	−11.3	−10.9	−4.1	−26.3	− 5.9	+20.4
9	−20.0	+ 2.7	−3.9	−17.0	− 8.2	+ 8.8
12	−45.0	− 9.0	+3.8	−50.2	−45.6	+ 4.6
15	−11.6	− 7.0	−4.5	−23.1	−12.8	+10.3
16	−12.0	− 3.6	−4.2	−19.8	−21.7	− 1.9

shows, for each accounting period used in the analysis, first, the expected deviation from standard cost which was estimated on the basis of the regression curves and second, the actual deviation of the reported cost of this accounting period from standard cost. The differences between these two represent the excess standard cost variance which has not been accounted for by this statistical study of presumably noncontrollable variances. It, therefore, indicates the amount of variance which is primarily attributable to managerial efficiency and for which some satisfactory explanation should be sought.

In addition to establishing a quantitative average relationship between cost variance and each of three important operating conditions an objective criterion of the seriousness of a departure was computed. Small departures from the curve norms may be attributed to chance and need not trouble the executives. But the large departures do warrant investigation. The dividing line may be drawn objectively by accepting the standard error of estimate as the criterion of significance. The chances are about two out of three that the cost variance of a particular accounting period will not depart from the estimated variance made from the curves by more than 18%. Departures in excess of this amount certainly require investigation.

Two measures of the success with which standard cost variance was explained by the three operating conditions used as independent factors were computed: the index of multiple correlation (.642), and the index of multiple determination (.411). The latter figure indicates that less than half of the squared differences in standard cost departures are accounted for by the three operating factors. The unexplained variance may be in part caused by incomplete removal of the effect of price changes; or it may be due to symmetrically distributed errors in setting the standard cost of particular articles of product. But, in large part, the low correlation is probably attributable to the unusual responses of management to kaleidoscopic changes in operating conditions—in short to the type of variation in managerial efficiency which we are interested in isolating for further investigation.

C. Evaluation

Certain objections may be raised to the method of analyzing standard cost variance which is proposed in this paper. It may be correctly contended that this procedure isolates numerically measurable variance, which may or may not be noncontrollable variance. This method does not pretend to completely separate controllable from noncontrollable variance. Instead it separates off that portion of variance which can be associated statistically with certain numerically measurable operating conditions; i.e., rate of output, size of lot, and number of new styles. The conditions themselves usually are beyond the control of the production department, but the adjustment which is made to these changed conditions is, of course, a function of management. Since the regression curves are based upon past reactions of management to these conditions, these curves measure the average or typical ability of management to cope with these changes during the years 1932 to 1934. Thus, they reflect the level of past efficiency of management in dealing with changes in volume, size of lot, and number of styles.

The so-called noncontrollable variance is thus really variance which has not been controlled in the past and which has fluctuated in a regular pattern in response to changes in the measured operating conditions. If this variance is brought under better (or worse) control in the period being studied, the fact will be recorded as a departure from the norms set by the curves. The fact that the norms are based upon the efficiency of the management in coping with these conditions in the immediate past should not detract from their usefulness as control devices, since past performance is a familiar and extremely practical criterion of efficiency. It may be asserted, as a second objection, that such an analysis merely measures but does not aid in controlling the variance which is attributable to output, size of order, and numbered new styles. This may be quite true although, incidentally, a knowledge of the costliness of fluctuations in exterior conditions such as these may prompt the executives responsible for these conditions to stabilize them; study is not designed to assist in controlling these conditions but rather to measure and remove their effect and then to isolate the variance which is due to other conditions which presumably can be controlled. A third objection, namely, that no attempt is made in this statistical procedure to trace the incidence of cost variance to the person responsible may also be conceded. The primary purpose of the statistical study is to clear the way for a further tracking down of cost variance to the responsible person. An extension of this procedure to an analysis of the cost variance of particular departments and of individual items of expense would, however, be of considerable assistance in tracing variance to individuals. By this means the portion of departmental cost deviation which is beyond the department manager's control and which is subject to statistical treatment could be measured and removed.

Fourth, it is stated that imperfections in the standard cost itself may be erroneously attributed to the influence of the independent factors used in the multiple correlation analysis. This is unlikely to be important for if error is persistent and present in all standard costs it is allowed for in the constant term of the multiple regression equation. Thus, its obscuring effect upon the appraisal of management is removed but this constant error is not attributed to the independent variables. Errors in standards which are not constant on the other hand are likely to be somewhat symmetrically distributed with respect to the independent variables and, hence, are likely to have little effect upon this relationship. Still another set of objections which may be better justified concern the limited accuracy of the adjustments suggested by the regression curves. It is true that these curves show only the average relationship, and it is important that the limited reliability of this average be recognized. The curves are frequently only tenuously sustained by widely scattered individual observations. It should further be pointed out that these relationships hold true strictly only for the three-year period used in their determination. To employ the results for control purposes for subsequent periods involves the assumption that fundamental conditions have not changed sufficiently to invalidate these established relationships. In the third place, the choice of independent factors may not have been entirely satisfactory. Certain cost factors of considerable importance may have been omitted and certain factors which were included may really measure a complex of causes. Hazards such as these are part and parcel of any statistical investigation. No brief is held for the statistical significance of the findings of this particular study. It is presented merely as a suggested method for supplementing the present methods of analyzing standard cost variance in the hope that controllable variance can by this means be better isolated and analyzed.

Certain important advantages may be claimed for the methods suggested in this paper. Isolation of that portion of standard cost variance which can be explained on the basis of known relationships with certain numerically measurable cost conditions, will, it is hoped, provide a better starting point for tracing the incidence of controllable variances to the individual responsible for them. It is clear, therefore, that this procedure is not an alternative to the methods now used in analysis of standard cost variance but is rather designed to supplement and sharpen these tools.

A second advantage enjoyed by this procedure is that it facilitates the visual administration of a standard cost system. By plotting the cost variance of a current accounting period on Chart 1–16 the variance which is in excess of that attributable to the three operating factors stands out clearly in the graph as the departure of this accounting period's cost from the graphically established norms.

A third advantage which may result from this method is that it may bring to light serious cost variances which may otherwise go undiscovered because of two opposing cost tendencies which counteract each other. That is, a favorable change in the price of raw materials may balance an unfavorable change in managerial efficiency and result in a negligible departure of actual cost from deflated standard cost. By using the proposed procedure the price variance, which has served, is removed by adjusting the standard cost to allow for it; this leaving the unfavorable efficiency variance clearly in sight.

Certain peculiar advantages may be realized by firms which have a very simple type of cost system. Such firms may find in this statistical analysis of variance a cheap substitute for the valuable control devices of a parallel current cost system and a comprehensive flexible cost budget. For firms which do not keep detailed current cost records to parallel the standard cost system a statistical analysis of variance provides a rudimentary causative explanation for standard cost variance not otherwise available. For those firms which do not have a comprehensive flexible budget, such an analysis provides an objective guide for modifying the fixed standard cost to take account of changes in rate of output and other cost conditions. The resulting flexible cost standards are rough substitutes for a detailed flexible expense budget.

Study 2

HOSIERY MILL

Contents

Chapter I

Introduction

The objectives of this study are, first, to develop, illustrate, and appraise statistical methods for determining the quantitative relation between the cost of operation

Reprinted from Joel Dean, *Statistical Cost Functions of a Hosiery Mill*, Studies in Business Administration, vol. 11, no. 4 (Chicago: University of Chicago Press, 1941), by permission of the publisher.

of an enterprise and those aspects of operation which influence cost and, second, to obtain, by these methods, estimates of the behavior patterns of the operating costs of a silk-hosiery mill.

Empirical determination of cost functions is useful for public policy and for economic theory, as well as for business administration. In periods of extraordinary extension of government controls, such as the present national defense emergency, reliable methods of determining the relation of volume of output to cost may be particularly valuable, since in a war economy knowledge of cost functions is important not only for securing maximum productive efficiency but also as a criterion for controlling or stabilizing certain kinds of prices.

The importance of the empirical analysis of cost for theoretical economists lies in the possibility that it may indicate the relative frequency of the various types of cost behavior which are postulated by economic theory. The usefulness of theoretical economics will, moreover, be increased if its premises can be made to correspond more closely to the facts of a given institutional framework. In view of the general preoccupation of economists with pricing problems and the significance of the internal economics of the individual firm, it is particularly important that cost theory be realistic.

There is a growing recognition of the importance of knowledge of cost behavior for the intelligent management of an enterprise. The increased use of budgets as instruments of control requires systematic prediction based on previous experiences and anticipated conditions. The statistical analysis of cost behavior furnishes a basis for estimating costs at various levels of production, which appears to be more reliable than the rule-of-thumb classification methods now commonly used. Statistical methods can serve not only to indicate reasons for the variation of actual cost from standard but also to give some quantitative indication of the importance of certain of these influences. Finally, statistical cost analysis provides estimates of short-run marginal cost (the cost of producing an additional unit with given plant facilities).[1] This type of cost information, although useful for short-run price and output decisions, is not provided by ordinary accounting procedures.

This present study is illustrative of a type of cost analysis that has yielded apparently reliable and useful findings in a number of similar investigations. It may, therefore, be hoped that in its general form the method is applicable to a wide variety of manufacturing plants. The emphasis placed on problems that may appear peculiar to the plant studied is believed justifiable because analogous issues will arise in the statistical analysis of the costs of any industrial process. Accounting data are never available in an immediately usable form and always require considerable correction and adjustment. Likewise, the findings of this investigation may have an applicability which extends beyond the particular enterprise investigated and may be typical of an entire sector of the economy in which comparable influences are at work.

Chapters i–v of this study deal with the nature of our findings. In chapter ii the characteristics of theoretical and empirical cost functions are discussed, and in chapter iii the statistically determined cost functions are presented. Chapters iv and v are concerned with an evaluation of the reliability of the findings and their practical application. Chapters vi–viii are devoted to various problems of statistical method, the selection of the sample, and the choice and correction of the dependent and independent variables. The appendixes deal with the results of some supplementary analyses and a description of certain technical procedures involved in the preparation of the data.

CHAPTER II

Nature of Theoretical and Statistical Cost Functions

A. The Meaning of Cost Functions

The analysis which is carried out in this section is based on the hypothesis that the cost of production of an enterprise is determined by certain specific operating conditions. The prices of input services are assumed constant in order to simplify the examination of the effects of the remaining variables on cost.[2] The basic proposition is, then, that there exists in the short run a functional relation between cost, the dependent variable, and a set of independent variables (other than prices of input services), which may include, for example, volume of production, size of production lot, and frequency of machine stoppage. The independent variables will be different for each type of manufacturing operation, although it can be stated that in general the most important variable is rate of output. The magnitudes of the independent variables are considered to determine the magnitude of cost. This relational dependence is shown symbolically by the cost function.

$$X_1 = f(X_2, X_3, X_4),$$

where X_1, the dependent variable, is cost, and X_2, X_3, and X_4 are the independent variables, for example, output rate, lot size, and shutdown frequency.[3]

A preliminary problem which is encountered in any attempt to determine an empirical cost function by statistical methods lies in the specification of the form of the cost function. There might be, for example, functions of the following types, each of which represents a different form of the cost function:

$$X_1 = f(X_2) + g(X_3) + h(X_4),$$
$$X_1 = f(X_2) + g(X_3X_4),$$
$$X_1 = f(X_2) + g(X_2X_4) + h(X_3) + l(X_4).$$

The most suitable form of the cost function is difficult to determine on a priori grounds in most empirical investigations. The procedure is to make as intelligent a guess as possible in the preliminary choice of a function form and later to alter the specification if necessary in the light of empirical evidence. The specification of the form of the cost function really amounts to a statement concerning the hypothetical universe (an infinitely large number of observations which is assumed to show the "true" nature of the cost function) from which the small sample of observations is drawn. The statistical method employed consists in obtaining an estimate of the constants in the "true" cost function from the available sample. We know, however, that there will be certain errors of sampling in the computed value of the coefficients. We wish to determine whether the numerical values of the coefficients are sufficiently large to be considered significant and not merely the result of errors of sampling. A cost function cannot be regarded as satisfactory if significant coefficients can be obtained for additional or higher-order terms, which have been omitted in the original specification. Consequently, by fitting functions with additional terms whose coefficients are not attributable to chance variation, improved specification of the cost function can be attained.

By abstracting from all independent variables except output, however, certain supplementary information is available which takes the specification of this restricted cost-output function out of the realm of pure guesswork. In the treatment of the economics of the individual firm customary in the analysis of perfect competition, the cost function refers to the functional relation between cost and rate of output.[4] It is also assumed, or should be assumed, that the particular firm is alone changing its output. That is, operating variables other than output are not included in the list of independent variables. This restriction of the cost function to its limited meaning as the cost–output relation makes it possible on theoretical grounds to impose a limit on the number of function forms which are suited to represent this relation. It is not markedly at variance with the facts, moreover, to emphasize the characteristics of the cost function in this restricted sense in view of the fact that rate of output is the most important determinant of cost.

B. Types of Short-Run Cost–Output Functions

It is necessary as a preliminary step to develop briefly certain considerations which are relevant to the behavior of cost in the short period. This is desirable for several reasons. First, theoretical analysis provides guidance in selecting the most likely statistical specification of the short-run cost function. Second, building the statistical superstructure on a theoretical foundation makes it possible to compare actually observed results with theoretical results and in this way to test the adequacy or generality of the underlying cost theory. Third, the theoretical models define a cost curve and an approach to the study of cost behavior which has considerable practical usefulness for managerial and predictive purposes. Once the basic relations between cost and output have been determined, estimates of the future cost of production at various levels can be obtained by adjusting the cost functions to reflect the effect of exogenous changes in wage rates, material prices, and possibly even technical methods.

In dealing with cost as a function of output, economic theorists have been accustomed to distinguish between short-run and long-run cost functions. The basis of this distinction is that in the long run all input factors (i.e., building, machinery, supervisory personnel, labor, and materials) are assumed to be completely adapted to the plant's rate of output, so that the output is produced at the minimum total cost. On the other hand, in the short run, one or more of the input factors, e.g., the plant, the machinery, or the supervisory personnel, is assumed to be physically fixed and not capable of immediate adaptation to changes in the rate of output. Thus, only part of the input factors are variable in response to changes in output rate. This situation is sometimes called partial adaptation in contrast to the long-run adjustment called total adaptation. Clearly, there exist a great many short runs, depending upon the number and importance of the factors which are fixed and how complete is the adjustment of the variable factors. Nevertheless, it is useful to maintain the distinction between the long run and the short run, imprecise though it may be. It is to be emphasized that in this study we are concerned with short-run cost functions only.

Not only are the cost functions limited to a description of short-run adjustments, but also they are restricted to static conditions. Such functions, therefore, show the relation between cost and output per unit of time when there exists a given fixed body of plant and equipment and when the following additional conditions are satisfied. First, there are no changes in basic operating conditions, such as technical improvements, changes in the skill of workers and in managerial efficiency. This implies also that technical methods of production do not vary for different levels of output. Second, current market prices of input factors are not variable—for example, wage rates, material prices, and tax rates are fixed. Consequently, no substitution among factors occurs as a result of price changes.

While it is apparent that the only admissible behavior of the static short-run cost function prescribes that total cost is an increasing function of output, there are three subcases to be distinguished. As a matter of terminological convenience, "degressive," "proportional," and "progressive" cost will be taken to refer to cost which increases less than in proportion to, proportional to, and more than in proportion to output. From combinations of these three kinds of increases of total cost, behavior patterns of cost can be distinguished which are applicable to different sets of operating conditions. First, there is the traditional case of an initial phase of cost degression, followed by proportionality, and finally cost progression as output increases. Second, costs are progressive throughout the whole range. Third, there is an original stage where cost is proportional, followed by eventual cost progression. Apart from the special cases to be discussed where the cost function reduces to a point or series of points, discontinuous cost behavior may be exhibited in any of these cases. Which of these types of behavior is to be found in a specific instance will be determined by the technical methods of production and the nature of the fixed plant and equipment.

The characteristics of the fixed equipment of an enterprise play a dominant role in the determination of its pattern of cost behavior. The particular quality of importance for short-run cost is the extent to which the flows of services emanating from a given plant can be varied or segmented. Segmentation, in this sense, therefore refers to the technical nature of the fixed equipment which allows the utilization of its services at variable rates. It is not an absolute concept, however, since there can exist varying degrees of segmentation. It is convenient to classify this quality on a scale varying from the absence of segmentation to discontinuous segmentation to perfect segmentation with a subcategory, imperfect segmentation.

The incorporation of segmentation into a fixed plant is dependent on both the technical nature of the equipment and the conscious managerial efforts in that direction. The first and most obvious situation where a high degree of segmentation is possible is one in which the fixed equipment can be divided into a complex of small units. Situations such as this will be referred to as "unit" segmentation. For example, in the hosiery mill whose costs are being analyzed, the knitting of the stocking legs is done on eighty-one approximately identical knitting machines.[5] Under such circumstances the successive introduction or withdrawal of the machine units permits wide variability in the services of the machinery despite its over-all fixity. It is further possible to introduce segmentation by varying the number and hours of the shifts per period that the fixed equipment is employed. It is convenient to designate this device by "time" segmentation. If the technical nature of the fixed plant is such that it can be used at varying speeds, "speed" segmentation can be obtained by more or less intensive use of the service of the equipment, i.e., by operating machines at faster or slower rates.

The means of attaining segmentation, as well as the technical structure of production, will determine the degree of segmentation which can be achieved in any instance. Unit segmentation is characterized by discreteness in the flow of services, as each successive unit of plant is brought into operation. The units may be so small in certain cases, however, that for practical purposes the discontinuities can be neglected. Time segmentation may also be characterized by marked discontinuities, for example, where the choice is between one or two shifts of eight hours rather than between an eight- and nine-hour shift. On the other hand, speed segmentation permits continuous variablity in plant use, although the range within which machine speed-up, etc., is a practical method of achieving segmentation is narrow.

The next question to consider is the variability or constancy of the proportions in which the services of fixed equipment and the services of the variable agents of production are combined at different levels of operation. It has been shown that the purpose of segmentation is to introduce variability into the flow of input services from physically invariable agents of production. Under the conditions assumed (constancy of factor prices, technical stability, etc.) the two aspects of the technique of the production process—the degree of segmentation and the variability of factor proportions—are sufficient to determine the short-run cost function of a plant. The different combinations of segmentation and variability of proportions yield the following eight cases:

1. Absence of segmentation
   a) variable proportions
   b) constant proportions
2. Discontinuous segmentation
   a) variable proportions
   b) constant proportions
3. Perfect segmentation
   a) variable proportions
   b) constant proportions
4. Imperfect segmentation
   a) variable proportions
   b) constant proportions[6]

Each of these cases will be considered in more detail in an attempt to isolate the combinations of operating conditions which are likely to be found in the technical processes which are being analyzed.

1a. *Absence of segmentation, variable proportions.* When the services which emanate from the fixed equipment cannot be varied, it may nevertheless be possible to combine different quantities of the variable-factor complex with this one rate of flow of services. If the technique of production described by the production function, $x = x(a, b)$, this circumstance can be represented by writing the production function as $x = x(a, b_0)$, where b_0 is some given level of b. This is the usual case taken to illustrate diminishing marginal productivity, for with such a production function the application of successive units of factor a yields first increasing, then diminishing, increments of product.[7]

1b. *Absence of segmentation, constant proportions.* When no segmentation is possible and the factor proportions are constant, we have the special case in which the production function reduces to a point, only one combination of factors and one amount of product being possible. This is an unrealistic case and need not be considered further.

2a. *Discontinuous segmentation, variable proportions.* When incomplete segmentation is the rule, the services of the fixed plant are available discontinuously at certain discrete levels. This may occur, for example, in unit segmentation where each machine can be operated at only one rate. The introduction of a machine therefore produces a discontinuity in the service flow, which then remains constant until the next machine is brought into operation. If the proportions are variable, the cost behavior over the range of output made possible by the introduction of an additional machine (i.e., its optimum capacity) will be a miniature of the cost function when no segmentation is possible and the proportions of the factors are variable. This introduces discontinuities into the production function and consequently into the cost function.

2b. *Discontinuous segmentation, constant proportions.* In the case of constant proportions, if segmentation is discontinuous for every level of the services of the fixed equipment, only one output can be produced as a result of the fixity of proportions between the factors. The production function and, consequently, the cost function are in this case represented by a series of points, no positions intermediate between the points being possible.

3a. *Perfect segmentation, variable proportions.* Perfect segmentation occurs when the services of the fixed equipment can be introduced in infinitesimally small quantities. These services are, in other words, continuously variable. The production function in this case can be represented by $x = x(a, b)$, where x is output, a the services of the variable agent of production, and b the service of the fixed equipment. If the proportions in which a and b are used are constant, the production function is homogeneous and of the first degree so that the marginal productivity of factor a will be constant within the range of variability of b. When the upper limit of b has been reached, however, the marginal productivity of a begins to decline. This pattern of behavior for the marginal productivity of factor a implies that the marginal factor cost is constant until the upper limit of variability of b is reached, when marginal factor cost rises.

3b. *Perfect segmentation, constant proportions.* The difference between this case and the previous one is irrelevant for the purposes here considered, for with complete segmentation it is always possible to achieve the optimum proportion between the factors, and this proportion will not change with changes in the level of output within the range of variability of the services of the fixed plant. Consequently, these circumstances are not essentially different from those where the proportion is constant because of technical considerations. Therefore, technical fixity of proportions need not be distinguished from the case where the proportion is fixed on economic grounds.

4a. *Imperfect segmentation, variable proportions.* There is a subcase to be considered in which the segmentation is a little less than perfect. Segmentation of plant which proceeds by employing small machine units may encounter some discontinuities with the introduction of new machine units owing merely to incomplete divisibility of the machinery. When, in turn, these machines can be operated at continuously variable rates above some minimum for each machine and when the proportions are variable, there will be discontinuities in the production function which are reflected in a discontinuous total cost function.

4b. *Imperfect segmentation, constant proportions.* The subcases of variable- and constant-factor proportions need not be distinguished for this type of segmentation since the optimum proportion, which is always attainable, will not vary with changes in output. In this respect it is similar to the case of perfect segmentation.

This is an exhaustive list of cases from the point of view adopted here, and it is necessary now to select the cases that may be relevant to the type of problem at hand. The case of complete absence of segmentation with constant proportions can be rejected summarily. The same is true for both cases of discontinuous segmentation, since neither is applicable to the type of industrial process that is being analyzed. There remain, therefore, the three cases: absence of segmentation with variable-factor proportions, perfect segmentation, and imperfect segmentation. It is convenient to designate these cases as Type 1, Type 2, and Type 3 behavior, respectively.

A complication is introduced by the fact that fixed equipment is not in general homogeneous. One may, therefore, find in practice that segmentation is not possible for certain parts of a fixed plant, while, on the other hand, a high degree of segmentation is possible for other parts. The derived cost behavior may, therefore, be a resultant of mixed forces. It is, nevertheless, convenient for purposes of analysis to treat each case in its simplest form.

The Type I behavior pattern results in the short-run cost function which is portrayed in Figure 1 of Chart 2–1, where output is measured on the horizontal, and cost on the vertical, axis. The course of total variable cost, i.e., that part of cost which fluctuates in response to output variation, is indicated by the cubical parabola *TVC*, which passes through the origin. The fixed costs which arise from the firm's positive inability to adapt certain input factors to prevailing output rates (i.e., because of the fixed equipment) have a monthly total which is constant, irrespective of output rate, and is designated by the horizontal line *TFC* in Figure 1. The vertical addition of the total fixed and total variable cost gives the curve of total combined cost *TCC*.[8] Since the total variable cost function rises first at a de-

Chart 2–1 Three Types of Short-Run Cost Behavior

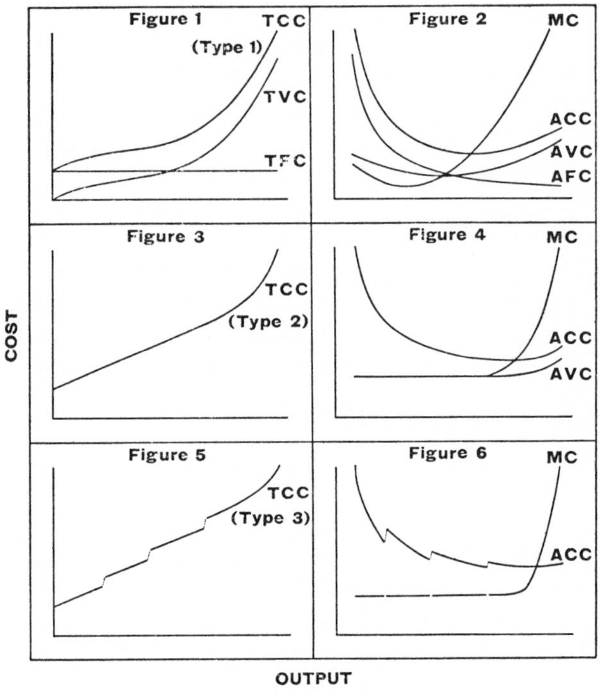

creasing, then at a constant, and finally at an increasing, rate, the total combined cost function obviously has the same shape, but an intercept higher by the amount of the fixed cost.

The corresponding behavior of unit costs is shown in Figure 2 of Chart 2–1. The fixed cost per unit will be represented by a rectangular hyperbola indicated by the curve *AFC*. Variable costs per unit (*AVC*) will be falling, constant, or rising, depending upon the phase of operations. Average combined cost, shown by the curve *ACC*, is obtained by vertical summation of the average variable and average fixed cost. The marginal cost function denoted by *MC* in Figure 2 represents the increment in total cost associated with a small change in the rate of output, or, alternatively, it shows the additional cost that must be incurred if output is increased by one unit. The marginal cost function, which is given by the slope of the total combined cost curve will, under the conditions postulated, decline at first and then bend upward, continuing to increase over the remainder of the output range.

In the case of Type 2 behavior, where the production function is homogeneous and linear over the initial range, the resultant total cost curve will also be linear over the same range. As a consequence of this characteristic of the production function, which results from perfect segmentation and fixity of factor proportions, total cost will be proportional to output below the level of the physical capacity of the equipment, after which point total cost will rise sharply. Therefore, the marginal cost will not change as the level of operations increases up to the critical point when it increases rapidly. This is illustrated by the cost curve Type 2 drawn in Figure 3 of Chart 2–1. The derivative cost functions—average variable, average combined, and marginal cost—are shown in the corresponding Figure 4. Marginal cost and average variable cost coincide over the operating range until the point is reached where no more fixed equipment can be brought into operation, when they diverge. It is highly significant that in this case marginal cost is constant at levels of output less than capacity.

The Type 3 case applies where there is imperfect segmentation of the given plant. The only difference between this and Type 2 behavior is that there are breaks in the cost function, where cost increases discontinuously. Even though there may exist discontinuities in the total cost of operations when, for example, successive machines are brought into operation, this will not influence the level of marginal cost. Discontinuities introduced into a cost function with an initial linear phase are pictured in Figure 5 of Chart 2–1. In the accompanying Figure 6 the behavior of the average and marginal cost curves for this discontinuous case are illustrated.

It may be that cost functions of this type are actually common in mechanized industry. It is convenient, however, to exclude the whole question of discontinuities by making the assumption (apparently realistic in this study) that the fixed equipment is rather highly divisible. Moreover, in view of the impossibility of finding a discontinuous function by the statistical methods employed, the Type 3 case can be omitted from further consideration. It is thus seen that the choice of the form of the cost-output function to be specified in this empirical study is limited to those illustrated by Types 1 and 2, namely, a linear function of the form

$$X_1 = b_1 + b_2X_2$$

or a cubic function of the form

$$X_1 = b_1 + b_2X_2 + b_3X_2^2 + b_4X_2^3.$$[9]

In making the choice between these two alternatives, two sources of information are particularly important. The first is an analysis of the technical process of production, with special attention being devoted to the degree of segmentation of the fixed equipment and the behavior of the technical coefficients of production of the variable factors. A description of the technique of hosiery manufacture, which is the production process analyzed in this study, is to be found in chapter vi, Section A.

One of the main points of relevance of the theoretical analysis is for the statistical problem of specification. This question is discussed in more detail in the following section of this chapter and in Section B of chapter iv, where the behavior of marginal and average cost is examined to see whether any further light can be thrown on the behavior of the total cost function.

C. Statistical Determination of Cost Functions

As indicated earlier, theoretical short-run functions are formulated on the assumption that productive equipment, managerial practices, and the prices of the inputs of productive services remain constant. Therefore, to determine a statistical cost curve which is the counterpart of the theoretical static cost function, it is necessary to select and adjust the data so as to approximate the theoretical model as closely as possible. To fulfil these prescribed conditions it was necessary, first, to select a sample firm and a period of observation in which dynamic elements, such as changes in the size of plant, technical production methods, managerial efficiency, and so forth, were at a minimum and, second, to rectify and adjust the cost and output data recorded in the firm's accounts in order to remove the effect of disturbing factors, such as changes in wage rates, prices of materials, tax rates, special accounting allocations, and so forth.

In addition to the influences allowed for in selection and rectification procedures, cost may be affected by other operating variables as well as the rate of output. Hence it is necessary to take account of additional independent variables which reflect operating conditions suspected to exercise an important influence on short-period fluctuations in cost, e.g., size of production lot, change in output from the previous period, style variety, and so forth.

Nevertheless, the main interest lies in determining the net effect of rate of output on cost when the influence of the remaining variables has been allowed for. If the statistical cost function can be determined in such a way as to eliminate the influence of all other independent variables apart from output rate, the resultant cost function will be the empirical counterpart of the theoretical cost function which has just been described.

The method of determining the nature of the statistical function is essentially mathematical, although there also is an approximative graphic method based on the more formal mathematical method. The graphic method of analyzing the influence of independent variables in multiple correlation problems is based on inspection and graphic approximation.[10] The use of this short cut results in such pronounced economy of time and technical skill that this procedure is likely to be more feasible than formal least-squares multiple correlation analysis when findings are to be used exclusively for administrative purposes. Graphic methods, furthermore, permit the preliminary selection of those independent variables that have a significant relation to cost and the provisional determination of the form of their relations. Thus, even if mathematical correlation analysis is to be used, graphic analysis is an aid in the selection of the relevant independent variables and the specification of the cost function. The functional relations can then be determined in a more objective manner by the method of least squares. Multiple correlation analysis by the method of least squares determines the representative functional relations between the dependent variable (cost) and each of the independent variables when the criterion of representativeness is that the functional form selected minimizes the sum of the squares of the deviations about this function. The method is complex, and only the briefest discussion of the statistical technique can be offered here.[11]

Suppose that it has been decided that the general form of the cost function is

$$X_1 = b_1 + b_2X_2 + b_3X_3 + b_4X_4 \, .$$

We have a series of observations of each of these variables which can be arranged in this way:

	X_1	X_2	X_3	X_4
January, 1935	X_1'	X_2'	X_3'	X_4'
February, 1935	X_1''	X_2''	X_3''	X_4''
. . . .	. . . .	. . . .	. . . .	. . . .
June, 1939	$X_1^{(n)}$	$X_2^{(n)}$	$X_3^{(n)}$	$X_4^{(n)}$

From these observations we must attempt to derive enough information to allow us to determine the values of the coefficients b_1, b_2, and b_3.

This procedure follows directly from the way in which the hypothesis on which the analysis is based is set up. As mentioned previously, the relation of primary importance and interest is the function showing the dependence of cost on output. Once we have determined mathematically the more inclusive cost function and expressed it as a multiple regression equation, e.g.,

$$X_1 = b_1 + b_2X_2 + b_3X_3 + b_4X_4 \, ,$$

it is possible to find the net influence of output on cost by allowing for the influence of the other independent variables and thus isolating the pure functional relation of cost to output. This relation as well as that determined by graphic methods will be referred to in the course of the discussion as the "partial regression" of cost on output.

The short-run cost-output function describes the relation between output and the least cost of producing that output with the given plant and production technique. Since the entrepreneur may actually fall short of minimization of costs for a given output and can never improve on the assumed condition, the statistical curve will tend to lie somewhat above the theoretical curve.

CHAPTER III

Statistical Findings

The enterprise whose cost behavior was analyzed is a hosiery knitting mill which is one of a number of subsidiary plants of a large silk-hosiery manufacturing firm. In the particular plant studied the manufacturing process is confined to the knitting of the stockings, that is, the plant begins with the wound silk and carries the operations up to the point where the stockings are ready to be shipped to other plants for dyeing and finishing. The operations in the mill are, therefore, carried on by highly mechanized equipment and skilled labor.

Cost functions were determined for combined cost and for its components: productive labor cost, nonproductive labor cost, and overhead cost. These functions were derived separately for monthly, quarterly, and weekly data. For the monthly and quarterly observations both simple and partial regressions of the various costs on output were obtained. In this chapter the statistical findings for the monthly data alone are presented, while the quarterly and weekly findings are given in Appendix A.

A. Monthly Observations

On the basis of a preliminary analysis of the technique of operations, the following three operating factors, in addition to output volume, were thought to have a significant influence on cost and were, therefore, considered as candidates for independent variables in the cost functions.

Replacements. Replacements consist of hosiery which is only partially manufactured and which is substituted for products found to be defective at any stage in the manufacturing process. The magnitude of replacements was considered significant both because of their direct cost and because of their indirect reflection of spoiled work.

Style changes. Despite the fact that the hosiery mill studied produced a relatively homogeneous product, some changes in styles did occur. Since manufacturing operations differ somewhat from style to style, it was decided that these style changes, which necessitate certain machine adjustments and wage changes, might influence cost significantly.

Change in output from previous month. If there exist rigidities in cost, the magnitude and direction of the change in output from one month to the next might exhibit a certain relation to cost. For example, if output in one month is extremely high, followed by a month with low output, the cost of the low output may be influenced if the mill is not able to adapt itself readily to the change.[12]

In order to obtain an index of their magnitude, "replacements" and "style changes" were measured by replacement labor and style-change labor, which were two of the nonproductive labor accounts in the company's records. "Change in output from previous month" was represented by the first differences calculated from the monthly output series, the direction as well as the absolute change in magnitude being taken into account. By means of graphic methods of multiple correlation analysis described earlier, the influence of these three variables on cost was investigated. For combined cost as well as for its three components (nonproductive labor cost, productive labor cost, and overhead cost) no significant net relation to the three independent variables was found to exist. These three factors which had been considered as independent variables were, therefore, eliminated from the cost function, leaving rate of output as the sole independent variable.

1. SIMPLE REGRESSIONS

Scatter diagrams were made between output and combined cost and its three components for the monthly data to indicate the form of the restricted cost function in which output is the only independent variable. The simple regression[13] indicated by the scatter diagrams appeared to be linear, so that a regression equation of the first degree with the general form $X_1 = b_1 + b_2X_2$ was fitted to the observations for combined cost and its three components. Furthermore, in view of the theoretical analysis of chapter ii, the technical organization of the hosiery mill substantiated the specification of linear total cost behavior,[14] the regression equations derived for the four categories of cost in the form of monthly totals, together with the statistical constants are shown in Table 2–1.

The results which are expressed in a mathematical form in Table 2–1 can also be shown by regression lines or scatter diagrams. The regression equations for combined cost and productive labor cost are illustrated graphically in Chart 2–2. Chart

Table 2–1*

HOSIERY MILL: SIMPLE REGRESSIONS OF COMBINED COST AND ITS COMPONENTS ON OUTPUT

(Monthly Observations)

	Combined Cost	Productive Labor Cost	Nonproductive Labor Cost	Overhead Cost
Simple regression equation	$_cX_1=2935.59+1.998X_2$	$_pX_1=-1695.16+1.780X_2$	$_nX_1=992.23+0.097X_2$	$_oX_1=3638.30+0.121X_2$
Standard error of estimate	6109.83	5497.09	399.34	390.58
Correlation coefficient (r)	0.973	0.972	0.952	0.970
Regression coefficient (b)	1.998±0.034	1.780±0.035	0.097±0.045	0.121±0.036

* The symbols have the following meaning:

$_cX_1$ = combined cost in dollars

$_pX_1$ = productive labor cost in dollars

$_nX_1$ = nonproductive labor cost in dollars

$_oX_1$ = overhead cost in dollars

X_2 = output in dozens of pairs

Chart 2–2 Simple Regressions of Total Combined Cost and Productive Labor Cost on Output (Monthly Costs)

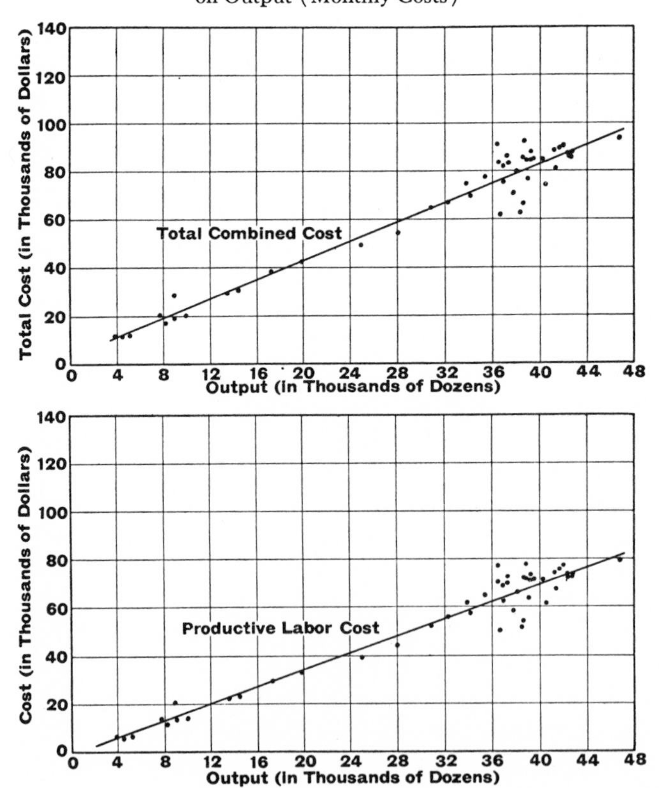

Chart 2–3 Simple Regressions of Non-Productive Labor Cost and Its Elements on Output (Monthly Costs)

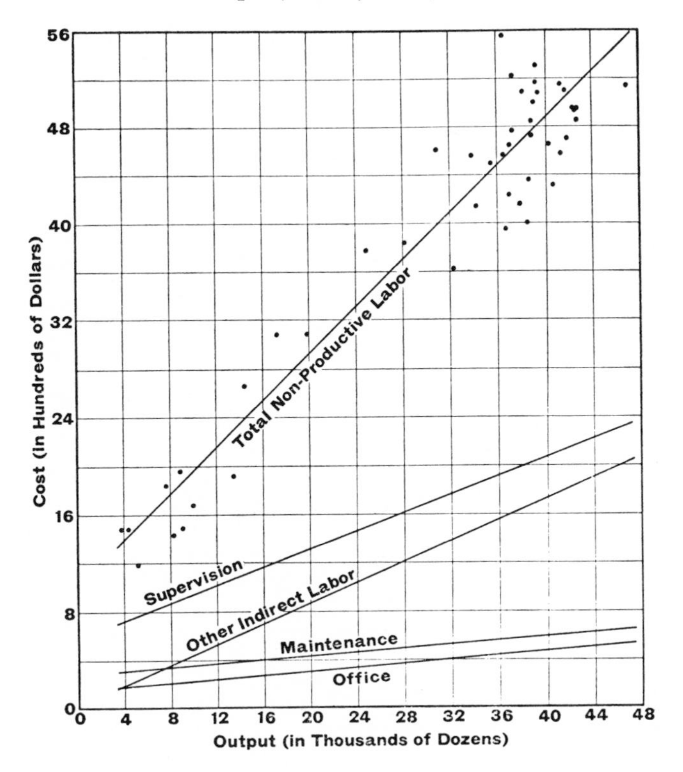

2–3 shows the simple regressions not only of the aggregate nonproductive labor cost but also of its principal elements: supervision, maintenance, labor, office staff, and other indirect labor. In order to show more clearly the nature of the individual cost functions, each of the cost elements and their total are measured from a common base, the X-axis, i.e., they are not cumulated. Simple regressions of total overhead cost and its elements are similarly presented in Chart 2–4.

2. PARTIAL REGRESSIONS

Graphic multiple correlation analysis showed that the deviations from the simple regression functions of cost on output were systematically ordered in time. This indicated that a correction for a time trend might be advisable. A time factor was, therefore, introduced explicitly into the least-squares multiple correlation analysis by the use of the variable X_3, which is a series consisting of the sequential numbering of the months in which observations were taken. In this way it was possible to isolate the systematic variation of cost as a function of time and to determine

Chart 2–4 Simple Regressions of Overhead Cost and Its Elements on Output (Monthly Costs)

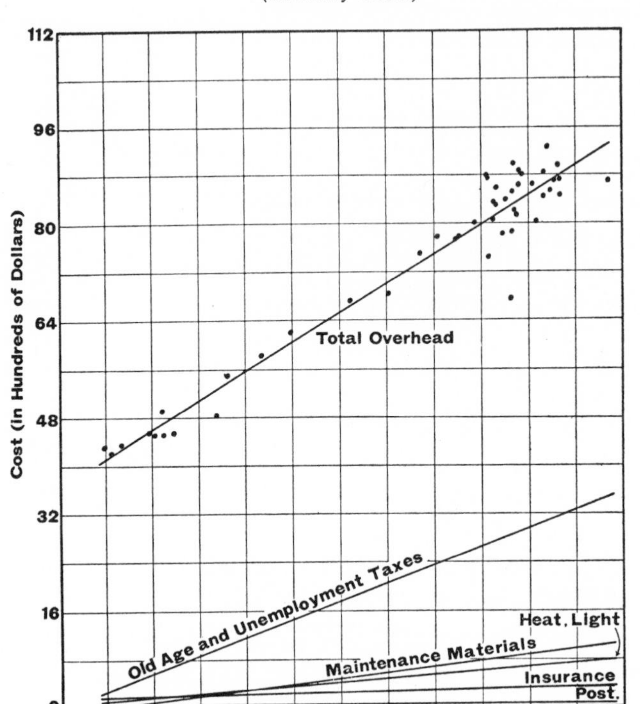

the net regression of cost on output. By allowing for the influences of changes in conditions through time which had not been taken into account by the rectification of the data, an estimate of the cost-output function which was possibly more accurate was obtained.

The graphic analysis showed a significant time trend for the three major cost components—productive labor, nonproductive labor and overhead—as well as for combined cost. The graphic partial regression of cost and time appeared to be curvilinear in the case of combined cost, productive labor cost, and overhead cost. A curvilinear multiple regression equation of the general form,

$$X_1 = b_1 + b_2X_2 + b_3X_3 + b_4X_3^2$$

was, therefore, selected as the most appropriate specification in these cases. This equation retains the linear specification chosen in the case of the simple regression, since this multiple regression equation is still linear with respect to output. In the remaining instance—nonproductive labor—a linear function, $X_1 = b_1 + b_2X_2 +$

b_3X_3, was fitted. In these equations X_1 is cost (in the form of totals per month), X_2 is output (in dozens of pairs), and X_3 is time (months numbered sequentially).

The results of the multiple correlation analysis of the monthly data for combined cost and its three principal components are shown mathematically in Table 2–2. These findings are also displayed in graphic form in the accompanying charts (2–5 through 2–8), in which the net or partial regressions of the various cost categories on output and time are shown.

In the upper section of Chart 2–5 the dots represent rectified monthly totals of combined cost that have been adjusted for the curvilinear time trend shown in the lower sections of the chart. Although the scatter is considerable, the distribution of the dots appears to substantiate the linearity of the partial regression of total cost over the observed range, from 4,000 dozen to 43,000 dozen pairs of hosiery. Beyond this range there is only one observation. The irregular line in the lower section connects cost observations which have been adjusted for output. They are deviations of the observations from the simple regression of cost on output arranged

Chart 2–5 Partial Regressions of Total Combined Cost on Output and Time (Monthly Costs)

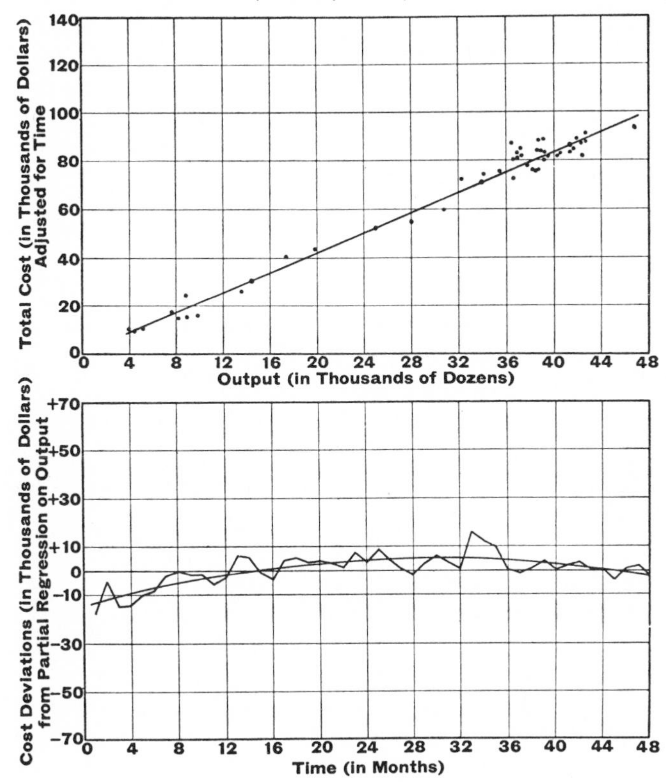

Chart 2–6 Partial Regressions of Productive Labor Cost on Output and Time (Monthly Costs)

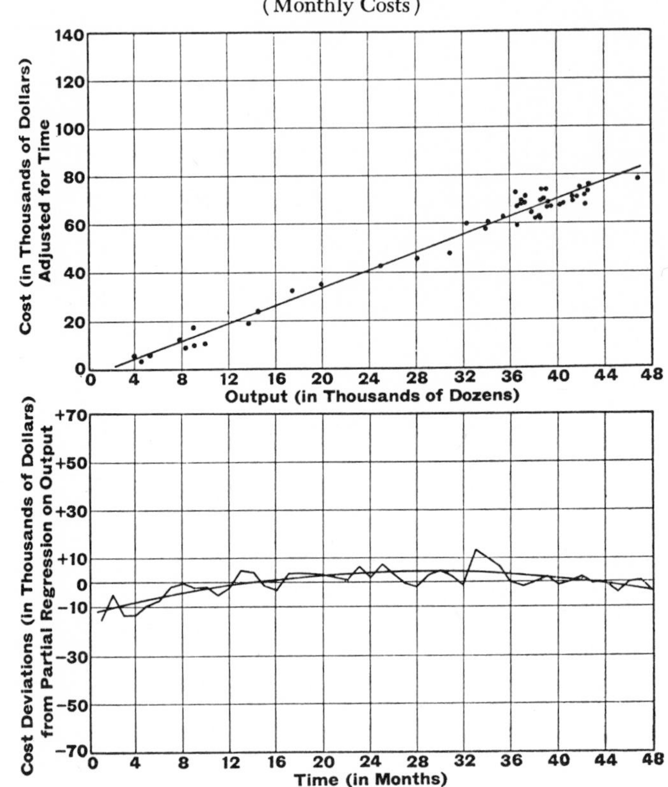

chronologically. The curved line fitted to these ordered deviations is the partial regression of cost on time, which is assigned a magnitude by the sequential numbering of the months.

A parallel portrayal of variations in productive labor cost with respect to output and time is found in Chart 2–6. The distribution of adjusted observations of monthly costs plotted against output in the upper section appears to be linear. As in the preceding chart, cost observations have been adjusted for the curvilinear partial regression of cost deviations on time shown in the lower section of the chart.

Chart 2–7 shows partial regressions for monthly totals of non-productive labor cost. In the upper section are plotted corrected cost observations adjusted for time trend. Again the amount of scatter and the character of the distribution of dots does not appear to justify specification of other than a linear partial regression. The deviations from this output regression, which were arranged chronologically and connected by an irregular line in the lower section of the chart, indicate a steady

Chart 2–7 Partial Regressions of Non-Productive Labor Cost on Output and Time (Monthly Costs)

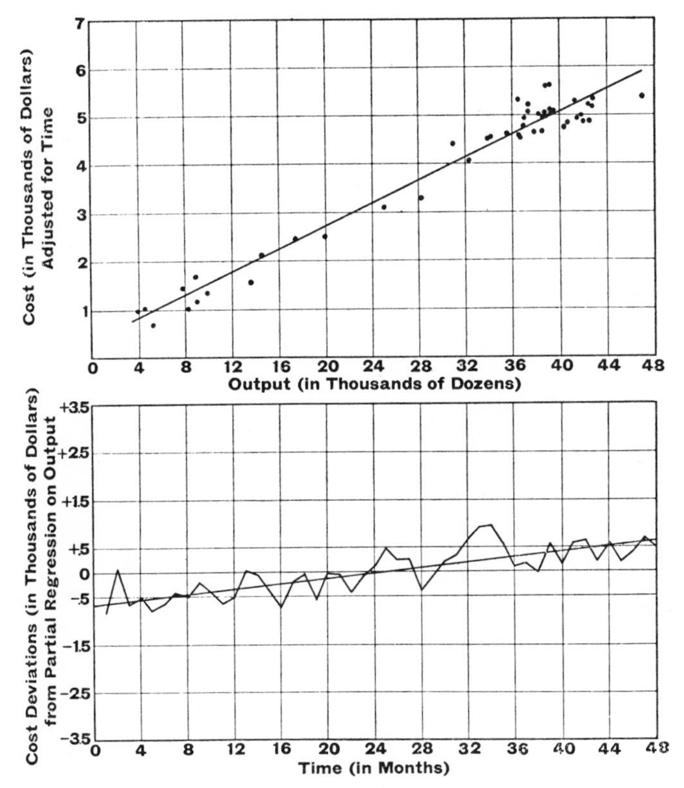

upward trend in nonproductive labor cost after allowance is made for the effect of output.

Chart 2–8 shows the partial regressions and adjusted observations of total monthly overhead cost. The scatter of adjusted cost observations plotted against output (shown by the dots in the upper section) is so wide and so approximately linear that fitting a cubic or parabolic regression curve does not appear to be justified. The linear partial regression shows that total overhead cost tends to increase with output at a uniform rate over the volume range studied. The lower section shows the time trend in overhead cost behavior, when allowance is made for the effects of output. The irregular line shows chronologically ordered cost deviations from the regression line appearing in the upper section of the chart. The trend is indicated by the curvilinear partial regression. There appeared to be a general tendency for overhead cost to increase during the first part of the period, to level off, and then to decline somewhat in the later months.

Table 2–2*

HOSIERY MILL: MULTIPLE AND PARTIAL REGRESSIONS OF COMBINED COST AND ITS COMPONENTS ON OUTPUT AND ON TIME
(Monthly Observations)

	Combined Cost	Productive Labor Cost	Nonproductive Labor Cost	Overhead Cost
Multiple regression equation	$_cX_1=-13{,}634.83+2.068X_2+1{,}308.039X_3-22.280X_3^2$	$_pX_1=-15{,}832.45+1.821X_2$ $1{,}205.593X_3-21.078X_3^2$	$_nX_1=-343.15+0.118X_2+$ $+27.668X_3$	$_oX_1=2451.60+0.130X_2+65.457X_3-0.987X_3^2$
Standard error of estimate†	3,983.31	3572.90	302.87	296.57
Coefficient of multiple correlation $\bar{R}$†	0.988	0.988	0.973	0.983
Coefficient of multiple determination $\bar{R}^2$†	0.977	0.977	0.946	0.966
Partial regression equation for output	$_cX_1=762.54+2.068X_2$	$_pX_1=-2993.03+1.821X_2$	$_nX_1=334.71+0.118X_2$	$_oX_1=3363.47+0.130X_2$
Partial regression coefficient for output	2.068 ± 0.071	1.821 ± 0.064	0.118 ± 0.005	0.130 ± 0.005
Partial regression equation for time	$_cX_1=-14{,}397.37+1308.039X_3-22.280X_3^2$	$_pX_1=-12{,}839.42+1205.593X_3-21.078X_3^2$	$_nX_1=-677.85+27.668X_3$	$_oX_1=-821.87+65.458X_3-0.987X_3^2$

* The symbols have the following meaning:

$_cX_1$ = combined cost in dollars
$_pX_1$ = productive labor cost in dollars
$_nX_1$ = nonproductive labor cost in dollars
$_oX_1$ = overhead cost in dollars
X_2 = output in dozens of pairs
X_3 = time

† Adjusted for number of observations.

Chart 2–8 Partial Regressions of Overhead Cost on Output and Time (Monthly Costs)

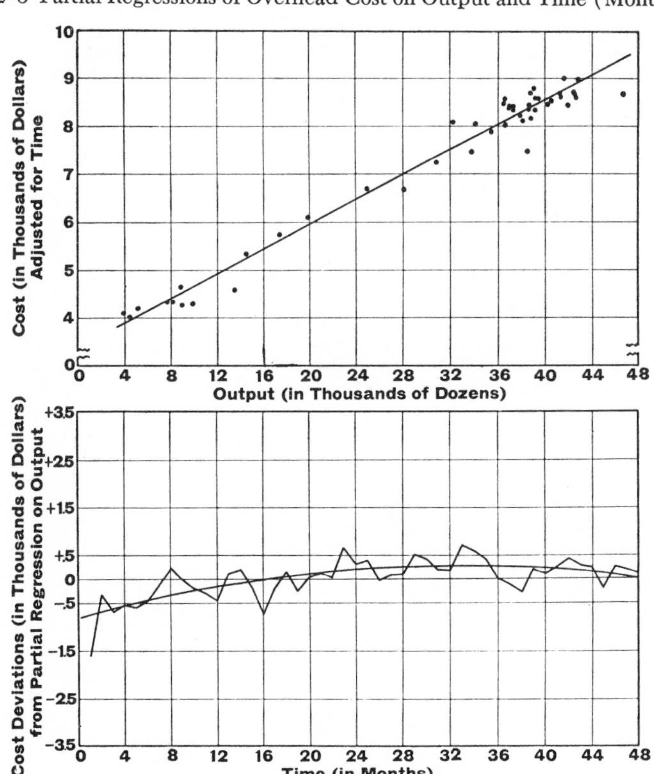

3. COMPARISON OF SIMPLE AND PARTIAL REGRESSIONS

In order to show the influence of the time variable which was introduced, a graphic comparison of the results of the simple and partial regression analyses is given in Chart 2–9. The simple and partial regression lines are plotted on the same axes for combined cost as well as for productive labor, nonproductive labor, and overhead cost. It is seen by inspection of the chart that there is a very close correspondence between the two regression lines.[15] The introduction of a time variable into a regression equation may in certain instances give very misleading results. For example, it might happen that a curvilinear simple regression function is transformed into a linear partial regression function, by removing the influence of a time variable with a spurious relation to the dependent variable. No such induced linearity has occurred in this case, however, and the only results of removing the time trend are to reduce the scatter of the observations about the regression line to some extent.

Nevertheless, the removal of the trend is an operation which must be performed

Chart 2–9 Comparison of Simple and Partial Regressions of Monthly Totals of Combined Cost and Its Components

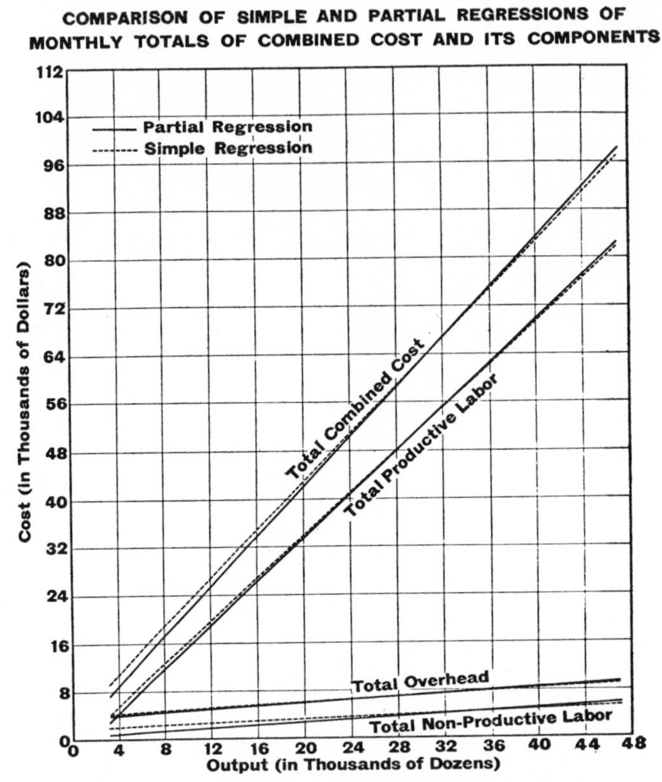

with caution. The opinion of the hosiery management was that the trend could be attributed to efforts to secure greater production control and efficiency which were initiated during the period under consideration. The reasons for its removal are not highly convincing, and in general it was found that this variation in cost of production of the period observed was difficult to account for satisfactorily. The trend may be attributable to fluctuations in managerial or labor efficiency or possibly to wage rate or other price fluctuations for which the data were not properly rectified. Because of this difficulty of knowing what a variation of cost through time actually indicates about changes in production conditions it was thought advisable to compute the regression equations shown earlier, expressing the relation of various costs to output without making any correction for a time trend.

4. MARGINAL AND AVERAGE COST

Both the simple and the partial regressions on output were determined for costs in the form of monthly totals. Both types of total cost functions were transformed[16] into average and marginal cost functions.[17] The equations for the derived average

Table 2–3*

HOSIERY MILL: EQUATIONS FOR TOTAL, AVERAGE, AND MARGINAL COST-OUTPUT FUNCTIONS OBTAINED FROM SIMPLE AND PARTIAL CORRELATION

(Monthly Observations)

	Total	Average	Marginal
	Simple Regressions		
Combined cost	$cX_1 = 2935.59+1.998X_2$	$cX_1/X_2=1.998+2935.59/X_2$	$dcX_1/dX_2=1.998$
Productive labor cost	$pX_1=-1695.16+1.780X_2$	$pX_1/X_2=1.780-1695.16/X_2$	$dpX_1/dX_2=1.780$
Nonproductive labor cost	$nX_1= 992.23+0.097X_2$	$nX_1/X_2=0.097+992.23/X_2$	$dnX_1/dX_2=0.097$
Overhead cost	$oX_1= 3638.30+0.121X_2$	$oX_1/X_2=0.121+3638.30/X_2$	$doX_1/dX_2=0.121$
	Partial Regressions		
Combined cost	$cX_1= 762.54+2.068X_2$	$cX_1/X_2=2.068+762.54/X_2$	$dcX_1/dX_2=2.068$
Productive labor cost	$pX_1=-2993.03+1.821X_2$	$pX_1/X_2=1.821-2993.03/X_2$	$dpX_1/dX_2=1.821$
Nonproductive labor cost	$nX_1= 334.71+0.118X_2$	$nX_1/X_2=0.118+334.71/X_2$	$dnX_1/dX_1=0.118$
Overhead cost	$oX_1= 3363.47+0.130X_2$	$oX_1/X_2=0.130+3363.47/X_2$	$doX_1/dX_2=0.130$

* The symbols have the following meaning:

cX_1 = combined cost in dollars
pX_1 = productive labor cost in dollars
nX_1 = nonproductive labor cost in dollars
oX_1 = overhead cost in dollars
X_2 = output in dozens of pairs

and marginal cost functions for combined cost and for its major components are found in Table 2–3.

The graphic counterpart of the results obtained from the partial regression equation for combined cost expressed in Table 2–3 are shown in Chart 2–10. The upper section shows the partial regression of total monthly cost on output, which is the same as that shown in the upper section of Chart 2–5. The marginal cost function which is pictured in the lower section is the first derivative of this total cost function. Since the total cost function is linear, its slope obviously remains unchanged; hence marginal cost is constant. From these results it is seen that the operating cost of producing an additional dozen pairs of hosiery (not including the cost of silk) is approximately $2.00 over the range of output observed.[18]

The average cost function, which lies above the marginal cost line in the lower section, was obtained by dividing the total cost function by output (X_2). This curve shows how cost per dozen varies with the number of dozens produced. Since the fixed cost is relatively small compared to the variable cost, the average cost function is only slightly curved and lies very close to the marginal cost function.

On the basis of the partial regression equations, estimates of combined cost and its components (productive labor, nonproductive labor, and overhead cost) were made for various levels of output. The equations which serve as a basis for these estimates are shown in Table 2–3. The derived estimates for monthly observations are shown in Tables 2–4 through 2–7.

Chart 2–10 Total, Average, and Marginal Combined Cost Derived from Partial Regression on Output (Monthly Costs)

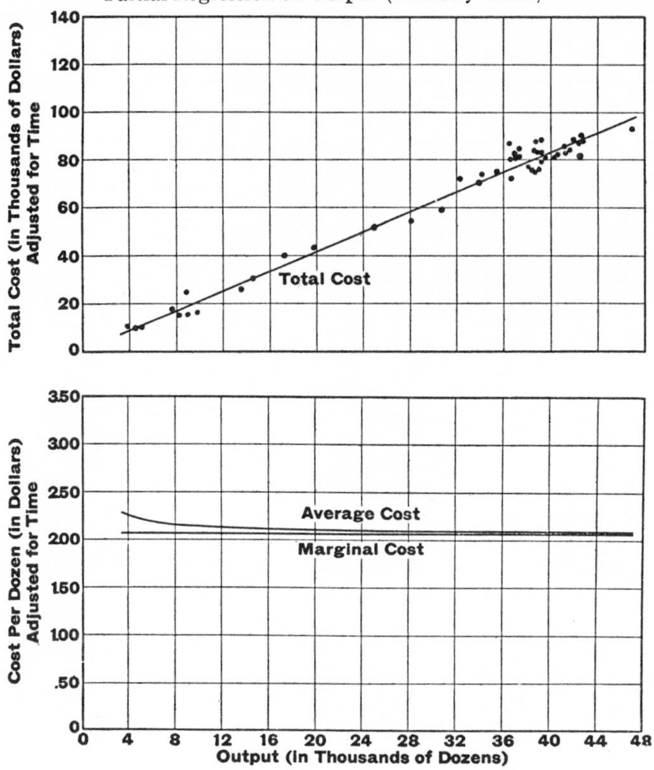

Table 2–4*

HOSIERY MILL: ESTIMATED COMBINED COST IN DOLLARS FOR VARIOUS LEVELS OF OUTPUT
(Monthly Observations)

Output (in Dozens of Pairs)	Estimated Total Cost	Estimated Average Cost	Estimated Marginal Cost
5,000..........	11,101.15	2.220	2.068
10,000..........	21,439.76	2.144	2.068
15,000..........	31,778.37	2.119	2.068
20,000..........	42,116.98	2.106	2.068
25,000..........	52,455.59	2.098	2.068
30,000..........	62,794.20	2.093	2.068
35,000..........	73,132.81	2.090	2.068
40,000..........	83,471.42	2.087	2.068
45,000..........	93,810.03	2.085	2.068

*The sources of the estimates are the partial regression equations for combined cost of Table 2–3.

Table 2–5*

HOSIERY MILL: ESTIMATED PRODUCTIVE LABOR COST IN DOLLARS FOR VARIOUS LEVELS OF OUTPUT
(Monthly Observations)

Output (in Dozens of Pairs)	Estimated Total Cost	Estimated Average Cost	Estimated Marginal Cost
5,000	6,114.45	1.223	1.821
10,000	15,221.94	1.522	1.821
15,000	24,329.42	1.622	1.821
20,000	33,436.91	1.672	1.821
25,000	42,544.39	1.702	1.821
30,000	51,651.88	1.722	1.821
35,000	60,759.36	1.736	1.821
40,000	69,866.85	1.747	1.821
45,000	78,974.33	1.755	1.821

*The sources of the estimates are the partial regression equations or productive labor cost of Table 2–3.

Table 2–6*

HOSIERY MILL: ESTIMATED NONPRODUCTIVE LABOR COST IN DOLLARS FOR VARIOUS LEVELS OF OUTPUT
(Monthly Observations)

Output (in Dozens of Pairs)	Estimated Total Cost	Estimated Average Cost	Estimated Marginal Cost
5,000	926.24	0.185	0.118
10,000	1,517.77	.152	.118
15,000	2,109.30	.141	.118
20,000	2,700.83	.135	.118
25,000	3,292.36	.132	.118
30,000	3,883.89	.129	.118
35,000	4,475.42	.128	.118
40,000	5,066.95	.127	.118
45,000	5,658.48	0.126	0.118

*The sources of the estimates are the partial regression equation for nonproductive labor cost of Table 2–3.

Table 2–7*

HOSIERY MILL: ESTIMATED OVERHEAD COST IN DOLLARS FOR VARIOUS LEVELS OF OUTPUT
(Monthly Observations)

Output (in Dozens of Pairs)	Estimated Total Cost	Estimated Average Cost	Estimated Marginal Cost
5,000.........	4,012.29	0.802	0.130
10,000.........	4,661.10	.466	.130
15,000.........	5,309.92	.354	.130
20,000.........	5,958.73	.298	.130
25,000.........	6,607.55	.264	.130
30,000.........	7,256.36	.242	.130
35,000.........	7,905.18	.226	.130
40,000.........	8,553.99	.214	.130
45,000.........	9,202.81	0.205	0.130

*The sources of the estimates are the partial regression equations for overhead cost of Table 2–3.

CHAPTER IV

Reliability and Significance of Findings

It is, of course, apparent that the reliability of any statistical analysis is limited by the quality of the original data. In this particular study this limitation is reinforced by the fact that it was necessary to perform certain rectification procedures by correcting for price variations and by reallocating cost and output. Consequently, in appraising the reliability of the conclusion that total combined cost is linearly related to output over the observed range of output, two questions must be raised in addition to that of the adequacy and quality of the sample. First, have the data been selected and manipulated in a manner which induces unwarranted linearity in the regressions? Second, is a straight line an adequate description of the cost regressions?

A. Limitations of Data and Methods

The difficulty of obtaining a sample which is satisfactory in other respects may mean that the observations represent an inadequate coverage of output variation, particularly at outputs approaching the physical capacity of the mill. There is no doubt that marginal cost would rise at some point as output moved beyond the range sampled, so that it may be emphasized that extrapolation of the cost–output functions at either end would be unjustified. The limited range of outputs observed does not, however, reduce the significance of the finding of constant marginal cost, if the observed variation in output includes the practical operating range. It is true that many enterprises operate at output rates short of maximum capacity much of the time. This may be the consequence of an imperfect selling market or of the economies to be attained by large-scale plants or of conscious managerial efforts to schedule production so as to avoid overloading of facilities. The sample may, therefore, be typical of operating conditions in an important sector of the economy.

The next limitations to be considered are those imposed by the method of selection and rectification of the data. These limitations will be considered under the following three heads: unjustifiable omission of costs, errors of allocation, and failure to stabilize irrelevant variations.

1. UNJUSTIFIABLE OMISSION OF COSTS

The difficulty which arises here is that the omission of certain cost elements, which bear some relation to output, may lead to an error in estimating the form and magnitude of the marginal cost function.

General firm overhead was excluded, although mill overhead was included in the cost estimates. Because of the arbitrary nature of allocation to any one mill, it was feared that the inclusion of general firm overhead would introduce greater error than its exclusion. Its effects upon marginal operating cost of this mill appeared, however, to be negligible since general firm overhead was not significantly correlated with the activity of this hosiery mill.[19] The conclusion was, therefore, that the omission of general firm overhead exerted no influence on marginal cost.

The cost of silk was also omitted, partly because the dominant importance of this variable cost would obscure minor departures from linearity in operating costs and partly because silk cost could be allocated to individual items with sufficient accuracy to raise no problem in the determination of marginal cost.[20] Since silk costs are assumed to be proportional to output, their elimination may prevent rather than induce an exaggerated appearance of linearity. The justifiable suspicion that silk costs might increase more than in proportion to output at extreme levels of operation as a result of wastage and spoilage is not borne out by the analysis or replacement cost which was presented earlier. The magnitude of replacements would reflect the extent of spoiled work, but no significant relation of replacements to output was found.

In addition to general overhead and silk cost a number of other mill costs were omitted. These excluded cost items, many of which cannot properly be regarded as production costs, amounted to only 2.2 per cent of corrected combined cost.[21] Moreover, some of these omitted elements were either allocated arbitrarily or were unduly influenced by accidental and irrelevant variation so that their inclusion would probably have led to more inaccuracy than would their omission. These miscellaneous omitted costs are so small that their inclusion is unlikely to affect the form of the observed cost function.

Thus, with the exception of silk costs there is little reason to suppose that the inclusion of these excluded elements of cost would have significantly altered either the slope or the form of the total cost function.

2. ERRORS OF ALLOCATION

Another general source of unreliability arises from the improper allocation of cost and output to time periods. Reallocation of output to the same period as the costs embodied in it is necessarily somewhat arbitrary.

The difficulty of allocating costs correctly to time periods is particularly serious for depreciation and maintenance costs, because of inability to pair these costs with the output that caused them. A machine can be repaired in advance of anticipated activity or subsequent to the activity. Accurate determination of the actual lags and leads is impossible. The arbitrary allocation of maintenance cost may mean that marginal maintenance cost is understated. But the alternative is to regard all maintenance cost as a function of output—a procedure which overstates maintenance cost. Since depreciation was recorded in the firm's accounts as a linear function of time, no depreciation is included in our marginal cost estimates. If a plant can be maintained in perpetuity and if maintenance cost can be broken down into that part attributable to use and that part attributable merely to the passage of time, no additional allowance need be made for depreciation. The magnitude of the error resulting from the omission of depreciation, therefore, depends on (1) the extent to which maintenance actually does maintain equipment unimpaired, (2) the extent to which use and time maintenance can be separated, and (3) the accuracy with which the true maintenance cost can be allocated to output.

With one exception no reallocations of cost proportionally to output were made,[22] and no accounts which represented proportional allocations were included. Therefore, errors arising out of allocation, while not strictly random, would scarcely induce linearity in the cost function.

3. FAILURE TO STABILIZE OR ALLOW FOR IRRELEVANT VARIATIONS

Another source of error may be in the inability to eliminate all irrelevant sources of cost variation. Even with careful selection of sample observations and rectification of the data, it may be possible to stabilize only imperfectly the environment in which cost-output behavior is observed. Certain small technical advances, unimportant individually, may cumulate to produce significant changes in the technique of operations. Managerial efficiency may deteriorate imperceptibly until checked by improved controls. The existence of a time trend in the cost residuals may reflect the influence of such cumulative changes or may merely show inadequate rectification of data. The use of time as an independent variable to allow for these influences may, on the other hand, merely obscure the meaning of the cost-output regressions. However, the comparison of these partial regressions with simple regressions shown in Chart 2–9 indicates that the discrepancies are unimportant.

B. Adequacy of Linear Regressions

An additional question which remains to be dealt with is the adequacy of the linear regressions which describe the cost–output relations. Any attempt to determine the functional relation between cost and output by statistical methods raises at once the problem of selecting a mathematical description of the relations in the hypothetical parent-population, of which the set of cost and output observations is a sample. In general where the functional form descriptive of a universe is unknown, this problem of specification is a difficult one. This discussion is intended to explain several devices which can be used as aids in the problem of specifying the short-run cost functions.

There are three approaches which are helpful in selecting a functional form of the universe to fit to the observed data. First, there are certain considerations resulting from the theory of cost which serve to place some restrictions on the admissible cost functions. Second, a preliminary statistical analysis of the marginal and average cost functions will afford evidence concerning the form of the total cost function.[23] Third, there are certain statistical tests available to strengthen one's confidence that the data are not in conflict with the hypothesis selected. The bearing of cost theory on the problem of specification has been discussed earlier in chapter ii and will not be considered further at this point. It is necessary, therefore, to examine the data only from the point of view of marginal and average cost behavior and certain statistical tests of significance.

1. ANALYSIS OF MARGINAL AND AVERAGE COST BEHAVIOR

In analyzing the cost behavior of the individual firm, it is convenient to use total cost functions and total cost curves. By simple calculation or geometric treatment the shapes of the subsidiary average and marginal cost functions can be seen immediately.[24] However, when a total cost function is used, the difficulty of specification is aggravated by the importance of the behavior of its first derivative, the marginal cost function.

It would be possible to fit a marginal cost function to first differences of the available cost and output observations and obtain the corresponding total cost function by integration. It is more convenient, in general, however, to proceed by fitting total cost-output functions and deriving the ancillary functions. Nevertheless, it seems advisable to carry out at least a preliminary analysis of marginal cost behavior in view of the information it affords concerning the shape of the total cost function. If the issue is between fitting a linear or a cubic total cost function, the choice will be simplified by the fact that the first will have a constant marginal cost function and the second a parabolic marginal cost function. Consequently, if the first differences of the cost observations divided by the corresponding first differences of output observations are plotted against output, the resulting approximation will yield some information concerning marginal cost behavior.[25] If the resultant scatter diagram gives the appearance of being best fitted by a horizontal straight line, one's confidence that the integral function is linear is increased. If, on the other hand, the regression in the scatter diagram is plainly parabolic, this is evidence that the total cost function is curvilinear. Such an analysis was carried out; it showed that the observed marginal cost approximation was constant with respect to output—a result which substantiated the hypothesis of linearity. A least-squares analysis of the relation of the approximation to marginal cost to output indicated that the correlation was not significantly different from zero. The coefficient of correlation was found to be 0.179, and its standard error, 0.152.

In addition a test was made to determine whether the size or direction of changes in output affected the magnitude of incremental cost. Chart 2–11 shows month-to-month first differences in total combined cost plotted against corresponding first differences in output. Since the relation is linear, this indicates that short-run changes in cost tend to be proportional to changes in output, regardless of the magnitude or the direction of the change. The least-squares regression line shown in this chart depicts the equation

$$(y_i - y_{i-1}) = K(x_i - x_{i-1}) ,$$

which was found to be $(y_i - y_{i-1}) = 1.99\,(x_i - x_{i-1})$. This provides a supplementary estimate of marginal operating cost of $1.99 a dozen, which is very close to that determined in other ways.

The linearity of this regression itself could be scrutinized more rigorously by making use of the technique of analysis of variance. This method is explained in more detail in the discussion of statistical tests. In view of other evidence, however, it was not considered essential to examine the regression of cost differences on output differences more objectively. The result of these two analyses of the behavior of the approximation to marginal cost is to substantiate the hypothesis of linear total cost regression and to indicate that the magnitude of marginal cost is not systematically related to the direction or magnitude of output changes.

If the total cost function has a point of inflection, i.e., a point at which cost progression sets in, average cost will rise from this point on. If, on the other hand, the

Chart 2–11 Relation of First Differences of Combined Cost to First Differences of Output

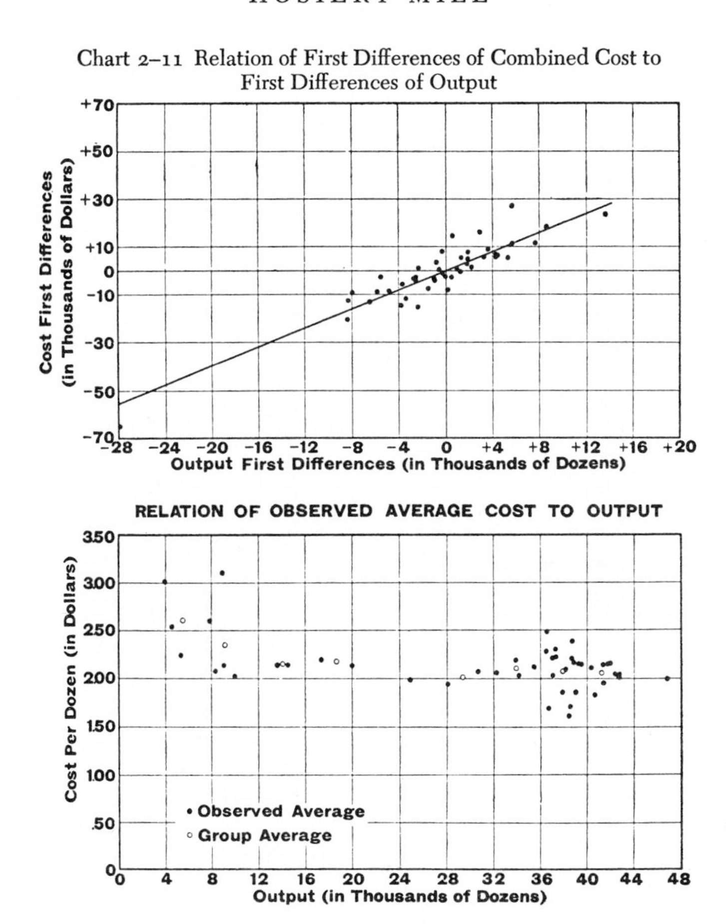

total cost function is linear, the average cost function will decline throughout the whole range because of the existence of fixed costs. In consequence, an examination of the relation of observed average cost to output will give some further evidence of the shape of the total cost function. In the lower panel of Chart 2–11, the behavior of average cost derived directly from the data is shown in the form of a scatter diagram. Neither the cost observations nor their group averages (which are marked with the small circles) show any consistent tendency to rise over the extremely high levels of output. One can conclude from this that the behavior of average cost is such that no curvilinearity of the total cost function which would be reasonable is indicated.

While this analysis is a valuable confirmation of the assumption of linearity, there are also available certain statistical tests of significance which may fortify the confidence to be placed in the specification chosen. Several of these tests will be considered in the following section.

2. STATISTICAL TESTS OF LINEARITY

If the universe is unknown or is known only to the extent that it is described by either the function

$$X_1 = b_1 + b_2X_2$$

or the function

$$X_1 = b_1 + b_2X_2 + b_3X_2^2 + b_4X_2^3,$$

it is possible to establish by reference to sample data only some range within which the parameters may lie. The only conclusion which one can reach by statistical inference is a negative one which expresses some degree of confidence that the parameters of the universe do not have certain assumed values, which, of course, does not establish the actual magnitude of the values. Suppose, for example, that a cubic function similar to the one represented above is fitted to the data. The procedure is to set up the null hypothesis that the universe value of the coefficients of both the squared and the cubed term are zero (i.e., that the function is linear). If the calculated coefficients are sufficiently small relative to their standard errors and if the probability attached to such ratios is greater than some reasonable or accepted level of significance, one can conclude that the data are not in conflict with

Chart 2–12 Parabolic and Cubic Regressions of Total Combined Cost on Output (Monthly Costs)

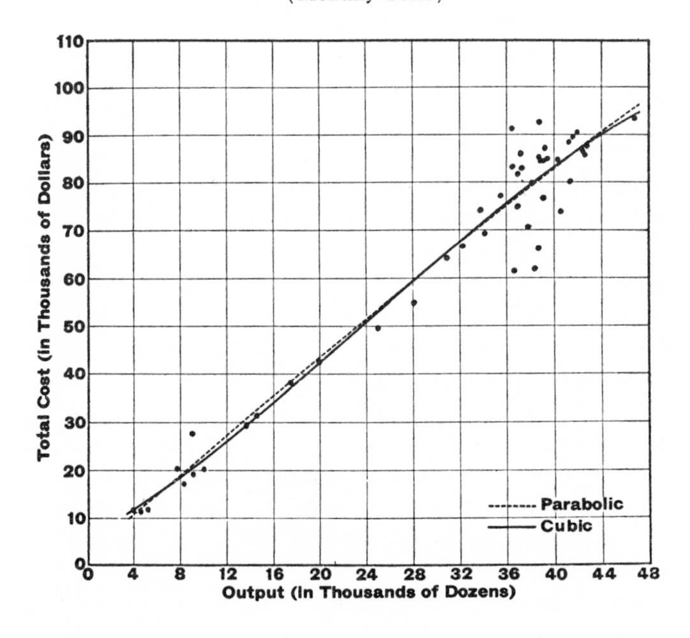

the null hypothesis. This does not, however, prove that the coefficients in the universe are zero. The data might, in fact, be consistent with some other hypothesis. It would, nevertheless, indicate that the observed results are consistent with the linear hypothesis.

In order to provide a basis for discriminating between linear and curvilinear regression functions, both second- and third-degree parabolas were, therefore, fitted to the monthly observations of cost and output. A graphic representation of these curvilinear regressions is given in Chart 2–12. The corresponding equations are: for the parabola,

$$_cX_1 = 2.387 + 2.068X_2 - .00145X_2^2\ ,$$

and for the cubic,

$$_cX_1 = 6.111 + 1.344X_2 + .0318X_2^2 - .00044X_2^3\ .$$

The ratio of the coefficient of the squared term in the parabola to its standard error was found to be .17. The significance of this ratio can be tested by reference to the distribution of t with $n = 45$, which indicates that a value such as the one observed could occur approximately eighty to ninety times out of a hundred as a result of chance errors (taking into account both tails of the t distribution). Similarly, the coefficients of the squared and cubed term in the cubic function would occur some fifty to sixty times in a hundred as a result of random sampling fluctuations, since the ratios of the coefficients to their standard errors are .605 and .641, respectively. Thus there is a high probability of obtaining such values of the regression coefficients by chance from a universe in which the parameters of the squared and cubed terms are zero. Therefore, nothing in these tests would lead to a rejection of the hypothesis of linearity.

An additional circumstance reinforced the rejection of the cubic hypothesis. The signs of the quadratic term in the parabola and the cubed term in the cubic are opposite to those postulated by the underlying cost theory. The fact that it is possible to test the regression coefficients for sign as well as for magnitude means that this is a particularly powerful test of significance. There are, therefore, on the basis of these considerations strong grounds for accepting the linear function in preference to some curvilinear function.

By the application of the analysis of variance technique it is possible to form some judgment concerning the relative "goodness of fit" of two functions.[26] In this particular instance the problem is whether or not the cubic regression fits the data significantly better than the linear regression. The test available is similar to the usual test of linearity of regression, in which the data are grouped into arrays and the relation of the variance of the array means about the linear regression is compared with the variance within arrays. If the former variance is unusually greater than the latter, there are grounds for rejecting the hypothesis of linearity in favor of the alternative regression function of higher degree. In the case of short-run total cost functions there is additional evidence that may narrow the choice of alternative forms to a specific type of cubic function. For such problems the test of the significance of the regression coefficients in the cubic function is the most powerful test available; hence it was not considered necessary to apply analysis of variance methods to the problem.[27]

The degree to which it is justifiable to apply these statistical techniques should not be overestimated. The conditions which are necessary for the validity of such tests are not met by our cost and output observations. First, the data are time series. Although sampling and correction operations may have reduced their serial dependence, the observations are still not independent. Second, the rectification and allocation procedures are so important in relation to the residual errors that it is dangerous to place too much emphasis on the behavior of residual errors. It may be that inaccuracies in the adjustment and correction procedures, which cannot be allowed for in tests of significance, far outweigh the actually remaining residuals. Nevertheless, these statistical tests do perform some useful function in supplementing other kinds of information used for choosing among different functional forms.

C. Indications of Reliability

Several measures of the reliability of the cost regressions—i.e., the standard errors of estimate, the correlation coefficients, the standard errors of the regression coefficients that were obtained in the correlation analysis—are shown in chapter iii in the various tables exhibiting the regression equations. In view of the nature of our data and sampling procedures, the various standard error measures should be regarded as indicating the minimum error and should be supplemented with additional indications of reliability, such as the correspondence of estimated cost to actual cost and the similarity among estimates of cost obtained by alternative methods.

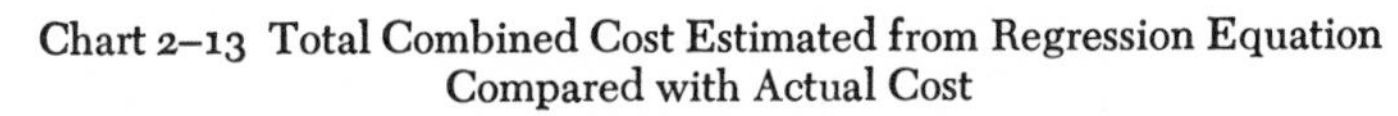
Chart 2–13 Total Combined Cost Estimated from Regression Equation Compared with Actual Cost

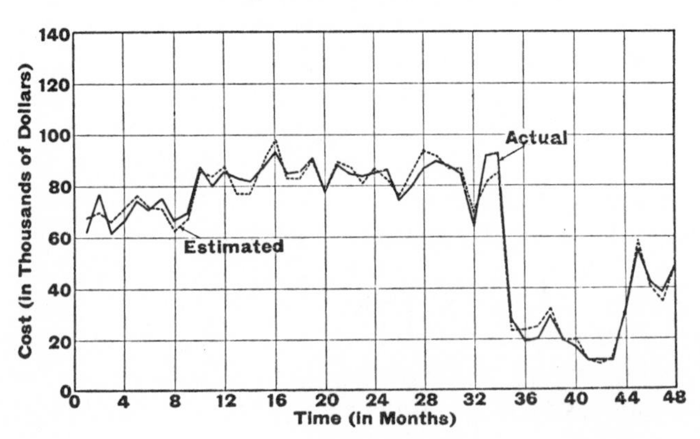

One such comparison is shown in Chart 2–13, where actual recorded cost is compared with estimated cost. In this chart the dotted line shows monthly combined cost estimated from the regression equation for each month of the analysis period, while the solid line shows the actually recorded costs for each month. The fairly close correspondence which exists during this period does not, of course, assure a continuation of such accuracy of prediction.

In Table 2–8 estimates of combined cost which were secured by summing the component cost estimates obtained from their partial regressions are compared with

Table 2–8*

HOSIERY MILL: ESTIMATES OF TOTAL COMBINED COST IN DOLLARS FOR VARIOUS LEVELS OF OUTPUT OBTAINED BY TWO METHODS
(Monthly Observations)

Cost	Output (in Dozens of Pairs)			
	10,000	20,000	30,000	40,000
Productive labor cost*	15,221.94	33,436.91	51,651.88	69,866.85
Nonproductive labor cost*	1,517.77	2,700.83	3,883.89	5066.95
Overhead cost*	4,661.10	5,958.73	7,256.36	8,553.99
Combined cost from summation of components	21,400.81	42,096.47	62,792.13	83,487.79
Combined cost from combined cost partial regression equation*	21,439.76	42,116.98	62,794.20	83,471.42
Difference	−38.95	−20.51	− 2.07	+16.37
Difference as percentage of summation estimate	0.18	0.048	0.0032	0.019

*The estimating equations are the partial regressions shown in Table 2–3.

those derived directly from the partial regression of combined cost on output.[28] Since the total cost observations were obtained by summing the components, very little difference in estimate would be expected from the two methods unless the fit of one or more of the component curves or the combined cost curve was bad.

There is a close similarity among estimates of marginal cost obtained by several alternative methods. Table 2–9 shows estimates of marginal combined cost obtained from monthly data and from quarterly data by differentiation of simple regressions and of partial regressions, as well as by summing the marginal cost of components. The correspondence of monthly and quarterly marginal cost indicates that errors of allocation among months were not serious. The similarity of marginal cost estimates derived from partial and simple regressions shows that use of time as an independent variable did not affect the slope of the total cost function to any great

Table 2–9*

HOSIERY MILL: ESTIMATES OF MARGINAL COMBINED COST IN DOLLARS OBTAINED BY ALTERNATIVE METHODS

Source of Estimate	Marginal Combined Cost	
	Monthly Observations	Quarterly Observations
Simple regressions:		
Marginal combined cost equation	1.998	1.972
Summation of marginal cost of components	1.998	1.972
Partial regressions:		
Marginal combined cost equation	2.068	2.042
Summation of marginal cost of components	2.069	2.044

*Estimates from marginal cost equations of Table 2–3 and Tables 2–16 and 2–17 in Appen. A.

extent. The closeness with which the sum of the marginal costs of components corresponds to that derived directly from the total combined cost functions themselves fortifies our confidence in the estimates of the marginal cost of the components.

D. Economic Interpretation

The constancy of marginal cost over the range of observed output is from the viewpoint of the economist the most striking finding of the statistical analysis. Although a marginal cost function of this type has not been considered to be the usual case in theoretical discussions of cost behavior, the technical conditions which would yield this result were outlined in the earlier theoretical discussion. It is necessary, therefore, to examine the technique of production operations to see if there is close conformity with the circumstances pictured in the theoretical model which portrayed a linear total cost curve over the range of output less than the technical capacity of the plant.

One technical aspect of hosiery manufacturing which would lead to constancy of marginal cost is the high degree of divisibility or segmentation of the fixed plant and equipment of the hosiery mill. Each machine unit constitutes almost a miniature hosiery mill in itself. The legging department, for example, consists of eighty-one substantially uniform legging machines, each approximating a self-contained production unit with its own balanced labor force and integrated into the total process by common supervisory and service functions. Other operations, such as footing and seaming, are similarly sectionalized into substantially identical machine units, each with its complement of labor. Variation in the number of shifts, the length of the working shift, and the number of days operated constitutes an additional method of segmentation, and alteration of the speed of machine may be a third means of segmentation. The existence of this highly segmented plant makes it possible to combine fixed and variable factors in optimum proportions in each segment rather than to employ fixed equipment at more than optimum rates.[29]

If increases in production could be accomplished by proportionate increases in all input factors, observed marginal costs would be constant, provided that the segments were of uniform efficiency and that more inefficient machines were not introduced progressively with increases in the rate of output. Since neither the machines nor the operators are of uniform efficiency, the employment of the more efficient units first and the use of the less efficient units only at high levels of output would result in a marginal cost function which would show an increasing phase as output increased. However, union regulations, repair programs, and a certain degree of specialization among machines deterred the management of the hosiery mill from taking full advantage of the different degrees of efficiency, so that rising marginal cost was not observable as a result of efficiency variation.

Since some of the fixed factors cannot be segmented—e.g., general supervisory and service functions—it might be expected that at some point marginal cost would rise because of more than optimum application of variable and semivariable factors to these fixed factors. In the case of executive personnel, however, there may be considerable excess capacity for normal outputs. In periods of high production the reserve of unused capacity and energy is drawn on, long-range tasks really allocable to future periods are postponed, and leisure is temporarily foregone, with the result that executive time and energy may be applied much more nearly in proportion to output than salary records would indicate.

It was shown in the theoretical discussion that for a given plant, sooner or later, a level of output must be reached beyond which marginal costs begin to rise. The statistical evidence, however, indicates that this critical level of operations was not reached in the hosiery mill during the period studied. In most real-world situations there are important deterrents to pushing output into that range of operations where marginal costs are rising. The economies to be attained by increases in the scale of plant may be so pronounced that it is cheaper to produce a given output by using an oversize plant intentionally at less than capacity than a small plant at full capacity. Furthermore, engineering and managerial conjectures may be that costs would rise so sharply beyond a certain well-defined limit that it is considered economical to overbuild a plant to take care of peak loads which might arise and thus to insure a certain operating flexibility. On the other hand, it may be the practice to manufacture for stock in periods of low demand in order to avoid the range of rising marginal cost which is encountered if operations are likely to be extended beyond the critical level during periods of peak demand.

Chapter V

Adjustment of Cost Functions for Practical Use

In order to make the findings of this study usable for practical administrative decisions, it is necessary to adjust the estimates of cost in three ways: by taking account of factor price changes, by allowing for omitted costs, and by adjusting cost estimates applicable to standard output units to the peculiarities of particular styles and specifications of hosiery.

A. *Adjustment for Input Price Changes*

To conform to the theoretical model the statistical cost functions are determined for the level of input prices prevailing in the base period. In order to use the cost estimates for decisions concerning present and future periods, the cost estimates must, therefore, be reflated to reflect current or anticipated input prices. Since the price trends of the components of combined cost may be diverse, it is desirable to correct the cost estimates of each component separately.[30] It is in this case necessary merely to compute the ratio of current or anticipated prices of input factors to the prices of the base period and multiply the relevant cost components by this ratio to bring the estimates up to date or to project them into the future. The reflation of marginal cost may be accomplished quite simply by expressing the input prices of each component for the month desired as a price relative of these input prices in the base period and multiplying this price relative by the deflated marginal cost of the component. This procedure is illustrated in Table 2–10. The results of reflating the marginal cost estimate to allow for changes in input prices are shown graphically in the upper section of Chart 2–14. Since this is primarily illustrative, the reflation is performed for the productive labor cost component only. The middle line shows the marginal cost function for productive labor for the level of input prices prevailing in the base period, and the other lines show marginal cost for levels of input prices prevailing in two other periods. The lower section of Chart 2–14 shows the reflated marginal cost of productive labor as a time series with the marginal cost estimate adjusted to reflect the input prices that prevailed at various times during the period of analysis. Correction of total or average cost involves an additional complication because, unlike marginal cost, the proportions of the major

Table 2–10
HOSIERY MILL: REFLATION OF MARGINAL COMBINED COST IN DOLLARS FOR A DIFFERENT LEVEL OF INPUT PRICES
(Monthly Observations)

Cost Components	Deflated Marginal Cost	Price Relatives for Base Period	Price Relatives for Different Period	Reflated Marginal Cost
Productive labor	1.821	100	90	1.639
Nonproductive labor	0.118	100	110	0.129
Overhead	0.130	100	105	0.136
Marginal combined cost	2.069	....	....	1.904

Chart 2–14 Related Marginal Productive Labor Cost (Monthly Costs)

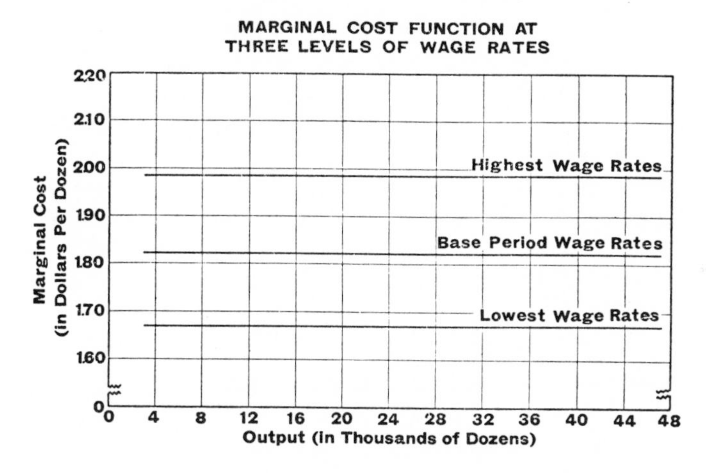

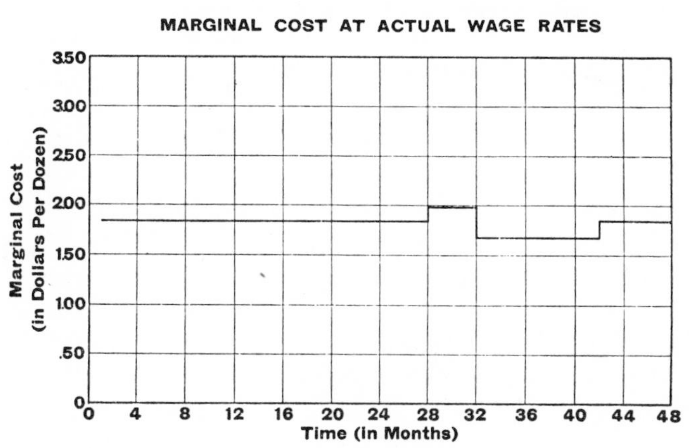

Table 2–11*

HOSIERY MILL: REFLATION OF TOTAL COMBINED COST IN DOLLARS FOR A DIFFERENT LEVEL OF INPUT PRICES
(Monthly Observations)

Cost Components	Cost at Output of 20,000 Dozen*	Price Relatives for Base Period	Price Relatives for Different Period	Reflated Cost
Productive labor....	33436.91	100	90	30093.22
Nonproductive labor	2700.83	100	110	2970.91
Overhead..........	5958.73	100	105	6256.67
Combined cost†	42096.47	....	....	39320.80

*Estimated from the partial regression equations of Table 2–3, at the level of input prices current in the base period.
†Estimated by summing the cost of the components.

components are a function of the rate of output. The adjustment must, therefore, take account of the output level as well as of the level of input prices. If the production in the current period is, say, 20,000 dozen, the cost of each component based on the prices prevailing in the base period can be read from the cost functions, and when they are multiplied by the current price relatives an estimate of current cost is obtained. This procedure is illustrated in Table 2–11. Estimates of average cost applicable to current or future periods could be obtained by the same method.

B. Adjustment for Omitted Costs

Certain elements of cost were not included in the "combined costs" used in the statistical analysis, and it is, therefore, necessary to allow for the effect of the omitted costs to obtain a usable estimate of marginal cost. In this study the omitted costs consist of the three groups—silk cost, general firm overhead cost, and miscellaneous costs. Silk cost was omitted from the analysis for two reasons. First, the inclusion of this one input factor of dominant importance, whose cost is almost certain to be proportional to output, might overshadow and outweight elements tending to produce curvilinearity in the combined cost function. Second, the company has sufficient knowledge of the amount and type of silk required for each style and specification of hosiery so that the adjustment of statistically determined costs to permit the inclusion of this omitted element is relatively simple. General firm overhead was omitted, first, because the particular plant studied constituted such a small part of the entire operations of the firm that the magnitude of firm overhead was unlikely to be affected by variations in the output of this one plant[31] and, second, because allocation of firm overhead to individual plants is necessarily arbitrary. The miscellaneous costs omitted include a number of trivial elements of cost which were excluded for a variety of reasons discussed later in Appendix B. Many of these omitted miscellaneous costs of the mill are not strictly production costs, so that no allowance need be made for them. Furthermore, the magnitude of this class of omitted cost elements was negligible, making up, approximately, only 2.2 per cent of the plant's total cost. Because of the trivial size of these omissions it seems unnecessary to adjust the marginal cost estimates to take them into account.

C. *Estimation of Marginal Cost of Particular Styles*

The estimate of marginal cost obtained is in terms of a composite product[32]—the average of the styles actually produced. In order to transform this estimate so that it is applicable to a unit of output of a particular style and specification, two alternative procedures may be followed. First, on the assumption that the degree of homogeneity of output permits us to neglect the cost of differences in operations among styles, a rough estimate of the cost of producing an additional dozen stockings of a given type can be obtained by adding the estimated cost of the silk required for the particular style to the marginal processing cost of the composite

Table 2–12*

HOSIERY MILL: ESTIMATES OF MARGINAL COMBINED COST OF PARTICULAR STYLES AT OUTPUT OF 20,000 DOZEN

(Monthly Observations)

Style	(1) Average Productive Labor Cost of Composite Product	(2) Average Productive Labor Cost of Particular Style	(3) Ratio of Col. 1 to Col. 2	(4) Marginal Cost of Composite Product	(5) Col. 3 × Col. 4	(6) Estimated Cost of Silk Requirements†	(7) Estimated Marginal Combined Cost of Particular Style (Col. 5 + Col. 6)
Style 102.....	1.672	1.70	1.017	2.068	2.103	2.30	4.403
Style 108.....	1.672	1.66	0.993	2.068	2.054	2.25	4.304
Style 116.....	1.672	1.65	0.987	2.068	2.041	2.15	4.191

*Estimates from partial regression equations of Table 2–3
†These figures are wholly imaginary.

unit. Second, if it is true that differences in operating costs are significant, the marginal cost of a particular style can be estimated by multiplying the marginal processing cost of the composite product by the ratio of the estimated productive labor cost of the particular style to the average productive labor cost of the composite product at the prevailing output rate. When the estimated cost of the silk requirements is added to this, an estimate of the marginal cost of a particular product is obtained. This procedure is illustrated in Table 2–12.

With this discussion of the adaptation of the statistical results the problem which was originally posed—the derivation of empirical cost functions—is essentially completed. However, in the course of the presentation of the statistical findings little attention was devoted to the practical problems encountered in the analysis of accounting records. It is necessary, therefore, to turn to the more prosaic details of the selection, adjustment, and manipulation of the data. Chapter vi deals with the selection of the primary data, with particular attention being given to the considerations which prompted the choice of the particular hosiery mill, analysis period, and observation unit. This discussion of the nature of the sample is prefaced by a brief description of the process of hosiery manufacture, in the hope that this may throw some light on the technical problems encountered. Following this, in chapter vii, the selection, classification, and rectification of cost data is discussed. Chapter viii deals with the selection and measurement of operating conditions whose influence on cost might be allowed for by introducing them as independent variables in the multiple correlation analysis. Special problems relating to omitted cost elements and to the measurement of output are discussed in the appendixes.

Chapter VI

Characteristics of the Plant and Observation Period

A. Description of Hosiery-Mill Operations

To make the methods and findings of this study more intelligible to the layman it is desirable at the outset to describe briefly some of the technical aspects of the process of manufacturing full-fashioned hosiery.

Raw silk is shipped to the hosiery manufacturer in skeins composed of long continuous threads. The skeins are first soaked in an emulsion of some vegetable or animal oil, for the dual purpose of softening the waxlike gum which coats the silk and lubricating the thread for subsequent operations. It is then wound on spools, doubled, and the doubled threads twisted together.[33] The number of threads twisted together determines the type of hosiery to be made: two and three threads being taken for the very sheer chiffons, and from three up to twelve for the heavier service weights. This set of operations is called "throwing." The next process is winding, where the silk is wound from paper cones to smaller wooden cones suitable for use on the knitting machines. It is at this stage of manufacture that the operations of the particular hosiery mill studied begin. It receives the silk already wound on wooden cones from the company's throwing mill and carries the process up to the point where the stockings are ready for dyeing and finishing.

Preparatory to the knitting operation the silk must be softened, which is done by passing it through an emulsion similar to the oil bath already mentioned. At the same time the yarn runs through knife-cleaners, which remove waste that might otherwise break the knitting needles or spoil the appearance of the fabric. From here the silk passes to the conditioning room, where it is kept at a high degree of humidity until it is needed.

The next step is knitting—the first actual manufacturing operation. Full-fashioned stockings are made on two machines, the "legger" which knits the leg and the "footer" which knits the foot.[34] The legging machines are composed of sections, each section knitting one leg. The leggers in the particular hosiery mill studied were eighteen- and twenty-four-section machines. The needles of the knitting machine are small and set close together—from twenty-six to thirty-eight to the inch, depending upon the gauge of the machine. Since each knitting head is approximately fourteen inches wide, there is a great number of needles in each machine. The distinguishing feature of full-fashioned hosiery is that the machine starts at the top with a full complement of needles, and as the knitting proceeds the fabric is reduced in width or "fashioned" by decreasing the number of stitches as the calf is reached, so that the stocking is narrowed in order to conform exactly to the shape of the leg. This part of the operation is called "fashioning." The footing machine itself is similar to the legger, the difference being that the sections are smaller and there is a smaller number of sections, each of which knits a foot at a time. The output of a footing machine will produce enough feet for three leggers. In the knitting of the feet there is also a fashioning operation which shapes the toe and narrows the sole so that it fits the foot.

The operation which unites the feet and the legs is called "topping," and consists of transferring the stitches that lie along the lower edge of the ankle fabric and along the inner edges of the heel fabric to the needles of the footing machine, with the result that the foot instep becomes a continuation of the leg instep and the two halves of the sole are attached to the inside of the heels. The transfer is accomplished by operators who first put each leg stitch on the points of a transfer stand, from here to the transfer bars, and then to the needles of the footer—an operation which requires great skill as each leg stitch must be exactly in place.

Topping completes the actual knitting operations, and the next process is "looping," during which the ends of the heels are joined and the point of the toe closed. The next operation is "seaming," where the two sole halves and the two edges of the leg are joined together.[35] The stocking is now substantially completed, the only remaining operations being examination before it is sent to another plant for dyeing and finishing.

In order to explain certain of the technical changes which occurred during the period studied, some mention should be made of the process which insures ringlessness in stockings. Natural silk, which is an animal product, is inherently lacking in uniformity, so that, unless special care is taken, unevenness is likely to appear in the knitted fabric. This irregularity is practically eliminated, first, by careful selection of the silk and, second, by the use of three-thread carriers in the knitting of the stocking leg, which is the so-called "ringless" process.[36]

B. Selection of Plant

1. CRITERIA OF SELECTION

The firm under consideration operated a number of technical units, several knitting mills, a throwing mill, and a finishing plant, from which it was necessary to select the one most suitable for the type of cost study projected. Each plant was accordingly examined with reference to the following criteria before the final choice was made.

a) *Homogeneity of product.* The diversity of products should be at a minimum in order to simplify the measurement of output and to make possible the conversion of cost estimates to usable practical terms. It is very desirable, moreover, from a theoretical point of view to be able to treat output as though it consisted of a single product, the fewness of actually diverse products reducing the error in such a simplification.

b) *Homogeneity of equipment.* Similarity of machine units is desirable in order to eliminate cost variations due to use of more efficient machines for some levels of production than for others. In the hosiery industry, selection of better equipment during times of slack production is relatively unimportant, because union restrictions prohibit allotment of work which discriminates against operators of inefficient equipment during periods of low or part-time employment.

c) *Degree of technical progress.* It is desirable to select for study a plant in which improvements in the technical methods of production are at a minimum. The existence of appreciable technical progress may exert an important influence on cost, and, if it cannot be measured or allowed for, an element of historical change is introduced in the description of the relation of cost to varying output volume. Since it is desired to determine the cost-output function for a given method of production, historical technical changes would obscure the nature of this relation.

d) *Length of production cycle.* The production cycle, which is the average length of time it takes the initial inputs in a process to emerge as finished products, should be short in order to minimize the reallocation of recorded costs and output necessary to place them in the accounting period which is appropriate, i.e., so that costs attributable to a certain output are actually matched with that output.

e) *Volume variation.* The nature of the operations should be such that the production volume varies from month to month in such a fashion as to include a wide range of volume and a fairly uniform coverage of this range.

2. CHARACTERISTICS OF PLANT SELECTED

After a survey of the various plants it was found that one knitting mill conformed most closely to all these criteria, and it was consequently chosen for intensive study.[37] The following special qualifications prompted its selection. First, it manufactured unbranded merchandise almost exclusively, which meant that the range of variations in quality and styles was smaller than that for mills producing differentiated hosiery of various brands. The equipment in operation was remarkably homogeneous. All legging machines in operation[38] were eighteen-section and either thirty-nine gauge or forty-two gauge, and all footing machines had twenty sections. Hence alteration of the type of equipment used as production volume increased was unlikely. Moreover, technical progress in the form of changed equipment was small and less than at the other mills during the period contemplated.[39] The production cycle was short (about seven days), partly because of the standardized character of the product and partly because of the management and organization of the mill. Finally, the volume of production per time unit fluctuated between zero and approximate physical capacity, although this range was not covered in so uniform a fashion as would have been desirable.

C. *Selection of Time Period and Observation Unit*

The data used for the analysis consisted of the monthly, quarterly, and weekly accounting records of the hosiery mill for the period beginning January 1, 1935, and ending June 30, 1939. In the selection of this observation period certain criteria were again set up, some of them similar to those used in the choice of a plant. The following conditions were regarded as desirable: (1) the existence of a wide range of output variability and fairly uniform coverage of the range; (2) constancy of size of physical plant; (3) minimum technical progress; (4) stability in management methods; (5) uniformity in cost records and adequacy of records of volume variation, cost variation, and change in other operating conditions; (6) recency of period, in order to make the results as useful as possible to the management; (7) a number of observations large enough to permit generalization and yet small enough to be conveniently manageable in statistical correlation analyses.

During the period selected, output varied from zero to a level approaching the maximum physical capacity of the mill. Although the range was not uniformly covered, some observations were available to represent most sections of this range. The size of the physical plant was unaltered, and the quantity and type of equipment remained approximately the same. Moreover, the technology was relatively stable as compared with earlier periods. Management methods were essentially unchanged, the same superintendent and most of the foremen being employed at the mill during the entire period. Records were reasonably adequate for analysis of cost and output variations and of other factors affecting cost. The recency of the period enhanced the value of the study to the management, made records more easily available, and rendered more reliable the information that had to be obtained from the memory of the firm's officials rather than from written sources. Finally, the length of the period provided a convenient and yet sufficient number of monthly observations.[40]

The calendar month was chosen as the best unit of observation for the following reasons. In the first place some of the required information was obtainable only in monthly figures. Again, the fluctuations within the month in output and in other operating conditions were apparently not very great. Finally, the month was the unit of time used by management as the basis of its operating plans.

Although major attention was paid to the analysis of the data available from monthly records, subsidiary analyses were also made using weeks and quarters as the units of observation, although these units appear to be less satisfactory than months. Many records, particularly those of certain cost items, were not kept on a weekly basis. Quarterly figures, on the other hand, although yielding somewhat more complete information on operating conditions than did monthly data, concealed month-to-month variations within the quarter. Furthermore, it was not possible to obtain a very large number of quarterly observations within a period suitable for short-run analysis, i.e., a period in which there were only negligible changes in such operating conditions as technical methods, equipment, management, and size of physical plant.

Chapter VII

Classification and Adjustment of Cost Data

This chapter is concerned with two problems which arise in the use of cost-accounting records to determine empirical cost functions. First, the available data must be classified and selected in such a way as to yield the maximum information about the behavior of different categories of cost. Second, it is necessary to adjust the amounts to remove the effects of changes in the prices of the input services in order to isolate the desired cost-output function.

A. Classification of Cost Data

The operating costs of the hosiery mill studied were classed by the company into the three categories: (*a*) productive labor cost, (*b*) nonproductive labor cost, and (*c*) overhead cost, noncontrollable and controllable. In this study major interest lies in the sum of productive labor, nonproductive labor, and overhead cost, designated by "combined" cost, which is the principal dependent variable. These subsidiary accounting categories were used in addition as dependent variables in separate analyses and, as explained before, are referred to as the "components" of combined cost. Some attention was also paid to the behavior of the individual constituents of the cost components, the elements of cost.

The analysis of the behavior of individual elements and components of expense is advantageous for several reasons. In the first place individual accounts may require different corrective devices, corresponding to the varying influences which give rise to the need for rectification. Irrelevant influences may differ in kind for different expenses and may also operate with varying intensity on the various categories of cost. If either of these conditions prevails, the analysis of the cost items separately will result in a more exhaustive statement of relations than will the treatment of cost in composite form only. The same considerations apply to the influence of independent variables as well as to the influences which are removed by rectification. An independent variable may affect only certain components of cost, and its effect on one of these components may be greater or less than on another or may even be different in character. In this case, therefore, the use of cost categories leads to more informative results.

Closely related to the advantages outlined above are certain practical benefits of component analysis arising when a cost study such as this is made the basis of a flexible budget. The use of component analysis means that derived relations, obtained on the basis of certain fundamental conditions existing during the period of investigation, can more easily be modified to take account of changes in these conditions. Such changes are likely to affect only some of the components and to affect even these in different ways. Therefore, a budget can be more detailed and more accurately adjusted to variations in operating conditions, and the estimates more precisely modified to accord with changed input prices if the analysis is so made that separate correction and reflation of the components is possible. Again, if underlying conditions affecting only a few cost elements change, it is possible to readjust the costs affected so that it is not necessary to scrap the entire findings concerning the cost–output relations.

It is desirable to investigate in more detail the content of the three components of cost—productive labor, nonproductive labor, and overhead cost. The accounts included under these three categories are as follows:

A. Productive labor

Legging	Looping
Footing	Seaming
Topping	Examining

B. Nonproductive labor

Supervision, knitting, general, other	*Samples*
Office expense	*Line-elimination labor*
Other indirect labor	*Chauffeurs and helpers*
Building maintenance	*Learners—all operations*
Minimum-wage guaranty	*Machine cleaners*
Machine maintenance and repair	*Replacements*
Style change	*Starting up machines*

C. Overhead
  1. Noncontrollable
     - *General administrative expense*
     - *Depreciation*
     - Taxes
     - Insurance
     - Unemployment insurance
     - Old age pension
     - Shutdown expense
     - Dues and subscriptions
     - Water
  2. Controllable
     - *Coal*
     - *Stationery and printing*
     - *Supplies and expense* (sundry, electrical, sanitary
     - Heat, light, electricity, and power
     - *Freight and express*
     - Telephone and telegraph
     - *Traveling expense*
     - *Suppers and overtime*
     - *Building repairs—material*
     - Postage
     - *Freight and express*
     - Knitting machines, accessories, and needles
     - *Line-elimination parts*
     - *Machine conversion*
     - *Oil*

Those costs which are italicized were omitted from the analysis. The general question of cost omissions has been discussed briefly in chapter v on the "Adjustment of Cost Functions for Practical Use," and a more complete discussion is offered in Appendix B, where the excluded accounts are discussed in detail. All recorded elements of cost, with the exception of those italicized, were included in one of the components of cost.[41] A number of these cost elements could not be used directly in their recorded form, however, since it was necessary to remove extraneous influences, such as price variations, by procedures of rectification. It is now necessary to consider in some detail the problems involved in obtaining from these recorded accounts costs which are strictly comparable to those of the theoretical model outlined in chapter ii.

B. Rectification of Cost Data

The process of cost rectification is intended to eliminate from the data two major influences which tend to obscure the nature of the short-run cost–output function. The principal sources of distortion arise from fluctuations in the prices of various input services[42] and from discrepancies between the time of recording cost and the time of recording output. The procedures devised to correct for these influences will be discussed in the remainder of this chapter.[43]

It is the purpose of this discussion to describe the adjustments which were made to the various elements of cost actually included in the correlation analysis. In general it can be said that adjustments in costs for price changes were accomplished by constructing special index numbers of individual prices to deflate the costs affected and thus remove the influence of price variations. When physical quantities of the input factors were recorded, they were simply multiplied by the price in the base period to determine their cost. In certain cases it is also necessary to make some adjustment for lags and leads in the recording of cost. A detailed discussion of the rectification of the cost data is offered under the three divisions corresponding to the firm's accounting designations—productive labor, nonproductive labor, and overhead cost.

1. PRODUCTIVE LABOR

The only adjustment necessary for the productive labor accounts was the deflation of recorded labor cost in order to eliminate the effects of changes in wage rates, a number of which occurred during the period of analysis. Recorded cost data were corrected for the effects of changes in labor prices by constructing index numbers of the plant's wage-rate payments and deflating the recorded cost to make the rectified figures comparable for the various months included in our study.[44] There were four different wage-rate schedules in effect during the period of analysis. Separate index numbers were constructed for each operation with January, 1939, wage rates as 100, and these indices were employed to rectify the labor costs by operation. The index numbers are shown in Table 2–13.

Table 2–13
HOSIERY MILL: INDEX OF PRODUCTIVE LABOR WAGE RATES

Period	Operation					
	Legging	Footing	Topping	Looping	Seaming	Examining
January, 1935—July, 1937....	105.0	96.9	86.3	121.0	108.9	93.3
August, 1937—February, 1938.	112.4	103.6	97.5	128.2	116.9	98.8
March, 1938—December, 1938.	98.4	88.8	80.3	97.9	87.4	79.7
January, 1939—June, 1939....	100.0	100.0	100.0	100.0	100.0	100.0

2. NONPRODUCTIVE LABOR

The methods of rectification of the nonproductive labor account[45] differed somewhat, as is shown in the detailed discussion which follows.

a) *Supervision: general.* The supervision account was stabilized at the 1939 figure of $450 per month. Since this account represents the salary of one man and since the same person held this position throughout the analysis period, any variations in the account were simply the result of changes in the rate of remuneration and were, therefore, eliminated.

b) *Supervision: knitting.* Knitting supervision consisted of the salaries of foremen in the legging and footing departments. This cost was plotted as a time series with final corrected production, and a scatter diagram was also made to show the relation between knitting supervision and legging production. For knitting supervision, even though a few months at the end of 1937 and the beginning of 1938 were rather high relative to output, there seemed to be no adequate basis for correction, and the account was used as recorded.

c) *Supervision: other.* This cost represented foremen's remuneration in operations subsequent to footing. The account was plotted as a time series with final corrected production, and a scatter diagram was constructed using other supervision and corrected production. In the last half of 1937 the cost of other supervision was found to be high relative to output. It was discovered, however, that the weekly amounts of other supervision, when totaled by months, were lower than the monthly control totals by an approximately constant amount. It appeared that some item, not previously included in this account, had been added to the monthly records in 1937. The weekly figures were, therefore, used instead of the monthly.

d) *Office expense.* The expense of running the office of the factory, when plotted against production, showed a definite, although rather widely dispersed, relation. The most serious part of this dispersion was removed by using the weekly figures rather than the monthly in those few cases where there was a difference between the two.

e) *Other indirect labor and* (*f*) *minimum-wage guaranty.* Other indirect labor—i.e., cleaning, matrons, watchmen, etc.—showed a high correlation with corrected production when the two series were plotted as time series. There was no record of changes in wage rates of persons in this category. Minimum-wage allowance also showed some relation to output, although the account was small and did not vary greatly. Neither of these accounts was adjusted.

g) *Building maintenance* and (*h*) *machine maintenance.* Building maintenance was likewise included in its original form. Machine maintenance, however, was exceedingly large in 1939 in comparison with its relation to production (both corrected and recorded) for the other years. Knitting machine accessories, to which machine maintenance is presumably related, showed a similar departure for 1939. It was discovered in this year that the accessory account was for the first time divided into two parts—needles and other accessories—and that, if only the needle expense was used, the previously existing relation to production continued. This gave rise to the supposition that the accessory and maintenance labor accounts before 1939 had only included needles and the labor of installing needles and that in 1939 some extraordinary repairs were undertaken. Therefore, in the accessory account the expense for items other than needles was omitted and the machine maintenance account was proportionately reduced.

3. OVERHEAD

a) *Real estate taxes,* (*b*) *water,* and (*c*) *dues and subscriptions.* Real estate taxes were obviously unrelated to the intensity of use of the plant, so they were rectified for changes in the tax rate by the use of the 1939 figure. Water and dues and subscriptions were found on preliminary examination to be independent of the level of production. These two accounts were accordingly also stabilized at their 1939 figure which was the most convenient rectification device.

d) *Postage.* This account was used simply in its recorded form.

e) *Telephone and telegraph.* A lag adjustment was made for telephone and telegraph bills to allow for the fact that, although the base rate is the same from month to month, the special charges for any given month pertain to the activities of the preceding month.

f) *Heat, light, electricity, and power.* The heat, light, electricity, and power account was adjusted for seasonal variation for the months of April through November and was apportioned on the basis of production for the remaining four months of the year. The seasonal index was constructed by taking, for each month, the average of the three central ratios of actual value to the values computed from the visually fitted regression of the account on final corrected production.[46] For the months of December through March the observations were so widely dispersed and hence the corresponding ratios so heterogeneous that a reasonably representative average could not be calculated. Therefore, since the values for the other months (when adjusted for seasonal variation) showed a close association with output, the total heat, light, electricity, and power expense for each period December–March inclusive was allocated among the fourth months on the basis of the volume distribution for the corresponding period. This procedure is shown in detail for 1936–37 in Table 2–15. The rectification of the heat, light, electricity, and power account for changes in rates was considered; but investigation showed that only small rate changes had occurred and that their effects on the total bill were negligible, so that apart from the seasonal correction no further adjustment was made in this account.

Table 2–14

HOSIERY MILL: SEASONAL INDEX OF HEAT, LIGHT, ELECTRICITY, AND POWER

Year	April	May	June	July	August	September	October	November
1935	....	....	90.0	79.2	86.3	90.8	96.9	109.4
1936	115.4	100.8	....	73.4	88.5	96.3	89.5	105.0
1937	101.5	107.8	93.1	86.3	93.0	97.1	100.0	111.4
1938	98.7	129.7	89.8	....	....	....	....	....
Average	105.2	112.8	91.0	79.6	89.3	94.7	95.5	108.6

g) *Maintenance materials.* Part of the rectification of knitting machine accessories has already been referred to in connection with the adjustment of machine maintenance labor. It was also necessary to remove the effects of speculative buying from January through June, 1937. This was done by the use of a three months' moving average, centered. An additional correction was needed for a two months' lead over production, which was apparent in 1935–37 from a comparison of the time-series charts of production and of maintenance materials.

Table 2–15
HOSIERY MILL: ALLOCATION OF HEAT, LIGHT ELECTRICITY, AND POWER EXPENSE

Year and Month	Corrected Production	Percentage of Production to 4-Month Total	Heat, etc., Book Value	Heat, etc., Adjusted Value
1936				
December	41,284	26.1	627.52	683.98
1937				
January	40,338	25.6	663.45	670.87
February	36,589	23.2	672.20	607.98
March	39,521	25.1	657.43	657.77
Total	157,732	100.0	2,620.60	2,620.60

h) *Insurance.* The insurance account includes building and equipment insurance, which is constant for each month, and insurance on inventory (primarily in-process), which is indirectly related to output rate. Adjustments to the recorded inventory insurance expenses were required because of variation in the stock-turn rate due to imperfect synchronization of receipts, production, and shipments. Adjustments were made simply by substituting for those few months that showed wide departures from the insurance-output relation the values obtained from the visually fitted regression. The most notable deviations occurred in 1935 and 1939, and in the latter years these were known to be caused by the storage of unusually large quantities of finished goods in the mill.

i) *Old age and unemployment taxes.* Taxes for old age insurance and unemployment compensation were recomputed on the basis of 1939 rates. This rate was 1 per cent for old age taxes and 3 per cent for unemployment; therefore, the corrected figure for the two together was 4 per cent of the corrected total of productive and nonproductive labor.

CHAPTER VIII

Selection and Measurement of Independent Variables

A. *Selection of a Measure of Output*

It has already been emphasized that the main aim of this analysis is to isolate the relation between cost and volume of output. The volume of output is known to be of dominant importance as a cause of variation in cost and is of central interest in any empirical study of the form of the cost–output relation. An attempt to measure output, however, involves an immediate practical difficulty. The theoretical cost functions discussed in chapter ii assumed that output consists of homogeneous units of a single product. Actually, however, despite the relative homogeneity of the output of this mill, its output consists of a group of allied products. A number of styles of hosiery are produced with some variations in specifications. Each of these products can vary independently, so that the proportion of any one product to the total is not necessarily constant. In addition to different styles and specifications of hosiery, seconds and partially finished hosiery are also produced. There are two alternative solutions of the problem of output heterogeneity. First, it is possible to construct an index of output from a weighted combination of the various types of product. Second, each important product type or each significant aspect of output can be introduced as a separate independent variable in the multiple correlation analysis.[47] This second alternative was not followed in the study. The principal independent variable is consequently the volume of production itself measured by a weighted index, the other independent variables representing a number of variable operating conditions.[48] Knowledge of the extent of the effects of these other causal factors is sought primarily in order to eliminate their influence, and in this way to determine the isolated effect of fluctuations in the volume of production upon cost, although quantitative determination of the cost effects of these subsidiary independent variables has both theoretical and practical usefulness.

The first part of this chapter will be devoted to the problems encountered in devising a satisfactory measure of output. In calculating the volume of production, several alternative measures were considered. The first alternative was the number of stockings in dozens of pairs, and the second alternative was a modification of this measure designed to take account of the following operating factors: (*a*) style variation, (*b*) differences in specification, (*c*) seconds, (*d*) backwinds, (*e*) transfers, and (*f*) replacements. Each of these six factors may be regarded as an aspect of the physical volume of production, whose effects it may be necessary to allow for in determining a measure of output. The preliminary question arises of deciding upon the relative importance of the effects of each of these variables. Then, the unimportant variables having been eliminated, the choice remains of introducing the remaining factors as independent variables or of employing them to modify recorded production. It is necessary, first of all, therefore, to explain in detail the nature of these influences.

a) *Style variation.* Although the choice of this particular hosiery mill was made partly on the basis of relative homogeneity of product, its output was not absolutely uniform. A variety of styles was produced during the period of investigation, and, since these styles differed with respect to machine time and relative amounts of various kinds of productive labor required, the monthly output figures may not be precisely comparable. An index of style variation might be constructed by finding the difference between the weighted average of the standard cost of the styles produced each month and the average for the entire observation period. Such an index, which would reflect month-to-month differences in the estimated costs of styles and in the proportion of expensive and inexpensive styles, could have been incorporated into the index of output by weighting each physical unit produced by its standard direct cost. A degree of comparability sufficient to permit the use of unweighted units exists if the cost differences among styles are negligible. Investigation revealed that this condition was satisfied—a conclusion which was reached by an analysis of the variation in standard cost among the styles produced during a sample three-month period.[49] In view of the approximate homogeneity of output and the apparent stability of proportions, it seemed satisfactory to treat all styles as equivalent in a production index made up of a simple count of physical production.[50]

b) *Differences in specifications.* Differences in specifications are minor variations in the characteristics of stockings of the same style. These differences were found to be so slight that this aspect of output heterogeneity could safely be ignored. Changes in the specifications of a particular lot of the same style of product affect cost to a smaller degree than style variations, since specification changes require only minor adjustment in the equipment and since the differences in machine time and wage rates for different specifications are small compared with the changes necessary for different styles.

c) *Seconds.* Imperfect stockings graded as "seconds" were not differentiated from first-quality hosiery in the production records. Yet, from the viewpoint of cost as well as of market value, seconds might well be regarded as a distinct aspect of output, the inclusion of which on a basis of equality with regular output impairs the comparability of different units of output.[51]

Weighting might be accomplished by taking into account the fact that seconds cost slightly more and sell for considerably less than first-grade hosiery. To weight seconds on the basis of cost would, however, result in an unjustified inflation of the output index. On the other hand, to weight them on the basis of market price would lead logically to the employment of relative market values of all styles as weights in the output index. This set of weights, however, is not a desirable one. A third alternative is to regard the loss in market value arising from the production of seconds as spoiled-work cost in the production of first-grade hosiery. Two variants of this procedure are (1) to include only first-grade hosiery in the index of output and deduct from the cost of producing it the reclamation revenue from the sale of seconds and (2) to include both firsts and seconds in the output index and add to the cost of production the revenue loss due to the difference between the price of first-quality hosiery and second-quality hosiery. However, it appeared likely that the error introduced by the inclusion of seconds in the production index was relatively slight. For all except the last four months of the period of analysis, this mill was producing under contract to a customer with low-quality standards, so that the proportion of seconds detected at the knitting mill was unusually small. Furthermore, the finishing processes were being performed by the customer rather than in the company's own finishing plant, and the seconds that were not detected before they left the knitting mill were not charged to the mill. In other words, the number of stockings classified as seconds due to damages during the knitting process was small because of low-quality standards and was further reduced because undetected damaged articles were sold in unfinished form as firsts.[52] In view, therefore, of the relative unimportance of seconds it was decided not to correct recorded production for this influence.[53]

d) Backwinds. "Backwinds" are stockings hopelessly spoiled at some stage in the knitting process. The process of backwinding is the unraveling of defective hosiery in order to recover the silk. Backwinds affect the cost behavior of this mill[54] because costs are incurred in the production of the stocking, yet the product is not reflected in the recorded production. Spoiled work of this type, which is an unavoidable concomitant of hosiery production, is properly to be regarded as a cost of unspoiled output completed rather than another dimension of output.[55] Special provision for this influence was consequently not attempted in the construction of an output index.

e) Transfers. Transfers of in-process inventories from plant to plant for further processing presented a serious problem. This hosiery mill performed operations subsequent to legging upon stockings which had been transferred to it from other mills which legged the stockings and, therefore, received credit for production. Consequently, this particular mill incurred the costs of these subsequent operations without having the output reflected in its production records. As a result the production of the plant was understated by the extent of the work done on products transferred for operations subsequent to legging.

A survey of the records indicates that transfers were made in large amounts rather continuously over the period of study. In consequence, although the proportion of the total cost incurred in the subsequent operations was not great, the aggregate cost involved was large, and the problem appeared to be important enough to require a solution. Omission from the analysis of the periods during which transfers were made was not feasible because transfers occurred in every month except one. Two alternative methods of correction were available. Either the reported production could be augmented to take account of the unrecorded production, or the cost of the additional operations could be deducted from the reported cost figures. The latter method would have involved difficulties because the effect of this additional production upon the various elements of cost, particularly the elements of overhead costs, would be extremely difficult to determine and, even if ascertainable, would have involved much labor. The procedure followed was to increase recorded production by the number of transferred stockings expressed in units of legged or completed product. Detailed description of the method of measuring transfer production is to be found in Appendix C on the "Construction of an Output Index."

f) Replacements. Replacements are partially processed hosiery held in reserve for substitution for units of regular production lots found defective at any stage. Such a reserve makes it possible to route work through all operations in round lots of ten-dozen pairs each, without interruption. As this reserve replacement stock was used up and as new styles were introduced, it was necessary to produce in-process stock specifically for this reserve. This practice affected cost behavior in two important ways. First, all operations on a rejected stocking prior to the operation which spoiled it had to be paid for even though no recorded output resulted.[56] Second, misallocation of reported production and costs among accounting periods resulted because it was difficult to assign nonproductive "replacements" labor cost to the output to which it contributed, since replacements were not necessarily produced at the same rate that they were being substituted for defective goods in the regular production process.[57]

Neither of these effects upon recorded production appeared to be serious in this particular mill for several reasons. First, the quality standards were less severe than in other mills so that the proportion of replaced hosiery was low. Second, the time disparity between the production and use of replacement reserves was at a minimum in this mill because changes in style were relatively infrequent and the replacement stock was replenished currently rather than in large lots at the time its exhaustion was imminent. It was decided, consequently, to omit replacement stockings from the output index, in view of the inaccuracy of allocation to production periods which their inclusion would entail.[58]

Of the original six aspects of output considered, apart from regularly recorded production, the first four—style variation, differences in specification, seconds, and backwinds—were found to be of negligible importance. Of the remaining two —transfers and replacements—it was decided to modify recorded production to take account of transfers and to use replacements in a multiple correlation analysis to test its importance.

As a result of these considerations the final index of production used as the output variable in the correlation analysis was a weighted composite of regular recorded (legged) production and transfer (partial) production, adjusted for a recording lead arising from the practice of recording production before the completion of all operations. The weights were determined on the basis of relative direct labor cost.

The correction of regular production for a recording lead was based primarily on the company's estimate of a seven-day production cycle and on the assumption that the total cost incurred during a given month attributable to the previous month's recorded production is proportional to the corresponding productive labor cost of the production which overlapped. An estimate of the proportion of a week's production done on each day had to be made, and this involved the further assumption that the productive labor cost per operation for each period was evenly distributed over the period and that the steady flow of production could be for practical purposes regarded as proceeding by discrete daily intervals.

The first step in the rectification procedure, therefore, was to make a correction for an accounting error that resulted in disagreement between monthly and weekly output totals for some periods. This correction was necessary both because a weekly analysis was contemplated and because weekly figures were used in removing the lead from monthly data. Recorded monthly production and monthly production derived from summing the appropriate weeks were both plotted as time series with total cost, the former being found much more highly correlated to cost. Adjustments were, therefore, made in the weekly data for the few months in which disagreement occurred. The actual procedure involved in the correction of production, which necessitates other tedious calculations, is explained in more detail in Appendix C on the "Construction of an Output Index."

B. Other Independent Variables

The second section of this chapter is concerned with the selection of those operating conditions apart from volume of production which were considered to have an effect on cost behavior and with the selection of series adequate to represent these conditions. It is convenient to begin by listing the apparently relevant variables and then to consider their interrelations and the general methods of eliminating or measuring them. The preliminary list of operating variables suspected to be influential includes the following:

a) Style variation
b) Differences in specifications
c) Seconds
d) Backwinds
e) Transfers
f) Replacements
g) Changes in output from previous months
h) Style changes
i) Specification changes
j) Differences in piece rates for a given specification
k) Technical progress
l) Shifts and hours of shifts
m) Labor turnover

The first six of these aspects of production—*(a)* to *(f)*—have already been discussed in connection with the measurement of output. Style variation was made the subject of a special subsidiary investigation, and it was found that differences in cost among styles were relatively slight and sufficient to permit the omission of this variable. The negligible influence of differences in specification, seconds, and backwinds has already been discussed and need not be treated further. Transfers were included in the measure of output and will not be considered further as a possible independent variable.

With reference to replacements, however, on the chance that its influence on cost behavior might be more important than its apparent direct effects, replacement labor cost was introduced as an independent variable in the correlation analysis,[59] notwithstanding the possibilities of its improper time allocation.

It remains, consequently, to investigate the remaining operative conditions which may be of sufficient importance to use as independent variables in the correlation analysis.

g) *Change in output from previous period.* The magnitude and direction of change in output from that of previous months may have an important bearing upon cost because of the existence of cost rigidities. It may be that it is not easy to adjust all costs when there is a change from a higher to a lower level of output because of contractual obligations of various kinds. This means that there may be minor discontinuities in the cost function. The importance of this problem, which was indicated by the wide range of change in output, justified the retention of changes in output from the previous month as an independent variable in the correlation analysis, where the size of the change was measured in percentage terms, both magnitude and direction being taken into account.

h) Style changes. There are two important aspects of style changes. One is style development, which did not occur in this hosiery mill and can therefore be ignored. The other aspect of style changes is the adaptation of labor and equipment to the production of different styles.[60] The effects upon cost of such shifts to different styles are clear. First, machine adjustments are required for every style change. Second, certain additional wage costs are incurred because, during the learning period, employees are paid their previous average earnings, which are usually higher than the compensation justified by their actual production of the new style. Finally, a certain volume loss is sustained, because of the slowing-down of operations. The simplest measure of the extent of this product modification is the account which represents the pay-roll costs attributable to style changes. This account was accordingly omitted in the measurement of cost and used as an independent variable in the correlation analysis. Alternative measures could have been either a simple enumeration of style changes, or the number of style changes weighted by volume or by the number of machines involved, preferably the latter. The labor required to obtain either of these alternative indexes, however, would have been considerable and was probably not justified by the importance of the effect of style changes upon cost.

i) Specification changes. A similar problem arose in connection with specification changes.[61] These were minor adaptations within the framework of a given style, and their effects upon cost were smaller in degree than those of style change, since they involved only small adjustments in equipment and in production technique. Hence both the adaptation-period loss and the volume reduction were not significant as compared with those incurred as a result of a style change. This factor appeared to be of negligible importance and was consequently omitted from further consideration.

j) Difference in piece rates for different specifications. The piece rate requires adjustment if it becomes necessary to produce stockings of a given specification on machines which are not entirely adapted to that specification. This complication rarely arose in the mill studied, however, because the variety of specifications was small and the equipment highly homogeneous.

k) Technical progress. Technical progress during the period of observation raises the problem of noncomparability of accounting periods. Improvements in equipment or in technical methods which reduce cost distort the static relation between cost and output which is to be isolated. The assumption is that the observations exhibit short-run variations in cost uninfluenced by long-run adjustments in the size of plant or by technical changes in the character of the production process.

Nevertheless, several technical improvements have taken place in the hosiery knitting equipment of the mill in the last few years.[62] First, three-thread carrier attachments were installed in a number of the knitting machines.[63] The use of three-thread carriers improves the quality of the stocking by making it possible to produce a ringless stocking;[64] but, aside from its trivial influence on depreciation and maintenance cost, it does not affect operating costs significantly. Second, half-speed shock absorbers were attached to thirty-five legging machines toward the end of 1937. This innovation was considered to have a negligible influence on cost and was therefore disregarded. Finally, during the period studied the number of sections in some of the knitting machines was increased.

In view of the absence of the type of equipment changes that affect cost behavior and the negligible effect on cost of the changes that were made, the necessity of making allowance for technical changes did not arise. It was possible, therefore, to neglect the influence of this aspect of operations.[65]

l) *Shifts and hours per shift.* Differences among accounting periods in number of shifts, in hours per shift, and in hours per week may affect costs over a certain range of volume variations. If a production level of 15,000 dozen pairs of hosiery were achieved by operating all equipment on a single shift, costs might be different from those incurred by operating half the equipment on a double shift. Multiple-shift operation might be expected to result in different costs because of higher wage rates and lower labor efficiency on night shifts and because of differences in the degree of utilization of equipment.[66] An analysis of the variation in hours and shift which occur within each month and a tabulation of the number of machines on single and on double shift for each month of the analysis showed that the month-to-month variation was small enough to be neglected. In consequence, shifts and hours per shift were not considered sufficiently important to warrant further consideration.

m) *Labor turnover.* Labor turnover has a direct effect upon cost because of the necessity for guaranteed minimum wages during the learning period despite the fact that the learner's production does not justify the wage on a piece-rate basis. This direct cost is segregated by a special overhead account, "Minimum-Wage Guaranty"; but this account reflects only the additional wage cost of learners and does not reflect losses due to spoiled work and to lower output than could have been obtained from the equipment by experienced workers. Examination of this situation indicated that this indirect cost was unimportant. The great reservoir of experienced hosiery workers in this area that can be drawn from in restaffing a mill made the learner problem unimportant, particularly during the observation period. It is the opinion of the executives that labor turnover is not great and that it does not vary significantly from month to month. It appeared clear that its probable importance did not justify a detailed analysis.

As a result of this discussion, the choice of influences likely to prove to be significant independent variables has been restricted considerably. The only variable operating conditions which were considered important enough to retain for mathematical analysis were: (*a*) replacements, (*b*) changes in output from previous months, and (*c*) style changes. The influence of these variables which were investigated by the methods of correlation analysis has been explained in chapter iii, which deals with the statistical findings.

Appendix A

Statistical Findings Based on Quarterly and Weekly Observations

A. *Quarterly Findings*

The analysis and findings of the quarterly data paralleled those of the monthly data closely. The simple regression equations of cost and output as well as multiple regression equations which made allowance for the effect of time were determined for combined cost and its components.

1. SIMPLE REGRESSIONS

The simple regression equations and certain of the statistical constants are shown in Table 2–16. A supplementary presentation of these findings is given in Charts 2–15, 2–16, and 2–17. Chart 2–15 illustrates the simple regressions of total combined cost and productive labor cost on output. Chart 2–16 shows the breakdown of nonproductive labor cost into its various elements, as well as the simple regression of nonproductive labor cost on output. Similarly, Chart 2–17 shows the simple regression of overhead cost and its constituent elements on output. These graphic findings are so similar to those in which monthly observations only were taken that an elaborate explanation of their significance is not necessary.

2. PARTIAL REGRESSIONS

As in the case of monthly costs, a significant relation between cost and time was determined by graphic correlation analysis after the influence of rate of output had

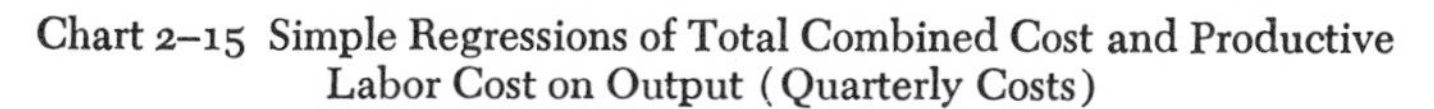

Chart 2–15 Simple Regressions of Total Combined Cost and Productive Labor Cost on Output (Quarterly Costs)

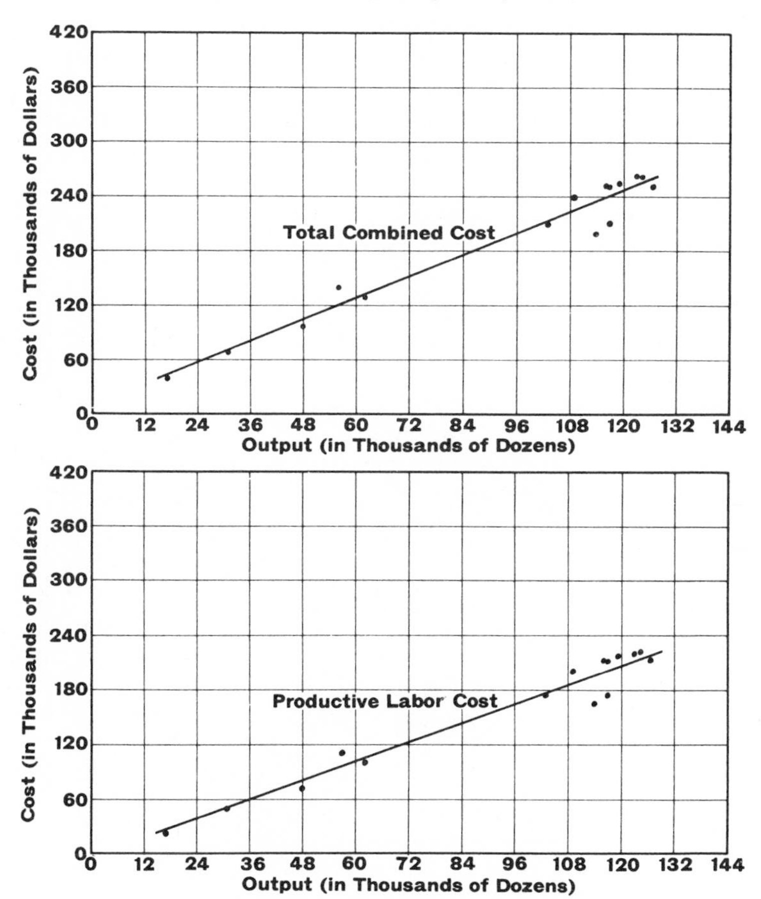

been allowed for. The cost-time function was curvilinear in the cases of combined cost, productive labor cost, and overhead cost and linear for nonproductive labor cost. Curvilinear multiple regression equations were, therefore, fitted in the three appropriate cases, and a linear multiple regression equation in the remaining case. The results of the multiple correlation analysis are presented in summary form in Table 2–17.

This same information is shown in graphic form in Charts 2–18 through 2–20. Chart 2–18 portrays the partial regression of total combined cost on output in the upper section and the partial regression of total combined cost on time in the lower section. Charts 2–19 and 2–20 show the analogous graphic illustration for productive labor and nonproductive labor. The chart showing the partial regres-

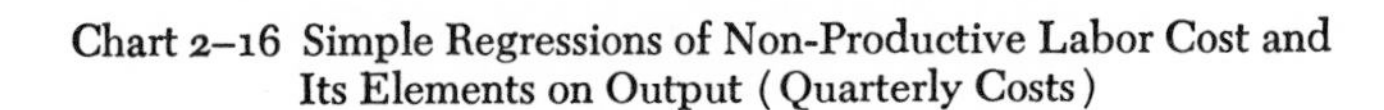

Chart 2–16 Simple Regressions of Non-Productive Labor Cost and Its Elements on Output (Quarterly Costs)

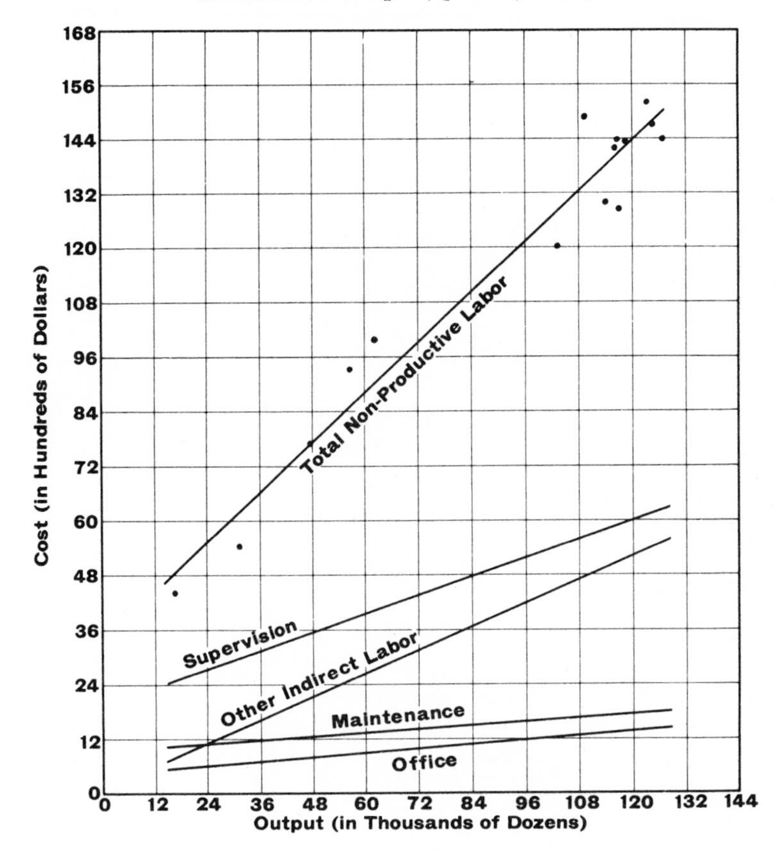

sions of quarterly overhead cost was so similar to Chart 2–19 that its inclusion was considered unnecessary.

3. COMPARISON OF SIMPLE AND PARTIAL REGRESSIONS

It is evident from a comparison of the simple and partial regression lines derived from quarterly observations that the linearity of the combined cost function is unaffected by the introduction of the time variable so that no essential difference in the form of the cost function results. The same considerations which were mentioned in the case of the comparison of monthly simple and partial regressions apply as well to the quarterly comparison.

Chart 2–17 Simple Regressions of Overhead Cost and Its Elements on Output (Quarterly Costs)

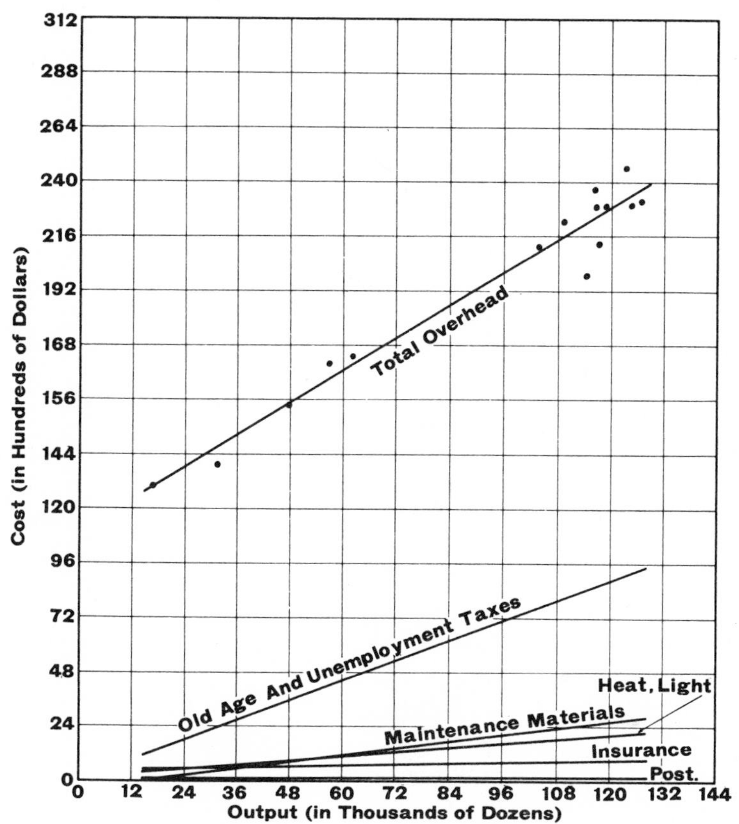

4. MARGINAL AND AVERAGE COST

Just as in the case of monthly observations, marginal and average cost functions were derived from the partial regression equations for combined cost and its components. The resulting equations are found in Table 2–17.

These findings are also shown graphically in Chart 2–21. This chart illustrates the results by portraying average and marginal combined cost derived from partial regression of quarterly combined cost on output. The upper section shows the linear total cost function together with the adjusted observations on which it is based. Since its slope is constant, the marginal cost function, which is shown in the lower section, is horizontal. Thus the analysis of the quarterly totals of operating cost (exclusive of raw material) indicates that the cost of producing an additional dozen pairs of hosiery is approximately $2.00 over the whole range of output

Chart 2–18 Partial Regression of Total Combined Cost on Output and Time (Quarterly Costs)

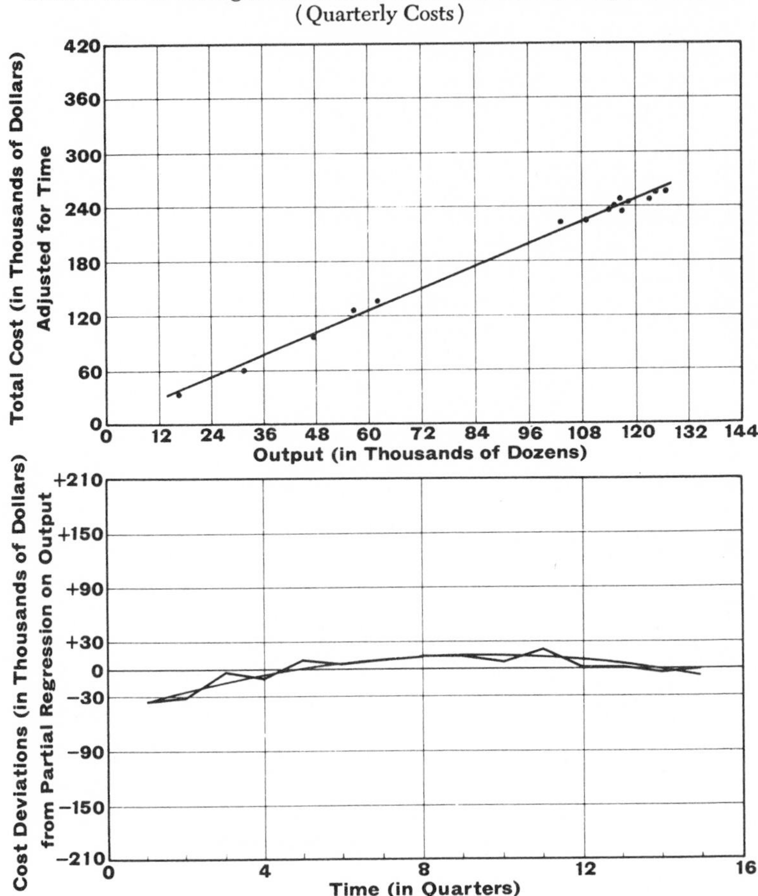

observed. The relation between cost per dozen and rate of output is shown by the declining curve in the lower section of the chart, which lies slightly above the marginal cost curve.

B. Comparison of Monthly and Quarterly Findings

A comparison of the monthly and quarterly findings for combined cost and its components at several rates of output is found in Table 2–18. To facilitate comparison with monthly totals, the quarterly figures have been divided by three. Examination of this table shows close correspondence over the observed range of output. This is to be expected since the quarterly data are merely sums of monthly data.[67]

The quarterly data, by averaging monthly variation within the quarter, tend to reduce the range of observed variation in cost and its determinants.[68] This process substitutes arithmetic averaging of variation within the observation unit for quad-

Chart 2–19 Partial Regressions of Productive Labor Cost on Output and Time (Quarterly Costs)

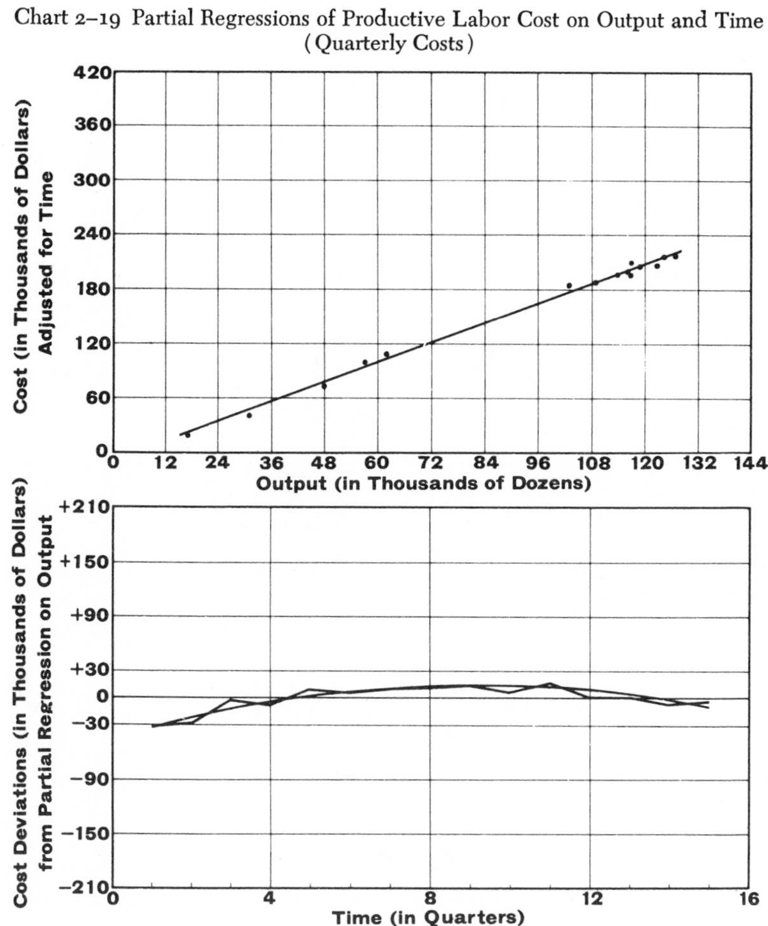

ratic averaging among observation units (accomplished by least-squares fitting of monthly observations), and in this way the extent of random variation tends to be concealed.

Errors of allocation among time periods may be reduced by using the quarter as the observation unit, since intermonthly, noncompensating errors will be offset. The close similarity of monthly and quarterly findings indicates that uncorrected errors of allocation are either unimportant or unrelated to rate of output.

C. Weekly Findings

A supplementary analysis of the behavior of weekly costs was also carried out. Unfortunately, the data available from weekly records were incomplete and could not be accurately allocated to proper periods. A major difficulty arose in attempting

Chart 2–20 Partial Regressions of Nonproductive Labor Cost on Output and Time (Quarterly Costs)

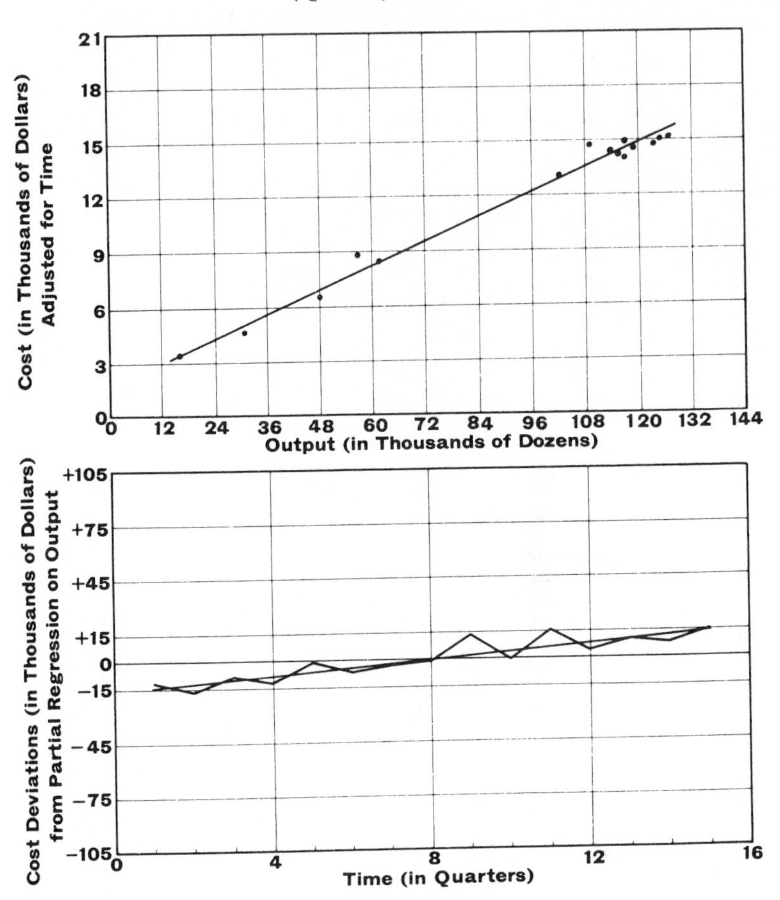

to secure an accurate measure of output. The weekly records of partially finished hosiery shipped in for further processing were incomplete, so that it was necessary to use "recorded" production as a measure of output without taking into consideration the transfer shipments. As a result it was not possible to make an analysis which would parallel the monthly and quarterly analyses described previously. Furthermore, certain cost items, recorded only as monthly totals, had to be arbitrarily allocated among weeks. The weekly analysis was confined to the year 1937. This gave a sufficiently large but not unwieldy number of observations and, furthermore, insured that possible changes in plant and equipment or technical methods would have a smaller influence than if observations scattered over four years had been used.

A curvilinear simple regression equation was fitted to these weekly data, since

Table 2–16

HOSIERY MILL: SIMPLE REGRESSIONS AND DERIVED FUNCTIONS OF COMBINED COST AND ITS COMPONENTS ON OUTPUT*

(Quarterly Observations)

	Combined Cost	Productive Labor Cost	Nonproductive Labor Cost	Overhead Cost
Simple regression equation	${}_cX_1 = 9{,}958.45 + 1.972X_2$	${}_pX_1 = -4{,}282.09 + 1.760X_2$	${}_nX_1 = 3{,}292.35 + 0.092X_2$	${}_oX_1 = 10{,}948.52 + 0.120X_2$
Standard error of estimate	15,059.74	13,734.33	743.082	923.58
Correlation coefficient, r	0.979	0.978	0.977	0.979
Regression coefficient, b	1.972 ± 0.054	1.760 ± 0.056	0.092 ± 0.057	0.120 ± 0.055
Average cost equation	${}_cX_1/X_2 = 1.972 + 9{,}958.45/X_2$	${}_pX_1/X_2 = 1.760 - 4{,}282.09/X_2$	${}_nX_1/X_2 = 0.092 + 3{,}292.35/X_2$	${}_oX_1/X_2 = 0.120 + 10{,}948.52/X_2$
Marginal cost equation	$d_cX_1/dX_2 = 1.972$	$d_pX_1/dX_2 = 1.760$	$d_nX_1/dX_2 = 0.092$	$d_oX_1/dX_2 = 0.120$

*The symbols have the following meaning:

${}_cX_1$ = combined cost in dollars

${}_pX_1$ = productive labor cost in dollars

${}_nX_1$ = nonproductive labor cost in dollars

${}_oX_1$ = overhead cost in dollars

X_2 = output in dozens of pairs

Table 2–17

HOSIERY MILL: MULTIPLE AND PARTIAL REGRESSIONS AND DERIVED FUNCTIONS OF COMBINED COST AND ITS COMPONENTS ON OUTPUT AND TIME*

(Quarterly Observations)

	Combined Cost	Productive Labor Cost	Nonproductive Labor Cost	Overhead Cost
Multiple regression equation	${}_cX_1 = -43{,}164.52 + 2.042X_2 + 12{,}979.091X_3 - 690.795X_3^2$	${}_pX_1 = -50{,}349.58 + 1.801X_2 + 12{,}089.699X_3 - 658.498X_3^2$	${}_nX_1 = -65.83 + .0111X_2 + 207.568X_3$	${}_oX_1 = 7{,}229.32 + 0.132X_2 + 591.703X_3 - 25.779X_3^2$
Standard error of estimate, S†	6,681.18	6,020.31	503.65	603.79
Coefficient of multiple correlation, R†	0.996	0.996	0.989	0.991
Coefficient of multiple determination, R^2†	0.992	0.992	0.979	0.982
Partial regression equation for output	${}_cX_1 = 3{,}562.46 + 2.042X_2$	${}_pX_1 = -8{,}067.84 + 1.801X_2$	${}_nX_1 = -1{,}594.72 + 0.111X_2$	${}_oX_1 = 9{,}831.85 + 0.132X_2$
Partial regression coefficient for output	2.042 ± 0.082	1.801 ± 0.074	0.111 ± 0.005	0.132 ± 0.007
Partial regression equation for time	${}_cX_1 = -46{,}726.98 + 12{,}979.091X_3 - 690.795X_3^2$	${}_pX_1 = -42{,}281.74 + 12{,}089.699X_3 - 658.498X_3^2$	${}_nX_1 = -1{,}660.54 + 207.568X_3$	${}_oX_1 = -2{,}602.53 + 591.703X_3 - 25.779X_3^2$
Average cost equation	${}_cX_1/X_2 = 2.042 + 3{,}562.46/X_2$	${}_pX_1/X_2 = 1.801 - 8{,}067.84/X_2$	${}_nX_1/X_2 = 0.111 + 1{,}594.72/X_2$	${}_oX_1/X_2 = 0.132 + 9{,}831.85/X_2$
Marginal cost equation	$d_cX_1/dX_2 = 2.042$	$d_pX_1/dX_2 = 1.8013$	$d_nX_1/dX_2 = 0.111$	$d_oX_1/dX_2 = 0.132$

*The symbols have the following meaning:
${}_cX_1$ = combined cost in dollars
${}_pX_1$ = productive labor cost in dollars
${}_nX_1$ = nonproductive labor cost in dollars
${}_oX_1$ = overhead cost in dollars
X_2 = output in dozens of pairs
X_3 = time

†Adjusted for number of observations.

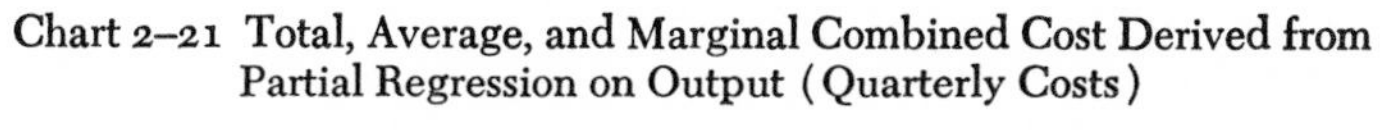

Chart 2–21 Total, Average, and Marginal Combined Cost Derived from Partial Regression on Output (Quarterly Costs)

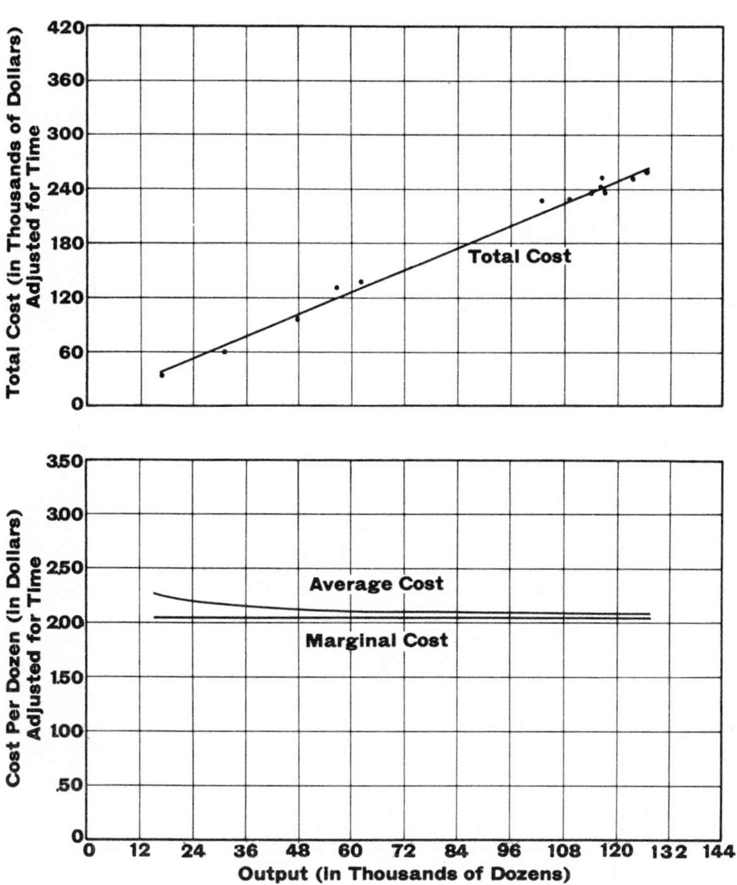

it did not appear necessary to take into account the influence of a possible time trend. The results of this correlation analysis are presented in Table 19. The weekly information available was more complete for productive labor cost, so that this is the only component of combined cost for which the regression equation is given.

The results summarized in Table 2–19 are presented graphically in Charts 2–22 and 2–23. In the upper section of Chart 2–22 are shown two alternative simple regressions of total combined cost on output. Although the observations may justify a parabolic regression, in view of their wide scatter, the incompleteness of the weekly data, and the clear linearity of monthly and quarterly regressions, the linear regression appears more reasonable. Marginal cost functions derived from both the parabolic and the linear regressions are shown in the lower section of Chart 2–22. In Chart 2–23, there is a graphic portrayal of total, average, and marginal produc-

Table 2–18

HOSIERY MILL: COMPARISON OF MONTHLY AND QUARTERLY FINDINGS FOR COMBINED COST AND ITS COMPONENTS AT DIFFERENT RATES OF MONTHLY OUTPUT*

	10,000 Dozen		20,000 Dozen		30,000 Dozen	
	Monthly	Quarterly ÷ 3	Monthly	Quarterly ÷ 3	Monthly	Quarterly ÷ 3
Combined cost	21,439.76	21,602.49	42,116.98	42,017.49	62,794.20	62,402.49
Productive labor cost	15,221.94	15,323.72	33,436.91	33,336.72	51,651.88	51,349.66
Nonproductive labor cost	1,517.77	1,641.57	2,700.83	2,751.57	3,883.89	3,861.57
Overhead cost	4,661.10	4,595.28	5,958.73	5,913.28	7,256.36	7,231.28

*Estimated from the partial regression equations of Tables 2–3 and 2–17.

Table 2–19

HOSIERY MILL: REGRESSIONS OF COMBINED COST AND PRODUCTIVE LABOR COST ON RECORDED WEEKLY OUTPUT*

(Weekly Observations)

Regression equation	${}_cX_1 = -10,484.69 + 6.750X_2 - 0.000328X_2^2$	${}_pX_1 = -14,059.40 + 7.102X_2 - .000367X_2^2$
Standard error of estimate S†	634.80	618.18
Coefficient of multiple correlation R†	0.946	0.942
Coefficient of multiple determination R^2†	0.895	0.888
Regression coefficient of output	6.750 ± 1.726	7.102 ± 1.680
Regression coefficient of output squared	0.000328 ± 0.000130	0.000367 ± 0.000127

*The symbols have the following meaning:
${}_cX_1$ = combined cost in dollars
${}_pX_1$ = productive labor cost in dollars
X_2 = recorded output in dozens of pairs
†Adjusted for number of observations.

Chart 2–22 Total and Marginal Combined Cost Derived from Linear and Parabolic Simple Regressions on Recorded Output (Weekly Costs)

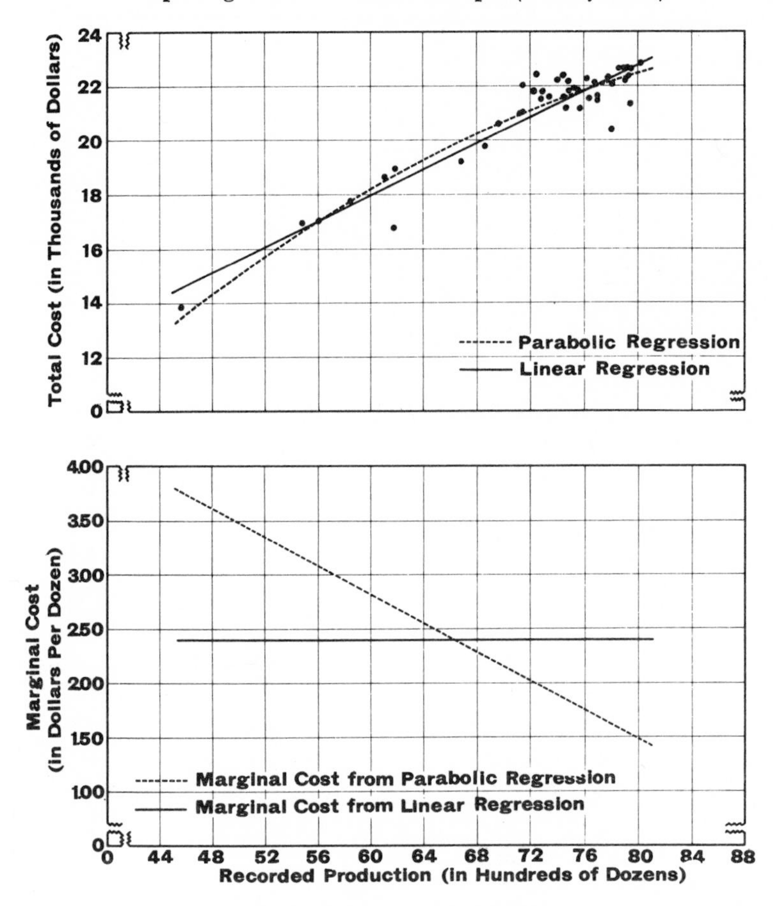

tive labor cost. The upper section shows three alternative simple regressions of cost in the form of weekly totals. The parabola appears to fit the observations but is subject to the limitation just mentioned for combined cost. The linear regression has a small negative intercept, which is unrealistic. Therefore, an alternative linear function was fitted subject to the logical condition that cost be zero at zero output. This third alternative sacrifices accuracy of the measurement of the slope (marginal cost) for a more realistic intercept. Since extrapolation to the cost axis is unjustified, this pseudo-realism may not be worth attaining. The marginal functions derived from these three regressions are shown in the lower section of Chart 2–23. The marginal cost corresponding to the unrestricted linear regression is probably more reliable than the others, but even it is not very satisfactory.

Chart 2–23 Total and Marginal Productive Labor Cost Derived from Linear and Parabolic Simple Regressions on Recorded Output (Weekly Costs)

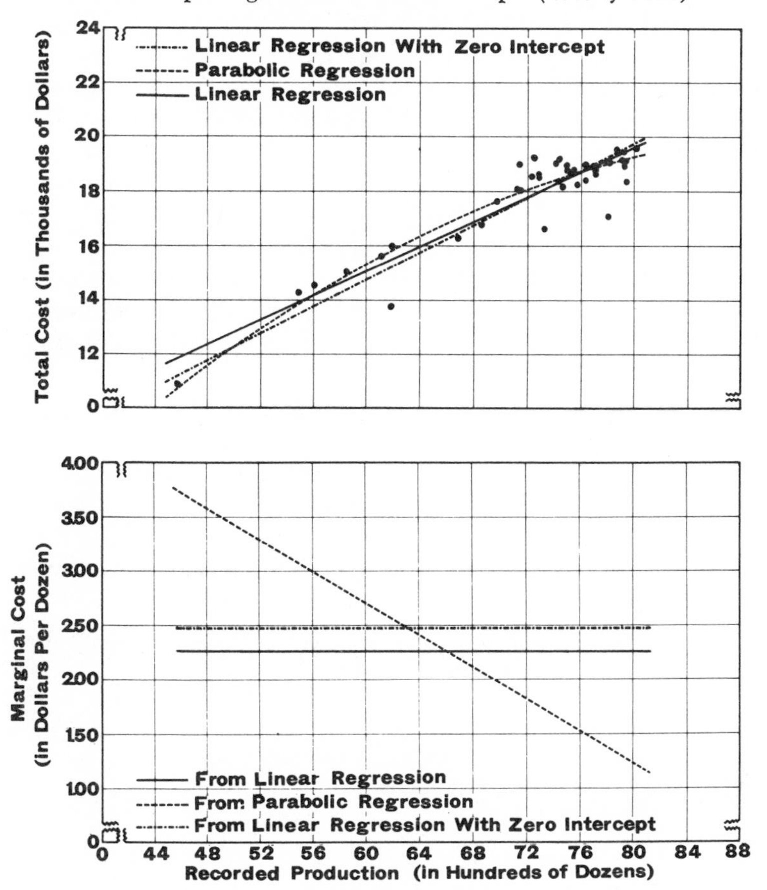

Appendix B

Omitted Costs

This Appendix is concerned with a description of the accounts which were excluded from the correlation analysis, together with the principal reasons for their omission.[69] The excluded cost accounts may be classified as follows on the basis of the principles governing their rejection.

1. ACCOUNTS THAT ARE NOT COSTS OF PRODUCTION IN A STRICT SENSE

Certain costs recorded for the hosiery mill, e.g., transportation to the firm's other plants, are not true production costs of the plant.

a) *Freight and express* was omitted both because it was arbitrarily allocated and because it was not truly a part of the cost of producing hosiery, production for our purposes ending with the appearance of the finished product.

b) *Building repairs—material* entries were not concentrated intentionally in periods of slackened production activity. Furthermore, the time at which the need for repairs became apparent did not coincide with the production of that output that may have been directly associated with the deterioration of the building. In addition to the fact that the cost of building materials is not properly a production cost, the insolubility of the allocation problem led to the rejection of this account.

c) *Chauffeurs and helpers* was the allocation to this plant of general trucking expense for trucks garaged at one of the other mills of the firm but serving all mills. It represented a transportation cost, as opposed to an actual production expense. The account was discontinuous, its first entry, which was small, appearing in June, 1939.

d) *Learners—all operations* recorded payments to learners who could not earn a certain minimum wage on the basis of their piece work and was a cost incurred only when a labor shortage necessitated the hiring of inexperienced workers. Entries in this account were intermittent and the expenditures apparently related to changes in the labor market but not to the level of output in the hosiery mill.

e) *Machine cleaners* was also a small and discontinuous account which appeared in the records only for a few months during which specialized machine cleaners were employed. During the rest of the period the cleaning of the machines was done by the knitters.

f) *Machine conversion* recorded the costs of converting eighteen-section knitting machines to twenty-four-section machines. This conversion did not take place until the last two months of our analysis period, and the altered equipment was not used until after the close of the period. Therefore, the machine-conversion account was omitted.

2. ACCOUNTS WHICH MEASURE INDEPENDENT VARIABLES IN THE COST FUNCTION

It was pointed out in the previous discussion of statistical methods that there are several independent variables apart from output which may exert an influence on the dependent variable (combined cost or its components). Since the aim of the analysis is to isolate the simple cost–output relation, some measure of the magnitude of the irrelevant independent variables is desired. In some cases it may be that a quantitative measure of an independent variable is afforded by a cost account. It is, therefore, convenient to exclude the account from cost, the dependent variable, and make use of it instead as an independent variable.[70]

a) *Replacement* was a direct labor expense incurred in making stockings that were placed in the reserve stock used for replacing spoiled production.[71] It was thus a waste cost, which is properly a part of the finished stockings. It would seem probable that such an expense would bear some definite relation to volume; but, if this relation existed, it was obscured by improper allocation of the replacement account. The proper lag could not be determined, possibly because it was of irregular length. At any rate, the omission of this item had little effect on the results, since it constituted only about 0.001 per cent of total corrected cost.

b) *Style change* was also used as an independent variable. This account, which represented the direct labor cost of style change, was the payment of earnings lost during the period of adaptation to a different style (regardless of whether or not the style had been produced in this mill previously). Since style change was not measured in the output index,[72] the omission of style-change labor served to remove partially the direct influence upon cost of a factor irrelevant to production volume. To remove its indirect effects this account was used as an independent variable. Since the account was small in size and highly variable, its omission, even if improper, did not greatly affect the findings concerning the cost–output relation.

3. ACCOUNTS WHOSE TRUE VALUE IS INDETERMINATE

Unfortunately, there were also accounts that had to be rejected, even though they were related to production, because they could not be adjusted to remove the variation arising from irrelevant influences or from accounting practices. In this category were included (a) expenses whose true magnitude was indeterminate, the book figures for which represented arbitrary allocations, (b) expenses that could not be allocated to the proper time period, and (c) expenses in which there was important variation, arising from extraneous influences, that could not be removed. The above classifications are, however, not mutually exclusive, and some cost elements may be characterized by two or more of these conditions.

a) *Depreciation,* although charged as a linear function of time, probably included use depreciation as well. To the extent that the use depreciation exceeded the time depreciation and was not fully reflected in maintenance, the inclination of the total cost function and the magnitude of marginal cost will be understated by this treatment. Since there was no way of determining use depreciation and since repairs probably made it minimal because moving parts had to be closely maintained to insure quality, inclusion of arbitrarily estimated use depreciation seemed inadvisable.

b) *General administrative expense* included the allocation to this mill of the firm's administrative expense. The true magnitude of such a cost was indeterminate, the book figures representing arbitrary accounting allocations. It was, therefore, omitted from the analysis.

c) *Starting up machines* was omitted from the analysis because it represented capital investment rather than an operating cost. This cost arises because when the mill resumes operations after a shutdown period, maintenance which amounts almost to rebuilding the machines is necessary.

4. ACCOUNTS WHICH ARE SO NEGLIGIBLY SMALL THAT THE LABOR OF RECTIFICATION IS UNJUSTIFIED

In general, very small accounts were omitted if doubt existed as to their content, the sources of their variation, the method of recording them, or if their correction would involve the construction of intricate rectification devices.

A discussion of the circumstances that prompted the omission of each account follows. First, those small elements of cost which were omitted on the basis of their negligible importance or for other reasons will be catalogued and their content discussed.[73] Prolonged consideration of such small accounts and the construction of refined corrective devices in order to rectify them did not seem worth while.

a) *Coal costs* were recorded on a purchase basis which approximated the timing of consumption. This expense was related solely to heating and was, therefore, a function of temperature rather than of hosiery output.

b) *Stationery and printing* was made up primarily of work tags and other stationery, the use of which was directly related to production. However, entries in this account were made on a purchase basis rather than on a use basis, while at the same time the inventory fluctuated violently. In order to smooth out some of the irrelevant variation and to make a necessary lag adjustment, a three months' moving average, centered on the last month, was computed. This series was then plotted against the production index, but the relation found was exceedingly vague and ill defined.

c) There were three *supplies and expenses* accounts: electrical, sanitary, and sundries. Together they constituted only about 0.7 per cent of total corrected cost. The electrical account included mostly materials allocated to the maintenance of the building. The largest single entry in the sundries expense account was the cost of pay-roll delivery. This account, which also included miscellaneous items too small and too uncertain in incidence to be readily classified, showed little relation to production volume.

d) *Traveling expense and* (*e*) *Suppers and overtime* were likewise too small and intermittent to merit rectification for inclusion in the analysis.

f) *Line-elimination parts* was an even smaller account than the corresponding labor expense. It was recorded on a purchase basis and thus presented an additional rectification difficulty, along with those mentioned for line-elimination labor.

g) *Oil* included purchases of oil for use in maintenance of machinery. It was not properly allocated to production periods and represented only 0.007 per cent of total corrected cost.

h) *Samples* represented the labor cost of small lots of hosiery produced irregularly for sample purposes only. The cost is negligibly small in this mill and unrelated to production volume.

i) *Line-elimination labor* represented machine maintenance required by the necessity of continuous needle alignment in order to eliminate vertical lines in stockings. It was a small account and primarily a quality-maintenance cost. This cost was related to production volume but was incurred so irregularly that there appeared to be no consistent lag or lead to provide a basis for redistribution to the proper time period. The assumption might have been made that each unit of volume contributed equally to maladjustment of the needles, thus indicating that line-elimination labor should be made directly proportional to output. The validity of this hypothesis, however, is dependent on the method of speeding up used in this plant and has not been verified. Probably the omission of this account merely lowers the level of marginal cost very slightly and does not affect the shape of the marginal cost function. The total line-elimination labor account was only 0.27 per cent of the total corrected cost for the entire period.

Appendix C

Construction of an Output Index

In the earlier discussion of the selection of a measure of output in chapter viii, it was pointed out that a correction in the available output figure was necessary because of the practice of recording output before its actual completion. It is also necessary to make certain adjustments for partially completed stockings, called "transfers," received at the mill for further processing. This Appendix is concerned with the details of the method devised for allowing for this recording lead and the way in which transfer production was incorporated into the output index.

A. *Correction for Recording Leads*

From the pay-roll records of March through May, 1939, the total number of dozens of pairs of hosiery passing through each stage of production and the total productive labor cost of each operation were obtained. The average productive labor cost of each operation was obtained for each of the operations—legging, footing, topping, seaming, looping, and examining. The detailed figures are presented in Table 2–20. Estimates were also made of the time, in days, consumed by each operation, the results of these estimates being shown in Table 2–21.

From this point the reasoning proceeds in the following way: Stockings legged on the last day of any given month (and therefore recorded as part of that month's production) and starting through the footing process the first day of the next month incur the cost of all operations subsequent to legging in the second month. The cost of the operations following legging constituted 55.6 per cent of the total labor cost, and, therefore, this proportion of the last day's legging production was included in the second month's production figures.

Table 2–20

HOSIERY MILL: AVERAGE PRODUCTIVE LABOR COST IN DOLLARS PER DOZEN

Operation	Average Productive Labor Cost per Dozen
Legging	0.74202
Footing	.25891
Topping	.31148
Looping	.10729
Seaming	.14870
Examining	.05962
Mending	0.04433
Total	1.67235

Table 2–21

HOSIERY MILL: TIME IN DAYS CONSUMED BY EACH OPERATION

Operation	Time Required (in Days)
Legging	1 2/3
Footing and topping	1 1/6
Looping	1 1/6
Seaming	1 1/6
Examining	1 1/6
Mending	2/3
Total	7

Similarly, stockings legged by the beginning of the last day of the month, that is, stockings started through the footing process at the beginning of that day, did not complete footing and topping until the next day and, therefore, incurred part of their footing and topping cost and all their subsequent operation costs the second month. In particular, footing and topping require 1 1/6 days. Assuming equal distribution of productive labor cost over this period, one-seventh of footing and topping cost was incurred the second month. That is, the cost incurred in the second month amounted to one-seventh of the footing and topping cost plus the cost of all subsequent operations the sum of which is equal to 26.4 per cent of total productive labor cost. Therefore, 26.4 per cent of the legging production of next to the last working days of the month was shifted to the next month's production.

An analogous treatment was used for each day at the end of the month back to the day whose legging production was recorded in the month in which all its costs were incurred. Table 2–22 shows the percentages which were used for each day. The foregoing procedure is based on daily production figures, and these could not be obtained directly but only estimated.

Table 2–22

HOSIERY MILL: ALLOCATION OF PRODUCTION TO THE MONTH IN WHICH COST WAS ACTUALLY INCURRED

Day of Legging in a Given Month	Percentage of Total Labor Cost Incurred in Subsequent Month
Last day	55.6
Next to last	26.4
2d to last	16.9
3d to last	10.0
4th to last	4.7
5th to last	1.3

For the period July, 1935, through December, 1938, the weekly production figures did not overlap months. If a month ended on Wednesday, the production for that week is reported in two figures—one for Monday through Wednesday and one for Thursday and Friday. It was thus possible to calculate the ratio of the first three days' production to the weekly production. Similar ratios were obtained for Monday, Monday and Tuesday, etc., and the average percentage for each combination of days was calculated. From these cumulative proportions which are shown in Table 2–23, the ratios that were used as a basis for allocating weekly production to days were derived. The figures are shown in Table 2–24, both for cases where production for the week in which the month ended was a single figure and for cases in which it was split at the month's end into two figures.

Their validity was checked by applying them to the 1939 weeks, where weekly production figures did overlap months. After the overlapping weeks were split between months on the basis of the estimated daily ratios, monthly totals were obtained, which tallied very closely with the monthly control totals. As an example of this procedure, Table 2–25 shows the final correction method for determining how much of January, 1935, recorded production should be shifted into February. The exact figures varied from month to month depending upon the day on which the month ended and upon the occurrence of holidays.

Table 2–23

HOSIERY MILL: CUMULATIVE ALLOCATION TO DAYS OF THE WEEK

Days	Average Percentage of Weekly Production
Monday	13.4
Monday-Tuesday	32.9
Monday-Wednesday	46.8
Monday-Thursday	73.5
Monday-Friday	100.0

Table 2–24

HOSIERY MILL: ALLOCATION OF WEEKLY PRODUCTION TO DAYS

Day of Week	Percentage of Weekly Production Legged on Various Days	Percentage of Monday and Tuesday Production Legged on Various Days	Percentage of Monday, Tuesday, and Wednesday Production Legged on Various Days	Percentage of Monday, Tuesday, Wednesday, and Thursday Production Legged on Various Days
Monday.....	13.4	40.7	28.6	18.3
Tuesday.....	19.5	59.3	41.7	26.5
Wednesday..	13.9	....	29.7	18.9
Thursday....	26.7	....	....	36.3
Friday.......	26.5	....	....	....
Total....	100.0	100.0	100.0	100.0

Table 2–25

HOSIERY MILL: EXAMPLE OF ALLOCATION OF MONTHLY PRODUCTION, JANUARY TO FEBRUARY, 1935

(Month Ends on Thursday)

(1) Day	(2)* Percentage of Week's Production Legged on Given Day	(3) Dozens Legged Week Ending February 1	(4) Percentage of Cost Incurred in February	(5)† Number of Dozens To Be Shifted to February
Thursday..	26.7	2,175	55.6	323
Wednesday.	13.9	2,175	26.4	80
Tuesday...	19.5	2,175	16.9	72
Monday...	13.4	2,175	10.0	29
		Dozens Legged Week Ending January 25		
Friday.....	26.5	1,034	4.7	13
Thursday..	26.7	1,034	1.3	4

*Columns (2) and (4) are taken from Tables 2–24 and 2–22, respectively.

†Column (5) is the product of columns (2) and (4), and its total is the number of dozens of January legging production that were transferred to February.

B. Measurement of Transfer Production

"Transfers" are partially completed hosiery transferred to the hosiery mill from other mills for the completion of the remaining operations. Operations are carried to the point where the legs and the feet have been joined. It has been explained that production is recorded at the time when hosiery is legged so that credit for production goes to the other mills. But, both because of the contribution of the particular mill to the finished hosiery and because of the costs incurred through transfers, some allowance must be made to include this aspect of production in the final measure of output.

The basic data on transfer production were the records of the number of unfinished stockings received by the hosiery mill for further processing from other mills. All these incomplete units were treated as though they had incurred their legging, footing, and topping cost before reaching the mill and, therefore, had only to undergo the remaining operations of looping, seaming, and examining. This is a realistic assumption in a large majority of the cases. Since the transfer data were weekly figures and since there was no basis for determining the day the transferred hosiery was started on the looping process, it was assumed that of the week's recorded transfer receipts, 20 per cent went into production on each of the five working days. Adjustment was made to allow for the fact that part of the transfers upon which work was started near the end of the month incurred costs in the subsequent month.

The method employed here was that weekly figures were allocated between months on the assumption mentioned that 20 per cent of the recorded weekly transfers were started into the looping process at the beginning of each day of that week. Then a procedure analogous to that used for allocating complete production was employed. Thus transfers started into production at the beginning of the last day of the month incur the next month one-seventh of their looping cost, as well as all the cost of the subsequent operations—seaming through mending—that is, they will incur 74.5 per cent of the latter operations' cost the next month. Therefore, 74.5 per cent of the last day's receipts were attributed to the second month's production. The corresponding percentages for other days are shown in Table 2–26.

When both complete and partial production had been adjusted so that they fell into the proper cost period, the only necessary correction remaining was the weighting of partial output before combining the two types of production. The weight chosen was the relative productive labor cost of transfer work as compared with that of complete production. Thus a uniform weight of 21.5 per cent was applied to adjusted transfers before they were added to adjusted total output to give the final corrected measure of output.

In an attempt to construct a simpler production index than the one used, this same weight had already been applied to unadjusted transfers in combining them with unadjusted legged production figures. The resulting series, however, as judged by the criterion of closeness of relationship to productive labor cost (adjusted for changes in wage rates, by operation), was much less satisfactory than the more refined measurement.

Table 2–26

HOSIERY MILL: ALLOCATION OF TRANSFER PRODUCTION TO MONTHS

Date of Receipt of Transfer Production	Percentage of Transfer Productive Labor Cost Incurred Subsequent Month
Last day	74.5
Next to last	40.7
2d to last	19.4
3d to last	3.1

Study 3

BELT SHOP

I. Introduction

The intimate relation between cost and exchange value directed attention to the theory of cost as soon as value theory became a subject of serious discussion. Because of economists' pre-occupation with the problem of value, cost has occupied a central position in theoretical economics. Although a great deal of subtle analysis has been devoted to cost, the notion is still attended with some ambiguity. Therefore, it is necessary to be certain that the meaning of "cost" in any given instance is clear. In this investigation "cost" is taken to mean the "expenses of production" an entrepreneur incurs in operating an enterprise.[1] The cost theory developed from this point of view is concerned with the magnitude of the cost associated with different levels of operation of a given enterprise. The simplicity of this relation should not be overemphasized,[2] nor its importance underestimated. This cost behavior has a crucial role in determining the most profitable adjustment of the individual enterprise to its economic environment. Consequently, the business executive is also vitally concerned with the response of cost to changes in output.

In this study an attempt was made to build upon the existing theoretical foundation by determining the empirical counterpart of the static cost–output functions for enterprises with immobile plant and equipment. This procedure is desirable as a means of comparing theoretical with statistically determined relations. Furthermore, since cost prediction requires a knowledge of the basic relation between cost and output, the estimation of future cost is much simplified by this approach. Such statistical analysis may also afford information that is not immediately available from the firm's cost accounts: e.g., (1) the expected total and average cost under a given set of operating conditions (adjustable budgets), (2) the additional cost that must be incurred if output is increased by a small amount. Nor is managerial interest in a statistical analysis of cost behavior confined to its immediate usefulness for forecasting purposes; the techniques used here have a wide applicability in the control of costs and in price policy.

Reprinted from Joel Dean, *The Relation of Cost to Output for a Leather Belt Shop*, National Bureau of Economic Research, Technical Paper 2 (New York, 1941), by permission of the National Bureau of Economic Research.

Rigorous empirical investigations of cost designed to compare statistical results with the cost behavior prescribed by economic theory have not been numerous. This is to be attributed to the difficulties of obtaining confidential cost data and of finding firms that meet the requirements specified in the underlying cost theory, i.e., single product, unchanging technical methods, unchanging equipment, etc. Although each industry and enterprise offers its own peculiar problems, the case study reported in this paper illustrates the results that can be obtained by statistical analysis of accounting records as well as the problems encountered in determining these results. It is hoped that this description and illustration of analytical methods that have proved valuable in several similar investigations will stimulate research in an area that has both scientific and practical importance.

II. Theories Concerning Static Short-Run Cost Functions

In the hope that this paper may reach some non-economist readers, we first summarize the fundamentals of short-run cost theory to clarify the basis of later discussion. Underlying the whole discussion of cost theory is the notion of a cost function —a function that shows the relation between the magnitude of cost and of output. The existence of such a function is postulated upon the following assumptions: (1) there is a fixed body of plant and equipment; (2) the prices of input factors such as wage rates and raw material prices remain constant; (3) no changes occur in the skill of the workers, managerial efficiency, or in the technical methods of production. It is the shape of this cost function that is of primary interest in both the theoretical and statistical parts of this paper.

Money expenses of production depend upon the prices and quantities of the factors of production used. Since prices are assumed to remain unchanged, the shape of the cost function will be determined by the physical quantities of the factors used up at different levels of operation. And since these quantities are functionally related to output, their relation to cost can be represented by a cost–output function. Thus the underlying determinant of cost behavior is the pattern of change in the factor ingredients as output varies.

This pattern of change will be determined by the technical conditions of production. In general, when there exist fixed productive facilities to which variable resources are applied, the law of diminishing returns is assumed to govern the behavior of returns. The mere presence of fixed equipment is, however, not sufficient to ensure the operation of this law unless in addition the productive services that are forthcoming are available only at a fixed rate. The essential question is, therefore, not the physical fixity of the plant but the invariability of the rate of flow of services from the plant.

On the basis of the nature of the fixed productive facilities two technical situations can be distinguished: (1) the services of the fixed productive facilities are available only at a fixed rate, (2) their services can be drawn off at varying rates by bringing these facilities into operation piecemeal. The behavior of cost differs radically as between the two situations. In the first, the law of diminishing returns is fully applicable: marginal returns increase at first and then diminish. The corresponding type of cost function possesses the greatest generality and has usually been considered typical by economists. The behavior of total or combined[3] cost, on the above assumption, is shown in Chart 3–1. Combined cost is, for analytical

Chart 3–1 Cubic Total Cost Curve

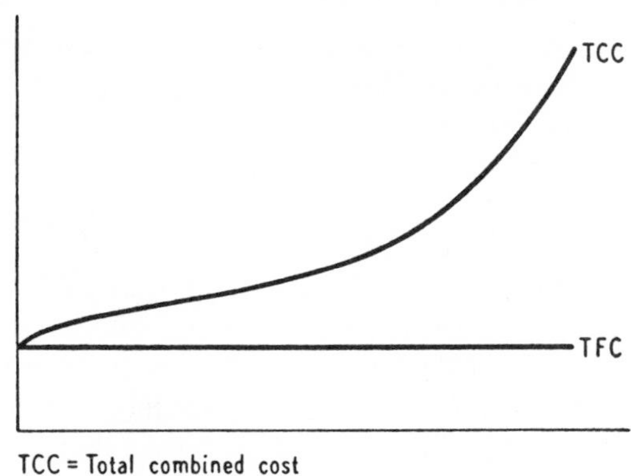

purposes, divided into fixed cost and variable cost. Fixed costs, which arise mainly from the firm's investment in fixed plant and equipment, are defined as those expenses whose monthly total is constant irrespective of the output rate. Total fixed cost is shown by the line *TFC* in Chart 3–1. Total variable cost, on the other hand, depends on the rate of output and is measured by the vertical distance between the curve of total combined cost, *TCC*, and of total fixed cost, *TFC*. In this case the total combined cost curve rises first at a decreasing rate and eventually at an increasing rate as output increases.

Fixed cost per unit declines steadily as output increases, the magnitude of average fixed cost being shown by the rectangular hyperbola *AFC* in Chart 3–2. Variable costs per unit will be falling at first and then rising as output increases. The curve *AVC* in Chart 3–2 describes the postulated behavior of total variable cost. The curve of average combined cost, *ACC*, derived from the total combined cost curve, is of the same general U-shape as average variable cost, first declining and then rising.

Strictly speaking, the marginal cost curve shows the rate of increase of total combined cost, but approximately it can be taken to indicate the additional cost that must be incurred if output is increased by one unit. The behavior of the marginal cost function is shown by the curve *MCC* in Chart 3–2. It has a falling phase in the low range of output and rises as marginal returns begin to diminish. The prominence of marginal cost in economic literature explains our emphasis upon the behavior of this particular function rather than on total or average functions.[4]

Chart 3–2 Average and Marginal Cost for Cubic Total Cost Curve

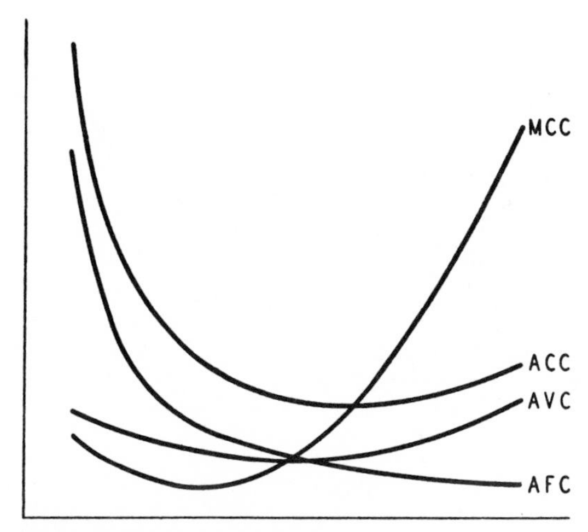

The total combined cost function associated with this shape of marginal cost curve can be represented by a great many types of function. Since the simplest functional form that possesses the necessary characteristics is probably a third degree parabola or cubic, it is convenient to refer to the cost behavior relevant to this case by specifying the cubic total combined cost function.

The second type of cost behavior mentioned above must now be considered. If the fixed equipment and machinery involved in a process can be broken down into small units, so that each segment can be combined with variable productive services, it may be possible to avoid diminishing returns until all fixed equipment has been brought into operation.[5] In other words, if each segment is equally efficient, new segments can be introduced whenever returns to the equipment already in use begin to diminish. This means that rising marginal cost can be postponed until all fixed productive facilities have been brought into use. Therefore, over the range of output less than capacity (when "capacity" means that all segments are in operation) marginal cost will be constant.[6] The resulting total combined cost curve is illustrated in Chart 3–3. The curve (*TCC*) is, over most of the output range, a positively sloping straight line whose intercept on the cost axis represents the total fixed cost. At some high level of operation it is assumed that the total combined cost curve ceases to be linear and bends upward. Marginal and average variable cost coincide until the level of output that utilizes all segments is attained; at this output, the two curves diverge. The average combined cost curve lies above the marginal cost curve over the low ranges of output and is eventually intersected by the marginal cost curve at its minimum point.

Chart 3–3 Linear Total Cost Curve

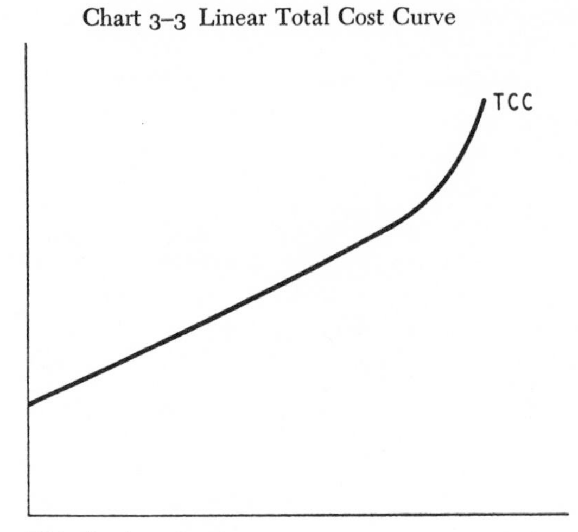

TCC = Total combined cost

Two alternative models of short-run cost functions have been considered: in one the total cost curve is a cubic curve; in the other the curve is linear until some extreme level of output is reached. The statistical analysis of cost data that follows is designed to indicate which type of theoretically postulated behavior is consistent with the cost behavior of the plant studied during the period of observation. Before discussing our methods and findings, we examine certain sources of divergence between these theoretical cost functions and their empirical counterparts.

III. Sources of Divergence of Empirical from Theoretical Static Cost Functions

In attempting to determine empirical cost functions that are the strict counterparts of those specified in the static theory of cost, several difficulties were encountered. First, it was not possible to include all costs in combined cost. The difficulty of allocating the cost attributable to jointly produced articles necessitated the omission of certain elements. Furthermore, the cost accounting information available may not be sufficiently accurate to represent faithfully the costs actually incurred. Second, the idealized conditions of production visualized in theory will not be fulfilled in any concrete situation. Not only is it sometimes impossible, for technical reasons, to make the continuous adjustment of the variable factors hypothesized, but also managerial inertia or other rigidities may be sufficient to hinder adjustment to changed conditions. It is perhaps to be expected that theoretical curves intended to be generally descriptive in a qualitative sense of all cost functions will neglect, for the sake of simplicity, the rigidities that may be peculiar to given industries or firms. Rigidities, however small, will be a continuous source of divergence whose influence cannot, practicably, be removed by statistical methods.

This divergence is accentuated by the degree of the entrepreneur's knowledge concerning market and technical conditions and his ability or willingness to adjust operations in order to attain for any output the minimum cost combination of factors, the basic assumption on which the theoretical model is drawn up. Nevertheless, an empirical function derived from data for one particular enterprise necessarily reflects the prevailing conditions of production in it, since these conditions represent a selection from among numerous possible forms of organization. Whether the influence of rigidities is sufficiently great to cause the shape of the empirical functions to differ essentially from the theoretical functions can be ascertained only after continued research in the field. Third, a set of problems arises from the attempt to approximate the static conditions assumed in theory. Any firm selected for study operates in a changing environment to which it continually adjusts. Most important are the variations in the prices of factors of production, which, unlike the influence of rigidities, can be removed from the cost data by a process of "deflation."

In general, the divergence between the production process analyzed and the theoretical situation assumed to exist can be minimized by three methods: (1) careful selection of a sample, with due regard to both the firm and the time period chosen, in order to reduce the influence of dynamic elements; (2) rectification of the data to remove directly the influence of disturbing factors that could be measured adequately; (3) multiple correlation analysis of the relation between cost and other variables whose influence was not directly measurable, in order to examine and remove their effects.

The application of these three methods is discussed in Sections IV, V, and VI. The statistical findings are presented in Section VII and their validity appraised in Section VIII. The concluding section attempts to interpret and qualify the findings, reconciling them with the results prescribed by the static theory of cost behavior.

IV. Collection of Data

It is much easier to isolate the relation between cost and output if the observations in the statistical sample are as free as possible from the influence of other variables affecting cost. We were fortunate, in this regard, to secure the active cooperation of a leather transmission belting manufacturer, one of whose plants met our sampling requirements admirably. In addition, the nature of the accounting data of the firm facilitated statistical analysis, since costs were kept in considerable detail as totals of expense for four-week accounting periods. Complete records of output measured in several different ways were available, as well as supplementary information concerning operating conditions affecting costs. All these records were comparable for several years.

The assumption, discussed above, under which short-run cost curves are drawn is that changes in the rate at which the plant is utilized are not accompanied by changes in the scale of the plant, in the technical methods of production, or in prices of input factors sufficiently great to induce substitution among the factors of production.[7] These conditions seemed to be approximately fulfilled by the firm during the period studied. Changes in the level of cost and output were frequent and large, as can be seen from Chart 3–9, and were unaccompanied by changes in scale or in basic methods of operation. Moreover, the technical processes were such as to make it unlikely that the technical coefficients of production would be altered significantly by changes in the price of input factors. As a rule, short-run cost functions are easier to determine for typical establishments in declining industries. Capital investment during the analysis period is usually small, output being less than plant capacity. Since, logically, capital is retired only when there is a probability that sales will not cover variable cost and return on the liquidation value of the equipment, changes in scale are minimized. Moreover, it is reasonable to suppose that technical progress is less likely to become a complication of any significance.

Analysis of cost is simplified when a single physically homogeneous product is manufactured. Difficult problems arise when cost must be apportioned among several products because of joint processes or common use of production facilities. In the leather belt plant the assumption of product homogeneity was not altogether justified, for not only were several grades of belting made but there were also certain highly dissimilar minor products. Nevertheless, the differences in grades and sizes of belting did not appear to affect the cost functions significantly. Since the products other than leather belting constituted a relatively small share of total output and were made in separate departments, the cost attributable to them was isolated and omitted from the analysis.[8]

A further difficulty in analyzing cost arises when there is a pronounced and irregular lag between the incurring of cost and the recording of the corresponding output. The concern here is not with the engineering problem of determining the length of the production cycle, but with the statistical problem of correcting for a lag arising from the practice of recording cost in the period in which it occurred, instead of at the time the output to which it contributed was completed. This lag-correction may become complicated when it is necessary to readjust each element of expense. Corrections for this lag become progressively less reliable as the production cycle lengthens. In the belt shop, however, the production cycle was short—a few days only.

Another reason for selecting this particular plant was that the operating conditions causing cost variation were few and those exerting the greatest influence on cost were measurable.

Selection of Sampling Period

The period January 1, 1935 to June 1, 1938 was chosen for study because it fulfilled the following conditions most satisfactorily:

1. The rate of output and other measurable cost determinants varied sufficiently to yield observations over a wide range and to afford fairly uniform coverage of this range.

2. Detailed and complete records of variations in cost and in the chief operating variables affecting cost were available.

3. The plant and equipment remained unchanged during the analysis period, permitting the observation of short-run adjustments uninfluenced by long-run changes.

4. Since technical methods of production were constant, the cost-output relation was not obscured by simultaneous changes in production methods.

5. The data were sufficiently recent to allow their use for cost forecasting and budgeting.

The four-week accounting period used by the company in its cost records and production plans was taken as the observation time unit for this study. During the three and a half years of observation, there were forty-five such accounting periods or "months." Since any theoretical analysis is properly in terms of rates of output, it is desirable in empirical studies of this kind to choose an observation unit in which the flow of output is uniform. In general the shorter the period the more likely is this condition to be fulfilled. In the plant studied there was evidence that fluctuations in the production flow within the four-week planning and recording period were small. Attempts to allow for the variation that did exist are discussed in Section VI.

Selection of Cost Elements

In selecting elements of cost for the analysis, we omitted expenses that were arbitrarily allocated to the leather belt shop and that bore no apparent relation to its operating conditions. For example, certain general overhead and head office and administrative expenses, which were distributed among the various plants operated by the firm, were excluded.

To ascertain whether the omission of overhead common to several plants led to an underestimate of the marginal cost of the leather belt shop, an exploratory graphic correlation analysis was made, using the common overhead as the dependent variable and the level of operations of the belt shop and the other plants as independent variables. The level of operations was measured by output and also by labor input. Since this analysis showed no significant partial regression of overhead on the activity of the belt shop, the absence of marginal general overhead cost was indicated. However, in the measurement of combined cost, certain elements of overhead cost as well as all elements of direct cost were included.

Although combined cost was of central theoretical and practical significance, it was found desirable to study also the behavior of individual constituents of cost. Combined cost was first broken down into two components, "direct" and "overhead" cost. The terms "direct" and "overhead" cost refer to accounting classifications and should not be confused with the economic categories of fixed and variable cost. The distinction between direct and overhead cost is based upon the ease of identifying cost with the particular units of output that give rise to it, whereas the distinction between fixed and variable cost depends upon whether cost varies with changes in output. In the plant studied none of the overhead cost was completely fixed.

The components of cost were broken down still further into their elements. Since the forces affecting cost vary in their impact upon different elements, separate correction of each element was necessary to remove irrelevant variations caused by changes in the prices of input factors and lags in recording. Furthermore, separate analysis of the behavior of the elements provides a basis for setting cost standards and flexible budgets in sufficient detail to be managerially useful. This detailed knowledge also makes possible more specific and exact allowances for changes in factor prices or minor alterations in the technique of production. Individual analyses were carried out, first, for direct cost and its elements: direct labor, leather, and cement; and second, for overhead cost and its elements: indirect labor, supplies, repairs, depreciation, taxes, insurance, salaries, sundries, dies and rings, heat, light and power, and water.

V. Rectification of Data

Rectification of the data is designed to eliminate influences that tend to obscure the true relation between cost and output. Since the influences on cost (apart from output) affect the various elements of combined cost differently, composite correction is unlikely to be accurate. For this reason it was found desirable to use specialized rectification devices for the various elements of cost.[9] There are two sources of distortion that necessitate rectification: (1) the time lag between the recording of cost and of the associated output, (2) variations in the prices of raw materials, labor, and other factors of production.

Time Lag

Rectification of the time lag between the recording of cost and of the output causing it ordinarily involves two steps: (1) the determination of the proportion of cost recorded in a period other than that in which the corresponding output is recorded; (2) the determination of the length of the recording time lag. Sometimes these magnitudes are readily found, but it is usually necessary to have recourse to estimates based on technical considerations and engineering opinions. These estimates can be supplemented by statistical analysis designed to test objectively the correctness of the engineering calculations.

The recording of the cost of machinery repairs, supplies, and cement is subject to a time lag sufficient to warrant attention.[10]

The amount of machinery repairs seems to bear a fairly definite relation to output. Information from operating executives indicated that in any given accounting period about one-fifth of machinery-repair expenditures were for minor replacements necessitated by current production, which, therefore, are likely to be recorded in the same period. The larger fraction, although to some extent attributable to the mere passage of time, was caused primarily by output and could be allocated to the production activity of approximately three months earlier. Since the influence of output on repair cost is cumulative and somewhat fortuitous, however, there are wide fluctuations in machinery repairs from period to period.[11] Accordingly, the corrected series of machinery-repair data is composed of one-fifth of the current figure and four-fifths of the cost three months later.[12]

The cost of supplies, in contrast to machinery repairs, is recorded in advance of the output for which they are destined, since usually supplies are not entirely used up within the period in which they are purchased. Analysis of the elements of the cost of supplies, supplemented by opinions of executives, showed that an average lag of one accounting period existed for approximately two-fifths of the recorded expenditure. Consequently, three-fifths of the book figure was combined with two-fifths of the cost in the preceding period to give a set of corrected data from which, it is believed, the recording error was largely removed.[13] The distortion caused by the lag in recording cement cost was recognized but could not be removed. The estimate of the cost that was used was the sum of the value of the inventory of the preceding month and the difference between the value of the inventory at the end of a given month and the total of the invoices for cement purchased in that month. An error in the allocation of this cost arises when invoices for cement consumed in one month are received in the following month. Its magnitude is indicated by the great variation in cement cost expressed in terms of cents per pound of finished product, a quantity that should remain fairly constant because actual cement cost may be expected to be approximately proportional to output. An attempt to offset the error by relating the output of each period to the cement cost of the preceding period yielded less satisfactory results than the use of corresponding months, probably because merely a fraction of the expenditure is incorrectly recorded. Since there seemed to be no way of segregating the wrongly allocated portion, rectification was abandoned.[14]

Part of the irrelevant variation in leather cost attributable to the lag in recording was removed by using the quantity of material charged out of the cutting room into production rather than the quantity charged into the cutting room. Since the cutting department constitutes a reservoir storing widely fluctuating amounts, the quantity of materials entering the cutting room is more remotely related to output than the quantity supplied by it.[15]

Changes in Wage Rates and Material Prices

To obtain empirical cost functions analogous to the static theoretical functions described above it is necessary to hold the prices of input factors constant at some base level. Two assumptions are implied in this formulation: (1) that substitution among the input factors did not take place as a result of changes in their relative prices; (2) that changes in the output rate of the enterprise exerted no influence on the prices paid for its factors.

Examination of the technique of production indicates that the first assumption is justified for the period studied. Whether the second assumption represents the actual circumstances depends upon the conditions under which the output of the particular enterprise fluctuates. Professor Viner has distinguished three kinds of change in competitors' output that affect the relation between the firm's rate of output and the prices of factors: (1) if the change in the firm's output is accompanied by offsetting changes in competitors' output, the industry's demand for input factors remains unchanged; (2) if there is no change in competitors' output, the industry's demands for input factors is increased only by the firm's increase; (3) if the changes in the firm's output are paralleled by changes in competitors' output, substantial increases in the industry's input demand accompany the firm's increased output. The third type of expansion seems most probable in a mature production goods industry of the type under consideration. However, the influence of changes in industry output on factor prices depends on the extent to which factors are specialized to the industry. Since the leather belting industry accounts for merely a small part of the total demand for the principal input factors, it is unlikely that its expansion would be sufficient to induce variations in input prices.

Whether or not it is correct to assume that the firm's rate of output is of negligible influence on factor prices, the only practical approach to the determination of the firm's cost-output functions is to exclude entirely the effects of industry adjustments. Although influences attributable to changes in the output of the industry may have been represented in the observations, these could hardly be disentangled and purged by multiple correlation procedures. It appears preferable to investigate the effect of factor price changes on marginal cost by a method explained in Section VII where the analysis of cost components is discussed.

For each accounting period variations in total cost arising from changes in the prices paid for factors were partly eliminated either by deflating the costs affected in order to render them comparable with costs at base year prices or by substituting for the prices actually paid an average monthly price for the years studied. The second procedure was applied when more precise rectification seemed unpractical or inadvisable.[16]

Direct and indirect labor, salaries, and cement were corrected by deflating the recorded expenditures to correspond to rates and prices of the base period. Leather cost, depreciation, insurance, and taxes, on the other hand, were rectified by charging them to production at an average rate for the entire period.

1. DEFLATION

For direct and indirect labor and salaries an index was constructed from records of actual wage rate and salary changes from the rate existing on January 1, 1935. Since all wage and salary modifications were general, with the exception of a few salary adjustments resulting from reorganization of the executive personnel, the computation was relatively simple.

Of the three constituents of cement cost—film, solution, and liquid cement—only the second and third varied sufficiently during the period studied to necessitate correction. The indexes used were weighted arithmetic averages of price relatives, with January 1935 as the base period and with the proportions of the average 1935 value for each element as weights.[17] The relatives were computed from monthly average prices for each element.

2. STABILIZATION

The cost of leather for each accounting period was computed on the basis of a uniform price per pound. Inventory price variations of leather reflect fluctuations in the price of hides, changes in the operating cost of processing departments preceding the belt shop, and alterations in the proportions of different qualities of leather going into output. Qualitative differences in this raw material apparently exerted no significant influence on the cost relations under consideration. The most appropriate method of rectification seemed to be the use of an average leather price for the period.[18]

Depreciation was held constant at the average monthly depreciation charge for the period. This procedure, by arbitrarily preventing depreciation from affecting the position or shape of the marginal cost function, may impair, to some extent, the validity of the findings. Ideally, use-depreciation should be separated from time-depreciation, since only that part of depreciation which arises from the actual operations of a plant is relevant in determining the cost occasioned by different levels of operation.[19] The shape of the marginal cost function depends upon whether use-depreciation is a linear, increasing, or decreasing function of intensity of utilization. This relation as well as the magnitude of use-depreciation depends upon maintenance standards and upon the effects of uninterrupted high speed utilization upon the deterioration of equipment. Depreciation caused by physical deterioration due to the passage of time and by losses in value as a result of technological progress or changes in product specification (obsolescence) affects merely the height of the intercept of the total cost function on the cost axis, not the shape of the function itself.

Unfortunately, from the accounting records, in which depreciation was charged on a "straight-line" basis, i.e., as a linear function of time, time- and use-depreciation could not be differentiated. The month to month differences in depreciation that did occur in the records arose from arbitrary annual changes in the depreciation rate made to correct for past errors and to adjust for past or expected profits.[20] The stabilization of the depreciation rate at its average monthly value was intended to remove these accidental and irrelevant variations. This procedure understates marginal cost only if, apart from time-depreciation, there occur significant losses of value arising from use that are not restored by repairs.

The costs of taxes and insurance were likewise stabilized at their average values for the entire period, again on the hypothesis that their variations are unrelated to the quantity of goods produced each month. The state excise tax, which constitutes merely a small portion of the tax bill, is alone proportional to output. Most of the change in monthly totals is a result of small changes in the annual tax rate, for which refined correction would not be worth while. Insurance cost also varied primarily because of changes in annual rates which were again unrelated to monthly changes in output or other operating conditions.[21]

Unrectified Errors

Several elements of cost were left wholly or partly unrectified, even though their magnitudes were influenced by some irrelevant variation.

The small relative importance of the cost of dies and rings and the difficulty involved in rectification justified the omission of any correction for this cost. Fluctuations in the cost of supplies arising from price changes were ignored both because of the minor importance of the cost and because of the labor involved in correcting for the great diversity of products recorded in the supplies account.

The book figures for water, heat, light, and power were also used. The water, heat, and light data did not appear to need correction, and only a small part of the variation in power cost could be considered irrelevant. It might have been desirable to remove the fluctuations in the cost of power caused by changes in temperature and number of hours of daylight in different periods, but the complexity of any suitable corrective device indicated that attempts at rectification would cost more than the increase in accuracy would warrant.[22]

VI. Methods of Analysis

Selection of Technique

Multiple regression analysis seemed most suitable for investigating the relation of the rectified cost to output and the other operating variables.[23] This approach yields measures of: (1) The relation of cost to each independent variable that influences its behavior after the effects of the other variables have been allowed for,[24] a relation displayed in the form of a curve or schedule showing cost for each of a series of values of each independent variable. (2) The importance of the combined effect of the several variables upon cost, a measure needed to indicate the degree to which cost behavior has been accounted for. (3) The reliance that can be placed on the derived curve or function as representative of cost behavior, subject to the limitations implicit in time series data, discussed below. Information such as this is of special value if the cost analysis is to be used as a basis for flexible budgets or to determine marginal cost.

Two methods of multiple regression analysis were employed. The graphic method was used for exploratory purposes, because of its economy and flexibility, and also because the net regression curves so determined serve admirably to present the statistical findings.[25] For the final analysis, fitting by the method of least squares was preferable because of its greater objectivity, the wider acceptability of its error formulas, and the fact that the order in which the variables are considered does not affect the results. The preliminary graphic analysis was intended to determine the various causal factors exerting an influence on cost sufficient to justify their inclusion in the least squares regression analysis and to aid in choosing the general character of the function that best represented the net relation. Although the independent variables were subject to error, the least squares curves and error formulas were computed on the usual assumption that the dependent variable alone is subject to error. This treatment can be justified on two grounds. First, the primary objective was to determine functions that enable prediction of cost from the values of the independent variables, rather than to discover the true functional relation or mutual regression function.[26] For this purpose the procedure is valid, despite errors in these independent variables. Second, it seemed probable that the independent variables finally chosen were subject to less error than cost. The cost data were defective because of recording errors and the possibility of improper rectification in several important respects, notably: (1) omission of certain allocated overheads, (2) stabilization of certain elements whose variation was considered irrelevant, (3) removal of dynamic influences by deflation, (4) reallocation with respect to time periods. The output data, on the other hand, were not subject to fluctuations in price levels, lags in recording, or arbitrary allocation.

Selection of Form of Cost Observations

Preliminary analyses were made first solely for combined cost, i.e., for the aggregate of the various cost elements included in the study. Additional, more detailed analyses covered not only the combined cost function, but also the cost functions for overhead cost, direct cost, and their constituent elements.

Combined cost was analyzed in the form of totals for the accounting period rather than in the form of cost per unit of product. Experimentation with these alternative approaches in previous cost studies has shown that analysis in the form of total rather than average cost yields more convenient and reliable findings.[27] The conversion of cost in total form to average and marginal form, which may be desired for interpretative purposes, is a simple matter. Marginal cost, for example, is the rate of increase in the total function or the slope of the net regression line of total combined cost on output.[28] When the total cost function is linear, marginal cost is simply the coefficient of net regression of total combined cost on output.

Variables Causing Cost Variation

In analyzing the relations of cost to the measurable causal influences, the variables selected for testing were those the management thought might affect cost in some degree. The tests required that the variables have marked independent influences on cost not reflected in other causal forces, i.e., it was necessary for the net regression of cost on the independent variable to be significant. Since certain variables might exhibit a quantitatively significant influence on some cost elements but not on others, this test was applied to various cost elements. Furthermore, the independent variables must account for a significantly large part of the variation in cost.[29]

Each influence selected as relevant was accordingly separately examined in order to ascertain: (1) the reasons for its influence on cost; (2) the best statistical series available for its measurement; (3) its net correlation with cost. The following operating variables were examined to determine their probable effects upon the behavior of cost in the leather belt shop:

1. Output (measured in square feet of single-ply belting)
2. Average weight per square foot of single-ply belting
3. Average width per square foot of single-ply belting
4. Magnitude and direction of change of output from preceding month
5. Percentage of single-ply belting in total output
6. Variability in rate of output within accounting periods
7. Size of manufacturing lot
8. Proportion of special orders
9. Rate of labor turnover

1. OUTPUT

The rate of output could normally be expected to exert a predominant influence on the magnitude of monthly cost because expenses incurred for materials and direct labor, which vary directly with it, are large. Square feet of single-ply equivalent belting was chosen as the measure of output primarily because the cost of operating the leather belt shop was more closely related to area than to weight, dollar value, or the standard cost of output, which were considered as alternative measures.[30]

Production of double- and triple-ply belting is the principal source of fortuitous variation in the area measure of output. To take account of it we converted the area of finished belting into equivalent single-ply belting.[31]

The independent effect of square feet of output on cost was tested by graphic correlation analysis for combined cost and for several of its major elements. In each instance a significant net relation was found. Output was therefore used as an independent variable in the final least squares correlation analysis of each aspect of cost.

2. AVERAGE WEIGHT

The average weight of belting output was believed to influence cost because of its clear relation to raw material cost and its effect upon the cost of certain processing operations. Average weight per square foot of single-ply equivalent was therefore tested as an independent variable in order to have a measure reflecting the effect of weight upon cost independently of the influence of output measured in square feet. The strength of the independent relation was examined by graphic multiple correlation and a marked net correlation was found in the case of both total combined and direct cost. Average weight was therefore selected as another independent variable for the mathematical correlation analysis.

3. AVERAGE WIDTH

For certain manufacturing operations, cost per square foot appeared to be affected by the average width of the belt. The influence of width alone was most accurately reflected by the average width of single-ply equivalent belting. Since graphic correlation indicated that this independent influence was of minor consequence, this variable was omitted from the least squares analysis.

4. MAGNITUDE AND DIRECTION OF CHANGE IN OUTPUT FROM PRECEDING MONTH

In order to detect the influence of two types of factors not already removed in the data rectification, the magnitude and direction of change in output from that of the preceding month were tested as independent variables, both separately and in combination. This procedure served as a rough test of the reversibility or continuity of the empirical cost function. Ordinarily static cost functions are assumed to be such that cost is a unique function of output regardless whether the output is attained by a large or small increase or decrease from the preceding period. The observations, however, may not fulfill the conditions assumed, since the cost associated with operating at 60 per cent of capacity after a period of operating at 40 per cent and after a period of operating at 80 per cent may not be the same. A concrete situation, consequently, may fail to conform to the conditions postulated in theory which assumes that adjustments are instantaneous and frictionless.

When the magnitude of the change, regardless of its direction, was used as an independent variable in a graphic correlation analysis, no significant net relation to cost was disclosed. Direction of change was then studied by separate analyses of the cost-output relation for periods of increased and of decreased output, but no noticeable difference was found. A combination of the two influences was then tried by using an independent variable ranging all the way from large increases to large decreases, again without definite indication of a relation. The magnitude of the departure from the preceding month's output was, therefore, excluded from the least squares analysis. In the case of one component, however, a noticeable relation between direction and magnitude of change in output was indicated by the graphic analysis. This variable was accordingly included in the formal regression analysis of overhead cost.

5. PERCENTAGE OF SINGLE-PLY BELTING IN TOTAL OUTPUT

The cost of both finishing and cement is greatly affected by the number of plies. In view of the marked variation from month to month in the proportion of single-ply belting in total output the influence of this factor was considered as an independent variable but was rejected because graphic analysis revealed no significant net correlation, the output measure chosen having adequately reflected the cost changes associated with this variation.

6. VARIABILITY IN RATE OF OUTPUT WITHIN ACCOUNTING PERIODS

Fluctuations in the rate of production were not fully reflected in the output data derived from the records which were kept for four-week accounting periods. These data, therefore, neglect intra-month variation. The same monthly output can be achieved by operating at full capacity for two weeks and then shutting down for two weeks as by operating at half capacity throughout the four weeks. By planning ouput in advance, however, and scheduling production at an even rate, the management had so reduced intra-month variation that they did not believe it affected cost greatly. A measure of this variation might have been obtained from the average deviation, standard deviation, or coefficient of variation of daily production, but daily output records for the entire period of analysis were not available. Since a satisfactory measure was unavailable and the effect of this variation was believed to be negligible, no attempt was made to include it in either the graphic or the least squares analysis.

7. SIZE OF MANUFACTURING LOT

The size of the manufacturing lot may markedly affect cost in processes for which the setting-up of machines is expensive. This is especially true when output is composed of diverse products and a different machine set-up is required for each product. Since neither condition existed in the leather belt shop, size of manufacturing lot was rejected as an independent variable.

8. PROPORTION OF SPECIAL ORDERS

Ordinarily special and rush orders cause a certain amount of confusion and inefficiency. It was the opinion of the executives, however, that production in the belt shop is so well scheduled and routed that operations are not significantly disturbed by special orders. Moreover, a suitable measure was difficult to construct because of the lack of data on special orders for the earlier periods. This factor was, therefore, rejected.

9. RATE OF LABOR TURNOVER

Because of the expense involved in the selection and assimilation of new personnel, labor turnover may exert a pronounced influence on cost. Inquiry indicated, however, that in this particular plant the influence of labor turnover had only a negligible influence upon month to month variation in cost. The reasons are first, that the labor turnover rate is approximately constant, and second, that increases in the labor force are achieved mainly by rehiring regular employees previously laid off, rather than by hiring inexperiencd workers. This variable, in view of its minor effects, was excluded from consideration.

The conclusion emerges from the preceding discussion of the operating factors affecting cost that only three satisfied the criteria of suitable independent variables for the least squares multiple regression analysis: *output,* measured by square feet of single-ply equivalent belting; *weight,* expressed as average weight per square foot of single-ply equivalent belting; *magnitude and direction of change of output from preceding month,* this last, however, being included solely in the analysis of overhead cost behavior.

VII. Findings

Examination of the process of manufacturing leather belting in the light of the theoretical considerations presented in Section II, together with the statistical distribution of the cost observations, led to the specification of a linear total combined cost function in the multiple regression analysis. Support for the hypothesis of linearity of total cost behavior, based on non-statistical considerations, is analyzed in more detail in Section IX, while the statistical aspects of the problem of specification are examined in Section VIII.[32]

The findings for combined cost, direct, and overhead cost, and the elements of cost are presented in this section in both graphic and tabular form. The total, average, and marginal aspects of cost behavior are shown for the following independent variables: (1) output, measured in square feet of single-ply equivalent belting (with the effect of average weight upon cost and upon output allowed for); (2) average weight (with the effect of square feet of output allowed for); (3) both output in square feet and average weight of belting.

The regression equations showing this cost behavior and certain other statistical constants are summarized in Table 3–1. To indicate how closely the various aspects of output used as independent variables are associated, particular attention was paid to the coefficients of multiple correlation and determination.

Analysis of Combined Cost Behavior

The total, average, and marginal combined cost curves derived from the least squares partial regression equation of total combined cost on output (after the effect of average weight has been allowed for) are shown graphically in Chart 3–4.[33] The total cost curve, which rises at a constant rate, yields a hyperbolic average cost curve, which falls continually within the range of observations. The derived marginal cost curve is a horizontal straight line, lying below the average cost curve. From this it is seen that the cost of producing an additional square foot of single-ply equivalent belting remains unchanged ($0.77) regardless of the rate of output.[34]

Chart 3–4 Partial Regression of Total, Average, and Marginal Combined Cost on Output

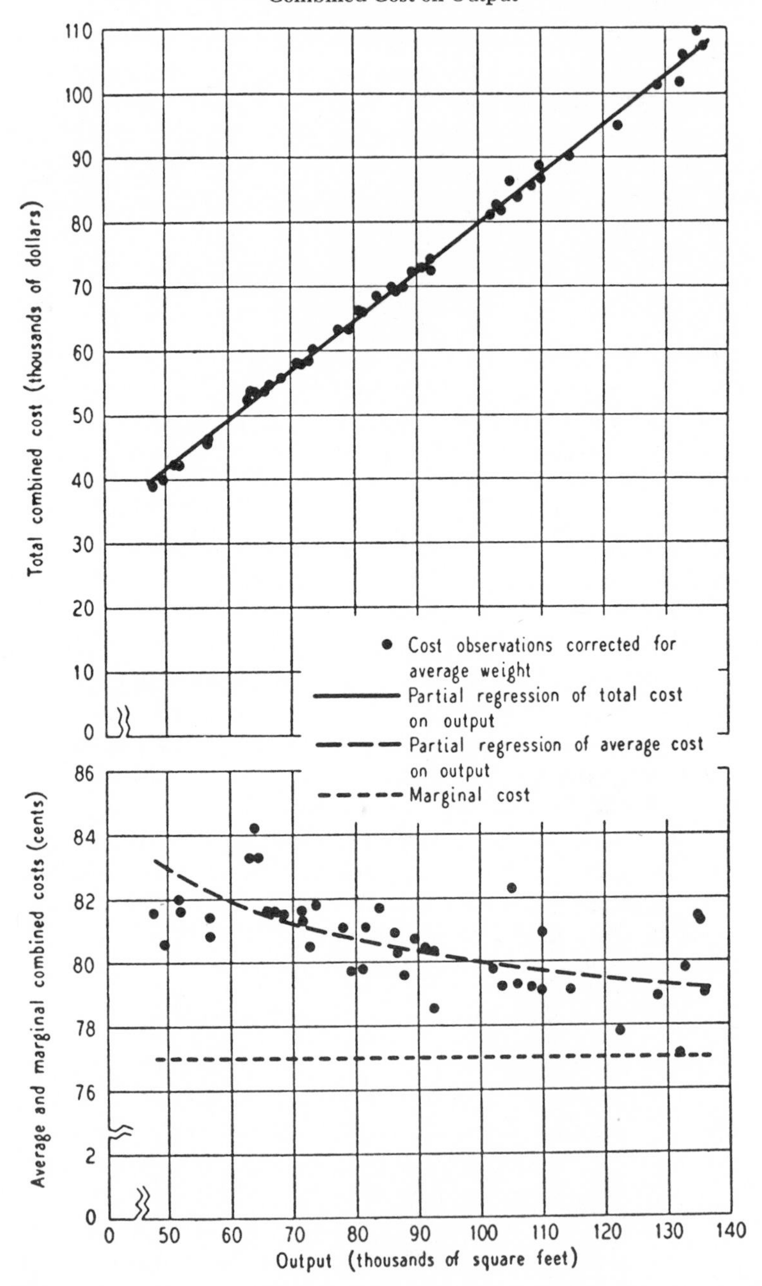

Table 3–1

SUMMARY OF FINDINGS OF CORRELATION ANALYSIS FOR COMBINED, OVERHEAD, AND DIRECT COST

Statistical Constants and Equations	Combined Cost	Overhead Cost	Direct Cost
Multiple regression equation	${}_tX_c = -60,178 + .770\ X_2 + 70,181.30\ X_3$	${}_tX_o = 2,157.78 + .0108\ X_2 - 4.828\ X_4$	${}_tX_d = -61,636.9 + .760\ X_2 + 69,324.130\ X_3$
Standard error of estimate	$\overline{S}\ 1.23 = 1,050.0$	$\overline{S}\ 1.24 = 164.4$	$\overline{S}\ 1.23 = 985.2$
Coefficient of multiple correlation	$\overline{R}\ 1.23 = .998$	$\overline{R}\ 1.24 = .842$	$\overline{R}\ 1.23 = .999$
Coefficient of multiple determination	$\overline{R}^2\ 1.23 = .997$	$\overline{R}^2\ 1.24 = .709$	$\overline{R}^2\ 1.23 = .997$
Partial regression equation for output (X_2)	${}_tX_c = 2,973.75 + .770\ X_2$	${}_tX_o = 2,108.72 + .0108\ X_2$	${}_tX_d = 662.54 + .760\ X_2$
Partial regression coefficient for output (X_2)	$b12.3 = .770 \pm .0063$	$b12.4 = .0108 \pm .00018$	$b12.3 = .760 \pm .0060$
Partial regression equation for average weight (X_3)	${}_tX_c = 7,394.2 + 70,181.30\ X_3$		${}_tX_d = 5,059.80 + 69,324.13\ X_3$
Partial regression coefficient for average weight (X_3)	$b13.2 = 70,181.30 \pm 643.4$		$b13.2 = 69,324.15 \pm 603.7$
Partial regression equation for change in output (X_4)		${}_tX_c = 3,107.69 - 4.828\ X_4$	
Partial regression coefficient for change in output (X_4)		$b14.2 = 4.828 \pm 1.760$	

${}_tX_c$ = combined cost in dollars
${}_tX_o$ = overhead cost in dollars
${}_tX_d$ = direct cost in dollars

X_2 = output in square feet of single-ply equivalent belting
X_3 = average weight in pounds per square foot
X_4 = change in output in square feet from preceding month

Chart 3–5 Partial Regressions of Total Combined Cost on Output and Average Weight

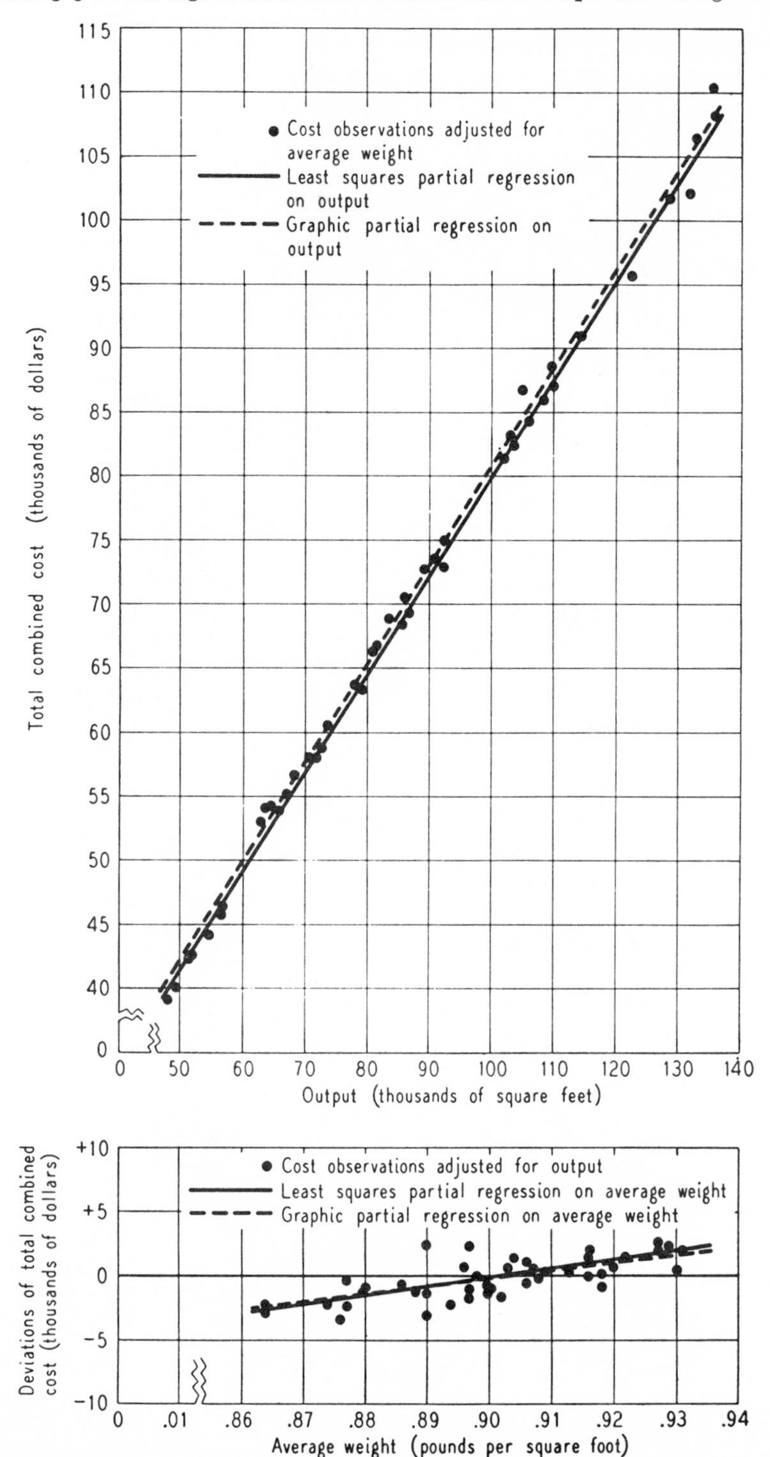

The statistical findings concerning the partial regression of combined cost are shown graphically in Chart 3–5 for the two methods of analysis, graphic and least squares. The lower panel of this chart shows the net relation of cost to average weight of belting when the influence of output measured in square feet has been removed.[35]

Table 3–2

ESTIMATES OF COMBINED COST IN DOLLARS FOR DIFFERENT LEVELS OF OUTPUT MEASURED IN SQUARE FEET

Output	Total Cost	Average Cost	Cost of Increment in Output
40,000	33,770	.844	....
60,000	49,168	.819	15,398
80,000	64,566	.807	15,398
100,000	79,964	.800	15,398
120,000	95,362	.795	15,398
140,000	110,760	.791	15,398

If both average weight of belting and number of square feet are introduced as independent variables, the cost function can be expressed in the form of a multiple regression equation which shows the value of total combined cost associated with various combinations of output and average weight. This equation, determined by the method of least squares, is:

$$ {}_{t}X_{c} = -60.1780 + 0.770\,X_2 + 70.181\,X_3 $$

By making use of this equation, it is possible to arrive at estimates of total combined cost that would be expected for different combinations of the variables, output and average weight. An example of the use of the equation for this purpose is shown in Table 3–3, in which a number of estimates have been drawn up.[36] The

Table 3–3

ESTIMATED TOTAL AND AVERAGE COMBINED COST IN DOLLARS FOR VARIOUS COMBINATIONS OF OUTPUT IN SQUARE FEET AND AVERAGE WEIGHT IN POUNDS PER SQUARE FOOT

	Average Weight					
	.86		.89		.92	
Output	Total	Average	Total	Average	Total	Average
40,000	30,974	.774	33,079	.827	35,185	.880
60,000	46,372	.773	48,477	.808	50,583	.843
80,000	61,770	.772	63,875	.798	65,981	.825
100,000	77,168	.772	79,273	.793	81,379	.814
120,000	92,566	.771	94,671	.789	96,777	.806
140,000	107,964	.771	110,069	.786	112,175	.801

standard error of estimate of cost in this equation was found to be $1,050. This figure must be interpreted subject to the limitations of time series data mentioned above. If variations from the equation are random, an estimate of cost based on the regression equation would, in approximately two cases out of three, lie within a range of plus or minus $1,050 of the actual cost for the period. The coefficient of multiple correlation (i.e., the simple correlation between actual cost and cost estimated from the net regressions) was 0.998, and the coefficient of multiple determination,

Table 3–4

ESTIMATES OF OVERHEAD COST IN DOLLARS FOR DIFFERENT LEVELS OF OUTPUT IN SQUARE FEET

Output	Total Overhead Cost	Average Overhead Cost	Cost of Increment in Output
40,000........	2,541	.064	216
60,000........	2,757	.046	216
80,000........	2,973	.037	216
100,000........	3,189	.032	216
120,000........	3,405	.028	216
140,000........	3,621	.026	216

0.997. If the sample were random, a coefficient of multiple determination of 0.997 would indicate that almost the entire variance in cost was accounted for by the variance of the independent variables.

Analysis of Cost Components

The two components of the combined cost of the leather belt shop, direct cost and overhead cost, were analyzed in the same way. The findings for overhead cost

Table 3–5

ESTIMATES OF OVERHEAD COST IN DOLLARS FOR VARIOUS CHANGES IN THE LEVEL OF OUTPUT IN SQUARE FEET FROM PRECEDING MONTH

Percentage Change in Output	Total Overhead Cost
−30	3,252
−15	3,180
0	3,108
+15	3,036
+30	2,864

($_tX_o$) are presented first. The two independent variables considered are output measured in square feet and the percentage change in output from the preceding month (X_4). The relation of overhead cost to output (after allowing for the effects of the percentage change in output from the preceding month) is illustrated in the upper panel of Chart 3–6 by net regression lines derived by both graphic and least squares multiple correlation.[37] The independent effect of the magnitude and direction of change from the output of the preceding month (when the effects of the current rate of output are allowed for) is shown by the net regression line in the lower panel of Chart 3–6.[38] From this chart it is clear that the relation of overhead cost to the two independent variables is linear, although considerable dispersion of the individual observations about the regression lines is evident.

Partial regression curves for total direct cost ($_tX_d$), showing its relation to output and to average weight, determined by both graphic and least squares analysis, are presented in Chart 3–7.[39] A numerical illustration of the meaning of these regression curves is presented in Tables 3–6 and 3–7.

Table 3–6

ESTIMATES OF DIRECT COST IN DOLLARS FOR DIFFERENT LEVELS OF OUTPUT IN SQUARE FEET

Output	Total Direct Cost	Average Direct Cost*	Cost of Increment in Output
40,000......	30,396	.760	15,198
60,000......	45,594	.760	15,198
80,000......	60,791	.760	15,198
100,000......	75,989	.760	15,198
120,000......	91,187	.760	15,198
140,000......	106,385	.760	15,198

*The intercept of the total direct cost curve on the cost axis is so small that there is no perceptible variation in average cost.

Table 3–7

ESTIMATES OF DIRECT COST IN DOLLARS FOR DIFFERENT AVERAGE WEIGHTS IN POUNDS OF BELTING

Average Weight	Total Direct Cost	Cost of Increment in Weight
.86	64,675	....
.88	66,062	1,387
.90	67,448	1,386
.92	68,834	1,386
.94	70,221	1,387

Chart 3–6 Partial Regressions of Total Overhead Cost on Output and Change in Output from Preceding Period

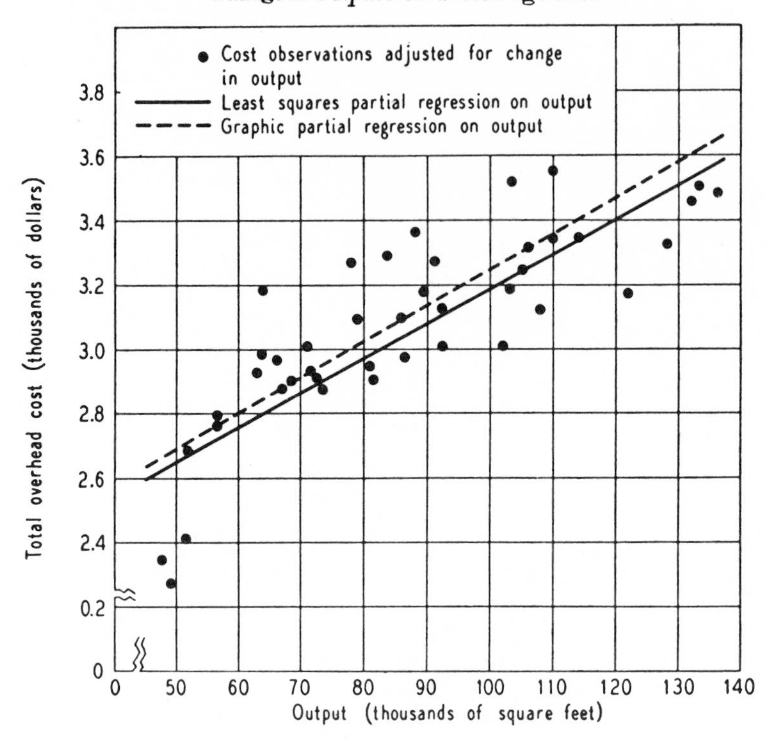

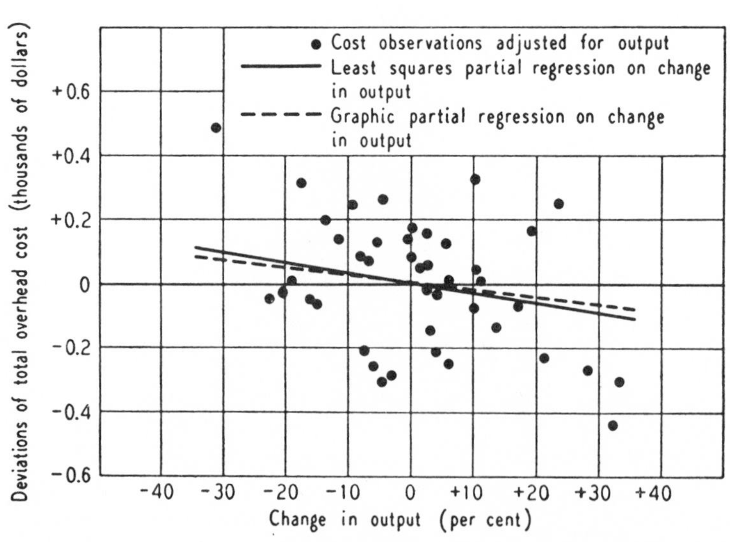

Chart 3–7 Partial Regressions of Total Direct Cost on Output and Average Weight

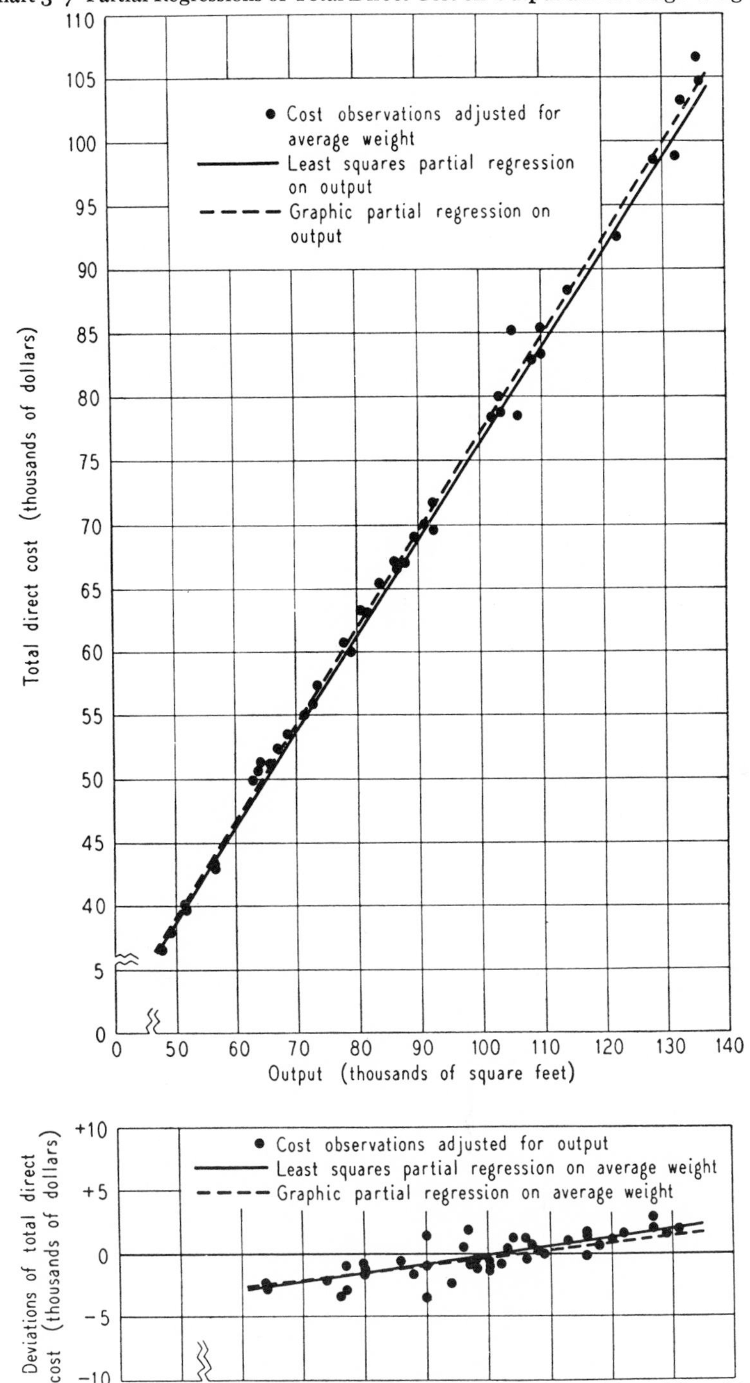

Analysis of Cost Elements

The elements of cost making up the cost components, direct and overhead cost, were examined separately by graphic methods to determine their behavior at different levels of leather belt shop activity. The relations established by this graphic analysis are summarized in Table 3–8. The various regression lines are illustrated in Chart 3–8.

Table 3–8

REGRESSION EQUATIONS FOR ELEMENTS OF COST ON OUTPUT DETERMINED GRAPHICALLY

Cost Element	Regression Equation
Direct	
Cement..........	${}_1X_d = -.06 + .0282\ X_2$
Direct labor......	${}_2X_d = .99 + .0538\ X_2$
Leather..........	${}_3X_d = .90 + .675\ X_2$
Overhead	
Fixed charges*....	${}_1X_o = .905 + .00143\ X_2$
Indirect labor....	${}_2X_o = .249 + .00207\ X_2$
Repairs..........	${}_3X_o = .0516 + .00120\ X_2$
Supplies.........	${}_4X_o = .098 + .00380\ X_2$

*Taxes, depreciation, insurance, power and water.

Particular interest attaches in this study to the behavior of marginal cost, defined above as the cost attributable to an increment in output of one square foot of single-ply belting. Estimates of this marginal combined cost derived from the regression equation for total combined cost, and from the regression equation for the individual elements of cost, are presented in Table 3–9. In addition, a comparison is made of the analogous marginal cost that results from a unit increase in average weight estimated by different methods. It is seen from the table that marginal combined cost and the marginal cost resulting from unit increments in average weight are approximately the same whether estimated directly or by summation.

Behavior of "Reflated" Cost

In establishing the functional relation of cost to output, the prices paid for materials and labor were held constant during the period of analysis. If, however, such statistical functions are to be useful for cost forecasting, as guides to price policy, and in determining whether the cost incurred in any period differs from the general pattern of behavior, prices of input factors appropriate to the period must be substituted for the "deflated" or stabilized prices used in the analysis. Fortunately, such a computation is relatively easy since if the cost of any group of elements for a given set of prices is known, the physical quantities of the factors can be deter-

Chart 3–8 Regressions of Elements of Combined Cost on Output

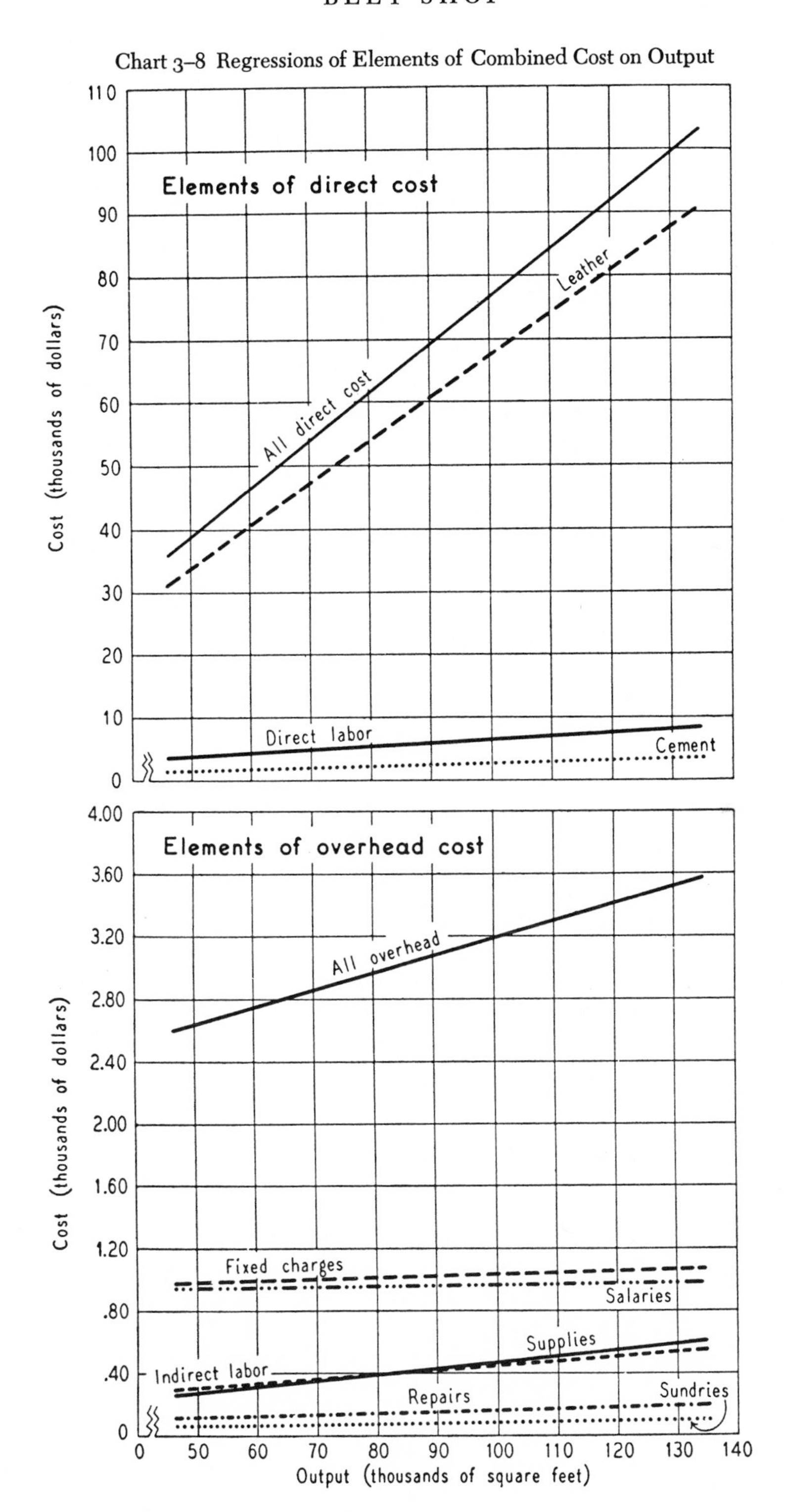

Chart 3–9 Fluctuations of Recorded Average Cost, Reflated Marginal Cost, and Output

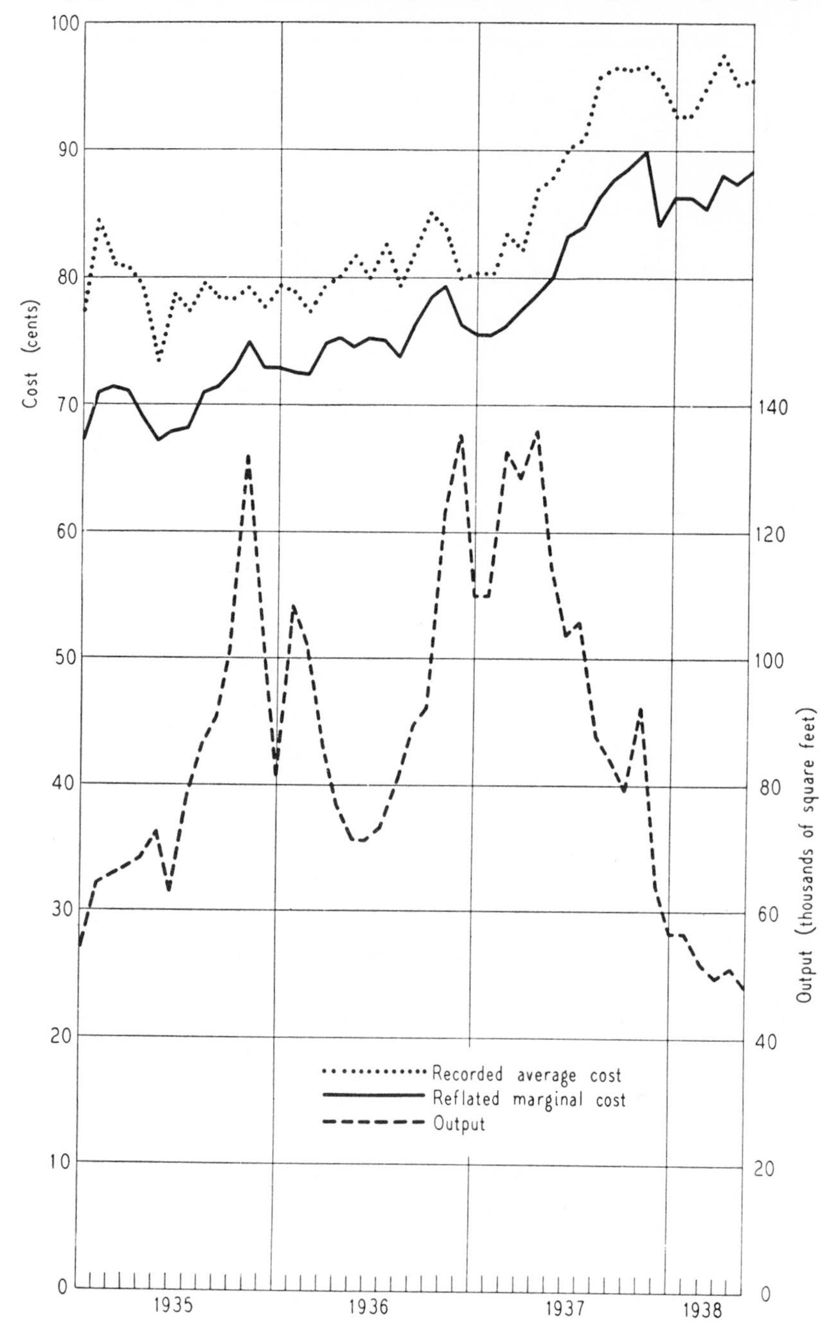

Table 3–9

ESTIMATES OF MARGINAL COST IN DOLLARS BY ALTERNATIVE METHODS

Cost Element	Marginal Cost of a Square Foot	Marginal Cost of a Pound*
Cement	0.28	.0251
Direct labor	.054	.0485
Leather	.675	.6063
Sum of elements of direct cost	.757	.6809
Direct cost (estimated independently)	.760	.6826
Fixed charges	.001	.0009
Indirect labor	.002	.0018
Repairs	.001	.0009
Supplies	.004	.0036
Sum of elements of overhead cost	.008	.0072
Overhead cost (estimated independently)	.011	.0099
Sum of all elements of direct and overhead cost	.765	.6881
Combined cost (estimated independently)	.770	.6916

*The marginal cost of a pound was obtained by multiplying the marginal cost of a square foot by a constant representing the number of pounds per square foot. This alternative index of output is not to be confused with the variable, average weight per square foot, included in the analysis.

mined. The magnitude of the elements of cost appropriate for another set of prices can then be found by multiplying the quantities by the appropriate prices.

Chart 3–9 shows marginal cost "reflated" to reflect the prices actually existing in the period. The rough similarity between the fluctuations of "reflated" marginal cost and those of recorded average cost arises from the predominant importance of leather cost in both. The departures from similarity, attributable mainly to fluctuations in output (also shown in this chart) reflect the inverse relation between output and the proportion of fixed cost to recorded average cost.

VIII. Validity of Observed Relations

Some potential sources of error that might influence the statistical results have already been discussed briefly. In order to appraise the validity of the statistical findings, we now examine in more detail their limitations, which may be attributable either to inadequacies inherent in the data or to the technique of analysis. The following considerations may conceivably have an important bearing upon the reliability of the findings of this investigation: (1) The sample may be inadequate, the observations not being representative, particularly for high output. (2) Certain cost elements that bear some relation to output were omitted, for example, allocated general firm overhead. (3) The rectification procedure may have errors and shortcomings, such as improper allocation of cost to time periods, elimination of price changes that may have resulted from variation in the plant's output rate, and the impossibility of eliminating non-random errors in the data. (4) Sufficient account may not have been taken of all operating conditions that influence cost; specifically, the rejection of certain independent variables in the multiple regression analysis that exert an appreciable influence on the behavior of cost may not have been justifiable. (5) The regression function may have been incorrectly specified, a possibility that arises when there is a large scatter in the observations so that there is some doubt concerning which function fits the data best. The first, third, and fifth of these limitations are of particular interest in this study.

Since the firm belonged to a declining industry, the sample was to a certain extent not representative of industrial conditions in general. Subnormal activity prevailed in the industry during the period studied, January 1935 to June 1938. Subnormal activity refers not only to a low level of output, but also to the aggregate of phenomena correlated with or directly due to a *prolonged* period of low output, including pessimistic expectations concerning future developments. This situation presents, however, certain advantages for statistical cost analysis since under such circumstances it is possible to avoid the difficulties arising from secular growth in scale of plant. Moreover, since in fact numerous industries do experience a secular decline in the demand for their products, the sample may not be valueless.

Price corrections and allocations of recorded cost to accounting periods, the latter being admittedly arbitrary, are relatively large adjustments that are likely to be attended with inaccuracy. Even price changes must be corrected for approximately, and the method of cost allocation may conceal a tendency for the marginal cost of repairs, cement, and supplies to rise. These elements of cost, however, constitute only 4.22 per cent of combined cost, so that the possible error is negligible.

There is some uncertainty whether regression functions of a linear form describe the observations as well as a curvilinear function form. The theoretical considerations underlying the alternative forms of the short-run total cost curve were examined in Section II. Two cases of cost behavior were examined: (1) that described by a curvilinear total cost curve, which rises first at a decreasing rate and eventually at an increasing rate as output increases, and (2) that in which the total cost curve is linear to the point of physical capacity, when it begins to bend upwards. Preliminary graphic multiple regression analysis of rectified observations of total combined cost supported the hypothesis of linearity. However, the strong preference of economic theorists for the curvilinear hypothesis made it advisable to test the hypothesis of linearity by methods more rigorous than graphic analysis.

Objective tests of the linearity of the total cost functions are especially desirable since approximate linearity of the curve is no positive assurance that the corresponding marginal cost is constant. Even though the discrepancy between the linear and curvilinear total cost curve is barely perceptible upon visual examination, the smallest degree of curvature in the total cost curve means that the marginal cost curve is curvilinear rather than constant. For this reason it is advisable that the cost functions be examined not only in total form but that the marginal and average cost functions be determined independently of the total cost function. The greater the confidence one has in the shape of the subsidiary marginal and average cost functions, determined independently, the more confidence one can have in the total cost curve consistent with them. In view of the importance of the distinction between the linear and the curvilinear total cost specification, careful attention must be devoted to any evidence that is of aid in choosing between the two. The remainder of this section is consequently concerned with statistical tests that may afford some useful guidance in making this decision:

1. Monthly total cost observations were classified into arrays corresponding to sub-groupings of output in order to obtain two independent estimates of the variance: (*a*) the variance of the observations about the means of the array; (*b*) the variance of the means of the array about the linear partial regression of cost on output. A linear functional relation may be regarded as an adequate representation of the cost-output regression if the variance of the array means about the regression line is not significantly greater than the variance within arrays. If the former variance is unusually greater than the latter, however, one is led to suspect that a curvilinear regression function is preferable.

2. Residuals from the linear multiple regression surface were computed and classified into ten groups according to output. The variance of residuals about their group means was then compared with the variance of the group means about the general mean. By testing residuals in this form it was possible to avoid the difficulty arising in the preceding test from the slope of the partial regression curve.

3. The relation of "incremental" cost[40] to rate of output and to other operating conditions was examined. If incremental cost is not significantly related to output, this is additional evidence that marginal cost is constant and consequently that the total cost function is linear.

4. Similarly, the relation of average cost (derived directly from the accounting records) to output yields further information concerning the shape of the total cost curve. If the average cost curve does not rise as output increases over the observed range, the existence of the rising phase of the cubic total cost function following a point of inflection is not substantiated.

5. A cubic cost–output regression function was fitted to the data by multiple regression analysis and the significance of its squared and cubed terms tested by Student's *t*-test.

Since there is some question concerning the validity of arbitrary, rule-of-thumb applications of statistical tests of significance to data derived from time series, it is necessary first to consider briefly the rationale of these tests. Suppose the residuals from the multiple regression of total combined cost are classified into arrays according to values of output, one of the independent variables. Even if these residuals were selected at random from a homogeneous, normal universe, some variation among the means of the arrays would be expected. Fisher's *z*-test enables one to decide, on the basis of a probability distribution, whether such variation is unusually large, when the hypothesis is that the original observations of cost, output, and weight were selected at random from a trivariate universe in which the regression is linear and the distribution of the arrays of the dependent variable normal and homoscedastic. The power or effectiveness of such a test, as a test of linearity, depends on the fact that curvilinearity of regression leads to unusually large or significant measures of dispersion about the means of arrays. Hence, the test is a good one if other hypotheses differ only in form of the regression function. This is the case here, since mere absence of linearity may be considered the alternative hypothesis.

The assumptions of random sampling, etc. on which the z-test is based are admittedly not satisfied by the data. In this study, however, these assumptions are not so unrealistic as they would be in most studies involving economic time series. Because of rectification to eliminate dynamic influences, evidence of parallel cyclical fluctuations in the form of positive serial correlation among residuals is absent. Indeed, the sign of the serial correlation coefficient is negative. Moreover, it was not even necessary to use time as a catch-all independent variable. Although the various rectification procedures almost certainly improved the data for the purposes of this investigation, the need for such procedures introduces some inexactness in the z-test, even if it were otherwise strictly applicable. The inadequacy of the technique of rectification has probably introduced sources of error additional to and more important than the errors tested; and since the rectification was not part of the least squares fitting process, its influence could not be allowed for by adjusting the number of degrees of freedom.

Despite these misgivings concerning the correspondence of the data with the specifications required for the application of analysis of variance tests, it seemed desirable to test the signifiance of the relations established by all available methods.

Analysis of Variance of Total Cost Observations from Linear Partial Regression[41]

In order to test the linearity of the relation of total cost to output the cost observations were classified according to output rate into ten groups. The variance of the observations about the mean of each group was then compared with the variance of these group means from the corresponding points on the cost regression. Since the cost-output function under examination is a partial regression representing the influence of only one of the two independent variables, it was necessary to correct the total cost observations for the influence of the other variable, average weight per square foot. This was accomplished by expressing the total cost observations as deviations from total cost estimated from the partial regression of cost on average weight, and then adding these deviations to the observed mean of total cost.[42]

The corrected total cost observations having been classified according to output, the mean output and the mean total cost were computed for each group. Comparison of the variance within groups and the variance among the several means of total cost and the total cost values estimated for corresponding means of output revealed a significantly greater intra-group variance. The value of z is found to be -1.1307, when n_1 is the degrees of freedom among the group means and n_2 the degrees of freedom within groups. To substantiate a hypothesis of curvilinear regression, a high positive value of z is required. The 5 per cent point for z when $n_1 = 7$ and $n_2 = 33$ is 0.4164. If a positive value for z higher than this had been determined, curvilinearity would be indicated. The value of z actually found, however, was negative and large (absolutely), -1.1307. Since the mode of z is zero, this value lies below the mode and farther away than would occur frequently by chance. In fact, the magnitude of z that would occur in random sampling once in a thousand is only slightly larger (absolutely), about -1.26. Therefore, the indication of linearity is unusually strong. The conclusions drawn from this test, when compared with those resulting from the analysis of variance of residuals from the linear *multiple* regression, afford interesting evidence of the distortion of the intra-group variance caused by the use of adjusted residuals from steeply sloping partial regressions.

Fisher's z-test, therefore, shows that the variance within arrays is much smaller than would be expected on the basis of random sampling according to the specifications described above, i.e., the sample regression is more nearly linear than one would expect it to be even if the observations on cost, output, and average weight were selected at random from a universe in which the regression is linear.

Analysis of Residuals about Linear Multiple Regression Surface

Since the large negative value of z in the preceding test seemed to be attributable primarily to intra-group variance caused by the slope of the partial regression line, an alternative form of the z-test was applied. Residuals were computed from the multiple regression surface instead of from the partial regression curve. These residuals were classified by output into ten groups. Their intra-group variance was then compared with the variance among the groups. In this case the value of z is $-.03089$, using n_1 for the degrees of freedom among groups and n_2 for the degrees of freedom within groups. In contrast to the preceding test, this value of z is not notably below the average value for random samples from a universe of the type specified. The 5 per cent point is 0.4164. The existence of curvilinearity of regression would be expected to cause unusually large variance within the arrays of the residuals about the multiple regression surface. However, since the observed variance within groups is not significantly unusual, this z-test also apparently indicates that the distribution of the observations is consistent with a hypothesis of linearity.

Analysis of Incremental Cost

The analysis of incremental cost was designed, first, to provide additional evidence concerning the hypothesis of the linearity of the total cost function; second, to explore an alternative method of studying the behavior of marginal cost, and third, to test by an independent estimate the magnitude of marginal cost found by differentiating a fitted total cost function.[43]

Incremental cost estimates were obtained directly from the cost observations by expressing the difference between the adjusted[44] total cost for each month and that for the preceding month as a ratio to the corresponding month to month first difference in output. This ratio represents the observed average increment in cost for the range of the specific increases in output and is designated, as noted above, as incremental cost. The behavior of incremental cost at various levels of output can be seen in Chart 3–10. On the assumption that the magnitude of incremental cost

Chart 3–10 Incremental Cost at Various Levels

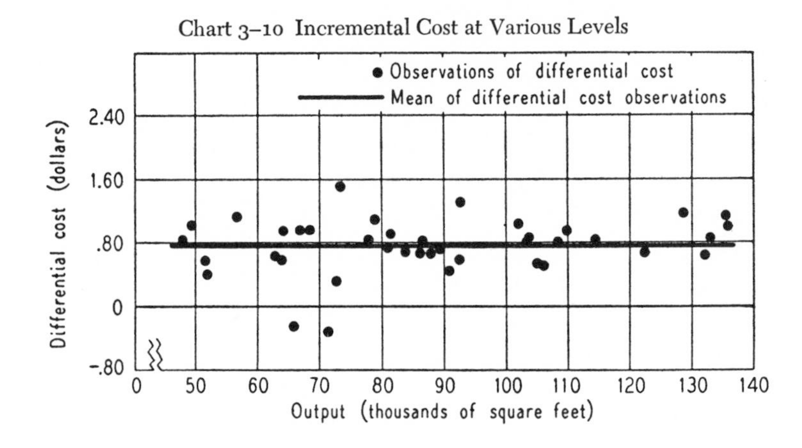

is unrelated to the level of output, the arithmetic mean of the incremental cost observations was computed and found to be $.767.[45] The wide scatter of the incremental cost observations, together with the assumption that it is independent of the output rate, restricts the confidence to be placed in this value of incremental cost as an estimate of marginal cost. Nevertheless, such an estimate is very close to the magnitude of marginal cost derived from the total cost function by differentiation ($.77).[46]

There are, however, certain essential differences in the nature of these two cost estimates. First, average incremental cost has reference to finite and sometimes large increments in output, whereas marginal cost, estimated by differentiating a fitted total cost function, is relevant for very small (theoretically infinitesimal) changes in output. Second, incremental cost was derived from scattered observations subject to much random error while marginal cost, estimated from a continuous function, is not influenced by random variation. Third, the total cost observations used in computing incremental cost were not corrected to remove the estimated average influence of average weight, whereas this distortion was removed in estimating marginal cost.

The hypothesis that the data in total cost form show no evidence that the magnitude of marginal cost is related either to output or to the other independent variables was tested by analyzing the effect on incremental cost of various independent variables, including some not used in the least squares analysis of the total cost observations. Since the primary objective was to ascertain the existence of a relation, rather than to determine its precise nature, the functional relation between incremental cost and output was examined by means of the analysis of variance. This validating device serves as a more rigorous test of the findings concerning constancy of marginal cost, being more objective than visual examination of the incremental cost observations. The reliance to be placed in this test is limited, it should be remembered, by the magnitude of the random variation in incremental cost and the difference between these observations and marginal cost in its more precise sense.

The procedure was to group incremental cost observations (omitting one case because of its unusualness) according to the corresponding values of the particular independent variable involved, and to test the existence of a relation between incremental cost and each variable by applying Fisher's z-test to determine the significance of the ratio of the variance within groups and the variance among groups. If the value of this ratio is found to be not significant, there is on this ground no reason to reject the hypothesis that no relation exists between incremental cost and the particular independent variable considered.

Since the method of grouping may influence the results, tests were applied for two groupings of each independent variable. The observations were first divided into three or four classes, then redivided into ten equal groups and retested in order to ensure that the use of too broad classifications had not obscured the relations.

Both tests for the existence of correlation between incremental cost and output demonstrated that no relation significant in a statistical sense existed. The value of Fisher's z is only slightly larger than its average value in random samples from an uncorrelated universe and might easily have occurred by chance. The value of z is 0.0726 when $n_1 = 9$ and $n_2 = 30$, while the 5 per cent point is 0.3925. This result, as well as the results obtained for the other independent variables, applies to the ten-group classification.

The investigation of the relation of incremental cost to average weight of belting revealed the same general situation. The value of z is -0.1465, the negative value indicating that this magnitude of z is below the average value expected in random samples from an uncorrelated universe.

The lack of a relation between total cost and direction and magnitude of change in output indicated by graphic analysis was substantiated by statistical tests using the analysis of variance. The magnitude of z is -0.7721 when $n_1 = 9$ and $n_2 = 30$, a negative value so far below the average value in random samples from an uncorrelated universe that it would be exceeded in more than 95 in 100 cases by pure chance. The analysis of variance test for incremental cost and absolute magnitude of change in output does not show the existence of any significant correlation.[47]

In general, the analysis of incremental cost substantiates the findings of the total cost analysis, both in indicating the lack of a relation between marginal cost and output and in providing a subsidiary estimate of the marginal cost almost identical with the marginal cost derived from the total cost equation. The form of the incremental cost observations, however, together with their great chance variability, restricts their reliability as a basis for the validation of relations established by the analysis of total cost observation and limits their usefulness in estimating marginal cost directly. Although direct analysis of first differences of cost and output is a more economical way to estimate marginal cost than correlation analysis of total cost, it is distinctly less reliable.

Analysis of Average Cost

The distribution of observations of average cost affords some additional information concerning the shape of the total cost function. A total cost curve of the conventional form, represented by a cubic parabola, leads to a U-shaped average cost curve. The scatter diagram of adjusted observations of average cost in the lower panel of Chart 3–4 does not suggest this sort of distribution. On the contrary, the scatter conforms closely to the average cost curve derived from the linear total cost curve. This curve, of course, differs from that which would have been found by fitting a curve to recorded average cost. The deviation to be minimized by the least squares fitting would differ for total and average cost observations, since the correlation of total cost and output was not perfect. The curve of average cost derived from the total cost curve nevertheless appears to describe with reasonable accuracy the behavior of recorded average cost. To determine the degree of this corespondence, the correlation coefficient was computed between recorded average cost and average cost derived from the equation

$$ {}_aX_c = .770 + \frac{2.974}{X_2} $$

(this equation was derived from the partial regression equation of total combined cost on output, after allowance for the influence of average weight). This coefficient was found to be 0.866; the multiple correlation coefficient for total cost is 0.998.

These four types of statistical test indicated that the cost and output data for the leather belt shop, for the range of output observed, are consistent with the hypothesis that the total cost function is linear. They showed, moreover, that a different approach to the determination of marginal cost yields substantially the same result as that obtained mathematically from the total cost function, and that the function for average cost obtained from the function fitted to the total cost observations explains most of the variation in recorded average cost.

Analysis of Fit of Cubic Function

To aid in discriminating more specifically between a cubic and a linear functional form, a fifth test was applied. A third degree regression function of the general form

$$_{t}X_{c} = b_1 + b_2X_2 + b_3X_3 + b_4X_2^2 + b_5X_2^3$$

was fitted by least squares multiple regression analysis and the significance of various regression coefficients was examined by applying Student's t-test. The mathematical analysis yielded the following partial regression equation of cost on output:

$$_{t}X_{c} = -12.995 + 1.330\,X_2 - 0.0062\,X_2^2 + 0.000022\,X_2^3$$

The behavior of this regression function is illustrated graphically in Chart 3–11, in which the derived marginal and average cost curves are also shown together with the partial regression of cost on the other independent variable, average weight. Neither the marginal nor the average cost function exhibits great variability. At the extreme of the output range, the parabolic marginal cost curve lies about 14 per cent above the level of constant marginal cost ($.77). At intermediate levels of output it lies a maximum of 4 per cent below. Because the cubic total cost curve has a negative intercept, the average cost curve behaves illogically even within the observed range. In the range between 45,000 and 65,000 square feet, the average cost curve rises; beyond this it falls until it is intercepted by the marginal cost curve, whereafter it rises slightly.

To test the suitability of the cubic function it was necessary to determine whether the regression coefficients of the higher-order terms are small enough to be attributable to errors of sampling.[48] Student's t-test was, therefore, applied by computing for the squared and cubed terms the ratios of the beta coefficients to their respective standard errors. These ratios are -2.01 for the squared term and 1.94 for the cubed term. Interpolating for $n = 38$ in a table of the distribution of t, it is found that the 5 per cent point is 2.025. However, the entries given in this table to determine the criteria of significance are based upon the sum of the tails of the t-distribution—a procedure that does not seem justifiable since the sign of the regression coefficient is specified in the theoretical model. Taking into account only one tail of the distribution, the higher order terms in the equation seem even more significant, since they lie near the 2½ per cent point.

In order to interpret the results of this test properly it is necessary to consider briefly its nature and determine whether the data comply with the statistical specifications implied in this type of test. The t-test applied to a cubic function tests the significance of the fit by setting up the null hypothesis that the universe value of the squared or cubed term is zero. This hypothesis is then examined by determining the probability of finding regression coefficients as great as those observed by random sampling from a universe of cost, output, and average weight in which the regressions are linear and the arrays of cost with respect to the independent var-

Chart 3–11 Partial Regressions of Total, Average, and Marginal Combined Cost for Third Degree Regression Function

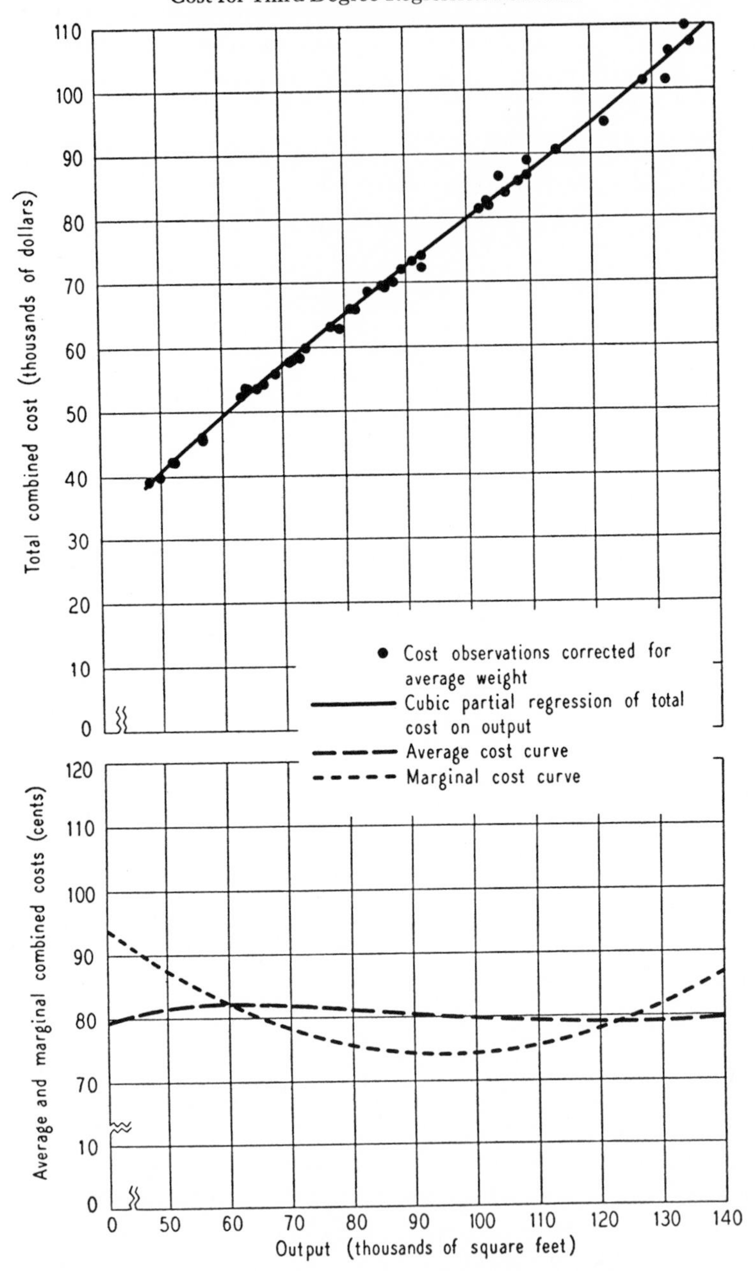

iables are normal and homoscedastic. If the probability of obtaining coefficients as great as those observed in random sampling from such a universe is higher than some arbitrarily established level (say 5 per cent), the hypothesis that the cubed term in the true function is zero is not disproved. If the probability is small, 5 per cent or less, the cubed term may be regarded as significant and the hypothesis of linearity rejected. This test is effective in discriminating between the hypotheses of linear and cubic cost behavior, provided the data conform to the requirements of the test and that the two types of behavior seem equally realistic in view of the technical methods of production encountered in the manufacturing process analyzed.

Table 3–10

SUMMARY OF STATISTICAL CONSTANTS FOR MULTIPLE REGRESSION OF TOTAL COMBINED COST ON WEIGHT, OUTPUT, OUTPUT SQUARED, AND OUTPUT CUBED

	Constant Term	Coefficient of X_2	Coefficient of X_2^2	Coefficient of X_2^3	Coefficient of X_3
Regression coefficients.......	−75.7730	1.3304	−0.0062	0.000022	69.7622
Regression coefficients in standard deviation units...		1.737	−1.5011	0.7810	0.0632
Standard error of regression coefficients in standard deviation units...........		0.3545	0.7470	0.4019	0.0091
Ratio of regression coefficient to standard error in standard deviation units...		4.90	2.010	1.943	6.945

The data, however, fail to meet the sampling specifications in several significant respects. The usual limitations inherent in time series are present, though they seem to be less serious for these data than for most, but there are other reasons for not expecting a normal, homoscedastic distribution of residuals. The magnitude of the rectification adjustments, which are unlikely to lead to a random distribution of errors, dwarfed the tiny residuals that are the basis for the *t*-test. These correction procedures, necessarily approximative, in themselves constitute sources of variation greater than those included in the test specifications. Moreover, it was not possible to include the rectification devices in the test by adjusting the degrees of freedom. If it had been possible to study the residuals in the original data, or to take account of rectification adjustments in the test, entirely different conclusions might have been reached.

In order to approximate a static competitive model, it was assumed that the prices of the factor inputs are independent of the output level of the individual firm. However, the activity in any firm is likely to be closely related to the activity in the industry, and it is not reasonable to assume that input prices are independent of the operating level in the industry. If this close association exists, it is to be expected that rising factor prices accompany high levels of activity in the firm; and falling factor prices, low levels of activity. Moreover, with expanding industry demand, recourse may be had to factors that are inferior in quality, while when industry demand is low, superior factors will be retained by the firm. These two considerations are especially relevant in explaining the cubic shape of the total cost function. If, for high outputs cost increases more than in proportion and for low outputs less than in proportion to output, the associated total cost function has a cubic shape. In the case under consideration the observations apparently responsible for the particular shape of the cubic function, i.e., the costs associated with high and low outputs, were likely to be inadequately rectified for price and quality changes, the bias being in the direction that would create a cubic total cost function.

The four observations lying above the fitted line at the highest output levels were in December 1936, and March, April, and May, 1937 (Chart 3–11). Thus they occurred at a cyclical peak, when defects in rectification would be expected to overstate deflated cost. The six observations lying below the fitted line at the lowest recorded outputs were depression months—January to June 1938—a period in which the rectification devices may not have accounted adequately for price and quality fluctuations.

The negative intercept of the total cost curve, and the consequent illogical behavior of the average cost curve within the range of observations, cast some further doubt on the validity of the cubic function. Moreover, as pointed out in Section II and discussed in Section III, there are indications that the technical structure of the production process does not correspond to that assumed in the cubic model.

In view of all these considerations, the curvature within the observed range does not seem to substantiate decisively the hypothesis that the total cost function is curvilinear.

IX. Conclusions

The statistical evidence presented in Section VIII gives some support to the conclusion that marginal cost is constant within the range of output examined in this study. The findings of such an investigation as this that are most significant for economic theory can be presented adequately by considering solely the behavior of marginal cost; for, if the course of the marginal cost function is known, the shape of the total cost function is apparent. (Supplementary information is needed to determine the magnitude of fixed cost and the behavior of average cost.) Some caution must be observed, however, in comparing the marginal cost function of a model firm under static competitive conditions with marginal cost function derived by statistical methods from empirical data. The observations that are the basis of the statistical estimate may not have been adequately purged of the influence of extraneous variables by the sampling, rectification, and correlation analysis procedures. To the extent that dynamic factors are present in the cost and output observations the empirical curves will not be a precise counterpart of the curves described in theory. It appears likely, however, that the most important dynamic influences were eliminated in the data adjustments.

On the assumption that our statistical techniques have successfully isolated the static marginal cost curve, it is desirable to attempt to account for the particular results observed. The explanations will be considered under the following headings: (1) possibility of excess capacity; (2) segmentized organization of plant; (3) conventional rigidities.

Possibility of Excess Capacity

The most obvious explanation for the constant marginal cost observed for the belt shop is that only one portion of the marginal cost curve was examined. A rising phase of marginal cost is, therefore, not disproved and, on the other hand, is almost certain to occur as physical plant capacity is approached. It is likely, however, that the major part of the possible variation of output was explored. The index of output ranged from 40 to 135, with four observations over 130. The largest output observed had not been exceeded during the preceding ten years of operation; plant managers expressed the opinion that this output represented "practical" plant capacity. This may be interpreted to mean that somewhere beyond the range of observations marginal cost would rise markedly.

Since excess capacity of an economic sort is indicated, an exploration of its causes may lead to a better understanding of the implications of the findings. Were it possible to assume that the firm was in long-run equilibrium, so that its long-run average cost curve was tangent to a falling demand curve, the firm would be operating on the descending phase of its short-run average cost curve. In such a situation, visualized in theories of monopolistic competition, excess capacity could be ascribed to imperfections of competition. Such an assumption, however, would be highly unrealistic and unjustified by any available information concerning the firm. More valid explanations might be found in a shrinkage of demand subsequent to the time investment in fixed plant was made, or over-investment as a result of optimistic expectations. These causes, the first of which seems important, would be operative in a non-equilibrium situation. The shrinkage in demand evident in the firm's sales records is explained by the encroachment of chain drives and rubber belts on a field previously dominated by leather belting.[49] Despite the general secular decline in demand, some further reasons must be sought to explain why, in view of the large seasonal and cyclical shifts in demand, the plant was not forced to operate in the area of rising marginal cost. Two possible explanations may be mentioned. First, the possibility of anticipating peaks in sales and manufacturing for stock in times of slack demand may have made it unnecessary to force production beyond the critical level. This is especially true of leather belting manufacture since sales peaks can be forecast and the product can easily be stored.[50] Second, the declining trend of industry demand may have shifted the firm's individual demand function to the point where sales fell so far short of original expectations that even seasonal and cyclical peaks do not strain the plant's capacity.

Segmentized Organization of Plant

Constancy of short-run marginal cost is, as pointed out in Section II, consonant with the technical organization of this plant and of similar mechanized operations. Segmentation of this plant into a number of similar operating units, each of which can be withdrawn from operation without influencing the efficiency of others, tends to result in constant marginal cost up to the point where all units are fully utilized. Variation in the time these units are operated, attained either by changing the number of shifts or the length of the work week, will accomplish the same end.[51] Two further questions are whether all the fixed plant and equipment can be segmented and whether the essential labor force is homogeneous. Concerning the first, it is likely, in practice, that segmentation of certain parts of a fixed plant is not possible, while a high degree of segmentation is feasible for other parts. If machinery and operating equipment are segmented, even though the buildings, etc. remain indivisible, all physically fixed plant relevant for marginal cost behavior may be considered to be segmented. The second question, concerning the homogeneity of the factors of production can best be treated as a part of the question of conventional rigidities.

Conventional Rigidities

If the management of this firm had been perfectly free to adapt the organization of production to variations in output by taking full advantage of the differences in efficiency in the segmented units, it is doubtful that the tendency to increasing marginal cost would be so completely offset as the statistical findings indicate. Although neither machines nor operators are of uniform efficiency, to select the units of the various factors with a view to employing them in the order of their efficiency was possible only within narrow limits. Seniority rights, repair programs, humanitarian and other considerations apparently deterred the management from taking full advantage of differences in efficiency. For these reasons, the hierarchy of efficiency that undoubtedly existed was not reflected in rising marginal cost with more intensive utilization of plant.

Restrictions of this kind therefore constitute a type of rigidity[52] that prevents the employer from selecting the minimum cost combination of factors for a given output.[53] As we know that conventional rigidities are present, we conclude that there is effective homogeneity of segments.

Thus the possibility of chronic underutilization of plant, the existence of segmentation, conventional rigidities and therefore effective factor homogeneity further substantiate the findings of constant marginal cost. The implication of this study, an implication of interest to industrialists as well as to economists, is that constant marginal cost within the usual range of output may be more prevalent under modern operating conditions than has been implied by much economic theory.

Study 4

DEPARTMENT STORE

I. Introduction

An empirical study of the operating behavior of the individual enterprise should supply factual information for verification of theoretical formulations. This is a counsel of perfection, however, and at the present level of development of economic science much of the value of this type of empirical study lies (1) in determining whether the assumptions underlying the economic theory of the individual firm are sufficiently typical, extensive, and realistic to have descriptive usefulness and to admit of empirical verification and (2) in developing statistical techniques which will in time permit the effective use of the results of theoretical research in the solution of economic problems.

The systematic theoretical analysis of the form of the cost function of the individual firm was a by-product of the attempt to clarify the relation between cost curves and industry supply curves under the assumption of perfect competition. This assumption, however, was not only unrealistic in an institutional sense, but it oversimplified the problem of cost and output determination by implying that the only part of the marginal cost curve relevant for output decisions is the rising section. The development of the economics of imperfect competition represented, therefore, a significant improvement in the theoretical background of empirical research, since by dispensing with the postulates of perfect competition it reduced the assumptions necessary to the theory of costs to the familiar technical laws of diminishing returns and variable proportions, and the psychological postulate of economically rational behavior.

When the assumption of perfect competition was dispensed with, however, it immediately became necessary to analyze that category of expenditure which was devoted to influencing the individual firm's demand function. The conditions of

Reprinted from Joel Dean, "Department-Store Cost Functions," in *Studies in Mathematical Economics and Econometrics*, ed. Oskar Lange et al. (Chicago: University of Chicago Press, 1942), pp. 222–54, by permission of the publisher. I am grateful to Mr. Walter Hoving, President of the Lord & Taylor Department Store at the time the study was made, and to Mr. Oswald W. Knauth, then President of the Associated Dry Goods Corporation, the parent company, for their generosity in making the data available for study and for permitting disclosure of the name of the firm.

equilibrium for the individual firm were restated to include selling cost as a variable category of expenditure exhibiting a functional relation to sales.[1] In the formal solution of the equilibrium condition, it is assumed that selling expense and expenses of production in the narrow sense can be separated and the amount of each determined independently. By definition, the two may be clearly distinguished on a functional basis; selling costs are accordingly expenditures designed to alter the shape or position of a firm's demand function, while costs of production proper consist only of those outlays technically necessary to produce a given physical quantity of a particular product.[2]

Empirically, it may in some cases be possible to perform the operations which are required to separate cost data into these two categories. When the business organization whose costs are being analyzed is specialized to supplying service, however, the basis of the distinction itself becomes uncertain. This is true because the assumption made in theoretical writings that a "given demand" for a good exists apart from the conditions under which the good is presented to the public is only realistic for very homogeneous and frequently purchased commodities. More usually, it is impossible to isolate "demand" from selling or promotional cost which create or modify it.

Promotional activities designed to crystallize and focus demand constitute a large part of the total operations of a department store. To some extent such promotion may quite properly be regarded as a utility-creating function, constituting an integral part of output of a retail store, namely, its sales service.[3] Moreover, promotional activities pervade every accounting category of department-store cost and contribute to every aspect of stores' sales-service product. Thus even if the basis of distinction were clear, the practical difficulties of extricating promotional costs from pure production cost in the data would probably be insuperable.

The solution adopted for this difficulty was to consider all merchandising costs as costs of production. A more general argument may be made out for this procedure on the ground that selling expenditure can only augment the sales of the affected commodity by increasing its utility to consumers; more strictly, selling expenditure converts a given physical commodity into a different and more desirable economic good, and may accordingly be regarded as a cost of production in its most rigorous sense.[4]

Whatever may be the merits of this argument, its acceptance points to an additional source of embarrassment and difficulty in this study. If we are to consider that selling expenditures change the character and utility of the product, the unit of output is no longer homogeneous as selling costs vary. A cost function derived from accounting data over a period of time during which selling expenditure constituted a varying proportion of cost would therefore relate to a product which was not homogeneous throughout the sampling period.

This nonhomogeneity of output arises from the fact that promotional cost does not bear the simple technical relation to output that pure production cost does. The amount of selling costs it is advantageous to incur presumably depends upon

(1) the position and slope of the firm's demand function and (2) the effectiveness of selling expenditure in modifying this demand function, rather than upon technical conditions of production. In the present case, however, since selling costs and costs of production are inseparable, the quantities of output and of selling expenditure cannot be determined independently, although some variation in the proportion of selling expenditure to strictly productive expenditure is undoubtedly possible. Especially in a department store dependent upon its prestige, promotional expenditures for spacious salesrooms, sales advice, and elaborate display are an essential requirement for the maintenance of sales volume. The situation approximates closely joint production in industry.

The possibility that the proportion of selling to production costs may be kept fairly stable by certain institutional conventions or standards should therefore not be overlooked.[5] In such a case, the output would be relatively homogeneous, and the cost function discovered would be more likely to possess the stability necessary for purposes of prediction.

To recapitulate, the ambiguity surrounding the concept "unit of sales service," which may properly be considered the product of a retail enterprise, made it impossible to separate selling costs from costs of production proper. The inclusion of a variable and perhaps indeterminate quantity of promotion results in heterogeneity of output through time. The cost function determined by our procedures cannot therefore be compared precisely with the static model of cost theory. All that can be claimed is that it shows the relation between a composite cost and some admittedly imperfect index of output.[6] Its heuristic value will, therefore, be confined to providing a description of merchandising costs when certain operating conditions have been stabilized in order to isolate the influence of output on cost.

As this suggests, with the exception of the inclusion of selling costs among costs of production, every attempt was made to make this empirical situation investigated conform to the static model for the short run. The investigator must wrest from his dynamic data something as nearly approaching a static function as possible if he wishes to find cost or demand functions which are conformable to those upon which so much valuable theoretical elaboration has been based. Such a procedure increases the usefulness of the findings for purposes of estimating costs and also makes their interpretation more clear. Once the basic relation between costs and output has been established, it becomes relatively simple to adjust for the effect of changes in operating conditions, such as shifts in wage rates, material prices, etc., when desired. Of course, changes which lead to factor substitution cannot be handled in this way, since the cost function itself will be altered.

The main problems in determining the cost–output relation arose from the necessity of breaking down available accounting data into the categories of fixed and variable costs as defined in theoretical writings and the selection of procedures for "rectifying" the data in order to remove the effect upon cost of changing conditions of production. No completely satisfactory solution could be found to these problems in some instances, even though no attempt was made to analyze the cost functions of the department store as a whole.[7] Only three carefully selected departments in a large metropolitan store were studied, and in these departments no analysis was made of the behavior of aggregate operating costs. Since the allocation of overhead costs to these units is necessarily arbitrary, it was obviously impossible to make any economically significant estimate of the department's true aggregate cost. Fortunately, the critical questions of cost theory center on marginal rather than average costs, although a knowledge of average costs is indispensable to the individual firm in computing its profits. In the present study, therefore, all allocated general store overhead costs were excluded. The analysis was confined to the following cost elements: (1) advertising, (2) salespeople's salaries, (3) other departmental salaries, (4) inside delivery, (5) outside delivery, and (6) direct departmental expense. Obviously some of these departmental costs are to a degree fixed, but the inclusion of such items does not affect the determination of marginal cost and yields average cost behavior useful for certain types of managerial decisions. The resulting estimates of marginal costs are subject to error because of the failure to include certain general store expenses which may bear some relation to the departments' volume. Moreover, some items which have been included may have little relation to volume but are administered, allocated, or recorded in such a fashion as to show a spurious correlation with sales volume. For example, several of the indirect advertising cost items which are allocated to departments on the basis of dollar sales may appear to possess a closer relation to volume than is in fact the case.[8] Similarly, the closeness of the relation of inside delivery cost to volume may be exaggerated by the method of computing this cost.

II. Methodology

The major methodological problems in deriving empirical cost functions from accounting records arise from the necessity of purging the data of the effects of dynamic influences which were at work during the observation period.

In order to get an adequate number of observations, monthly data were analyzed for sixty consecutive months, covering the years 1931–35 inclusive. In attempting to isolate the static components of these data, three chief methods of freeing the cost data from the distorting influences of extraneous variables were employed. First, by careful selection of the establishment, departments, cost items, and time period investigated, certain factors which would have obscured the cost-volume relation were held constant. Second, the influence of another group of factors was allowed for by using them independent variables in the multiple-correlation analysis. Third, the effects of several disturbing forces were removed directly from the data by familiar rectification procedures.

A. Collection of Data

1. *Selection of establishment.* Among distributive enterprises department stores are of particular significance because of their quantitative importance in distribu-

tion and because of the variety of merchandise carried. In searching for an establishment to co-operate in a study of this type, a large metropolitan department store was found which had kept comparable and unusually comprehensive accounting and statistical records over a period of years. Furthermore, the management was sufficiently interested in research to understand the importance of this study to economics and to see its practical usefulness in business administration.

2. *Selection of departments.* Selection of the selling departments which would be most suitable for statistical cost analysis was made after an examination of each department with reference to the following criteria:

1. The sales volume should vary from month to month in such a fashion as to give a wide range of volume and a fairly uniform coverage of this range.
2. The heterogeneity of the merchandise sold in the department should be at a minimum, so as to simplify the problem of finding a suitable output index.
3. The character of the merchandise and the nature of the transaction should be relatively uniform from month to month, in order to minimize changes in the meaning of the output index through time.
4. The department should be relatively large, to maximize the managerial significance of the study and minimize the effects of indivisibility of input factors.
5. The changes in layout, general method of operation and managerial personnel during the period of analysis should be at a minimum, in order to approximate short-run cost functions by holding technology and plant scale as nearly constant as possible.

On the basis of these criteria, three departments were selected for study: hosiery, women's shoes, and medium-priced women's coats.

3. *Selection of the time period for analysis.* Since the objective of this investigation is to determine statistically the behavior patterns of short-run selling cost, it is desirable to study the effects upon cost of changes in operating conditions in a situation which precludes the possibility of bringing certain of the input factors into optimum adjustment—a situation in which some cost elements offer effective resistance to adjustment to the prevailing operating conditions. Analyses of monthly or weekly observations of cost and operating conditions satisfy this condition, since many input factors in the department store cannot be adjusted to these relatively rapid changes in output and other circumstances. Monthly observations appeared more suitable than weekly observations for this purpose since fewer arbitrary cost allocations are involved because most of the statistical and accounting records of the firm were in terms of months. In addition, random and irrelevant fluctuations are more likely to be averaged out by taking the longer period.

In determining the years to be selected for study, the following criteria were applied:

1. Changes in the space occupied, in layout, in general methods, and in the management personnel should be at a minimum.
2. Accounting records should be comparable throughout the period of analysis.

3. A sufficient number of months should be included to permit valid statistical analyses.
4. There should be sufficient independent fluctuations in demand, and consequently in output, to give a fairly wide range of observations of both cost and output.

Using these criteria as a basis for selection, the best years for analysis appeared to be 1931, 1932, 1933, 1934, and 1935, which made available sixty monthly observations.

B. Direct Rectification of Data

Changes in operating conditions which could be removed directly are:

1. Time lag between the recording of cost and of the volume of output to which the cost contributed
2. Variation in the number of selling days in a month
3. Changes in salary rates and material prices
4. Seasonal variation not associated with sales volume

1. *Time Lag.* The time lag between the recording of cost and of operating conditions which give rise to the cost was found to be appreciable for advertising only. During the years in question, however, this lag had been foreseen by the store and to a certain extent removed from the data in the process of recording the cost. Any advertising which occurred on the last day of the month was charged to the following month, on the supposition that it was to the sales of the second month that such expense contributed. This procedure assumes that the largest part of the effect of advertisement is felt on the day following its appearance and that the part which is carried over to the subsequent day is so sharply diminished as to be negligible for practical purposes. Consultation with store executives indicated a consensus that this assumption is correct. The removal of a recording lag, therefore, appeared unnecessary.

2. *Number of selling days.* Variation in the number of selling days per month undoubtedly exerts an influence on the magnitude of cost, but the relation is neither proportionate nor constant because of the differential effect of length of month on the several subdivisions of total variable cost. Salaries of buyers are constant regardless of the length of the month. Salaries of the salespersons, stock persons, and clerical help are paid on a weekly basis and allocated to each month on the basis of the proportion of the overlapping week falling in the month. These costs, however, are not directly proportional to number of selling days, because of holidays, for which full salaries are paid. Other items, such as inside delivery, outside delivery, and departmental direct expense, vary somewhat with number of selling days, but this variation results indirectly from the influence of the number of selling days on the volume of sales rather than directly from the length of the work month.[9]

It may be concluded, then, that the number of selling days is a factor affecting both the dependent variable, cost, and certain independent variables. But since none of these relations can be considered proportional, the number of selling days should, therefore, be employed as an independent variable in the multiple-correlation analysis, in order to avoid the distortion and spurious correlation which would arise from proportional correction in the original data.

3. *Changes in input prices.* Changes in salary rates and material prices are also likely to cause variation in cost behavior over a period of time. In determining whether correction for these changes is necessary and, if so, how it should be effected, several important questions arise:

1. Were there significant changes in wage rates and material prices during the period under study?
2. Are the observed changes in rates separable from the changes in cost which are due to variation in volume of sales?
3. Can an index of wage changes be found that will reflect changes in the labor market alone and not the fluctuations associated with improvement of the skill of the individual?[10]

An analysis of the variations in the salaries of salespersons over the period under consideration revealed that the average salary in each of the departments declined throughout 1931 and 1932 and most of 1933, then remained fairly constant, with a slight tendency to rise during 1934 and 1935. The magnitude of this fluctuation, however, was not great. Furthermore, it was not a blanket reduction but was highly individualized. The desirability of removing the effects of this change in average salary from the cost data depends essentially upon whether the salary cost per unit of sales was significantly affected. If this cost has not changed during the period, then the adjustments in salary rates have been approximately proportional to the changes in sales volume and have thus brought about what is, in effect, a constant piece rate. A study of the average salesperson's salary cost per dollar of sales for each of the three departments showed that this rate did remain substantially constant. On the basis of this evidence of approximately constant factor prices per service unit, it was concluded that no correction was needed.[11]

Inside delivery cost figures were not corrected because a large proportion of their total was made up of salaries which were so administered as to defy index-number correction for the same reason discussed under salary rectification and because the materials used in this activity were so diverse that construction of an accurate price index would have been more costly than the improvement in accuracy would justify in view of the relatively small magnitude of these expenditures.

Correction of outside delivery, which was all handled by an independent delivery company during the period of study, was unnecessary because the package rate had remained unchanged over this period.

Advertising cost was found to vary mainly in response to changes in newspaper linage rates. Of advertising expenditure, 70 per cent was for newspaper advertising, and an average of 83 per cent of this proportion was paid for actual space (the remainder being for copywriting and illustration); so that about 60 per cent of advertising expense was proportionately dependent upon linage rates. The three departments in question were remarkably similar to the store as a whole with respect to the distribution of advertising expense.

Since there were significant changes in newspaper rates over the period of analysis, it was necessary to devise a correction index to eliminate the effect of these variations. A monthly index of the composite space-rate fluctuations for all the newspapers in which advertising was purchased was constructed and applied to 60 per cent of the original advertising cost data. These corrected data were then combined with the uncorrected 40 per cent to give a new series of cost figures from which the influence of rate changes had been removed.[12]

From the above discussion it is clear that rectification of most cost items for changes in wage rates or prices was found to be either unnecessary or unpractical so that only advertising cost was rectified for rate changes.

4. *Seasonal variation.* Although seasonal variation in cost may be regarded as deriving primarily from fluctuation in the volume of sales, it is to a lesser degree attributable to changes in other conditions. Since in this study volume is the principal causal factor to be associated with cost, only the seasonal variation not arising from changes in physical volume and in average value of sale should be removed from the data. It was decided, therefore, first, to study the net relation of cost to number of transactions and average value of transaction, and then to isolate and remove that part of seasonal cost not correlated with this relation.

The familiar Bean-Ezekiel technique of establishing the net regression of cost on number of transactions and average value of transaction was employed.[13] Cost residuals from these net relations exhibited a seasonal pattern. In order to correct for this, an additive type seasonal index was computed,[14] which was used to correct the costs of the coat and shoe departments. The seasonal pattern of the hosiery department was not sufficiently clear to warrant such correction.

Because of the regularity of seasonal changes in many aspects of department-store merchandising, the seasonal index arrived at in this way appears to correct the cost data for a number of distorting influences which could not be successfully removed individually. Differences in the amount and character of sales service performed, for instance, appear to have for some departments a regular seasonal pattern which accords with the observed net seasonal variation in costs.[15]

Although variation in the number of selling days in a month was not a strong enough influence to show significant correlation with cost, its effect may be partly reflected in the residual seasonal pattern. February, a short month, shows higher than average costs in all departments. July and August, short months because the store closed Saturday, show high seasonal costs. Lags between the month in which sales effort was expended and the month in which the transaction is recorded appear to differ from month to month in a fairly regular pattern. For example, February, being the beginning of the style season for coats, carries missionary work and even actual sales completion for transactions not recorded until the following month. Similarly, November is loaded with some expense of training new girls or carrying over trained personnel preparatory to the holiday peak. Vacations, which caused cost distortion difficult to remove directly, appeared also to be included in this blanket rectifier, for they occur only in the months of July and August, and in these months costs are seasonally high. Thus the blanket correction of seasonal variation appears, happily, to cover a variety of specific causes for cost distortion which could not be removed individually.[16]

C. Analysis of Relations

1. *Selection of index of output.* In choosing the measurement unit that best represents output of retailing service, both physical volume of sales and dollar volume of sales were considered. The index that most accurately reflected the effect of physical volume upon the cost of the departments appeared to be the number of transactions. Number of units seemed less acceptable since sales-service effort was not believed to increase proportionately in selling several units in one transaction. Sales service, moreover, seems to be directly associated with the number of transactions when a department has a fairly homogeneous "product"; and some items of cost, such as inside and outside delivery, are so incurred and allocated as to be closely related to transaction volume.

An alternative measure of output, dollar value of sales, possessed some advantages because some items of cost, such as advertising, appeared more directly related to value of sales than to number of transactions. On the whole it was felt, however, that value of sales did not represent as acceptable a measure of output as number of transactions, especially as it was possible to reduce monthly figures for number of transactions to a comparable basis by holding average value of transaction constant. While a complete correlation analysis was made using value of sales as the output index, the results were not as satisfactory as those obtained by using number of transactions with average value of transaction held constant. These results will not be presented in detail here.[17]

The size (money value) of the transaction had a clearly defined influence upon cost, even though its variation arose both from the number of units sold and from the size (money value) of these units. Records for the period of study did not permit a precise distinction between the two sources; but it seems evident that for the coat and shoe departments most of the fluctuations in cost came from differences in the value of units, whereas for the hosiery department most was attributable to differences in the number of units sold. Regardless of which of these influences predominates, the size of transaction probably affects cost materially. Even when the number of units is the same, more time is likely to be required for the sale of an expensive item than of a cheaper one.

The only available index of size of transaction was the average dollar value per sale. Since this measure reflects the effect of variation in dollar sales when number of transactions is held constant, it indirectly introduces value volume into the multiple-correlation analysis. This index is, however, subject to the same defects as dollar sales, namely, that it reflects changes in the retail price level, shifts in popularity among price lines, and changes in the nature of the article sold. Nevertheless, it appears to be a fairly satisfactory measure of an important cost influence.[18]

2. *Selection of additional independent variables.* In addition to output as measured by volume of transactions, and average value of transaction, the following cost influences were tentatively selected as independent variables for the correlation analyses on the ground that they appeared to influence cost significantly and that they could be measured and treated statistically, but that their effect could not be safely removed by rectification of the original data.[19]

1. Difference between actual and anticipated volume.
2. Intramonth variability in volume of transactions.
3. Number of selling days having unfavorable weather.
4. Number of selling days per month.
5. Change in volume from previous month.
6. Fluctuations in business conditions.

a) Difference between actual and anticipated volume. The departure of actual sales from anticipated sales seems to have an important effect upon the costs of a department largely because of the difficulty of adjusting the working force to a violently fluctuating sales volume. The existence of a versatile contingent sales force which can be shifted among departments and of a trained alumni corps available for part-time work makes possible fairly accurate adjustment of salespersons and stock personnel to the sales volume to the extent that sales can be correctly forecasted. But if actual sales fall below planned sales, the adjustment is defective, and cost per sale rises.

The difference between budgeted sales and actual sales was considered as a possible measure of inaccurate forecasting but was subsequently found unacceptable because of incomplete records. The most promising of the alternative indexes of forecasting inaccuracies appeared to be the ratio of each month's sales to the sales of the corresponding month of the preceding year, since the earlier year's volume is usually a most important consideration in making plans for the current year's sales.

b) Intramonth variability in volume transactions. Another important cause for variation in cost was thought to be fluctuation in sales volume within the time period selected as the unit of analysis. This fluctuation is of three main types: week-to-week variation, day-to-day variation within the week, and hour-to-hour fluctuation within the day. Fluctuation in volume affects cost partly because of the difficulty of adjusting the sales force to changes in need for its services. Intramonthly fluctuations may influence month-to-month differences in cost in two ways: (1) by departure from a predictable pattern of variation or (2) by such extreme irregularity of variation as to make adjustments impossible even when correctly anticipated.[20] An examination of each type of fluctuation indicated that variability of sales within the month would probably fail to account for differences in cost between months.[21] Nevertheless, an index of variability was computed for one department to test this conclusion. Day-to-day fluctuation was chosen as most likely to be important and easiest to measure. An index of the variability in number of sales transactions was constructed by expressing the difference between the highest day and the lowest day of the week as a ratio of the lowest day.[22]

c) *Weather.* Weather was thought to influence cost not only through its effect upon sales volume but also through its effect upon the predictability of sales. Bad weather, by suddenly driving sales below estimated volume, may make unnecessary the services of salespersons and stock persons previously employed.

Two kinds of adverse weather conditions can be distinguished: unseasonable weather and disagreeable weather. Since the former's disturbing effect upon cost was probably not great,[23] and since it was difficult to obtain an objective measure of unseasonable weather for the period under study, measurement of this phenomenon for use as an independent variable was abandoned.[24]

Disagreeable weather appears to affect costs by causing large day-to-day variations in sales which are not accurately predictable early enough to permit adjustments of selling and stock force. An index was constructed by tabulating for each month the number of days with extreme temperature (above 85° or below 15°) or with rainfall during store hours.[25]

d) *Number of selling days per month.* Direct rectification of data for length of work month was, upon analysis, considered both unnecessary and unpractical. As a check upon the conclusion that its effect was negligible, number of selling days per month was tested as an independent variable.

e) *Changes in volume from previous month.* The position of a point on a static cost function is assumed to be unaffected by the position of previous observations; that is, the cost-output relation for one period is not supposed to be influenced by the output of the previous period. Our cost function, however, may not correspond precisely to this model, since rigidities of various sorts may cause the cost associated with a given output to be different when this output has been attained by an increase from the previous level than by a decrease. To examine the reversibility of the empirical cost function, the magnitude and direction of change in output from that of the previous month was tested as an independent variable.

f) *Fluctuations in business conditions.* It might be supposed that fluctuations in general business conditions would influence costs in the present study mainly through changes in the various input prices and through variations in number of transactions and in the average value of transactions. By rectifying the data to remove the influence of changes in wage rates and prices, it was hoped to remove the greatest part of the "irrelevant" variation in costs attributable to what will be called, for lack of a better term, "the business cycle." The remaining effect of cyclical influences upon cost behavior was roughly tested by plotting the cost residuals of a graphic correlation analysis in chronological order to observe periodic fluctuations.

3. *Testing the influence of independent variables.* In the preceding section

certain cost influences were tentatively selected as independent variables for the correlation analyses on the grounds that they appeared to influence cost significantly, that they could be measured and treated statistically, and that their effect upon cost could not be safely removed by rectification of the original data. Number of transactions and average value of transaction were chosen as measurement units for physical output. In addition, the following sources of cost variation were analyzed:

1. Difference between actual and anticipated volume.
2. Intramonth variability in volume of transactions.
3. Number of selling days having unfavorable weather.
4. Number of selling days per month.
5. Change in volume from previous month.
6. Fluctuations in business conditions.

Two criteria were applied in selecting from this list the independent variables for the least-squares multiple-correlation analysis: (1) the factor must have an independent influence upon cost not accounted for by some other variable and (2) the factor must not be highly correlated with any other independent variable.

By a preliminary graphic analysis it was possible, first, to ascertain whether any net relation existed between cost and each of the tentative causal elements; second, to define the general character of this relation; and, third, to determine the degree of intercorrelation among the independent variables.

The net relations were tested by employing each factor as an independent variable in graphic multiple-correlation analyses of total departmental cost. Introduction of the independent variables in the order of their believed importance[26] made it possible to establish by successive approximation the net relation between cost and those items which proved most important and to test the correlation between each factor and the cost variation not attributable to a more closely correlated variable.[27] A clear net relation to cost was found for both number of transactions (X_2) and average value of transaction (X_4).[28] Cost deviations from these net regression curves were plotted against each of the remaining independent variables, to determine whether the factor was significantly correlated with the residual variation in cost not accounted for by the net regressions of X_2 and X_4.

The number of selling days per month was found to have no net correlation with cost in any of the three departments after the cost variation associated with X_2 and X_4 had been removed. Although the effect of the factor upon sales volume was clear, its net effect upon cost was not; and it was, therefore, not included among the independent variables.

The percentage change in number of transactions from the corresponding month of the previous year likewise showed no net correlation with cost for any department studied and was rejected as an independent variable.

Although unfavorable weather (as measured by the number of selling days per month that were rainy or uncomfortably hot or cold) appeared to have a clear effect upon sales volume, this index showed no statistically significant net relation to cost for any department. Apparently its only effect upon cost was through its effect upon sales volume; it was therefore not used in the least-squares analysis.

Daily variability in number of sales transactions likewise failed to show a statistically significant net relation to cost in each department and was, therefore, not included among the independent variables.

Change in volume from that of the previous month showed no net relation to cost, thus roughly showing the continuity or reversibility of the cost function. No cyclical pattern was found in the cost residuals from the X_2 and X_4 net regressions. This indicated that no additional correction for cyclical changes in supply prices and wages was needed and that the observed cyclical fluctuations of cost were primarily accounted for by fluctuations in physical and dollar volume of sales.[29]

To summarize, by determining graphically the net regression of each prospective independent variable on cost, a significant relation was found between cost and two factors: number of transactions (X_2) and average gross sale (X_4). Dollar volume of sales (X_3) showed a strong gross relationship but was adequately represented in the multiple-correlation analysis by X_2 and X_4. A well-defined residual seasonal pattern of cost variation was discovered and was removed from the data by a correction index. No net relation to cost, however, was found for the following factors:

1. Change from corresponding month of previous year.
2. Number of selling days per month.
3. Number of selling days per month having unfavorable weather.
4. Day-to-day variability in volume of transactions.
5. Business conditions.

4. *Determining the intercorrelation of independent variables.* The degree of intercorrelation among the independent variables was first explored by means of scatter diagrams, which provided a sufficiently accurate indication for these purposes.

Number of transactions (X_2) and average gross sale (X_4) showed no correlation for the coat and shoe departments. For hosiery no correlation was found for the years 1932, 1933, 1934, and 1935, but for the year 1931 a clear relation appeared to exist.

The high correlation which existed between some of the rejected independent variables and number of transactions (X_2) accounts, in part, for lack of any relation between these variables and cost after the effect of volume variation had been removed.[30]

III. Findings

The preceding sections have dealt with problems of collecting, rectifying, and analyzing the data in order to find the net relation between cost and output, with other influences held constant. In this section the findings of the study are presented.

Three departments of a retail store were studied: the women's medium-priced coat department, the women's hosiery department, and the women's shoe department. The cost analyzed for each department excluded general store allocated expenses and was confined to an aggregate of the monthly expenses of advertising, salespeople's salaries, other department salaries, inside delivery, outside delivery, and direct expenses. This aggregate cost, hereafter referred to as "combined cost," was studied in three forms: as a total of the monthly expense (total cost), as the average expense per unit of sale (average cost), and as the increment in total cost associated with an additional unit of sale (marginal cost).

Previous experiments in methodology have indicated that more useful and accurate estimates of cost behavior can be obtained by analyzing cost in terms of total expense for an accounting period than in terms of expense per unit of sale.[31] Cost behavior was, therefore, analyzed in terms of totals before converting the findings into average and marginal terms. Empirical cost functions were obtained by least squares multiple regression analysis of corrected monthly observations.

A. *Cost as a Function of Number of Transactions*

We shall first summarize the findings concerning cost behavior associated with variations in output as measured by number of transactions (the influence of average gross sale being allowed for) since this was accepted as the most useful measure of output.[32] The total, average, and marginal cost functions will be shown for the coat, hosiery, and shoe departments.

In order to keep absolute cost magnitudes confidential, both cost and output measures were transformed into index numbers. Number of transactions will henceforth refer to number of units of transaction index, and average and marginal costs must be understood to refer to the index unit. Average gross sale will also refer to index units rather than to dollars.

1. *Coat department.* The following partial regression equation for total cost was obtained for the coat department:[33]

$$X^{C}_{T} = 16.835 + 1.052X_2 - .00194X_2^2 .$$

It shows total cost increasing in a convex curve which rises at a declining rate as physical volume increases, as portrayed in Chart 4–1.

The behavior of average cost was determined by conversion of the corresponding total cost function rather than by direct correlation analysis of unit cost observations. It shows transaction cost declining in a hyperbolic curve as physical volume increases, as may be observed in Chart 4–1, and seen from the following equation:

$$X^{C}_{A} = 1.052 + 16.835/X_2 - .00194X_2 .$$

The marginal cost per transaction unit declines at a constant rate, as physical volume increases, in a function described by the following equation:

$$X^{C}_{M} = 1.052 - .00388X_2 .$$

Chart 4–1 Coat Department: Total, Average, and Marginal Combined Cost Derived from Partial Regression on Transactions

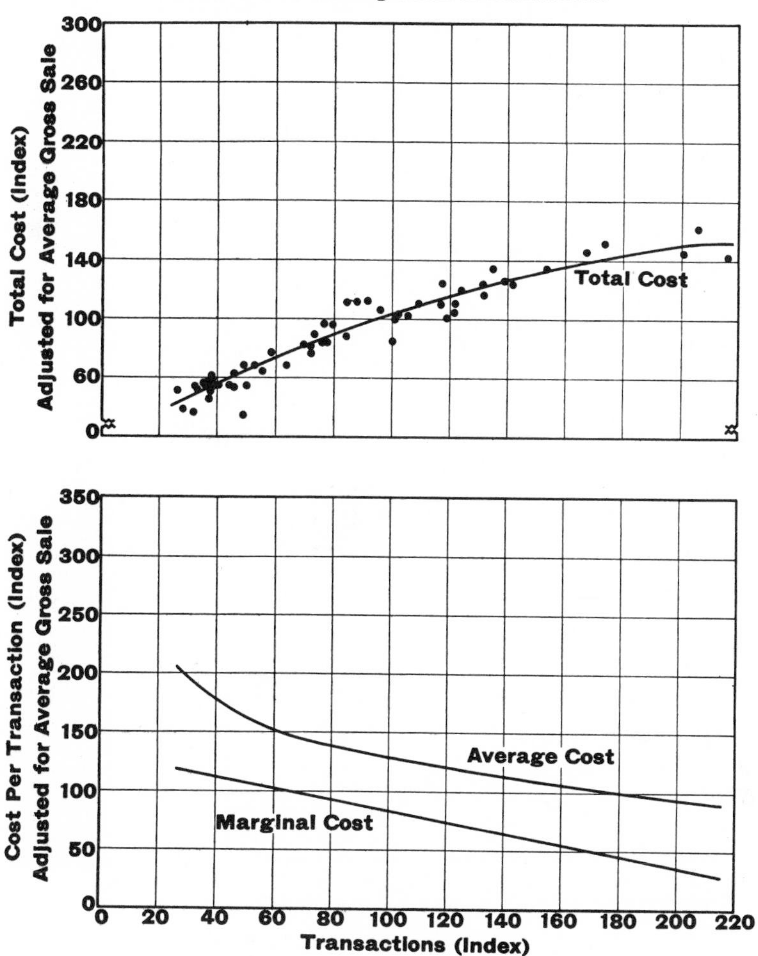

2. *Hosiery department.* For the hosiery department the least-squares equation for the net relation between cost and number of transactions, while holding the average gross sale at its mean, is

$$X^H{}_T = 55{\cdot}554 + {\cdot}347X_2 \,.$$

The graph of the equation, which indicates that total cost rises at a constant rate as number of transactions increases, is shown in the upper half of Chart 4–2. Inspection of the latter figure reveals that the observations upon which the curves are based are so unevenly distributed that the relation was well established only

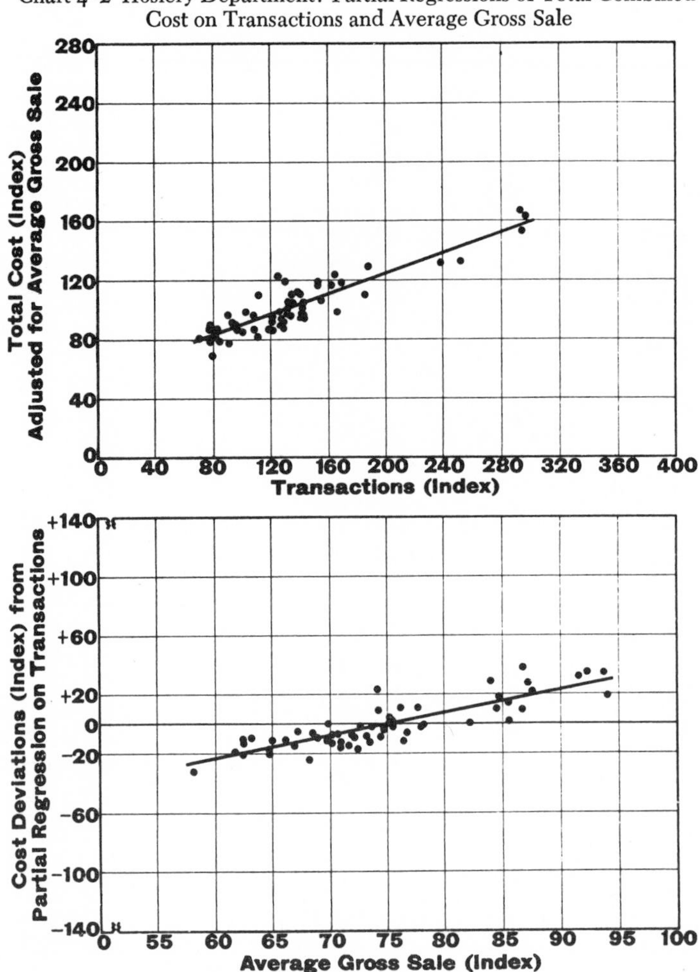

Chart 4–2 Hosiery Department: Partial Regressions of Total Combined Cost on Transactions and Average Gross Sale

between the transactions index figures of 80 and 170, and only tentatively defined for the range between 170 and 300.

The equation for average cost per sales transaction as related to number of transactions, when the average gross sale is held constant, is

$$X^{H}_{A} = .347 + 55.554/X_2 \,.$$

In tabular form this function is presented in Table 4–1. Examination of the table shows that average cost per transaction unit falls at a declining rate as physical volume of sales increases and tends to approach a constant at the extreme range of cost observation.

Table 4–1
HOSIERY DEPARTMENT: ESTIMATED TOTAL, AVERAGE, AND MARGINAL COST* FOR VARIOUS UNITS OF TRANSACTIONS (X_2) WITH THE INDEX OF AVERAGE GROSS SALE CONSTANT AT ITS MEAN

Transactions (Index)	Estimated Total Cost (Index)	Estimated Average Cost (Index)	Estimated Marginal Cost (Index)
60	76.374	1.273	.347
80	83.314	1.041	.347
100	90.254	.903	.347
120	97.194	.810	.347
140	104.134	.744	.347
160	111.074	.694	.347
180	118.014	.656	.347
200	124.954	.625	.347
240	138.834	.578	.347
280	152.714	.545	.347
320	166.594	.521	.347

* Derived from the equations:

$$X^H{}_T = 55.554 + .347\ X_2$$
$$X^H{}_A = X^H{}_T/X_2 = .347 + 55.554/X_2$$
$$X^H{}_M = .347$$

The marginal cost of one additional hosiery transaction was found to be constant at .347.

3. *Shoe department.* For the shoe department the net relation between total cost and number of transactions (X_2), when the average value of transaction is held constant at its mean, is depicted by the following partial regression equation:

$$X^S{}_T = 32.137 + .925X_2\,.$$

The graph of this function (Chart 4–3) indicates that total selling cost of the department increases at a constant, although not proportionate, rate as the physical volume of sales rises.

The average cost per transaction unit for women's shoes declines, as would be expected, when the number of transactions increases. This relation, when average value of sale is constant at its mean, is shown in the following equation:

$$S^S{}_A = .925 + 32.137/X_2\,.$$

The graph of the function is shown in Chart 4–3, where it may be contrasted with the total cost function.

The estimate of the marginal cost of an additional transaction index unit for the women's shoe department appears to be constant, throughout the observed range of physical volume, at .925.

B. *Cost as a Function of Average Gross Sale*

The findings for cost functions when number of transactions is accepted as the index of the output (with average gross sale held constant) which were presented

Chart 4–3 Shoe Department: Total, Average, and Marginal Combined Cost Derived from Partial Regression on Transactions

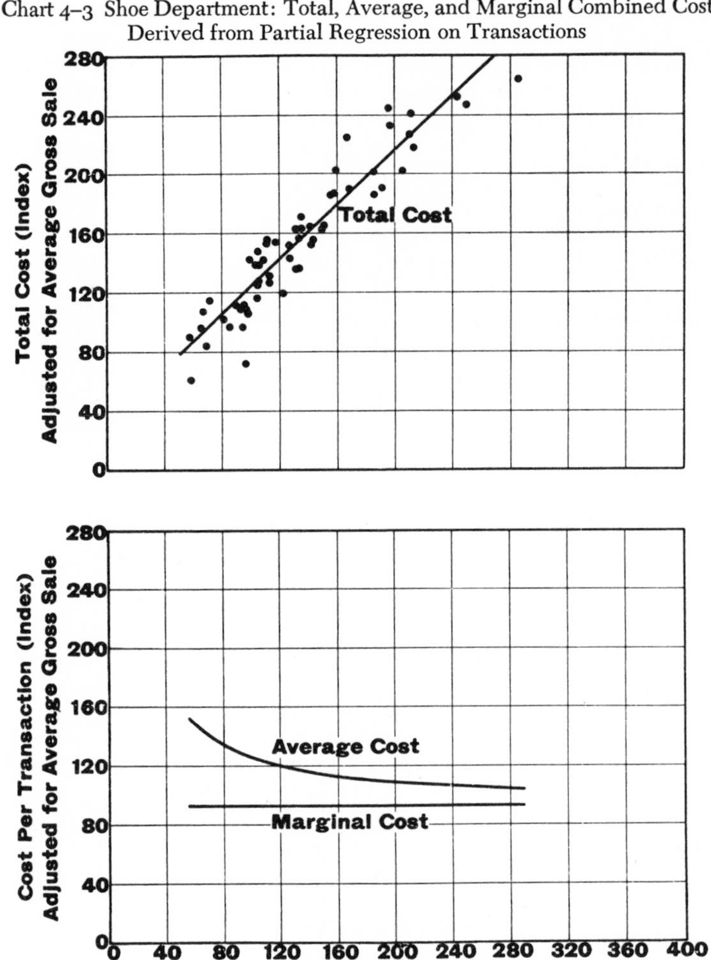

in the preceding section need to be supplemented by examination of the influence of another aspect of output. The significance of the net relation between the average gross sale and cost indicates that output should be considered as two dimensional and may increase either in the direction of more transactions or greater average size of the transaction. Accepting size of transaction (average gross sale) as the measure of output, and holding number of transactions constant at its mean, we obtain the results summarized in this section.

1. *Coat department.* The net functional relation between total cost of the coat department and average gross sale (with number of transactions constant) is shown by the following least-squares partial regression equation:

$$X^C{}_T = 35.942 + .787X_4\,.$$

Total cost tends to increase at a constant rate with the size of the average gross sale. Table 4–2, which was obtained by substituting in the above regression equation, shows this relation in the form of estimated total cost for various values of average gross sale.

Table 4–2

BUDGET COAT DEPARTMENT: ESTIMATED TOTAL, AVERAGE, AND MARGINAL COST* FOR VARIOUS MAGNITUDES OF THE INDEX OF AVERAGE GROSS SALE (X_4) WITH INDEX OF TRANSACTIONS CONSTANT AT ITS MEAN

Average Gross Sale (Index)	Estimated Total Cost (Index)	Estimated Average Cost (Index)	Estimated Marginal Cost (Index)
50	75.292	1.506	.787
60	83.162	1.386	.787
70	91.032	1.300	.787
80	98.902	1.236	.787
90	106.772	1.186	.787
100	114.642	1.146	.787

* Derived from the equations:

$$X^C{}_T = 35.942 + .787\,X_4$$
$$X^C{}_A = X^C{}_T/X_4 = .787 + 35.942/X_4$$
$$X^C{}_M = .787$$

The average cost per index unit of average gross sale, as average gross sale varies and number of transactions remains constant, is shown by the following expression:

$$X^C{}_A = .787 + 35.942/X_4\,.$$

The cost per unit increase in the monthly average value of transaction is constant at .787 over the observed range.

2. *Hosiery department.* For the hosiery department the equation obtained for the net relation between total costs and average gross sale, when the effect of number of transactions is allowed for, is

$$X^H{}_T = -13.889 + 1.557X_4\,.$$

The curve of this equation found in the lower half of Chart 4–3 shows that cost increases with the size of the average gross sale at a constant rate. As in the case of number of transactions, the wide dispersion of the observations casts some doubt on the validity of a linear relation.

The average cost per index unit of average gross sale, when the number of transactions is held constant, is described by the following equation:

$$X^H{}_A = 1.557 - 13.889/X_4\,.$$

The tabular representation is shown in Table 4–3.

Table 4–3

HOSIERY DEPARTMENT: ESTIMATED TOTAL, AVERAGE, AND MARGINAL COST* FOR VARIOUS MAGNITUDES OF THE INDEX OF AVERAGE GROSS SALE (X_4) WITH INDEX OF TRANSACTIONS CONSTANT AT ITS MEAN

Average Gross Sale (Index)	Estimated Total Cost (Index)	Estimated Average Cost (Index)	Estimated Marginal Cost (Index)
50	63.961	1.279	1.557
60	79.531	1.326	1.557
70	95.101	1.359	1.557
80	110.671	1.383	1.557
90	126.241	1.403	1.557
100	141.811	1.418	1.557

* Derived from the equations:

$$X^H{}_T = -13.889 + 1.557\ X_4$$
$$X^H{}_A = X^H{}_T/X_4 = 1.557 - 13.889/X_4$$
$$X^H{}_M = 1.557$$

The additional cost of increasing average gross sale by one index unit for the month was found to be 1.557.

3. *Shoe department.* The functional relation of total cost and average gross sale, with transactions constant, for the shoe department, is shown by the equation:

$$X^S{}_T = -23.427 + .837X_4 .$$

Total cost increases at a constant rate with the value of average gross sale, when physical volume is held constant at its mean.[34]

The average cost per unit of average gross sale, with physical volume constant at its mean and size of gross sale varying, is shown by the following expression:

$$X^S{}_A = .837 - 23.427/X_4 .$$

The cost of a dollar increase in average value of transaction is constant at .837.

C. Cost as a Function of Number of Transactions and Average Gross Sale

The findings concerning the combined effect upon cost of the two aspects of output—number of transactions and average gross sale—will be briefly presented in this section.

For the coat department the combined effect of number of transactions (X_2) and average gross sale (X_4) on total cost is shown by the following multiple-regression equation:

$$X^C{}_T = -35.440 + 1.052X_2 - .00194X_2{}^2 + .787X_4 .$$

The standard error of estimate, adjusted for degrees of freedom, was 7.901. The close relation of cost to the independent variables is shown by the value of the coefficient of multiple correlation, .980, and by the coefficient of multiple determination which indicates that about 96 per cent of the variance in rectified cost was accounted for by the independent factors.

The combined effect of number of transactions (X_2) and average gross sale (X_4) upon total cost of the hosiery department is described by the following equation:

$$X^H{}_T = -60.764 + .347X_2 + 1.557X_4 .$$

The confidence which can be placed in the relation described above is indicated by the standard error of estimate of 8.140, the coefficient of multiple correlation of .957, and the coefficient of multiple determination of .91.

For the shoe department the combined effects of number of transactions (X_2) and average gross sale (X_4) upon total cost is shown by the following multiple-regression equation:

$$X^S{}_T = -146.776 + .925X_2 + .837X_4 .$$

Provided that the basic conditions of the sampling period remain unchanged, considerable confidence can be placed on the above relations, as is evidenced by a standard error of estimate of 16.624, a coefficient of multiple correlation of .965, and a coefficient of multiple determination of .93.

IV. Interpretation of Findings

The findings of an empirical study of the individual firm should constitute some kind of evaluation of what theorists have been saying with regard to its economic behavior. However, such an appraisal would imply that the situation described by the statistician is strictly comparable to that generally postulated by the theorist. The actual situation investigated, in fact, deviates from the theoretical norm in two important respects, namely, (*a*) by the inclusion of selling costs in cost of production and (*b*) by the failure to purge the statistical data of all sources of dynamic nonconformity to the static model. Some brief comments will be made on these limitations of the findings in Sections A and B before an attempt is made in Sections C and D to explain and rationalize the results observed.

A. Inclusion of Selling Costs

As was mentioned above, the costs examined in this study were not confined to what is usually designated as cost of production proper. A priori, of course, there are equally good, if not better, reasons for supposing that marginal selling expenditure increases with intensified selling activity as there are for expecting that marginal cost of production rises with output. First, with more intensive utilization of fixed equipment returns per unit of input presumably decrease. Second, each additional unit of promotional cost has diminishing effectiveness since it becomes necessary, in order to increase sales, to detach customers with increasing degrees of personal affiliation and loyalty to rival firms, or to make present customers spend an ever increasing share of their incomes.[35]

Although our findings may be strongly influenced by the presence of costs designed to modify the existing demand functions, the relation of such costs to sales may not be revealed in our data. A spurious correlation between selling expenditure and sales could result from independent fluctuations in demand and from market-sharing, since it obviously is not possible to attribute all shifts in the store's demand functions during the period of analysis to its promotional activities. It is conceivable that selling costs in a mature competitive retail market have the effect of maintaining the store's share of the total demand—a demand which shifts seasonally and cyclically in response to changes in custom, in tastes, and in income.

It would be impossible to establish the relation between number of sales transactions and expenditure on advertising without eliminating variations in sales resulting from varying effectiveness of advertising, from changes in demand, and from other irrelevant factors. The same holds true of other elements of promotional expenditure. Consequently, our results cannot be interpreted as measuring the influence of the diminishing effectiveness of promotional expenditures as a firm's market is expanded at the expense of rivals.

The inclusion of both selling costs and pure costs of production in the empirical cost function, furthermore, indicates that the product, a unit of sales service, is not entirely homogeneous, since selling activity which increases sales may be considered to have enhanced the utility of commodities offered. However, there may be considerable stability in the proportion of selling costs to total cost.[36]

An additional reason for suspecting that the unit of output, sales service, may not be homogeneous over the observation period and may be correlated with the physical transaction rate is the fact that sales service per transaction is likely to diminish during busy seasons. As a consequence of this, a decline of marginal cost as the volume of transactions increases may merely reflect deterioration of sales service standards.[37] If this is the case, and if this deterioration has a tendency to decrease future marginal revenue, the management may regard falling marginal cost as a danger signal, indicating that sharply rising marginal promotional expenditure may be necessary in the future.[38]

B. Statistical Sources of Nonconformity

Besides the problems involved in promotional activity, there also exist certain statistical sources of possible nonconformity of the empirical situation to the theoretical model in addition to those mentioned in the sections dealing with methodology.

It is possible that, had a smaller observation unit been chosen (e.g., days or weeks instead of months), a more conventional cost function would have been obtained. If uncorrected data had been used, the expectation would be that the longer the time unit chosen the less would be the slope of the marginal cost curve, since marginal costs are a function of the time period allowed for adjustment to changed rates of output. If rectification of the data has been successful, however, adjustments of a long-period or quasi-long-period character have been eliminated, so that there should be no discrepancy between a function based on monthly findings and one based on a shorter time period.

From the purely technical viewpoint, it should probably be mentioned that the total cost functions are not defined beyond question as being straight lines or parabolas. The unexplained scatter of observations is great enough to permit a cubic of the traditional form to be fitted in each case. However, the curvature would be so slight as to be insignificant from a managerial viewpoint, so that it could scarcely affect any economic conclusions which might be derived from the linear and parabolic functions. In each instance a higher-order function than that selected was fitted and subjected to critical ratio tests, which indicated that the more complex function did not fit the data significantly better than that chosen.

C. Excess Capacity

If the plant of these three departments were systematically underutilized during the observation period, marginal costs might appear to be constant or falling over the whole length of the curve because of an inadequate range of observations.

Overbuilding as a result of errors of optimism in estimating the position and inclination of demand is not as elsewhere a plausible explanation for overcapacity, since department-store layouts are relatively easy to modify when such errors become apparent. This explanation, moreover, does not take account of the possibility that selling expenditures may eventually shift the demand functions so that capacity is fully utilized.

The equilibrium position of a department may be one of apparent excess capacity only. The fact that the magnitude of selling expenditures affects both the cost function and the demand function results in complex interrelations that may yield equilibrium at a point on the demand surface that appears to represent underutilization of capacity. Delays in the responsiveness of demand to various kinds of promotional expenditures, together with complexities of polyperiodic production, may accentuate this conditin. Store prestige may, furthermore, force some departments to operate at less than their departmental optimum capacity. An overelaborate layout and stock in particular departments could also result from the effort of the store to maintain comparable standards of sales service, variety of merchandise, display, etc., in all departments of the store. Store "good will" may thus necessitate a minimum size for the departments studied, greater than that justified by the average position and slope of their demand functions during the period examined.

We have no evidence that the observed outputs were those which would have maximized profits under these complicated conditions. In view of the complexity of the relations between cost and revenue, it is not surprising that this store conformed to the general pricing policy of department stores which is based on the "cost plus a given percentage" principle rather than on an attempt to calculate optimum price on the basis of interrelated cost and revenue functions. The "cost-plus" policy would yield the price and output which was "optimum" for any momentary situation only accidentally and might result in a systematic under-utilization of plant which would confine observations to the constant section of the marginal curve.

D. Technical Explanations for Constant Marginal Costs

A variety of technically plausible explanations can be given for findings of constant marginal costs in the hosiery and coat departments and falling marginal costs in the shoe department. Marginal costs may be constant if it is possible to "sectionalize" the activities of a firm; that is, if the factors employed can be grouped into small operating entities, each of which can be utilized to equal advantage, and each of which represents the optimum combination of factors.[39] In practice, of course, a plant cannot be completely sectionalized, since some factors, if only the management and organization, are fixed in the short run. As long as fixed factors are present, the law of diminishing returns will cause costs to rise over some portion of the marginal cost curve.

Second, the declining phase of the marginal cost curve will be particularly important if the fixed factors are not completely adaptable to varying inputs of the variable factors and if, in addition to "lumpiness" of the fixed factors, there are organizational economies available when larger amounts of the variable factors are used. The inclination of the curve will depend on the degree of technical "flexibility" to be found. An inflexible plant would have a U-shaped marginal cost function with a sharply defined minimum; greater flexibility would be present if outputs less or greater than the optimum could be obtained without entailing rapidly rising marginal cost.[40]

It is probable that the type of marginal cost function found in empirical studies can be explained by investigating (1) the degree to which sectional divisibility of the plant has been attained and (2) the degree of technical rigidity which exists.

In a department store technical rigidities are probably not so great as would be found in a manufacturing enterprise. The fixed elements of cost include, from the standpoint of the individual department, the general standards and reputation of the store, the location, in the present instance the size of the departmental layout, and to a certain extent the services of the buying and managerial staff. However, it is probable that buying and managerial service varies with output, even when no recognition of such variation is given in salaries or number of people employed. To this relatively small component of fixed factors might be added units composed of the optimum combination of selling service, advertising, delivery expense, etc. The degree of segmentation thus attained might offset, for considerable variations in output, the tendency to diminishing returns. Moreover, since there are few highly specialized input factors in a technical sense the degree of flexibility is probably great.

In conclusion, since the demand and cost functions are not independent when promotional expenditures are being made, it would be necessary in order to get a determinate solution to the problem of the optimum price and output not only to separate promotional costs from cost of production proper or to determine a functional relation between them but also to determine (1) the relation between selling costs and the demand function, (2) the time lag between the incurring of the selling expenditure and its effect on demand, (3) the relation between the output-price adjustment in one period and the demand function in successive periods, (4) the interrelation among the demands for various products both within and outside this department (e.g., "loss leader" and "ensemble" buying), and (5) the reaction of rivals to the firm's price, output, and promotion policies.

PART TWO

Cost and Plant Size

Introduction to Part Two

Part Two concerns statistical determination of the long-run cost curve, or curve of returns to scale, which shows the relationship between cost and size of plant. Two statistical studies designed to measure this relationship are "Finance Chain" (Study 5) and "Shoe Chain" (Study 6).

I. Measurement Problem

The hypothesis of the relationship between cost and size of plant that is tested in these two studies is diagramed in Chart Two–1. The three short-run average cost curves (*SAC*–1, *SAC*–2, *SAC*–3) show how average unit cost varies with the rate of use (measured by output) of a fixed plant of three alternative sizes. The minimum-cost size is represented by *SAC*–3. When all these short-run curves are drawn, an envelope curve tangent to the family of short-run curves can be plotted. This curve (*LAC*) shows, for each output rate, the level of minimum-cost and the size of plant that would produce it. The envelope curve of average unit cost first declines with increased size of plant as a result of net economies of scale, then flattens as savings of size are exhausted and/or offset by diseconomies of scale. As the size of the plant (or firm) is increased further, average unit cost ultimately rises because most or all of the economies of scale have been attained and the disadvantages of size have more than offset them.[1]

The long-run cost curve does not pass through the minimum points of the short-run curves (except for *SAC*–3), because on its downslope a slightly bigger plant would produce that output at lower cost, even though the plant was underutilized. (The reverse is true for *LAC*'s upgrade.)

What we seek in empirical studies is the statistical counterpart of this envelope curve, which shows what cost would be ideally—i.e., under complete adaptation of plant size to output. Time needed for adjustment is a proxy for completeness of adaptation, hence the long-run label.

The classic rationale for a U-shaped long-run average unit cost curve was developed for the single-plant firm and has been carried over to the units of a multiplant firm where it may not apply. The relationship of cost to size for the branch units of a multiplant firm is bound to be different from the relationship of

Chart Two–1 Relation of Cost to Plant Size

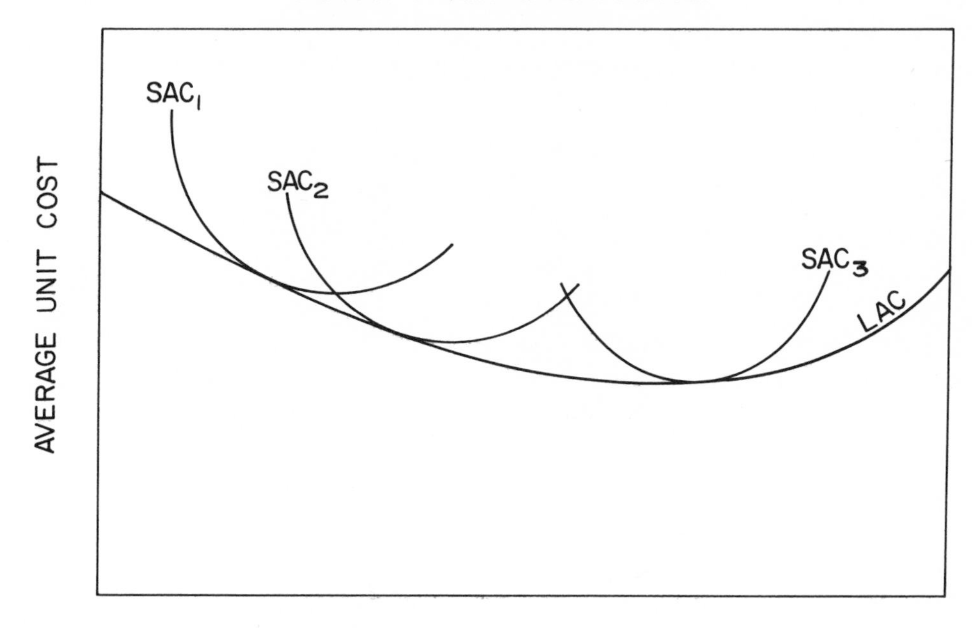

cost to size for a single-plant firm because the mix of activities is different. Those functions for which savings of size are believed to be greatest are usually centralized. Nevertheless, the branch units retain activities in which there are indivisibilities and savings of specialization, which are the main source of savings of plant size. The causes of the rising phase of the cost–size curve, if it occurs, will also be different. Branches (stores, loan offices, factories) do not perform head office functions such as policy guidance, overall coordination, capital rationing, and capital sourcing, where disadvantages of size are asserted to occur in large firms.

If the cost–size curve does turn up at large branch units, its rise cannot be attributed to diminishing returns to entrepreneurship, because the vital entrepreneurial functions are performed by the central organization. The managers of individual chain stores have sharply limited operational responsibilities that do not change appreciably with the size of the unit. Nor can a rising phase of the combined cost curve be attributed to higher advertising costs per unit for the large stores. Advertising outlays by or in behalf of individual chain units are usually small (partly because of lack of suitable media). Even if promotional costs did rise with size of market share, this would be inapplicable, since the firm's share of the local market is not systematically related to the size of the branch unit.

II. Alternative Methods of Measurement

To isolate the cost–size relationship, we want, ideally, to remove the effect upon cost of changes in technology, product mix, rate of use of plant, and prices of input factors. This is a tall order. Four alternative methods are:

1. Engineering predictions of costs of blueprinted plants that differ in size but not in technology.
2. Historical study of the path of cost during the growth of a single plant.
3. Statistical analysis of cost of plants of various sizes operated by different firms, studied and observed at the same time.
4. Statistical analysis of simultaneously observed costs of different-sized plants of a single corporation.

III. Method Used

Both the finance chain and shoe chain studies used the fourth method—observation of different-sized plants in the same firm—to isolate the cost–size relationship. The finance chain study is a statistical determination of the long-run relationship between cost and the size of individual loan offices of the Household Finance Company. The study was made five years before the shoe chain study; its technology is more primitive, though the measurement problems were equally difficult. Size was measured by the number of accounts in a loan office. Other cost determinants, used as independent variables in the analysis of average unit cost per account, were percentage of accounts delinquent, percentage of new accounts, percentage of renewal accounts, and population of branch office city. Combined cost and its major components (rent, salary, and general expenses) were studied separately for 1932 and 1933. Graphic multiple-regression analysis was used to determine net regression curves for cost in terms of unit cost per account. Total cost per year was also studied. The long-run incremental cost of expanding the size of a loan office was found to be substantially constant.

The theory of the relationship between cost and plant size was developed for manufacturing and therefore may be viewed as applying only to production costs. A finance company's costs are a mixture of persuasion costs (designed to induce or hold patronage) and production costs (designed to take care of that demand). The two kinds of costs cannot be separated, and their proportions may change with the maturity of the loan office, its rate of growth, and possibly its size.

The shoe chain study is a statistical determination of the relationship between operating cost and the size of retail stores of the Thom McAn shoe chain, a division of Melville Shoe Corporation. Of the several hundred stores operated by Melville, a sample of 55, all located in New York City and under one supervisor, was studied. The sample was representative of the size range of the chain's stores and of various operating conditions, such as sales mix and rate of utilization. Stores recently opened and stores selling women's shoes were excluded. Annual data on retail operations and costs for each store were studied for 1937 and 1938. Head office overhead costs, which were fixed in relation to shoe sale volume and which were allocated to the stores in an inevitably arbitrary fashion, were omitted.

Size of store was measured by output of sales service: the help of the shoe salesman in searching, fitting, and offering admiration and advice. Sales service was proxied by shoe sales in number of pairs. Cost components as well as combined cost were studied statistically.

IV. Usefulness of Findings

Findings of studies of the relationship between cost and plant size have three important managerial uses. One is to find the least-cost size for the operating unit (store, loan office, or manufacturing plant). This information is valuable even though it alone usually does not determine plant size. There are exceptions: one large grocery chain, for example, has sought to adopt a single, standard, least-cost store size in all locations, selecting sites with potential sales of this store size. Rivals of this chain usually permit store size to vary over a wide range, tailoring size partly to the predicted sales potential of the store's accessible market. Thus a second use of knowledge of the effect of size on cost is to make an intelligent trade-off between the losses of departing from a minimum-cost size and the benefits of better market orientation, i.e., superior location and more precise tailoring of store size to sales potential. Quite often the effect of plant size upon cost is not the only nor even the most important influence governing the profitability of a chain store or a loan office. Hence a third managerial use is to set tailored standards for controlling store expenses. Cost criteria that are adjusted for the effects of store size upon costs are, in that respect, more alibi-proof.

Knowledge of the relationship of cost to the size of plant (finance office, chain store, branch factory, etc.) will play an increasingly important role in capital budgeting decisions of large firms. Research in managerial economics, which is grounded on a realistic understanding of how business decisions are made, is also being continuously refined by advancing mathematical and statistical techniques of management science. Managerial economics will, in the future, in its application to decisions about plant size, be more sharply aware of four managerial dimensions of futurity: (1) that forecasted future costs alone matter for decisions and that past cost behavior is pertinent only as a basis for quantifying expectations; (2) that decision-making cost analysis needs to be structured by alternatives; (3) that uncertainty, being a critical feature of the future, should be explicitly recognized in the forecasts that compare alternatives; and (4) that forecasts and plans must be continuously revised as experience accumulates and the future unfolds. This kind of futurity setting for cost forecasting enhances the usefulness of cost–size studies for capital management decisions.

Knowledge of the effects of size of plant upon cost is also useful to academic economists. The interest of economists in savings of scale centered initially on the size of the economy as a whole (wealth of nations). This interest was refocused during the early 1950s on the size of the industry (growth economics) and, more recently, on the savings of size of the firm (structure of competition). Savings of plant size underlie all three kinds of economies of scale.

As to size of economy, econometric determinations of U.S. production functions indicate constant returns to scale, suggesting that least-cost size has been achieved by the U.S. economy. As to size of industry, econometric findings based on Cobb-Douglas production functions developed for major industries in the United States generally concur in constant returns to scale, indicating again that no further size savings in our mature major industries can be expected to result from growth of our economy.[2]

V. Criticisms and Limitations

Ten criticisms and limitations of the methods and findings of statistical estimation of the relation of cost to size of plant are of particular interest. The criticisms will be appraised first, then the limitations will be examined.

1. The findings of statistical cost–size functions are incompatible with the long-accepted theoretical model of the behavior of firms under perfect competition. The assumptions of atomistic competition require that cost functions, both average and marginal, be U-shaped. In the short-run, costs must ultimately rise with output in order for the firm to have a determinate output. In the long run, costs must rise with size in order to preserve perfect competition.

A horizontal short-run marginal manufacturing cost curve of the kind found in most statistical cost studies has been decried as providing no logical optimum for the firm's output under perfect competition. Hence there could be no equilibrium in the short run. Similarly, it is argued that a long-run average cost curve of the statistically determined shape, i.e., one that declines continuously over the observed range size or that is L-shaped, fails to set a cost-behavior limit to firm size. This failure implies that the logical consequence of competition is monopoly, not the preservation of atomistic competition. Thus both long-run and short-run statistical cost curves are incompatible with the equilibria that are needed for perfect competition.

Competition in the real world, however, is not atomistic but monopolistic. The optimum (equilibrium) rate of output is determined in the short run not by rising marginal production costs but by rising marginal selling costs. And in the long run the optimum size of a branch unit is determined not just by rising unit production costs but also, and dominantly, by rising long-run persuasion costs—i.e., costs of expanding the sales share of a geographically limited market at the expense of rivals who are few but competent. The assumption of perfect competition is an analytically useful model but not an attainable ideal. Thus the incompatibility of empirical findings with the theoretical model of cost behavior is not a tenable criticism.

2. Statistical cost findings are contradicted by facts observed in the marketplace: plants (and firms) that differ greatly in size survive despite the empirical evidence of scale economies.

Again, the assumption of perfect competition in the U.S. economy gives rise to this criticism, and the unreality of this assumption answers it. The market is assumed to be so big and so accessible under atomistic competition that size of market has no influence on the *size* of plant or firm. Instead, market size is assumed to determine only the *number* of suppliers that survive. In the real world, competition is monopolistic. The size of market that is accessible to a particular plant (e.g., chain unit) depends on the spatial distribution of demand. Size of accessible market is the main determinant of optimum size of a chain store because customers' costs of access (in terms of nuisance and time) usually confer decisive locational monopoly power on the convenient store. This, combined with uneven spatial distribution of demand, makes the optimum size of a branch unit differ widely from one location to another.

3. The observed declining phase of the long-run average unit cost curve is caused by survival of the fittest; it occurs because efficient firms grow big, not because big ones are efficient.

In some industries this assertion could be correct for size of firm. In the U.S. automobile industry, for example, a thousand firms have been killed off in the competitive struggle, and the handful left have survived because comparative competence conferred economies of size at a critical stage and has kept their costs competitive since. As the monistic explanation for economies of scale in existent firms, however, it is not a convincing criticism. An efficiently managed small firm does not always retain its competitive advantage as it grows larger: computer-assisted management requires different talents than records-in-hat management. Finally, there is ample support for savings of size in many industries, both in common observation and in theory. Insofar as the low costs of production of the firm are the main reason for its survival and growth, this criticism has an element of truth. But it applies only to size of firm: it is not correct for size of units of a multiplant firm. Differences in the size of loan offices or of chain stores are clearly not caused by their comparative success in reducing unit costs. Instead, size is planned at the head office and is dominantly the result of differences in the density of demand and in customers' cost of access.

4. The findings of statistical cost studies overstate economies of scale because firms (and plants) that are smallest, as measured by output, are likely on average to be

operating at unusually low rates of use of capacity; conversely, those with greatest output are most likely to be running at full blast.

This criticism, which has been ably stated by Milton Friedman[3] and illustrated in a detailed numerical example by George Stigler,[4] points up one frailty of output as the index of size. Its main thrust is the failure of empirical studies to remove the short-run cost effects of differences in rate of utilization. This criticism is valid only if size of plant (or firm), as measured by output, is positively correlated with rate of use of capacity. Negative correlation (which is also plausible) would cause savings of size to be *understated.* The existence of such a correlation and its quantitative importance for cost–size studies depends on whether causes of differences in rate of utilization are common and similar among branches of different size. One cause is cyclical fluctuations in business conditions, which should cause about the same degree of underuse of capacity for all units in a chain. Seasonal fluctuations can cause big changes in unit-cost but these are averaged out because the year is the observation period. Only those departures from designed output that are peculiar to individual branches and that are also correlated with measured size can cause this distortion.[5] This bias can, moreover, be avoided by cross-classifying firms on the basis of plant size (as was done in the finance chain study). Alternatively it might be avoided by introducing rate of use of capacity as an independent variable in the multiple-regression analysis, which would require that capacity be measured in terms of input.[6]

5. Accounting data on costs of firms (and plants) of different sizes do not supply valid information for measuring scale economies; omissions and non-market valuations embodied in the acounts distort true economic costs.

This criticism is based on the following analysis: if all accounting data were revalued continuously at the prices that would prevail in perfect markets, if all "economic costs" omitted by accountants were included, and if the industry (and the economy) were perfectly competitive and had no specialized input factors, then long-run average unit cost would be the same for each firm. The curve would be U-shaped and all firms would be of the same minimum-cost size.

Differences in size of firms could be caused only by errors or by transition to the optimum size. Mistakes in size of firm would quickly become visible to a perfect capital market and should result in changes in accounting valuations, so that correctly reported costs per unit would be the same for all firms and plants. Any specialized units of differing quality—e.g., superior technology and managerial talent—would be fully reflected by appropriate rents, if the capital and talent markets and the accountants were doing their jobs. This model of perfect competition also assumes that the market is sufficiently large and accessible and uniformly distributed geographically so that size of market plays no role in determining the optimum size of plant or firm. And, in this model, costs are defined to include all rents and all profits, so that they, by definition, equal total receipts.

This criticism is correct, of course, at its own level of abstraction. But we do not live in a world of perfect competition. Accountants do not continuously revalue input services and assets on the basis of perfect markets for talent and capital. In the real world of monopolistic competition demand is not evenly distributed, either spatially or in respect to product preferences. Rather, it is so unevenly distributed that this is a major determinant of the optimum size of branch unit. Differences in the size of a shoe store (or loan office) are caused by planned minimization of production costs together with business-getting costs, in response to extremely uneven demand distribution. Similarly, under conditions of monopolistic competition, differentiation of the firm's product-package makes it profitable to have branch units of different size. These size differences in an imperfectly competitive world cannot be viewed as "mistakes" that require continuous revaluation of all costs, which would make all unit costs of all firms the same.

It is true that accounted costs do leave out some things that economists view as costs. These omissions include the cost of equity capital, rent on company-owned buildings and equipment, and the potential market value of services and of superior management and of patents, know-how, market acceptance and other specialized inputs. As a consequence, these omitted "costs" are dumped into profits, so that accounted costs understate complete economic costs.

It is also true that the valuations embodied in accounted costs do not continuously reflect those of perfect markets (or even imperfect ones). These defects of valuation sometimes make accounted costs additionally understate economic costs, particularly during inflation. Mismeasurements arising from these valuation "errors" (departures from those of perfect markets) can differ widely among firms. Errors are particularly important for capital consumption allowances, for costing inventory withdrawals, and for long leases.

Among firms, these omissions and inadequacies of accounting valuations might cause substantial differences in accounted costs. But this would distort economies of scale measured from accounting cost data only if the size and direction of this mismeasurement were correlated with size of firm. Only if mismeasurement were biased as to firm size would it not average out.

Among plants (stores, loan offices) of the same firm, distortion of the measured cost–size relationship by the omission and valuations of accounting conventions is much less likely. One reason is that most cost omissions impinge more on central-office costs, which are excluded from the analysis, than on costs of branch unit operations. Moreover, any size-distortion of the accounted cost of branches can be removed by sloughing allocated fixed overheads. The impacts of some kinds of omissions (e.g., noncosted superiority-rents such as patents and brand acceptance) tend to be spread over branch units proportionately to sales-size. There is no reason to expect these errors of valuation and of omission to be correlated with size of branch. However, underpriced locational superiority of the site and underpayment of a superior manager of a store or loan office do distort branch cost data and these understatements may tend to be greater, proportionately, for the bigger units.

Another reason that valuation inadequacies are likely to distort measurement of cost–size relationships less for the units of the chain than for separate firms is the imposed uniformity of accounting valuations within a chain. This uniformity makes it easier for the investigator to detect egregious noncomparability of cost from this source and either to rectify the data or to cast out the observation. For example, accounting data on cost–size studies of units of a chain usually include depreciation charges that embody any differences in the original cost of depreciable equipment that are caused by disparate price levels at the time of purchase. These disparities can be corrected by restating assets and depreciation in constant dollars. Space rentals may be similarly distorted by differences in rent levels on the dates when long-term leases were signed. Such rentals could likewise be restated at replacement market value. Plants, branches, or stores may differ widely in technology. Units that are manifestly obsolescent could be omitted, so that the study is confined to units that are approximately homogeneous as to technology.

These circumstances and corrective actions can mitigate the distortion of statistical estimation of scale economies of plant size for units of a chain. But they do not answer satisfactorily the criticism of the incomplete content of accounted costs for studying size of plant, and they fail totally to meet this criticism for size of firm. It is true that in statistical estimation of cost functions we count as costs only those bookkept outlays that are required to obtain the services of input factors. We do not include imputed "economic" costs, such as cost of equity capital and market worth of owned patents, know-how, brand acceptance, underpaid managerial competence, or other sources of competitive superiority. We cannot include these elements of economic cost in statistical cost studies because they are unknown to us (and essentially unknowable). We must, therefore, rely on the valuations of the accountant and must confine the analysis to outlay costs (as opposed to opportunity costs). Rectification of cost data in these studies adheres to the spirit of outlay cost.

Accounting data for branch units of the same firm supply a moderately good measurement of the total flow of input services, including the capital consumption portion of capital input (assuming that differences in original cost of equipment caused by price level fluctuations have been removed, and a corresponding correction made for long-term leases). Within this framework of adjusted accounted cost there remain imperfect valuations and arbitrary allocations of overhead. Hopefully they are not systematically related to plant size and hence will widen the statistical scatter rather than distort the shape of the fitted function.

6. Statistical cost–size studies are expensive and of limited durability because their findings quickly become obsolete.

This criticism is correct. The limitation is of particular importance when there are rapid changes in technology, in customer preferences, or in the geographic distribution or accessibility of demand.[7] Statistical estimation of the cost–size function in a way that is satisfying to academics and useful for economic science is expensive, compared with the cost of primitive measurements that would probably suffice for most managerial uses. Nevertheless, to be confident of this and to know which short cuts least sacrifice precision usually requires a thorough analysis first. Quite properly, the roughhewn statistical estimation that can be built on this research foundation would rarely be made available to academicians, who in any case would not regard it as a valid empirical test of economic theory about economies of scale.

7. The size measure used in the studies was not satisfactory for testing the hypothesis. Mismeasurement may have distorted the statistical estimation of the relation of cost to plant size.

In "Finance Chain," size was measured in terms of output, as represented by number of loan accounts. An alternative approach would have been to use some measure of input, such as floor area, rental outlay, or number of employees. These were rejected, mainly because the theory to be tested calls for some dimension of output as the measure of size. Several dimensions of output were considered. An obvious one was dollar volume of loans. It was rejected because the amount of financial service, which is the underlying output of a finance chain, is essentially the same in kind and in quantity for a small personal loan as for a large one. Consequently, the dollar volume was considered a poorer proxy for the output of this underlying finance service than was the number of loans outstanding.

In "Shoe Chain," size of store was also measured in terms of output, by sales in pairs. Here again input measurements of size of store, such as floor area, number of seats, and number of salesmen, would have conformed to popular notions of size and were appealing in their simplicity. But they were ruled out as incompatible with the theory we sought to test. Alternative output measurements of size were numerous: dollar sales of shoes, dollar sales of shoes plus auxiliary items, and shoe sales in pairs. The last, being a close approximation to number of transactions, was a better proxy for the underlying output of a shoe store—namely, sales service—than were dollar sales of shoes or total dollar sales.

Though the calibration of size that we used was better than the alternatives, it still proved unsatisfactory for testing the hypothesis because of economic theory's simplifying abstraction of a single homogeneous product. For all firms and plants observed in the real world, multiproduct output is the rule. Consequently, developing an index of output which correctly weights and combines heterogeneous products is the single most important and most difficult problem in statistical estimation of cost functions, whether they be cost–output relations or cost–size functions. Quite properly, therefore, the way size is measured is a focal point of criticism. Unfortunately, there is no single correct measure of output for a real-world, multiproduct unit; hence no measure of size is beyond cavil.

The output of a shoe store is implacably heterogeneous. The sales-service supplied in support of a transaction differs among salesmen and, for any one salesman, among kinds of shoes and among customers. All that we can hope for is to average away this diversity. In choosing among the alternative output measures, all of which have frailties, we must be content with the least unsatisfactory proxy for the ideally weighted composite of heterogeneous sales-service output. If we are fortunate, the departures of the imperfect size index from the true size—which we cannot measure—will not be systematically related to size of plant. And in this case, no mismeasurement of the shape of the cost–size relationship results directly from the inadequacy of the measure of size.

8. The cost effects of low rates of utilization of capacity of branch units cause the statistical cost curve that is fitted to the cost observations to overstate the unit cost of the idealized envelope curve at all plant sizes.

Failure to remove the cost effect of plant utilization rate, or to make sure it was not inversely correlated with size, is an inadequacy of the statistical research methods in these two studies. Averaging out the cost effect of different average yearly use rates among stores of various sizes produces a statistical cost function that lies above the envelope curve of the theoretical model. However, it need not, for this reason, have a different shape.

This defect will not distort the measured shape of the cost–size function unless the overstatement is systematically different on average for large branch units than for small. Bias toward bigness will result only if (*a*) big stores are systematically more underutilized than small stores, or (*b*) the cost effects of underutilization are (because the cost curve is irreversible downward) more serious for big stores than for small. Neither of these influences is envisaged in the theoretical model. Hence the test of the hypothesis by the statistically estimated cost function is likely to be particularly imprecise in the large size range, where it may undermeasure the savings of store size.

This limitation on the precision with which the statistically determined cost–plant-size curve tests the hypothesis was particularly important for the shoe chain. The rate of utilization of capacity may have been systematically lower on average for the largest stores in the shoe chain during the analysis period. A lower use would cause higher unit cost, largely because the salary cost, which is the biggest component in many chain operations, is irreversible downward.

In theory the long-run average unit cost curve abstracts from cost behavior caused by different rates of utilization. Removal of the cost effect of rate of utilization was particularly difficult in these two statistical cost studies partly because both size and rate of use were measured in terms of output and partly because data were available for two years only. Over a longer period, the degree of underutilization might have been measured by the ratio of current output to some past high-water mark or to a cyclical peak if the data had been available. Alternatively, a hybrid utilization-rate measure could have been carpentered up by ratioing output to some input measure of size that served as proxy for capacity (e.g., floor area or number of seats). Neither of these refinements was undertaken.

The two main causes of pronounced underutilization are newness of the branch unit and fluctuations in intensity of competition that the unit faces. As to the first, we excluded from the analysis new and recently expanded stores and loan offices. This diminished but did not solve the problem. As to the second cause, wider fluctuations in utilization rate and a lower average rate for large branch units could be caused by their more vigorous and volatile competitive environment.

In "Shoe Chain," for example, the largest stores were located at the center of a transportation network in mid-town Manhattan, where they were in frontal competition with New York's largest (and at that time solely centrally located) department stores. In contrast, the smaller shoe stores, which tapped narrow markets bounded by walking convenience, faced, in those days, few and stodgy competitors. Most of these neighborhood stores had a limited locational monopoly and hence smaller volatility of sales and higher average use rates. Thus there is some evidence that the extent of fluctuation in sales and the average degree of underutilization depend upon the intensity of competition, which was generally greater for large, centrally-located Thom McAn stores than for small neighborhood units.

9. The observed size range of branch units in the two studies was narrow and the size-category of large units was thinly populated, hence the findings are unreliable for the largest store or loan office size.

This limitation detracts from the interest of the findings to economic theorists. Controversy centers on whether the curve turns up for largest units, i.e., the eventual appearance of net diseconomies of scale. This defect also has practical importance. In the trade-off of size economies against locational advantage, good forecasts of operating costs for units bigger than those as yet warranted by the size of the local market are needed.

10. Differences among plants (stores or loan offices) as to technology and quality of management were incompletely removed in the studies, creating a bias toward exaggeration of economies of scale.

One of the major drawbacks in cross-sectional statistical studies of firms is that differences in the quality of management and the modernity of plants cause differences in cost that are contaminants for purposes of determining cost–size functions. In the hypothesis that we seek to test, the economic theorist assumes perfect management and technology that is either optimum or uniformly nonoptimum among firms (or plants) or is valued in a manner that dispels the cost impacts of differences in technology, in locational advantage, in brand acceptance, etc. Among firms, these contaminating differences are great and probably bias cross-sectional measurements toward overstatement of scale economies, because the largest firms tend to be the best managed and most modern, and because the equity capital invested in cost-saving equipment is treated in the accounts as a free good.

This bias is likely to be less serious in cross-sectional analysis of branch units of the same firm. But it probably still exists, since the most talented manager and the most advanced equipment are likely to be assigned to the largest units of a chain. It will cause overstatement of scale economies to the extent that talent is undercompensated, or (inflation aside) the equipment that embodies the technological advances is priced below its worth to the user, or the capital costs of this investment are understated by the books. All three probably occur in some degree.

Study 5

FINANCE CHAIN

To determine statistically the long-run relationship between cost and size of plant, the annual costs of a chain of personal finance offices were studied. Long-run marginal and average cost curves were fitted to simultaneous observations of 143 offices, which covered a wide range of plant sizes. Each office performed most of the functions of an independent personal finance company and was regarded as a proxy for an independent establishment for purposes of the study.

The relationship between cost and plant scale was a statistical approximation to the neoclassic economic concept of complete adaptation to output rate of all costs, including the size of plant, that is labeled "the long-run cost curve." This label is misleading. The concept of complete adjustment of cost to size of plant need have nothing to do with time. Investment, not time, is usually the critical requirement for a bigger plant. The economic theory of cost–size relationship is essentially timeless. It permits all the adaptation needed to adjust the size of plant. But it excludes all but this one type of change, ruling out as irrelevant other time-related changes such as advances in technology, higher wage rates, and prices of other factors.

I. Methods

The total adaptation of cost to plant size assumed in the idealized envelope curve of long-run relation of average unit cost to output is unattainable in a statistical cost study because short-run changes in the intensity of use of fixed factors occur simultaneously with long-run adjustments. The factors of production (cost ingredients) usually form a hierarchy of fixity. Some can be changed cheaply and quickly to adapt to the level of business activity; others involve investment in adaptation and may take a long time. Adjustments of a short-run character (i.e., differing

This study was made in 1934 and appeared as "Appendix E" in Joel Dean, "A Statistical Examination of the Behavior of Average and Marginal Costs," Ph.D. dissertation, University of Chicago, 1936. I am indebted to the executives of the Household Finance Company for providing confidential cost data in 1934 and for helping me to understand their meaning. I am also grateful to Mr. G. R. Ellis, the present chairman, for his generous permission to publish the findings and reveal the name of the company.

intensity of use of a loan-office plant of given size) may average out and thus not distort the *shape* of the statistically determined long-run cost curve; but they will raise the overall *level* of the curve. The result is that the cost observations used in a statistical cost study (and the statistical curve fitted to them) lie above the idealized envelope curve (portrayed in Chart Two–1).

This study attempted to minimize the distortions caused by short-run adjustments by selecting a kind of business in which fixity of output rate was great compared with fixity of most cost components.

A chain of personal finance offices proved, in this respect, a particularly suitable choice. The uniformity of records, layout, personnel standards, operating policy, and management methods among branch offices gave the data an otherwise unattainable homogeneity.

Loans, the output of a finance chain, are made for a relatively long period (twenty months on the average) compared with the commitment period of some costs. Output, therefore, is rather inflexible downward. Loans can be expanded more quickly than they can be contracted, but even this upward flexibility is not mercurial. In contrast to the sluggish adjustments in the number of open accounts, most components of loan-office cost can be adjusted fairly quickly. As a consequence, the scale of plant, except for rented space, can be accommodated relatively easily to the remarkably stable rate of operation.

In a further effort to remove the distorting effects of short-run cost adjustments, loan offices that were less than twelve months old and offices that had grown rapidly in the last year were excluded from the analysis. Finally, to allow for unpurged short-run cost departures, an additional independent variable, ratio of new accounts to total accounts, was introduced in the multiple-regression analysis. All these measures minimized but did not eliminate the distortion caused by the inclusion of short-run, as well as long-run, adjustments in the cost data. In the case of some loan offices, this distortion may have been serious.

No rectification of the cost data for changes in price level or for output lag was made. Cost data were annual summaries, and the comparison among loan offices was made at the same point in time.

To isolate net relationships between cost and size of loan office, graphic multiple-regression analysis was employed. Each of 143 mature loan offices was used as an observation. Cost was the dependent variable (X_1). As independent variables for the analysis, five cost determinants were selected:

X_2 Size of branch office, expressed as the total number of open accounts.
X_3 Delinquency, represented by the ratio of delinquent accounts to the total number of open accounts.
X_4 Size of city, reflected by the population of the city in which the branch office is located. (This variable is expressed in logarithmic terms.)
X_5 New business, represented by the number of loans made to new borrowers plus the number of loans made to former customers, expressed as a percentage of the total number of open accounts.

X_6 Renewals, represented by the number of renewed loans made during the year, expressed as a percentage of the total number of open accounts.

To validate the findings of entity analysis, and to develop a new technique, a second kind of graphic multiple-regression analysis was carried out. Cell averages obtained from cross-tabulation were substituted for the 143 individual loan offices as observations. Loan offices were cross-classified according to three of the major cost determinants listed above: size of branch office (X_2), ratio of delinquent accounts (X_3), and size of branch office city (X_4). The resulting cell averages were used for several of the cost curves developed in the study.

Costs of the loan offices were studied not only in terms of combined cost but also in terms of major cost components.[1] For a loan office, the main components of cost include rent, salary, bad debt loss, contributions and general expenses, local taxes, and depreciation. The components selected for detailed investigation were (1) rent, (2) contributions and general expenses, and (3) salary. Two separate bodies of data were used: costs for the year 1932 and costs for the year 1933.

The relationship to plant size was determined statistically for cost in two different forms: (1) average unit cost (per account) and (2) total cost for the year. Long-run marginal cost was estimated from the statistically determined cost behavior patterns of both total cost and average cost per account.

II. Findings

Average Unit Cost

The behavior of combined cost per account for the year 1933 as determined by graphic multiple-regression analysis is portrayed by the partial regression curves in Chart 5–1. This chart compares the findings of the two multiple-regression analyses. The dotted curves in the figure were established by the first analysis, which used each of the 143 loan offices as an individual observation. The solid curves were determined by the second analysis, which used as observations the cell averages of the individual loan offices obtained by cross-classification. Differences between the two net regression curves are caused in part by the averaging of cross-classification and in part by the omission of two independent variables—X_5 (% new accounts) and X_6 (% renewal accounts)—from the cell-average regressions.

The top section of Chart 5–1 shows the effect of size of branch office (X_2) on cost. As expected, cost per account declines as plant size increases. The sharper drop of the regression curve obtained from individual observations may more truly indicate the long-period average cost behavior, since the effect of irrelevant cost factors may have been more completely removed and short-period adjustments more fully allowed for when all five of the independent variables were used and when each branch office was plotted individually.

One orthodox explanation for net internal economies of large-scale production, such as appear to be present here, is the limited divisibility of some factors of pro-

Chart 5–1 Average Combined Cost for Finance Company: A Comparison of Net Regression Curves Secured from Analysis of Individual Branch Offices and from Group Averages, 1933

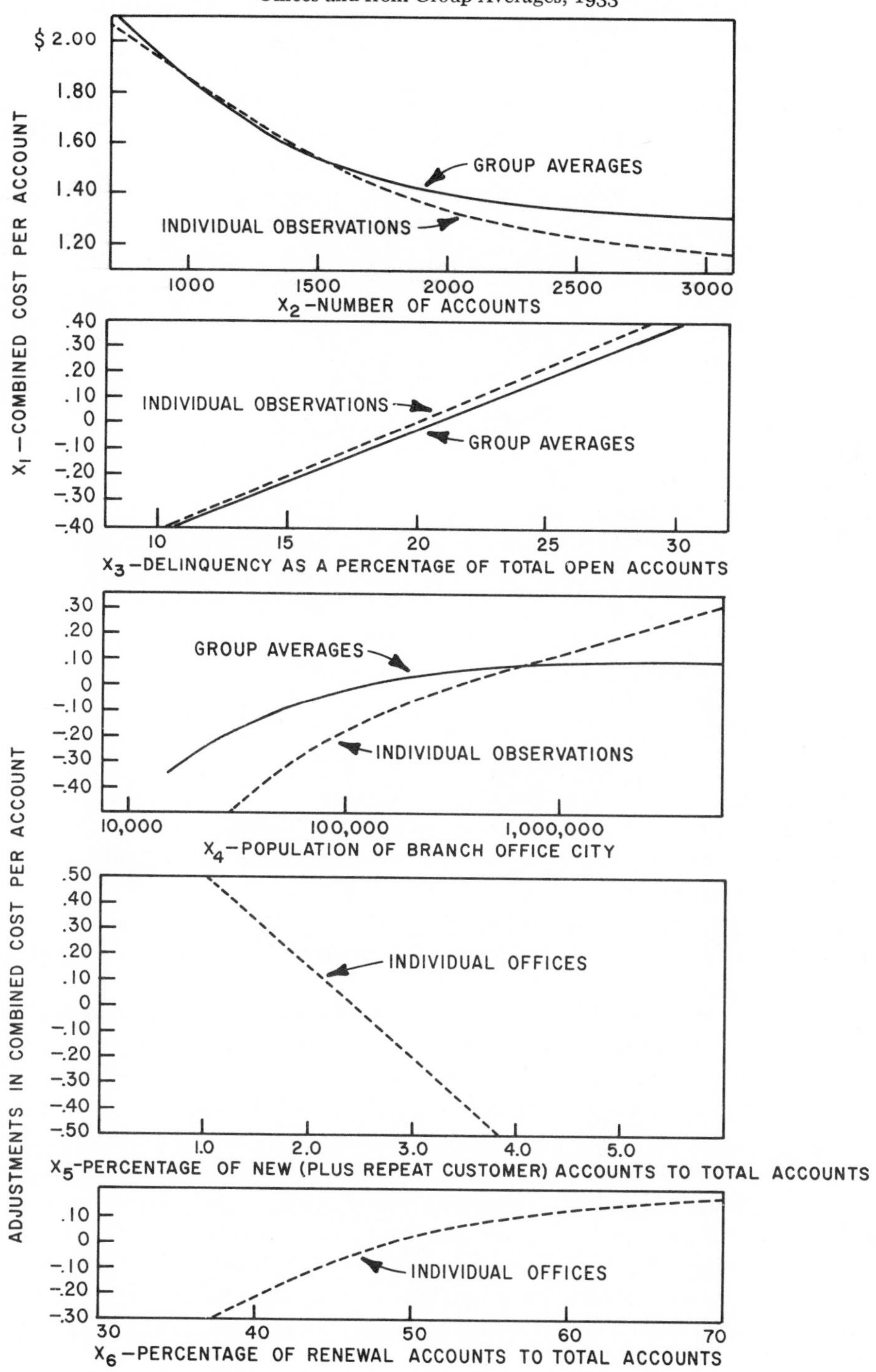

duction. Indivisibility prevents the advantageous proportioning of factors in a small-sized plant that can be achieved on a large scale. The pronounced scale economies in salary costs indicated in Chart 5–4 suggest that specialization of labor, as well as the essential indivisibility of the human being, is an important determinant of savings of branch-office size. Possibly, something akin to the inherent economies of a large machine (Edgeworth's steamship) produces a more efficient layout in a large loan office. Note the decline in rent cost per account, even after correction for size of city in Chart 5–2.

The theoretical hypothesis of the relationship between unit cost and size is a curve that first declines, then flattens, then rises after the optimum size is passed. The statistical cost curve obtained for the finance chain differs from the theoretical curve in that it does not turn up. Within the size range of loan offices in 1932 and 1933, unit cost flattens but does not subsequently rise and long-run marginal cost is constant; there is no evidence of net diseconomies of scale.

Economists have assumed that there must come a point where the gains from further increases in size would peter out and then be more than offset by the increasing cost of coordination and control by management whose efficiency shrinks as it is spread thinner. When the economic theory of the relation of cost to scale was formulated a century ago, each firm had but one plant: plant and firm were the same. Since then economic theorists have usually failed to make the distinction, which is important in 1934, between size economies of the firm and those of the plant (or loan office). The forces that cause scale economies are not the same and the size where they are exhausted is different. The source and likelihood of net diseconomies of size for a plant such as a loan office cannot be taken over bodily from early doctrine about size of firm.

The competitive implications of this declining long-run average cost curve are interesting. Under pure competition a decisive cost advantage conferred by size of loan office would forecast cutthroat competition for personal loans; competitive survival of the fittest would cause a drift toward monopoly. But under the kind of imperfect competition that characterizes the personal finance business—with locational monopolies, transportation barriers to increased loan office size, and extremely imperfect consumer information—the observed net relation of unit cost to size of loan office has a different and more limited significance. Big loan offices do not, because of savings of size, drive little ones off the map. The small office, despite its higher cost per account, is still the cheapest and most effective way for a chain to meet the competition of independents and of rival finance chains in serving a small, locationally circumscribed market.

The effect of slow payments—the net regression of delinquency (X_3)—on cost is shown in the second section of Chart 5–1. As expected, higher delinquency raises the average cost per account.

Bigger population of the branch office city (X_4) also has the expected effect of raising cost, as indicated by the regressions in the third section of Chart 5–1. Loan offices in large cities have higher salary rates and apparently (Ricardo's doctrine

Chart 5–2 Branch Office Rent Cost Per Account, 1933—Net Regression Curves for Three Influential Factors: Number of Accounts (X_2), Delinquency (X_3), and Population of Branch Office City (X_4)

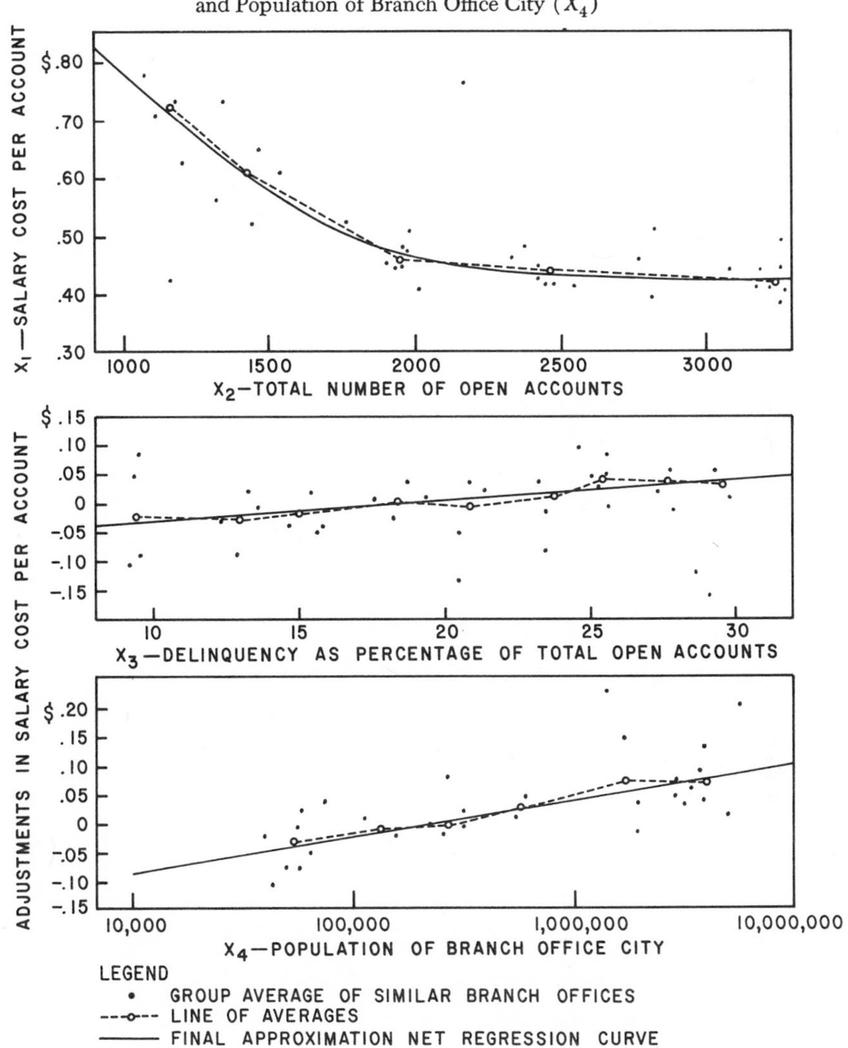

notwithstanding) higher rentals per account. A steeper regression curve for city size is obtained from individual loan office observations.

As indicated by the fourth section of Chart 5–1, a higher new-loan ratio (X_5) seems to be associated with lower costs. This regression is hard to understand. Possibly the new-loan ratio reflects the cost effects of growth, in addition to the cost of getting new business. Loan offices with a high new-loan ratio are also growing relatively fast. Hence this variable may actually measure the more advantageous use of fixed facilities (which were generally in plethora in 1932 and 1933) in growing loan offices as compared with stable or retreating offices. Also, a high ratio of new accounts during periods of depression indicates low collection costs. New borrowers are screened more rigorously in such periods. Moreover, it is relatively cheap to investigate and to collect from newly repatriated customers (included as new borrowers) if wise use is made of their previous repayment performances.

A higher percentage of renewals (X_6) is associated with high cost per account, as seen in the bottom section of Chart 5–1. This is reasonable. Particularly in 1932 and 1933 renewal often masked incipient delinquency, in that some renewals were simply a continuation of bad accounts that were especially costly to collect.

Three of the most important components of combined cost—rent, general expenses, and salary—were also investigated. The net regression curves obtained for 1933 rent cost are shown in Chart 5–2. Net regressions for general expenses in 1933 are shown in Chart 5–3. Salary-cost regressions for 1933 are shown in Chart 5–4. Examination of these regressions shows that the behavior of the components of combined cost was, in the main, in accordance with expectations. Each of the three component regression curves for size of branch office slopes downward to the right, indicating economies of scale in the behavior of these cost elements. Delinquency appears to increase salary cost and general expenses, as would be expected in the administration of slow collections. However, the delinquency regression curve for rent cost (the third section of Chart 5–2) suggests that delinquency has little to do with rent expense. The unanimous upward slope of the net regression curves for size of branch-office city indicates that operation in large urban centers brings increases in each of the constituent expenses investigated. This too might reasonably be expected.

The measures of the closeness of the relationship between cost and the several independent variables are summarized in Table 5–1. They were determined graphically; hence they do not have quite the same significance or precision as those obtained from a least-squares multiple-correlation analysis.[2]

Total Cost

Chart 5–5 shows the findings of the analysis of combined cost of loan offices in the form of total cost per year. A straight line was fitted visually to net-regression-adjusted cell averages in the form of totals for the year. Although the scatter of cell

Chart 5-3 Branch Office General Expenses Per Account, 1933—Net Regression Curves for Three Independent Variables: Number of Accounts (X_2), Delinquency (X_3), and Population of Branch Office City (X_4)

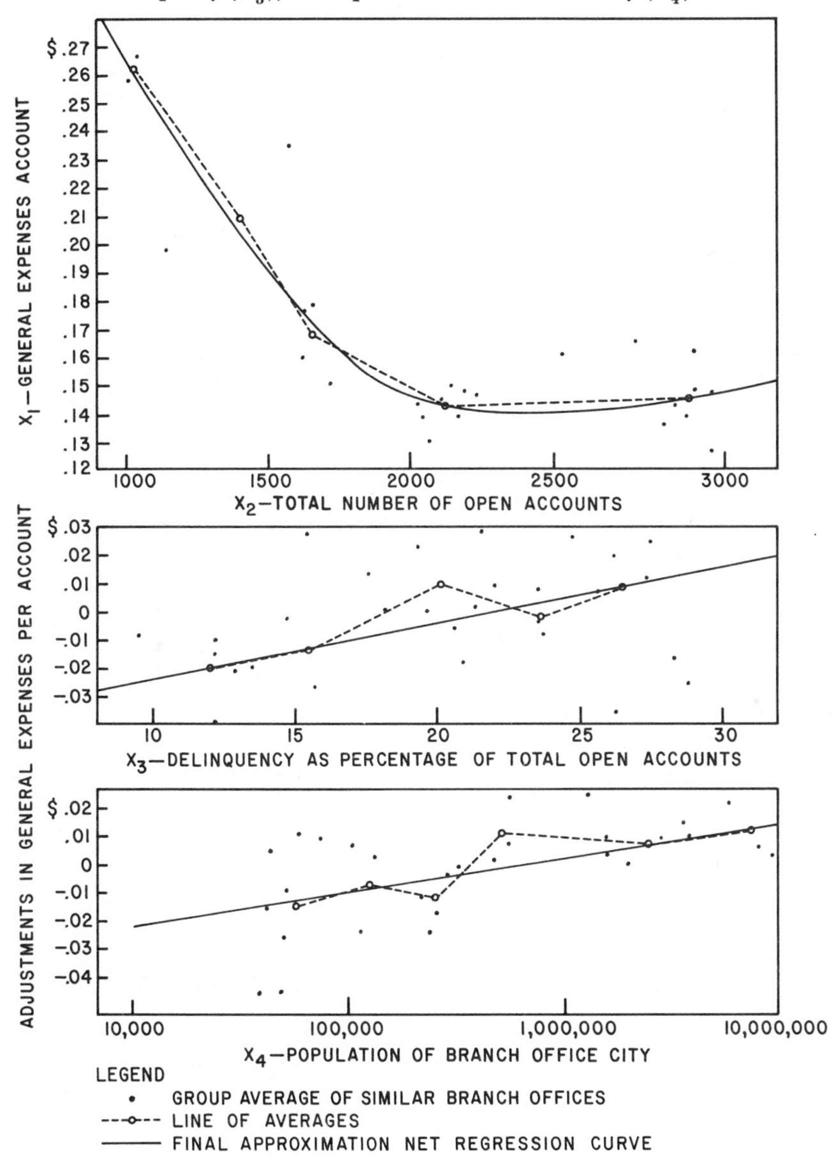

Chart 5–4 Salary Cost of Branch Offices, 1933—Net Regression Curves for Three Influential Factors: Number of Accounts (X_2), Delinquency (X_3), and Population of Branch Office City (X_4)

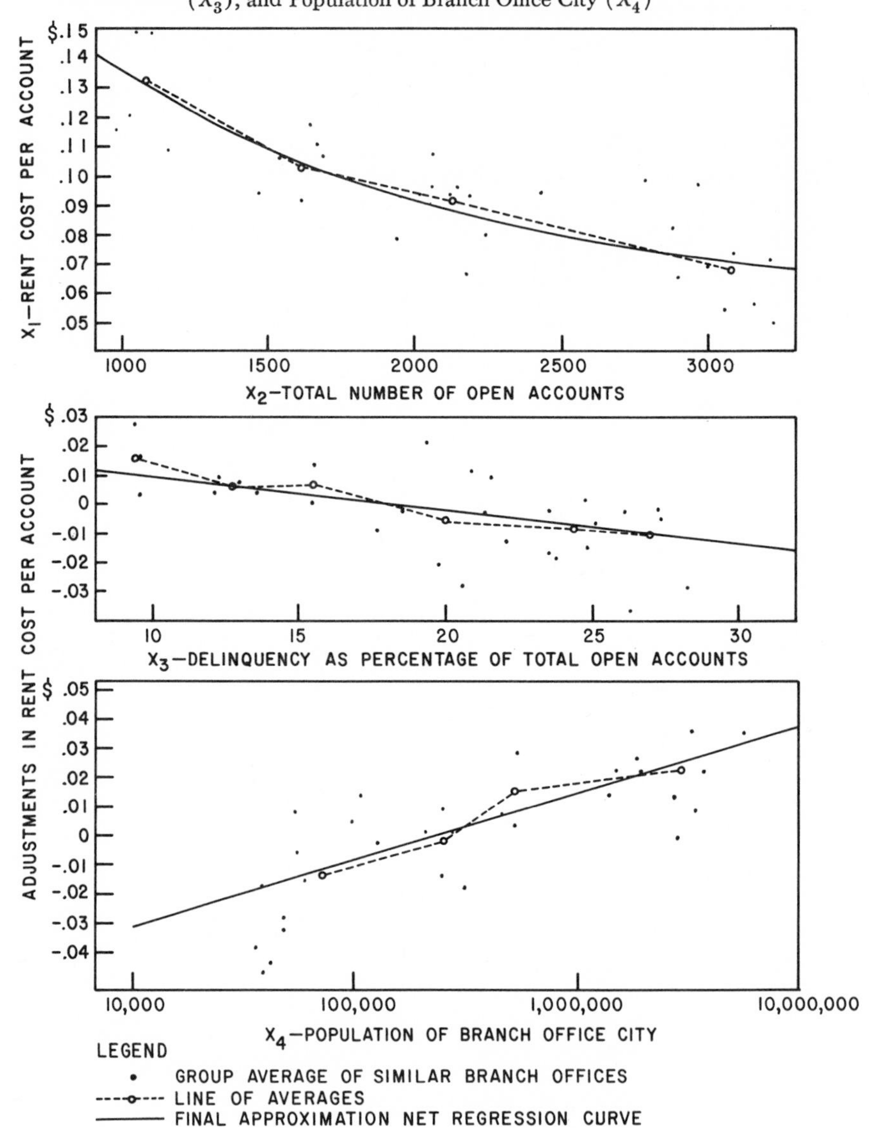

Table 5–1

MEASURES OF CORRELATION: REGRESSION ANALYSIS OF BRANCH OFFICE COMBINED COST OF FINANCE CHAIN, 1933*

Cost	Standard Error of Estimate	Index of Multiple Correlation	Index of Multiple Determination
Combined cost	.141	.942	.887
Salary	.0283	.700	.490
General expenses	.0271	.667	.445
Rent	.0161	.769	.591

*Independent variables: size of branch office, X_2; delinquency ratio, X_3; size of branch office city, X_4.

Chart 5–5 Marginal Cost Determined by Fitting a Straight Line to 1933 Adjusted Total Cost Observations

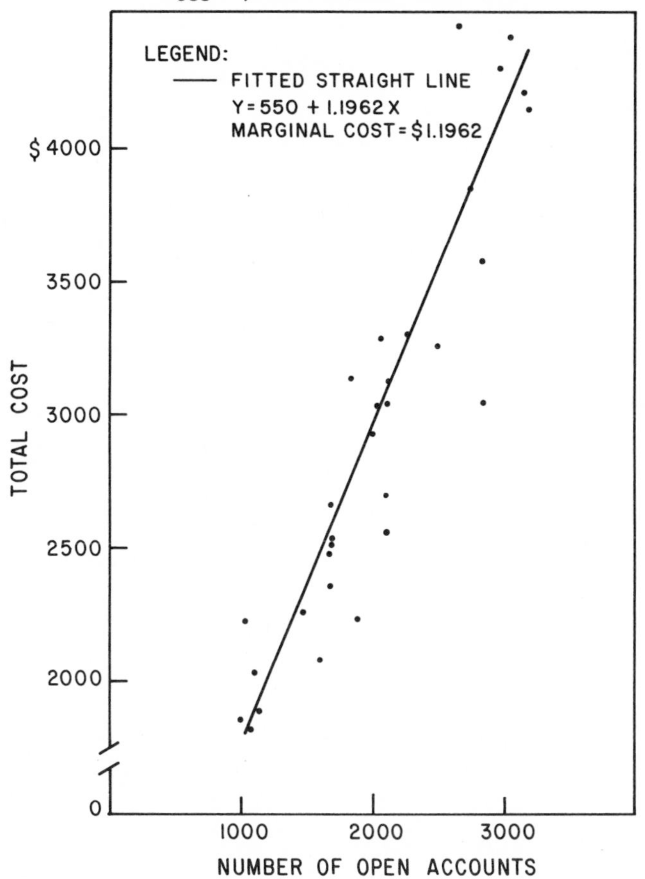

averages is wide, the distribution of these total cost data in relation to size of loan office (measured by number of open accounts) supplies empirical support for constant long-run incremental cost over the observed range of size.

Marginal Cost

Estimated long-run marginal cost was derived from the total cost findings. In addition, two marginal cost estimates were obtained from the net regression curves of average combined cost related to size of loan office. They were derived from the average unit cost curves obtained from (1) the multiple-regression analysis of the 143 individual observations (the dotted curve in the top section of Chart 5–1); and (2) regression analysis of the cross-classification cell averages, the third approximation cost–size regression (the solid curve in the top section of Chart 5–1). Both were derived from third-approximation net regressions.[3]

The estimates of long-run marginal cost obtained by each of these methods are shown for the observed range of loan-office size in Table 5–2. The most reliable estimate of long-run marginal cost is probably the one obtained from the total cost curve fitted to cell averages, shown in column 4. The marginal cost estimate shown in column 3 is derived from third-approximation average cost curve for cell averages. The curve is virtually flat: its mild U-shaped departure from constancy is not statistically significant. The declining marginal cost schedule shown in column 2 was obtained from the third-approximation average cost curve, using individual branch offices (rather than cell averages) as observations. Its continuous decline shows the sensitivity of marginal cost estimates to small differences in the shape of average unit cost curves: the two in the top section of Chart 5–1. This is a frailty of using regressions of cost per account to estimate incremental cost. Declining marginal cost is not supported by the net regressions of total cost (Chart 5–5).

Little reliance can be placed in the undulations of the two marginal cost curves derived from average unit cost net regression curves; minor changes in the slope of the average cost function are magnified in the differential curve. Nevertheless, the general downward slope of the marginal cost curve derived from cost per account of individual offices (column 2) casts some doubt upon the estimate of constant marginal cost obtained from the total cost curve. Both marginal cost curves derived from average cost per account concur in a declining phase up to a loan-office size of about 1,500 accounts. The declining portion of the various average cost curves appears to be the most reliable sector.[4] All the cost components studied exhibit an early decline in their derived marginal cost curves.

In summary, more confidence can be placed in the estimate of constant marginal cost derived from the total cost function than in the undulations of the two marginal cost curves derived from net regressions of average cost per account. The only possible exception is the initial falling phase of the marginal cost curve derived from both average cost regressions, which may be reliable to a size of about 1,500 accounts.

Table 5–2

RELATION OF LONG-RUN MARGINAL COST TO SIZE OF BRANCH OFFICE OF FINANCE CHAIN, 1933

	ESTIMATED MARGINAL COST*		
(1) Number of Accounts	(2) Average Cost (Individual Offices) †	(3) Average Cost (Cell Averages) ‡	(4) Total Cost (Cell Averages) §
800	$1.57	$1.10	$1.20
900	1.45	1.06	1.20
1000	1.32	1.02	1.20
1100	1.20	.98	1.20
1200	1.07	.95	1.20
1300	.95	.93	1.20
1400	.82	.93	1.20
1500	.73	.94	1.20
1600	.66	.95	1.20
1700	.64	.97	1.20
1800	.63	.99	1.20
1900	.70	1.02	1.20
2000	.80	1.03	1.20
2100	.87	1.05	1.20
2200	.89	1.07	1.20
2300	.90	1.09	1.20
2400	.89	1.11	1.20
2500	.87	1.12	1.20
2600	.85	1.13	1.20
2700	.83	1.14	1.20
2800	.81	1.15	1.20
2900	.78	1.15	1.20
3000	.75	1.16	1.20

*Derived from net regression curves obtained by graphic multiple-regression analysis of the combined cost.

†Derived from third-approximation net regression curve of average combined cost per account as determined by multiple-regression analysis, using individual branch offices as observations.

‡Derived from third-approximation net regression curve for average cost per account, using cell averages of branch offices as observations.

§Derived from the straight line fitted graphically to total cost residuals adjusted by multiple-regression analysis, using cell averages of branch offices as observations.

Estimates of long-run marginal cost have particular manageral importance for a personal finance chain. Most of a finance company's competitive actions require long-run rather than short-run adjustments, e.g., convenient locations, continuing outlays on advertising, and a believable history of competitive interest rates. Moreover, increases in size can be made by relatively small increments, so that the scale of operation can be adjusted moderately closely to fit changes in business volume. Finally, personal loans run for a relatively long period compared with the time over which some costs are flexible; thus the cost of taking on additional business may be approximately equal to the long-run marginal cost.

Study 6

SHOE CHAIN

CHAPTER I

Introduction

The major purpose of this study is to determine, by statistical analysis, the relation between the size and the operating costs of the retail stores of a large shoe chain. The size of a given retail store is an important influence governing profitability, although it is evident that the earning power of a store is affected by several other conditions as well, such as the nature and size of the tributary market area, the type and quality of the merchandise handled, and the character and ability of the store management. Since store size is subject to managerial control, a knowledge of the relation of cost to size is essential in order to formulate a rational policy of store size and store location and to set standards of operation.

The results of such an analysis, in addition to their administrative usefulness, are also of interest to economists. The empirical evidence afforded concerning the influence of size on cost has a direct bearing on the theory of the long-run cost behavior of the individual firm. This long-run cost behavior is described by the cost–output function in the case that the firm is able to adapt all its equipment and plant facilities to suit any selected output. Here the output of a firm is considered to measure its size, so that, in effect, the long-run cost–output function describes the course of cost as size varies.

Reprinted from Joel Dean and R. Warren James, *The Long-Run Behavior of Costs in a Chain of Shoe Stores: A Statistical Analysis*, Studies in Business Administration, vol. 12, no. 3 (Chicago: University of Chicago Press, 1942), by permission of the publisher. I am indebted to Ward Melville, Chairman of the Melville Shoe Corporation, for making the data available to me at the National Bureau of Economic Research in 1939, for encouraging the research as it crept along, for permitting its previous publication, and now for allowing me to reveal that the company was Thom McAn.

Two alternative methods are available to obtain the empirical counterpart of such a theoretical construction. A series showing the relation between cost and output can be derived either by (1) successive observations of one firm which has adapted itself to various outputs over a period of years or by (2) simultaneous observations of a number of similar firms adapted to different output levels, taken to represent alternative long-run adjustments of one firm. The first approach is subject to serious objections. The observations of an actual firm are likely to represent an expansion path which traces only growth or stationary positions but seldom large-scale decreases in size. In such a case the practical difficulty of abstracting from historical evolution, i.e., changes in technical methods, managerial skill, relative prices, etc.—the "other things" assumed to remain constant in the theoretical model—is usually insuperable. The second approach, on the other hand, raises the problem of finding a group of firms which produces a homogeneous product and which are sufficiently similar in other respects to be considered as observations of a single firm at different levels of output under conditions of total adaptation.[1]

Data of this second type were used in this study. Approximately integral units of a multiple-plant firm were studied, instead of attempting to find comparable independent firms. These units were observed simultaneously in 1937 and 1938. Although it is true that the individual stores of a shoe chain do not correspond precisely to the firms of economic theory,[2] each store individually performs almost all the functions of an independent retail store and might, therefore, be expected to have similar patterns of cost behavior. In addition, such units possess the important advantages of being almost identical with respect to records, merchandise, layout, personnel, and management methods. If it were not for the remarkable homogeneity of the retail outlets, comparability would be lacking, and even a rough approximation to the theoretical model would be impossible.

In view of the fact that problems of economic interpretation as well as problems of statistical measurement are raised in the analysis of these data, this monograph is divided into two parts. The first part (chap. ii) is devoted exclusively to the statistical treatment of the data, the problems of measurement encountered, and the presentation of the statistical results. The second part (chap. iii) deals with certain questions involved in the nature of retail selling and the theory of cost of the individual firm, in so far as they are of help in reconciling the statistical results with the underlying economic theory.

Chapter II

Statistical Analysis

The objective of this statistical analysis of a group of retail shoe stores is to determine the influence of the size of a store on its cost of operation. Before the statistical results are presented, however, it is first necessary to consider the nature of the information available and the problems involved in measuring size and cost. Section A of this chapter is consequently devoted to a description of the sample, while Sections B and C are concerned with the measurement of size and cost, respectively. Section D deals with the actual results of the statistical analysis.

A. Nature of the Sample

From the several hundred stores operated by the parent shoe corporation a sample of fifty-five stores was selected for the purposes of this analysis. Annual data on the retail operations of each store were made available for the two years 1937 and 1938.[3] The sample of stores was selected in such a way as to provide a fairly uniform coverage of various operating conditions. An attempt was made, therefore, to choose a group of stores selling different proportions of the various types of shoes which were, at the same time, representative of the sales volume of the stores in the chain as a whole.

In order to secure as high a degree of comparability as possible the sample was further restricted on the basis of the following considerations. First, only those stores situated in the metropolitan area of a large city were included. It was thus possible to minimize regional variations in wages, rentals, and public utility rates. Moreover, since this area is under the control of one supervisor, differences in managerial efficiency attributable to supervisory skill were eliminated. Second, although the parent shoe corporation operates several shoe chains, each of a different price class, stores from only one of these chains were included in this sample, in order to obtain the homogeneity of sales service associated with uniformity of selling price. Third, the sample was restricted to those stores selling men's and boys' and girls' shoes only; stores selling women's shoes being excluded. Fourth, stores recently opened were excluded from the sample, on the grounds that they might exhibit marked peculiarities because of immaturity. "Immaturity" in this case signifies not only that systematic underutilization may exist but also that sufficient time has not elapsed to allow the store to adapt its merchandise to the special characteristics of its market. Fifth, some attempt was made to select the sample on the basis of a uniform time distribution of sales, but it cannot be claimed that this was rewarded with any success.

The individual stores of the shoe chain are confronted with a market situation which can be described as monopolistic competition with elements of oligopoly. This situation leads to the inference that retail shoe stores of the type under consideration do not generally compete with one another on a price basis but confine their competitive efforts to attempts to alter the volume of their sales.[4] The price inflexibility of shoes in the range of quality under consideration is remarkable, in view of the violence of fluctuations of prices of the principal raw materials. While the brands of shoes sold by rival stores are differentiated products, the physical differences of comparable shoes of distinct brands within the same price range appear to be small to the inexpert eye. This similarity of products and the tendency of shoe stores to cluster geographically prevent the attainment of any considerable degree of monopoly by any one retail store. The total demand in the market area is divided among individual stores in proportions which tend to remain relatively constant. Expenses intended to maintain or increase one firm's share in the market will, therefore, be incurred continually. The share of each shoe store depends upon customers' habits, personal affiliation, advertising, and the sales-promotion activities of the firm rather than upon price concessions. Open price-cutting appears to be an ineffective means of enhancing a shoe store's share of the total market for shoes of a given price class. This is partly because the differentiation of products is so slight that close rivals will be forced to meet the concession immediately, which may precipitate a price war. The practice which shoe stores often follow of advertising a specific price ($10.00, $5.55, or $3.25) means that manipulation of these prices may impair customer good-will and nullify the value of the advertising which directed attention to the fixed price. Moreover, for shoe purchasers who judge the quality of shoes by their prices, price-cutting may demote a brand of shoes to a lower quality position.

As far as the factor market which confronts the firm is concerned, it appears that the demand of the individual stores is such a small proportion of the total that the prices of input factors are independent of the action of the individual store. Even though there is centralized purchasing by the shoe corporation of some supplies and some centralized hiring, which may affect factor prices, the influence of any one store can be neglected.

B. The Measurement of Size

The concept of the size of the individual firm is not a simple one. Size, however measured, cannot be dissociated from capacity in some sense. It is necessary to make a primary distinction between "technical" and "economic" capacity. Technical capacity refers to a maximum rate of operations, an upper limit set by the technical character of the plant and equipment. Economic capacity, on the other hand, refers to some rate of operations which is optimum from the point of view of cost, i.e., the rate at which average cost is a minimum. "Normal" capacity, in addition, may refer to some average or modal rate of operations which is not associated with either technical or economic capacity.[5]

Of the many size-concepts possible, it is necessary to select one which is practically measurable, relevant to cost, and which, at the same time, permits comparisons of the sizes of different firms. There are available several concepts of size which can be measured and which are not without significance for cost. The choice can be narrowed to the following three types of measure: (1) physical size of plant and equipment, (2) input capacity, and (3) output capacity.

1. THE AMOUNT OF FIXED EQUIPMENT

Size may refer to the aggregate of productive resources that are not alterable in the "short run." The amount of fixed equipment, however, is difficult to measure in terms that permit interfirm comparison, unless the equipment is highly standardized and in homogeneous units. If equipment is standardized, significant aspects of size for comparative purposes could be obtained—for example, by counting the number of spindles in a cotton mill, the number of beds in a hospital, the number of knitting machines in a hosiery mill, or some such similar unit.[6] In the case of shoe stores, one can obtain physical measures of plant size from the number of seats in a store, the area of floor space, or the number of standard lighting units (if electrical fixtures are sufficiently standardized). All these measures are subject to such a high degree of accidental variation, however, that none of them gives a trustworthy measure of size which permits interstore comparability.

2. INPUT CAPACITY

It may be possible in some instances to devise a measure of size based on input capacity. A commonly used index of size is the number of employees in an enterprise (in census classifications, for example). Raw-material inputs may also serve this purpose—for example, the crude-oil input of a refinery, the size of the charge of a blast furnace, or the number of tons of ore per day a crushing machine can handle. It may, however, often be extremely difficult to select from the complex of inputs that aspect which is most relevant. Especially when human services are involved, the concept of input capacity lacks the precision which is desirable when measurement is required. In the case of retailing, where the dominant input is labor service, measurement presents especial difficulties, so that input measures were rejected as a possible index of size.

3. OUTPUT CAPACITY

Another alternative is to look on the size of a firm as defined by its output. Here again, however, some important difficulties are encountered. First, when output consists of variegated products, a complex measurement problem is introduced.[7] It may, therefore, be necessary to employ index numbers to deal with cases of multiple products. Second, it is necessary in this case, as well as in the case of input capacity, to distinguish whether output represents the maximum possible rate of production, the optimum rate (for a given body of fixed plant), or some modal rate. In one of these cases the measure of size by output is clearly superior to any other. This occurs under the circumstances in which each output is produced at the minimum total cost for that output. Here there is a one-to-one correspondence between output and size, although it is true that this case may represent idealized behavior. Despite the fact that firms rarely operate under these minimum-cost conditions, the use of actual output as representative of size is the most satisfactory measurement device obtainable.

The next question to be considered is the information which is available concerning the output of the shoe stores. The total sales of each retail outlet in the sample are composed of the following types of merchandise: (*a*) men's shoes, (*b*) boys' shoes, (*c*) girls' shoes, (*d*) hosiery (men's, boys', and girls'), (*e*) rubbers, (*f*) sundries. These categories are self-explanatory except for sundries, which include bedroom slippers, shoe laces, shoe polish, and other small accessories. The dollar volume of sales of each category was available for 1937 and 1938, and information concerning the physical volume of sales of men's and children's shoes was also provided.[8] The assortment of shoe styles stocked differed somewhat from store to store, but information concerning the composition of shoe sales with respect to the various types and styles was not available.[9]

With this information at hand, the next step was to derive from it some satisfactory index of output. The analysis of the nature of shoe merchandising, which is discussed in the theoretical section (chap. iii, Sec. A), indicates that retail selling output is best considered as the production of sales service. Sales service is the contribution of the shoe salesman to the prospective purchaser in the form of help, advice, and expert knowledge. Without too much violence to the facts, this sales service can be treated as a homogeneous product.[10] Although the quantity and quality of sales service per transaction may differ somewhat from type to type, from style to style, from store to store, and from time to time in the same store, the variability of sales service is nevertheless limited. There are several influences making for uniformity of service standards. For example, all the stores in the sample are in one city, under the same supervisor, and have standardized recruiting and training policies. Moreover, the employment of extra salesmen drawn from a trained reserve force tends to prevent serious impairment of service standards during rush periods. Since marked deterioration of service would result in loss of customer good-will and patronage, it may be presumed that, with good management, efforts will be made to confine variation in service quality within narrow limits. The assumption is probably correct that the intangible output of sales service is highly correlated with some aspect of retail transactions which is capable of measurement. The immediate problem is to select some one composite measure of the volume of shoe sales in either physical or value terms which will most reliably reflect output in terms of sales service.[11]

To secure the best index of output, the components of physical sales should be weighted on the basis of the quantity of selling time and effort which they require. The amount of sales service rendered in the sale of a pair of shoes is closely associated with the quality and quantity of the effort made by the salesman. In an index of output, therefore, shoes which are difficult to sell should be weighted more heavily than easily sold shoes. Any measure of transactions will automatically assign weights to various aspects of retail sales, and it is the nature of the weights assigned by the several measures of sales which is the most important consideration in this selection. With this in mind, several alternative indexes were considered: (*a*) sales in dollars, (*b*) shoe sales in dollars (excluding sales of rubbers, hosiery, and sundries), (*c*) shoe sales in pairs weighted by selling price, (*d*) shoe sales in pairs weighted by cost of selling, and (*e*) shoe sales in pairs.

a) *Sales in dollars.* The dollar volume of sales represents the most comprehensive index of volume, since it includes, besides the sales of men's and children's shoes, the sales of hosiery, rubbers, and sundries. In this index, products are weighted according to their price, hence this measure may not accurately reflect the amount of service associated with the sale of each product. There is reason to believe that lower-priced children's shoes may necessitate greater sales service than the more expensive men's shoes. The error in the weighting of products other than shoes may be even more serious, since items of small unit price, such as hosiery, may take a disproportionately large amount of selling time and effort. However, there is evidently a highly complementary selling relation between men's shoes and sundry sales which may rectify the undue weight of the value of men's shoe sales.

b) *Shoe sales in dollars (excluding sales of rubbers, hosiery, and sundries).* This measure is defective in two respects. First, men's and children's shoes are again weighted on an undesirable basis. Second, sales of merchandise other than shoes are given zero weight, in view of their exclusion from the index.

c) *Shoe sales in pairs weighted by selling price.* The use of the average price of various types of shoes as weights would yield an index identical with shoe sales in dollars and is open to the same objections.

d) *Shoe sales in pairs weighted by cost of selling.* If an independent measure of the normal input required for each type of transaction had been available, this might have formed a basis for weighting. For example, the net influence on cost of men's and children's shoes might serve as weights. It is, however, impossible to achieve any satisfactory allocation of cost among the different types of shoes sold, since reliable subsidiary information about relative sales service or cost contributions was lacking. Consequently, the use of a specifically weighted combination of shoe sales was abandoned.

e) *Shoe sales in pairs.* This type of index weights all types of shoes equally.

Therefore, of all these measures of output, physical volume of shoe sales affords the last objectionable weighting of any of the measures considered. This measure does not take into account the sales of hosiery, rubbers, and sundries, which is an important omission. While there is some justification for this index in the probable existence of a highly complementary selling relation between shoe sales and other sales,[12] there is no positive assurance that the complementarity is great enough to validate this procedure.

Despite its omission of sundry sales, it was decided to employ number of pairs of shoes sold, including both men's and children's shoes, as the best available index of output. This choice was made on grounds of convenience and because a more satisfactory measure was lacking. Although the physical index is preferred as affording better weighting of the components of sales than the alternatives, it was decided to use dollar sales also as a subsidiary measure of output, so as to include the sales omitted from the physical index.

It may be a matter of indifference, as far as the statistical results are concerned, which one of the three indices—dollar sales, dollar shoe sales, and shoe sales in pairs—is used, in view of the close relation of all these measures. The coefficients of simple correlation between each of the series were calculated and found to be, for each pair, as follows:

dollar sales, dollar shoe sales: 0.9988
dollar sales, shoe sales in pairs: 0.9980
dollar shoe sales, shoe sales in pairs: 0.9979

The data available for the two measures of output which were retained, namely, shoe sales in pairs and dollar sales, needed no adjustment or corrections.

C. The Measurement of Cost

The primary cost data made available by the shoe corporation show the details of the annual expenditure of each store in the sample. The total operating cost is broken down into twenty-three accounts, for each of which the annual total is known for the two years 1937 and 1938. Each of these elements of cost is described in detail subsequently. [Their sum is here called "total cost," which is the same as "combined cost" in the other studies. "Total" also designates the form of the cost: total cost per year versus average cost per unit.]

Two of the elements of cost listed deserve special mention, however. These are general indirect expense, which is an allocation to each store of its share of the expenses of the chain as a whole which cannot be accurately charged to one account, and administrative expense, which is the share of the corporation's central administrative expenses charged to each store. In view of the arbitrary method of allocation, it was considered that the inclusion of these two elements might obscure the relations which it is desired to investigate. Consequently, in the calculation of operating cost these two elements were omitted.

In order to obtain an insight into the behavior of broader categories of cost, the elements of cost were classified into the following groups:

1. Selling expense
   a) salaries
   b) reserve salaries
   c) employees' discount
   d) hosiery award
   e) shoe bonus
   f) advertising
   g) taxes
2. Handling expense
   h) delivery (including inbound, outbound, and interstore)
   i) insurance
   j) postage
   k) supplies
   l) miscellaneous
3. Building expense
   m) rent
   n) light
   o) heat
   p) window display[13]
   q) repairs
   r) depreciation
   s) water, ice

A word of explanation concerning the attention which is given to the individual elements and categories of cost should be added. While it is true that primary interest lies in total cost, the separate analyses are valuable for the following reasons. First, for purposes of control it is important to trace the sources of total-cost variability, since this affords a clearer insight into the causes of abnormal behavior of the costs of an individual store. Cost groups or elements may be subject to influences peculiar to them. Second, separate analysis provides a partial check on total-cost estimates, since estimates of the subsidiary classifications of cost can be combined in order to arrive at total cost.

To give a clearer picture of the nature of costs of shoe selling, there follows a brief description of each element of operating cost.

1. SELLING EXPENSE

a) Salaries. The expenditure of a retail store on salaries is composed of the weekly salaries paid to full-time salesmen, managers, assistant managers, and other employees. Although the salary of an employee does not bear a direct relation to the volume of his personal sales, it reflects his length of service, occupational aptitude, appearance, and general promise and thus bears close relation to his sales-service capacity. In part, the proportionately higher salary costs of larger stores is explained by the personnel organization in the larger stores. They employ an administrative, nonselling store manager and also include on the pay roll such employees as a cashier, who also has the special function of selling hosiery to "walk-in" customers, and occasionally a porter.

A general increase in salary was put into effect in June, 1937. The rate of increase, however, was not uniform for all employees but varied from individual to individual, discrimination being based upon a reappraisal of each salesman's present and potential usefulness to the firm. Furthermore, throughout the period of observation there were continuous adjustments in salary rates. Apprentices, hired at low rates for a six-months' trial period, were subsequently given salary increases. These changes, however, were so small and applied so generally to the stores studied that it was not considered necessary to attempt to improve the comparability of stores by rectifying for salary adjustments within the year. Instead, a separate analysis was made for both 1937 and 1938 to obviate the question of noncomparability.

b) Reserve salaries. Although generally small, this is a significant element of cost. It is composed of payments to part-time salesmen, who are employed during the weekly and seasonal "peaks." Its importance is that it may give some indication of the extent of the irregularity of the time distribution of demand.[14]

c) Employees' discount. This is a discount allowed to employees on merchandise purchased for themselves or their dependents. In a sense it is part of their remuneration, although its magnitude for a particular employee is determined by irrelevant circumstances.

d) Hosiery award. In addition to shoes, the chain sells hosiery in a relatively low-price range, a special attempt being made to link shoe and hosiery sales. The hosiery award represents a salesman's commission of one cent per pair on all hose sold. It is a relatively small cost item which may, however, bear only an indirect relation to volume of shoe sales, since large quantities of hosiery are sold to "walk-in" customers, who wish to purchase hosiery only. Since hosiery sales were not included in the physical measure of the volume of output, the inclusion of a cost item pertaining specifically to sales omitted from this volume index may appear questionable. However, shoe sales are likely to be accompanied by hosiery sales, since a complementary selling relation exists between them, both because of the nature of the merchandise and because of the efforts of the stores to strengthen this relation. On the other hand, hosiery sales were included in the alternative index of output—dollar volume of total sales.

e) Shoe bonus. The shoe-bonus account represents a salesman's commission of two cents per pair on all shoes sold.

f) Advertising. Advertising, which consists mainly of a radio program and newspaper space, is focused on the trade-name and not on individual retail outlets and is consequently paid for by the chain as a whole. The cost is allocated to the stores within a region on the basis of dollar-sales volume.[15]

g) Taxes. A wide variety of taxes is included in this account: unemployment compensation taxes, old age pension taxes, a gross-receipts tax, a city occupancy tax, and an occupation tax. Some of these are proportional to salary, some proportional to sales, and some constant for each retail outlet. The separate tax items are allocated to the stores on the same basis on which the tax is assessed, so that their incidence is correctly reflected by the allocation.

2. HANDLING EXPENSE

h) *Delivery* (*inbound, outbound, and interstore*). The cost of inbound and outbound delivery is based upon the store's actual expenditure for the transportation of merchandise between the warehouse and the store, while interstore delivery refers to shipments from one store to another. The distance of the store from the central warehouse does not affect delivery cost, since the transportation contract is on a tonnage basis, regardless of the location of the store within the metropolitan area. Differences in delivery costs per store are largely explainable by rush deliveries and variation in weight per unit of sales. Outbound and interstore deliveries are small items of expense, ranging from $1.00 to $18.00 per year. They are unrelated to volume of sales but may be a function of the merchandising policies of the individual store manager. Their inclusion introduces a certain small random error, but their retention was decided upon in order not to remove one possible source of volume-cost relation.

i) *Insurance.* The expenditure for insurance premiums covers the insurance of fixtures, stock, and all store improvements. The chain has a self-insurance plan, according to which each store contributes to the general insurance pool on the basis of the volume of its sales. In addition, a few of the large stores had some commercial insurance policies.

j) *Postage.* This expense represents the cost of mailing correspondence and reports to the central office of the shoe corporation.

k) *Supplies.* The cost of supplies includes expenditures for wrapping paper, twine, electric-light bulbs, and other small items. This cost is obviously closely related to sales volume, but there may be some unexplained variability as a result of erratic consumption of electric-light bulbs, because some stores, subject to greater traffic vibration, burn out bulbs quickly.

l) *Miscellaneous.* The costs grouped in the miscellaneous category are small store expenses of negligible importance which are not easily classified in the other groups.

3. BUILDING EXPENSE

m) *Rent.* Rent arrangements vary widely from store to store. Some leases specify a minimum guaranty *plus* a percentage of sales, which is usually about 6 per cent. To the rent actually paid for the store's location is added $120 per year per store to compensate the real estate department of the chain for its expenses. The variation in rent per dollar of sales is affected by errors in the selection of locations or in estimating probable sales volume of a particular site, by differences in the time at which leases were made, by the length of the lease period, and by the ease or difficulty of adjusting contractual agreements.

n) *Light.* The light account represents payment for electricity to public utility companies. In 1937 a careful survey of every store was made, and alterations were made with a view to reducing light expenses by standardizing the lighting fixtures and the wattage of each light and by scheduling the timing of lighting the store and its windows. Some regional variation in public utility rates for different sections of the metropolitan area exist, but the differences are not significant. Variability of light expenditure from store to store is, therefore, primarily related to the floor space of the store, the display windows, and the hours of peak operations.

o) Heat. The cost of heating is not incurred directly by all stores as a specific expense but is included in the rent in many stores. A scatter diagram of this cost element against physical sales of shoes indicates no apparent relation, yet this cost was included in order not to underestimate the occupancy cost for stores for which heat is not included in the rent.

p) Window display. The window-display account includes only supplies in the form of material, posters, specially selected merchandise, and window fixtures used in the individual stores. It is not predominantly an allocated account, although it does include an apportionment of general office cost for special decorators.

q) Repairs. Repairs are necessary from time to time in the stores to take care of small breakages and the wearing-out of fixtures. The magnitude of the repairs account is small.

r) Depreciation. This element of cost is composed of depreciation on the fixtures and improvements of stores. Its amount depends upon the expenditure on necessary structural changes and fixtures and on the terms of the lease. Depreciation is calculated on a straight-line basis over the period of the lease or the probable length of the life of the asset, whichever is shorter. No apportionment of central-office depreciation is included. The relation of this element of cost to sales volume displays extreme variability—a fact which is accounted for by differences in the amount of equipment, the length of time over which it is being written off, and the proportion of the equipment which has been completely written off.

s) Water, ice. This is another very small account, nonexistent for some stores, which represents the cost involved in maintaining water-coolers.

In order to avoid having to employ rectification devices to allow for changes in prices from 1937 to 1938, each year was studied separately. It might have been thought preferable to combine the 1937 and 1938 observations in order to make a single summary analysis. It was learned, however, that input prices were higher in 1938 because of increased social security taxes, commissions, and salary rates. Furthermore, examination of the data in scatter diagrams of cost and output indicated a consistent and substantial increase in costs in 1938, which could not be attributed to changes in output. It was decided, therefore, to study 1937 and 1938 separately. The necessity for the adjustment of the cost data was thus avoided, apart from the omission of two arbitrarily allocated overhead items. Apparently, no important variation in salary rates, rent, or prices of supplies existed within the area selected. Thus the restriction of the sample to stores located in the same area obviated the necessity of deflating the cost figure to make them comparable from store to store. It is possible that prices changed during the year, but, because only annual data were available, it was not possible to rectify recorded cost within the observation period.

D. Statistical Findings

As pointed out already a separate analysis of the relation of cost to output in the fifty-five shoe stores was undertaken for each of the years 1937 and 1938. This was advisable not only because of the lack of precise comparability in the data for the two years but also because of the additional confirmation that would be afforded by agreement in the two sets of results. Preliminary examination of the relation between cost and output indicated that the pattern of cost behavior was essentially the same in both years.

This preliminary analysis of the data by means of scatter diagrams showed the observations to be heteroscedastically distributed in a fashion which indicated logarithmic dispersion. This was substantiated by the appearance of the scatter diagrams when plotted on double logarithmic paper. Consequently, except in the analysis of average cost behavior, the logarithms of the cost and output variables were employed in the calculations throughout the investigation.

Correlation analysis was used as the basic statistical technique, either linear or curvilinear functions being fitted to the data by the method of least squares. The fitting procedure was carried through, first, for total cost, which is the aggregate of reported expenditures of the store, excluding allocated administrative expense and general indirect expense. Two measures of output were used: (1) number of pairs of shoes sold and (2) dollar-sales volume. Second, the behavior of average cost, derived directly from the data, was examined by means of correlation analysis. Third, the cost behavior of each of the three components of total cost—selling, handling, and building expense—were analyzed by the same technique. Finally, certain of the individual elements of cost of particular interest were investigated separately.

1. TOTAL COST

The relation of total cost to physical volume of shoe sales for 1937 and for 1938 is shown in the scatter diagrams in the upper and lower panels of Chart 6–1. A detailed examination of the position of the individual stores indicated that the relation between cost and output remains fairly stable during these two years. Second-degree regression functions were fitted for both years, the general form being,

$$_tX_1 = b_1 + b_{2c}X_2 + b_{3c}X_2^2,$$

where $_tX_1$ is the logarithm of total cost and $_cX_2$ the logarithm of output in pairs of shoes. A mathematical description of the regression functions which were thus determined is shown in Table 6–1.

Chart 6–1 Simple Regressions of Total Cost on Output Measured by Shoe Sales in Pairs

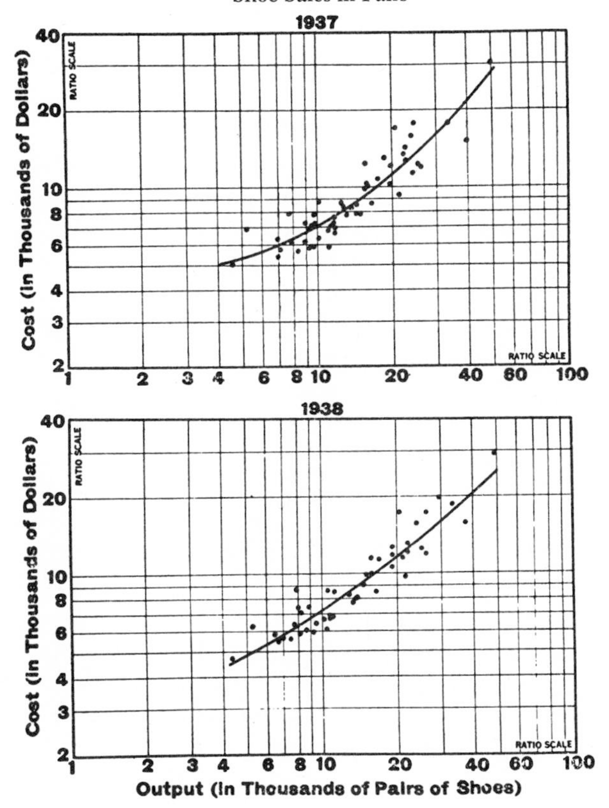

Table 6–1

SHOE-STORE CHAIN: SUMMARY OF STATISTICAL CONSTANTS OF TOTAL COST-OUTPUT* REGRESSIONS

$({}_tX_1 = b_1 + b_{2c}X_2 + b_{3c}X_2^2\dagger)$

Statistical Constants	1937	1938	Statistical Constants	1937	1938
b_1	9.029	7.617	b_3	0.458	0.380
b_2	−3.128	−2.460	S_{b3}	0.143	0.118
S_{b2}	1.194	0.982	$\frac{\beta_3}{S_{\beta 3}}$	3.197	3.220
$\frac{\beta_2}{S_{\beta 2}}$	2.619	2.505	R	0.920	0.948

* Output measured by pairs of shoes sold.

† ${}_tX_1$ is the logarithm of total cost, and ${}_cX_2$ the logarithm of output.

The use of logarithms of the variables in the analysis of the total-cost function has certain advantageous properties apart from the reduction of the scatter and heteroscedasticity of the observations. The principal virtue of logarithmic analysis is that changes in the magnitude of the logarithms of the variables in which interest lies represent relative changes in the variables. Consequently, the slope of a function which is in logarithmic terms gives the elasticity of the function of the variables themselves.[16] From a knowledge of the behavior of the elasticities it is immediately possible to deduce the shape of the cost–output functions. The parabolic shape of the fitted regression lines shows that the relative change in cost is increasing, compared to the relative change in output. In other words, since the slope of the regression line is always becoming steeper, the elasticity of total cost is increasing over the whole range of output. On the basis of this knowledge of the behavior of total-cost elasticity, it is easily possible to translate the relations portrayed on the double logarithmic scatter diagram into arithmetic terms.

For example, in order to analyze the behavior of average cost, it is necessary only to make use of the proposition that average-cost elasticity is equal to total-cost elasticity *minus* 1.[17] It is, therefore, apparent that unitary elasticity of total cost corresponds to zero elasticity of average cost. If total-cost elasticity moves from values less than unity to values greater than unity, it is seen that the corresponding elasticity of average cost is at first negative, then zero, and finally positive. This means that the average-cost curve displays the familiar U-shape with a minimum point where average-cost elasticity is zero. An inspection of Chart 6–1 indicates that for both 1937 and 1938 the minimum point of the average-cost curve is reached at an output of approximately 33,000 pairs, since it is at this level of output that total-cost elasticity (the slope of the regression curve) is unity.[18]

It is unfortunate that only three stores in the sample have outputs in excess of this critical level of operations, for one can, therefore, have little confidence in the precise magnitude of the optimum rate of operations. This does not mean, however, that the significant change in the elasticity of total cost is attributable only to the influence of the large stores. Even if they are excluded from the analysis, an upward bend occurs in the regression line—in fact, it becomes more marked, and the parabolic shape of the average-cost curve is even more clearly defined.

There are, indeed, some grounds for the exclusion of the three largest stores from the sample. Those stores which sell more than 30,000 pairs of shoes annually may have special characteristics which set them apart from the smaller stores in the sample. The three large stores are located in the downtown area and therefore are faced with selling conditions which are different from those stores in the outlying regions. While downtown stores have generally marked daily peak loads at noon and in the late afternoon, peripheral stores are more likely to be subject to heavy peak loads on different days of the week, e.g., Saturdays. This influence, which depends on location, may be sufficient to explain the atypical behavior of these stores. If one could be sure that these stores are not homogeneous, they could be omitted from the analysis altogether. If this were done, there would be stronger confirmation of the existence of an optimum-size shoe store for outlying metropolitan markets. However, since, even without the retention of these large stores, the existence of an optimum size is clearly shown, it was decided to retain the large stores in the statistical analysis.

The behavior of the total-cost function, in which dollar sales was used as the index of output, was very similar to those functions shown in Chart 6–1. The regression function fitted was of the general form,

$$ {}_tX_1 = b_1 + b_{2a}X_2 + b_{3a}X_2^2. $$

The scatter diagrams and the regression lines fitted by the method of least squares for 1937 and 1938 are shown in Chart 6–2. The same situation observed in Chart 6–1 exists in this case also. The fact that the regression line does not bend upward more sharply over the extreme range of output is a result of the influence of the three large stores. Nevertheless, there is a point on the curve where the elasticity of total cost is unity (i.e., average cost is a minimum). This can be determined graphically or from the regression equations. The regression equations and the relevant statistical constants are shown in Table 6–2. In the same way as before

Chart 6–2 Simple Regressions of Total Cost on Output Measured by Dollar Sales

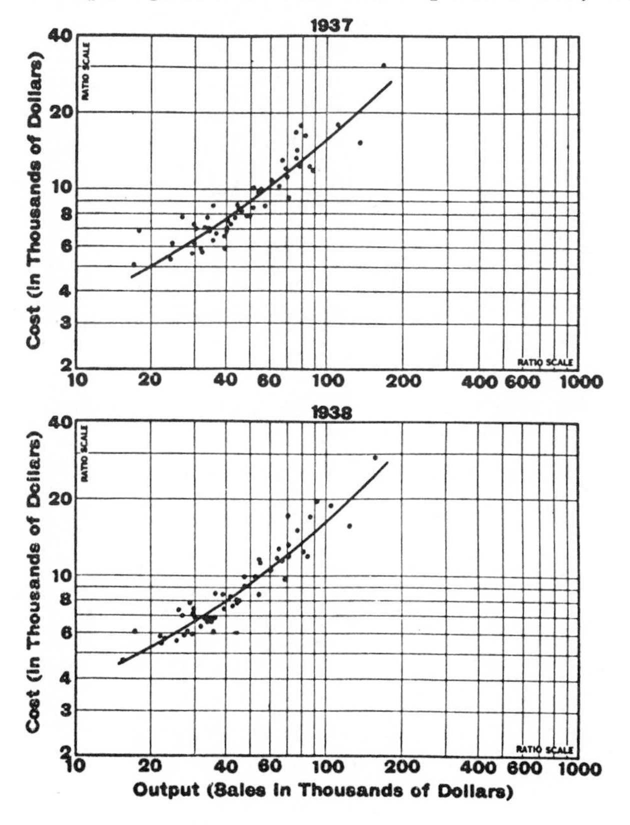

Table 6–2

SHOE-STORE CHAIN: SUMMARY OF STATISTICAL CONSTANTS OF TOTAL COST-OUTPUT* REGRESSIONS

$({}_tX_1 = b_1 + b_{2a}X_2 + b_{3a}X_2^2\dagger)$

Statistical Constants	1937	1938	Statistical Constants	1937	1938
b_1	11.322	9.383	b_3	0.483	0.402
b_2	−3.837	−3.038	S_{b3}	0.147	0.122
S_{b2}	1.387	1.147	$\frac{\beta_3}{S_{\beta_3}}$	3.274	3.278
$\frac{\beta_2}{S_{\beta_2}}$	2.767	2.650	R	0.924	0.950

* Output measured by dollar sales.

† ${}_tX_1$ is the logarithm of total cost and ${}_aX_2$ the logarithm of dollar sales.

Chart 6–3 Simple Regression of Average Cost on Output Measured by Shoe Sales in Pairs

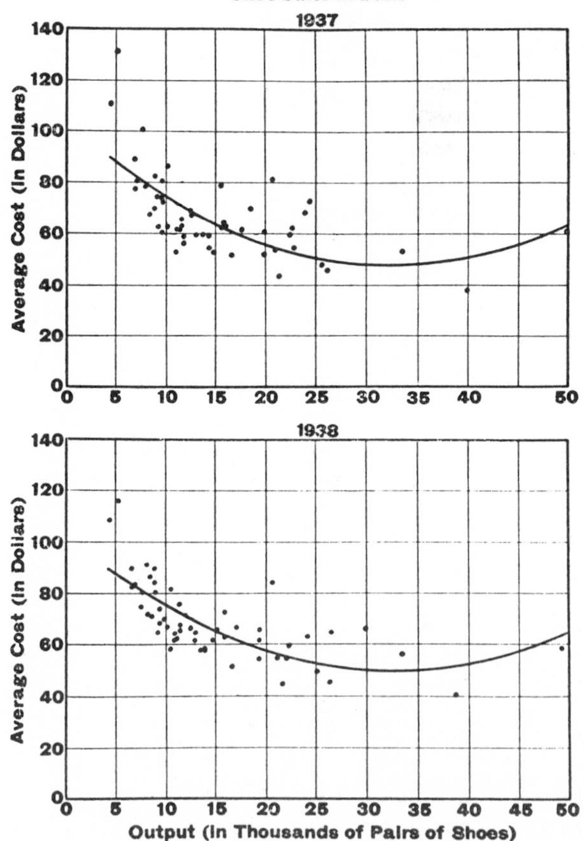

Table 6–3

SHOE-STORE CHAIN: SUMMARY OF STATISTICAL CONSTANTS OF AVERAGE COST-OUTPUT* REGRESSIONS

$({}_aX_1 = b_1 + b_{2c}X_2 + b_{3c}X_2^2\dagger)$

Statistical Constants	1937	1938
b_1	1.036	1.026
b_2	−0.03441	−0.03258
S_{b2}	0.006520	0.005289
$\frac{\beta_2}{S\beta_2}$	5.279	6.159
b_3	0.0005295	0.0005025
S_{b3}	0.0001352	0.0001128
$\frac{\beta_3}{S\beta_3}$	3.918	4.456
R	0.667	0.738

* Output measured by thousands of pairs of shoes sold.
† ${}_aX_1$ is average cost and ${}_cX_2$ is output.

the optimum output is found from the regression equations to be roughly $102,500 for 1937 and $104,900 for 1938.

2. AVERAGE COST

Apart from the information concerning average cost derived from the regression functions fitted to the logarithms of the cost and output data, supplementary confirmation was obtained directly by the analysis of average-cost behavior. Scatter diagrams were made of average cost per hundred pairs of shoes sold for the two years, the pictorial representation of this being shown in Chart 6–3. Regression functions were also fitted to the average-cost data, the mathematical description of which is given in Table 6–3. The scatter of the extreme observations is so wide that the behavior of average cost is not well defined in that region where the optimum output was found to be by other methods. Consequently, the upward trend of average cost is not apparent when the three large stores are excluded.

3. COMPONENTS OF COST

In addition to the foregoing analysis, the behavior of the components of cost—selling expense (${}_sX_1$), handling expense (${}_hX_1$), and building expense (${}_bX_1$)—were also studied. As explained earlier, these components of cost are made up by the following grouping of the cost elements: (*a*) selling expense: salaries, employees' discount, reserve salaries, hosiery award, shoe bonus, advertising, taxes; (*b*) handling expense: delivery (inbound, outbound, interstore), insurance, postage, supplies, miscellaneous; (*c*) building expense: rent, heat, light, window display, repairs, depreciation, water, ice.

The relations between output and each of these components of cost (in the form of annual expenditures for each category of expense) are analyzed for 1937 and for 1938 by methods similar to that used in determining the total cost–output relation. The only measure of output used in the determination of these subsidiary cost functions is the number of pairs of shoes sold. In both the graphic and the mathematical analyses of these relations, logarithms of the variables were employed. The results are shown in Charts 6–4, 6–5, and 6–6 and in Table 6–4.

a) *Selling expense.* A curvilinear regression function was fitted to the selling-expense and output data, the graphic illustrations of which appear in Chart 6–4. The curvilinearity of these regressions is quite marked and the dispersion of the observations much less than in the scatter diagram of total cost and output. The form of the regression function fitted is

$$_{s}X_1 = b_1 + b_{2c}X_2 + b_{3c}X_2^2,$$

Chart 6–4 Simple Regressions of Selling Expense on Output Measured by Shoe Sales in Pairs

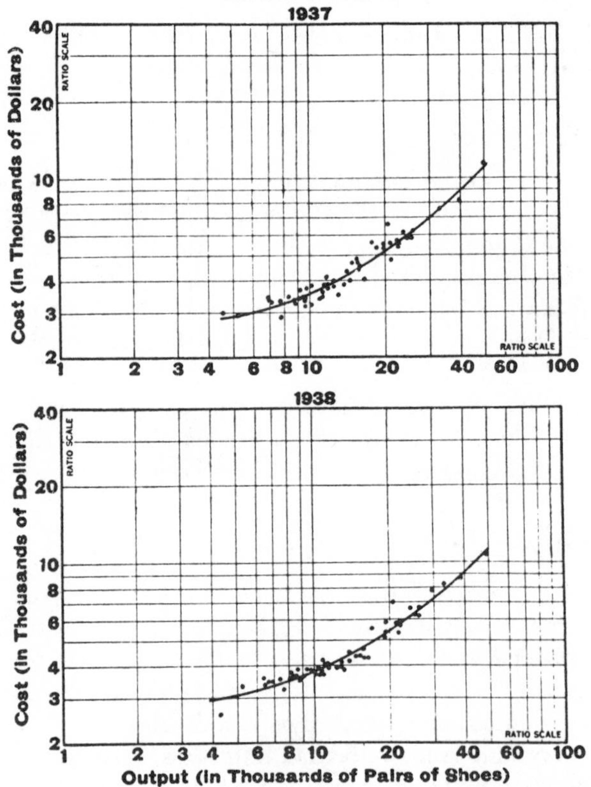

Chart 6–5 Simple Regressions of Handling Expense on Output Measured by Shoe Sales in Pairs

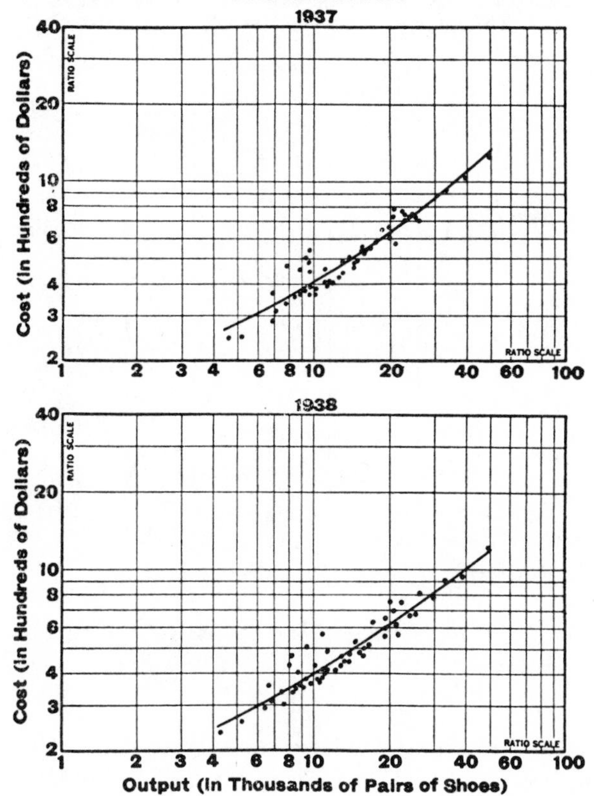

it being remembered that ${}_sX_1$ and ${}_cX_2$ are the logarithms of selling expense and output in pairs of shoes, respectively. The regression equation which was derived is shown at the end of this section in Table 6–4.

b) Handling expense. The scatter diagrams of handling expense and output indicate that the relation between them is not markedly different from the selling-expense–output relation. Therefore, the same form of regression function was fitted, namely,

$$ {}_hX_1 = b_1 + b_{2c}X_2 + b_{3c}X_2^2, $$

the graphical illustration of which is shown in Chart 6–5. The mathematical constants pertaining to this equation are shown in Table 6–4.

c) Building expense. A much less well-defined relation between building expense and output was found than for the two other components of cost. The scatter of the observations is very wide. The explanation of this scatter lies in the variability of rent, which makes up a large proportion of handling expense. The reasons for the erratic behavior of rent cost were discussed briefly in Section C of this chapter.

Table 6–4
SHOE-STORE CHAIN: SUMMARY OF LEAST-SQUARES ANALYSIS OF COST COMPONENT–OUTPUT REGRESSIONS

Description of Cost	Constant Term (b_1)	Coefficient of ${}_cX_2$ (b_2)	S_b	$\frac{\beta_2}{S_{\beta_2}}$	Coefficient of X_2^2 (b_3)	S_{b_3}	$\frac{\beta_3}{S_{\beta_3}}$	R
1937:								
Selling expense, ${}_sX_1$	8.942	−3.103	0.580	5.352	0.439	0.070	6.309	0.969
Handling expense, ${}_hX_1$	3.985	−1.269	0.885	1.433	0.231	0.106	2.178	0.949
Building expense, ${}_bX_1$	0.120	0.841	0.085	9.894	......	......	......	......
Total salaries	9.431	−3.334	0.603	5.531	0.463	0.072	1.404	0.963
1938:								
Selling expense, ${}_sX_1$	8.047	−2.652	0.525	5.051	0.384	0.063	6.084	0.974
Handling expense, ${}_hX_1$	2.104	−0.350	0.830	0.422	0.118	0.100	1.188	0.953
Building expense, ${}_bX_1$	−0.086	0.890	0.070	12.714	......	......	......	......
Total salaries	8.435	−2.831	0.565	5.013	0.401	0.068	5.917	0.967

Since no curvilinearity in the regression is evident, a linear function was fitted to the data,[19] its form being

$$ {}_bX_1 = b_1 + b_{2c}X_2 \,. $$

Chart 6–6 shows the scatter diagrams and the regression functions fitted by the method of least squares for the two years 1937 and 1938.

The mathematical results of the fitting procedures are summarized for the three components—selling, handling, and building expense (as well as for the cost element, total salaries)—in Table 6–4.

4. ELEMENTS OF COST

The most striking result of the analysis of total-cost behavior was the upward bend of the regression function in the range of high outputs. Partly in order to trace the cause of this and partly because of their intrinsic interest, the components of cost were analyzed separately. It is fairly clear from consideration of the relative importance of the components[20] and from a comparison of their regression functions that the major part of the curvature of the total-cost function is to be attributed to the curvilinearity of the selling-expense component. The various elements of these cost components were analyzed individually, with particular attention being paid to the elements making up the selling expense.

The form and clarity of the relations differed considerably; and, since the behavior of most of the elements of cost is of slight importance and interest, with one exception the statistical results are not presented here. The exception is the cost

Chart 6–6 Simple Regression of Building Expense on Output Measured by Shoe Sales in Pairs

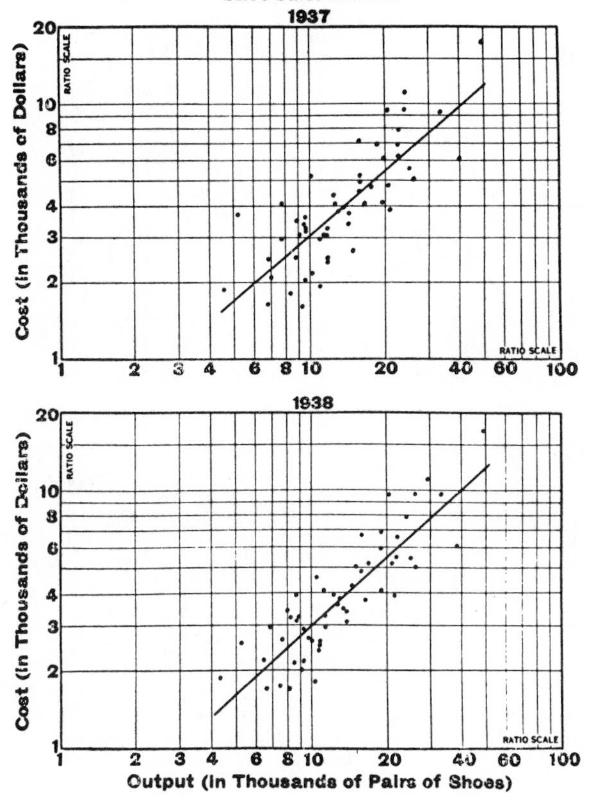

element—total salaries—which is of dominant importance in the selling-expense component (78 per cent in 1937).

Total salaries includes regular salesmen's salaries, reserve salaries, employees' discount, hosiery award, and shoe bonus. An analysis similar to that carried out for total cost and for the components of cost was made of this group of cost elements. The scatter diagram indicated a curvilinear relation between total salaries and output, so that a second-degree parabola was fitted. The graphic representation of this function is shown in Chart 6–7, and the relevant statistical constants are given in Table 6–4. The importance of this particular regression function is that it underlies the curvilinearity of the selling-expense output regression, which in turn is of dominant importance in producing the curvature in the total cost–output functions.

5. MARGINAL COST

Nothing so far has been said concerning the behavior of marginal cost, which usually has a central role in any discussion of cost behavior. The meaning of the

Chart 6–7 Simple Regression of Salaries on Output Measured by Shoe Sales in Pairs

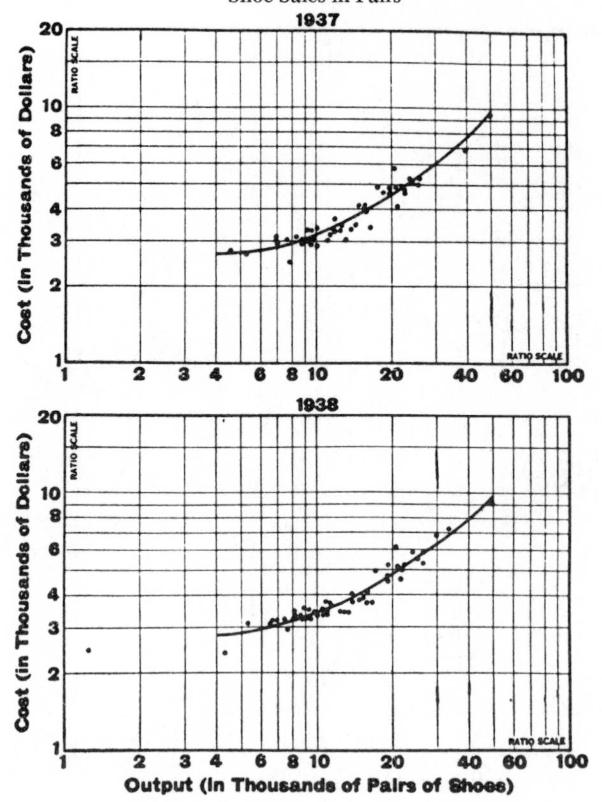

long-run, marginal-cost function is highly ambiguous in the case under consideration. To adopt Bowley's definition, long-run marginal cost is "the additional cost of producing one more unit after adapting the organization of the factors of production."[21] This concept has clear meaning where excess capacity or underutilization do not exist.

Chapter III

Theoretical Analysis

The preceding chapter was concerned with the statistical analysis of the relation between cost of operation and size for a group of fifty-five stores of a retail shoe chain. Size was measured by annual output expressed as volume of sales in terms of pairs of shoes. In this theoretical part of the study Section A will be devoted to a number of problems associated with the use of output as an index of size and to the difficulties of obtaining an adequate measure of retail output. Section B will consider the statistical results in the light of the theory of long-run cost of the individual firm. Although in the course of the last fifteen years the behavior of the costs of the individual firm has been analyzed in an extensive body of theoretical literature, there have been surprisingly few attempts to compare the theoretical formulations with the results of statistical analysis.[22]

A. The Nature and Measurement of Retail Output

A primary difficulty is encountered in attempting to define precisely the character of the output of a retail enterprise. The traditional theory of the individual firm considers the output produced by the firm to be units of physical commodities —gallons, bushels, or tons, etc. This treatment involves a special conception of "production," i.e., "a physical process by which certain physically measurable goods or services are combined in order to produce a physically measurable product or products."[23] On the other hand, Professor Knight has argued for a less material conception of production, preferring to think of "consumption as the enjoyment of consumption, and . . . . of production as the rendering of consumption services."[24]

Physically, retail commodities as a rule undergo no change in being placed in the hands of the consumer. Therefore, the treatment of production as a physical process of transformation is not appropriate to retail selling. Nevertheless, the retail store makes an essential contribution to the completion of this transformation. Even though the production process is incomplete until the actual initiation of consumption, the actual physical manufacture of retail commodities is accepted as a datum, the implication being that the physical nature of retail goods sold is irrelevant. It is, of course, perfectly legitimate to select one aspect or one stage in the process of production and abstract from preceding stages in studying it, since the same economic principles apply to the parts as to the whole.[25] By this approach it is possible to avoid the difficulties involved in the treatment of selling and promotional cost as a category of cost separate from the cost of production of the physical goods. Costs of retailing can be considered an integral and inseparable part of the cost of production, in view of the specialized conception of production adopted here.

The assumption upon which this study is based is that the output of a retail enterprise consists of the services which it renders to purchasers. These services place the physical commodities conveniently at the disposal of the consumer, widen his range of choice, and add to his knowledge of quality, grades, and prices.[26] The next question to consider is the extent to which retail output can be considered a homogeneous service. In general, the services associated with the sale of different goods—for example, in a department store—differ not only in quantity but also in quality. In the case of shoe retailing of the type considered, however, it is doubtful whether the quality of the services involved in the sale of different types of shoes is subject to marked variation. This is a result of the fact that sales of different sorts are made by identical individuals, and it is in the individual salesmen themselves that the sources of sales-service variability are to be found. The problem is rather that different quantities of service are associated with sales of different kinds of shoes. In particular, one may find that the quantity of sales service necessary to sell a pair of children's shoes is quite different from that embodied in the sale of a pair of men's shoes.

The embodiment of different quantities of service in the sales of different kinds of shoes was discussed earlier in connection with the weighting of different kinds of sales in order to obtain an index of output. Although there was some reason to believe that children's shoes deserve heavier weighting than men's shoes, it was not possible to devise suitable weights, in view of the lack of the precise information necessary.

If production had been treated in physical terms, one could have considered men's shoes and children's shoes as multiple products. Multiple production in this case refers to the output of several goods, where the proportions between the different goods are freely variable. This is to be distinguished from joint production, where the proportions in which the various goods are produced are dictated by technical necessity. Since, however, output is treated as an intangible service homogeneous from one type of shoe sale to another, it is not possible to employ multiple or joint production analysis in this fashion.

It was indicated in Section B of chapter ii that from a theoretical point of view the most suitable measure of the size of industrial enterprises is rate of output. It is necessary now to consider the assumptions underlying the measurement of size by rate of output and to investigate the bearing of the special features of the type of retail enterprise under discussion on this identification. The simplest case is the representation of size by "normal" or planned output, where the technical equipment of the enterprise is adjusted ideally to the originally planned output. The rate of output is assumed to be uniform and synchronized to a repetitive demand situation for each future period or "week." Moreover, equipment is assumed to be at all times utilized at its planned capacity. This is, however, an idealized case, and some account must be taken of the fact that all firms subject to daily, seasonal, and cyclical fluctuations in demand and using, none the less, equipment which outlasts successive waves of high and low demand, face a recurrent problem of unused capacity. It is not a question of errors of planning in many cases, rather it is foreseen by the entrepreneur that the existing demand situation gives rise to marked peaks and troughs.[27] It will be essential for the entrepreneur to secure some degree of flexibility in operations. As Stigler has pointed out, there are two obvious methods of securing flexibility: "the first is based on divisibility of fixed plant, which will reduce variable costs of suboptimal outputs . . . . the second . . . . is to reduce fixed plant relative to variable services, i.e., to transform fixed into variable costs."[28]

When the extreme variability of output which characterizes retail selling is introduced, there is no longer any clear-cut relation between output and size. Plants will be constructed so that peak loads can be handled in some manner, but it may well be true that under such circumstances the minimum-cost combination cannot be achieved for any output. Rather, the fixed plant may be designed to achieve sufficient flexibility so that it is possible to produce for demand in both the peaks and the troughs, operation being, however, supra-optimal in one case and sub-optimal in the other. As a result of such an uneven time distribution of demand, therefore, the precise correspondence of output and size which is relevant in the case of uniform and repetitive demand does not exist.

There are two ways out of this difficulty, neither very satisfactory. First, it is possible to select as a measure of size some arbitrary percentage of maximum-capacity output or alternatively an average of outputs over a period long enough so that both peaks and troughs are adequately represented. The second possibility involves a reclassification of the output of the firm. Because of the unevenness of the time distribution of demand, there are good grounds for believing that the output of a retail enterprise is different during peaks than during troughs. The quality of the service may deteriorate, despite the efforts of management to preserve uniform standards, when salesmen are forced to work under pressure.

There is, therefore, some scope for the application of joint production analysis in a different sense, since there may be some difference in the *quality* of sales service in peak periods and slack periods. One can then look on the output of retail shoe stores as the two grades of product—peak-period services and slack-period services. Output could then be treated as a vector quantity which may be so constituted as to allow a vector of given composition to be considered the normal output of a firm. This vector would then serve as an index of size. While this approach may have some value conceptually, to make use of it statistically would require detailed knowledge of the time distribution of sales for each store, which is lacking. Even with this information it is difficult in practice to decide on any dividing line separating different qualities of service.

B. The Theory of Long-Run Cost of the Individual Firm

The purpose of this section is to outline certain theoretical aspects of cost behavior which may be helpful in understanding the relevance of the statistical findings for the theory of long-run cost of the individual firm. First, a general treatment of cost will be presented, and, second, the modifications necessary to suit the actual situation encountered in the retailing of shoes will be introduced. It is necessary to interject at the outset some preliminary discussion of the theory of production which underlies the theory of cost.

In undertaking the production of a good the individual producer is faced with two types of decisions: the total output to be produced in a given period of time and the allocation of productive resources which will minimize the total cost of producing that output.[29] Suppose that an entrepreneur has decided to begin the production of a commodity in certain quantities or, strictly speaking, quantities per unit period of time. It is evident that, when he formulates his production plans, various methods of production are open to him. By "method of production" is meant the technical characteristics of the actual process of transformation.[30] Some qualitative combination of productive resources will be chosen in the light of existing and available technical knowledge. On the other hand, the quantitative relation which exists between the amount of product and variable quantities of the productive resources, once the qualitative composition of the resources co-operating in production has been selected, is the production function of the technical unit. The entrepreneur, having selected the method of production, constructs his plant and equipment in such a way that the forthcoming services are quantitatively ideal for production on the anticipated level.[31] That is to say, he chooses from the complex of combinations of available productive resources one set which, for the output selected, will be called the "minimum-cost combination."[32] Under the assumption that the share of the individual firm's demand for productive resources is not a large enough part of the total demand in the market to influence their prices, this minimum-cost combination will be given by the condition that the marginal productivities of the individual productive resources are proportional to their prices.

The scale of output chosen originally by the entrepreneur may, however, be altered in the face of extraneous influences. The manner in which the entrepreneur adapts the structure of his enterprise to external changes and the laws governing adaptation are of extreme significance for this investigation.[33] Interest here lies in the course of the total-cost curve in a strictly static sense, where it is assumed that changed circumstances will persist for a specified period of time and that adaptation to the new circumstances are instantaneously affected. Whether the type of adaptation is of a long-run or short-run character will depend upon whether deviations from the original planned output, when they occur, are considered by the entrepreneur to be sufficiently important or lasting to justify reconstruction of plant and equipment, i.e., adjustment of all the components of the factor complex.

For changes in output volume that promise to endure for a long period the typical behavior of the entrepreneur is to adapt his firm completely by changing the size of his fixed plant and equipment, as well as the rate of input of more easily variable factors, with a view to obtaining the minimum-cost combination of inputs for the new level of output. In conformity with Schneider's terminology, this type of adjustment will be denoted by "total adaptation."[34] Total adaptation is likely to be encountered in the case that a secular change in demand for the product of a firm is the cause of the change in output volume.

On the other hand, the entrepreneur confronted by changes in output of a temporary nature will alter only the quantities of easily varied factors, such as raw materials and labor, leaving the plant and equipment of the enterprise unchanged. Such a course of action can be designated as "partial adaptation." Output variability arising from seasonal shifts in demand is likely to cause only short-run adjustments or partial adaptation. The main theoretical issues in this study deal with total adaptation.

Total adaptation of an enterprise may also involve shifts in the production function. This means that the whole function changes, not that there is merely a movement to a different point on the function. There are three cases to be considered. First, new and independent technical advances may be incorporated in the production methods when the plant is adapted. These changes are not a result of increases or decreases in the scale of operations. Second, with increased output the larger scale of production may make the introduction of different, already known techniques feasible. Third, the original production function may be retained.[35] The assumption is made, although it is to some extent a matter of definition, that in the case to be considered no significant changes in productive technique are available with changing levels of output. This treatment may, therefore, be restricted to the case in which total adaptation is unaccompanied by a changing production function.

Attention must now be directed to the behavior of the total cost–output function under the assumption that all productive resources are substitutional and freely variable from zero to indefinitely large amounts. This is the traditional case of the "long run," which is considered to mean a state of affairs in which free variation of all ordinarily fixed productive resources is possible. This is the Marshallian "long run," his usage being shown in his comment:

> In long periods . . . . all investments of capital and effort in providing the material plant and the organization of a business, and in acquiring trade knowledge and specialized ability, have time to be adjusted to the incomes which are expected to be earned by them: . . . .[36]

Under the assumptions stated, there is no reason for the optimum proportion of factors, established for one output, to be different for any other level of operations. Under these circumstances, for total adaptation, the enterprise exhibits constant returns to scale, and total cost is, therefore, a linear function of output.[37] The total-cost curve is, therefore, a straight line, passing through the origin of the co-ordinate system whose axes are cost and output. This is illustrated in Chart 6–8. For this specialized theoretical model the shape of the long-run cost function has been derived. The model, however, is strictly static, assuming instantaneous and perfect adjustments. It is now necessary to attempt to bring the model closer to the real situation which is to be studied, i.e., the retailing of shoes. The problem encountered in the uneven distribution of sales, which was discussed in the previous section, must be given particular attention. Complete incorporation of the aspects of production theory which depend upon planning and expectations would lead

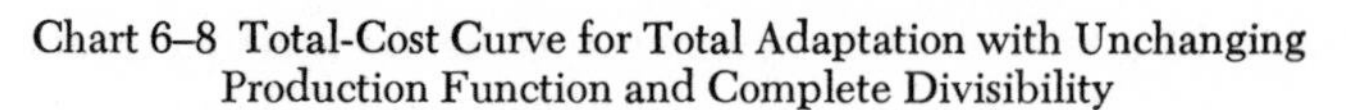

Chart 6–8 Total-Cost Curve for Total Adaptation with Unchanging Production Function and Complete Divisibility

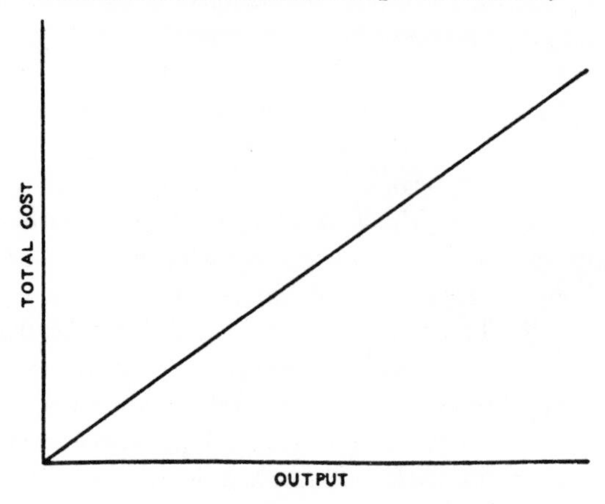

to a very complex model, whose nature cannot be discussed in this place. We can merely hope for a static model, which is sufficiently flexible to serve as a basis for discussion of the actual time problems encountered.

In developing a model to describe the behavior of the costs of a retail shoe store when it is in a position to adapt all its facilities (including salesmen) to changes in output, it is necessary to take into account the indivisibility of the agents of production.[38] It is evident that, actually, the capacity of a salesman is subject to wide variation. Nevertheless, it is convenient to make the assumption that each salesman possesses a specific sales capacity—an assumption which does not affect the substance of the argument.[39] As a consequence, as the capacity of the existing salesmen in a store is fully utilized, it becomes necessary to employ an additional man. These circumstances introduce a significant modification into the cost function which was previously derived. The change, which consists of the introduction of discontinuities, can be shown diagrammatically in Chart 6–9. The ordinate represents total cost, and the abscissa output. The total curve cuts the cost axis positively at a height representing the minimum costs which must be incurred before any operations are possible. Ox_1 represents the capacity of the first salesman, x_1x_2 the capacity of the second, and so on. When operations exceed the level Ox_1, therefore, an additional salesman must be employed, whose selling capacity is underutilized until the output Ox_2 is reached, at which point no idle capacity exists. The length of the line *RH* represents the salary (excluding commissions) of the new salesman. But now under these new conditions, as long as the second salesman is retained, the cost curve becomes *BRG*. Similarly, *DTE* becomes the cost curve for the enterprise operating in the range x_3x_4, again provided that the adjustment is not reversible. When there is complete reversibility, of course, the cost curve for total adaptation becomes the broken line *AHRGSFTE*.

There is good reason to believe that this discontinuous behavior of cost is rep-

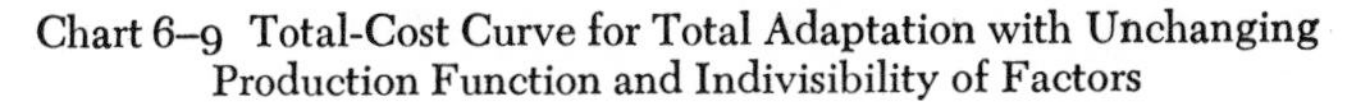
Chart 6–9 Total-Cost Curve for Total Adaptation with Unchanging Production Function and Indivisibility of Factors

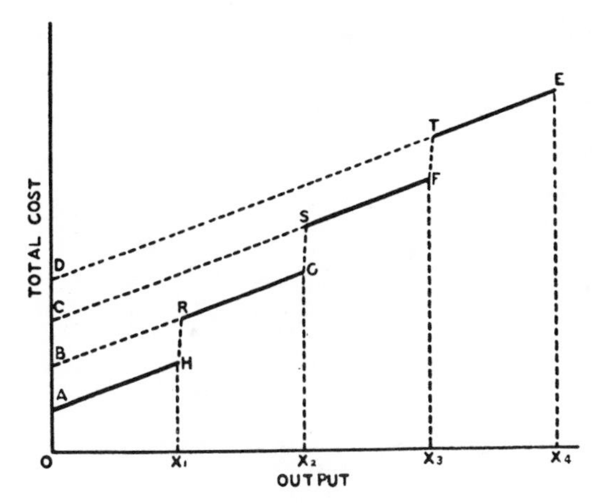

resentative of the retail selling of shoes.[40] The dotted sections of the cost curves in Chart 6–9 become of particular significance for this investigation. For example, suppose a store operates during its rush hours at or near its maximum capacity with the retention of the same quality of service. Then during slack hours in the morning or at other times the store will move along the cost curve on to the dotted section. This is a result of the fact that the store can anticipate the future occurrence of rush periods and will therefore retain the unused selling force during the intervening periods.[41]

The next question to consider is the contribution which such theoretical formulations can make to the understanding and interpretation of the empirical cost curves which have been determined.

C. Economic Interpretation of Statistical Findings

In contrast to the two theoretical models of cost behavior discussed in the previous section, the empirical cost function possesses a branch which rises more than proportionally to output when high levels of output are reached. The purpose of this discussion is to try to explain this shape of the empirical cost function.[42]

Customarily, in the English literature of the theory of cost, the treatment of long-run cost behavior is given in terms of the long-run average-cost curve, the so-called envelope or interplant curve. This curve is shown usually to be parabolic or U-shaped. In other words, it first declines, then rises after a certain level of output is reached. Chart 6–10 portrays the long-run average-cost curve as the envelope of a series of short-run average-cost curves. Points on *CC*, the envelope curve, show the level of average cost for the corresponding output when the complex of factors of production are harmoniously adjusted, i.e., the plant and equip-

Chart 6–10 Long-Run Average-Cost Curve with Complete Divisibility

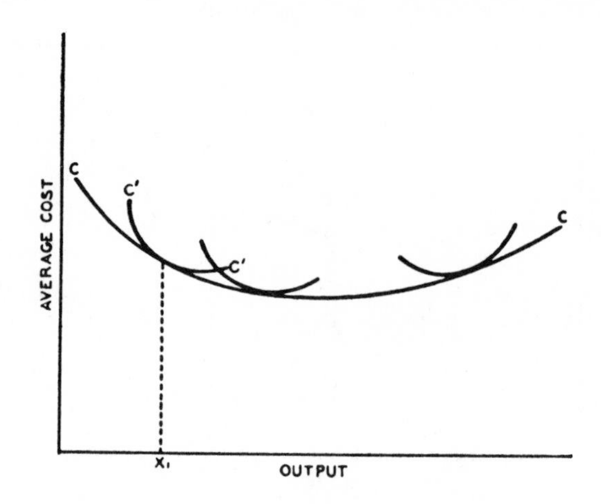

ment are ideally adapted to produce the given output. The short-run curve $C'C'$, which is tangent to the envelope curve at Ox_1, depicts the course of average cost, as the output for which the plant and equipment is suited is departed from. Attention is here restricted to the long-run curve. It was emphasized earlier, however, that the nature of shoe retailing is such that a harmonious adjustment of factors for any output is inconceivable because of demand irregularities. In consequence, the average-cost curve which is relevant in the case under consideration is one derived from the discontinuous total-cost curve depicted in Chart 6–9. This average curve is illustrated in Chart 6–11. In this case average cost at the outputs x_1, x_2, x_3, and x_4 corresponds to points on the continuous envelope curve in the traditional case.

Chart 6–11 Average-Cost Curve for Total Adaptation with Unchanging Production Function and Indivisibility

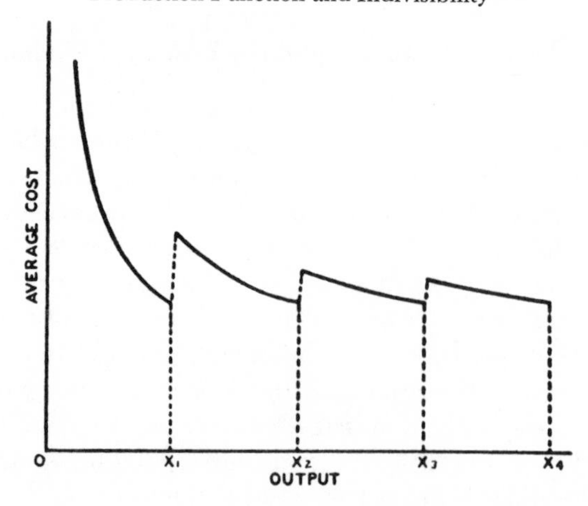

In actuality, the statistical observations which are available are not estimates of average cost for harmonious combinations, and there is no reason to believe that the empirical cost function indicates the shape of the envelope curve. It seems advisable, therefore, to drop the envelope curve as a theoretical foundation intended to elucidate the statistical results.

Nevertheless, in the discussions of the shape of the envelope curve in economic literature, there are certain considerations which may be relevant to the problem at hand. In particular, the explanation of the parabolic shape of the curve is of especial interest here. This behavior of average cost is sometimes attributed to internal economies and diseconomies of scale, which arise because of the existence of some indivisible fixed factors. In the long-run case by definition, fixity is possible only for some aspect of the entrepreneurial function. This is Kaldor's interpretation, which is a widely accepted one, illustrated in his comment:

> .... The costs of the individual firm must be rising owing to the diminishing returns to the other factors when applied to the same unit of entrepreneurial ability. The fact that a firm is a productive combination under a single unit of control explains, therefore, by itself why it cannot expand beyond a certain limit without encountering increasing costs.[43]

Such an explanation, however, is of little value in the case of the retail shoe stores. The entrepreneurial functions performed by shoe-store managers are almost nil, in view of the positions of the stores under the control of a central office. The managerial and supervisory activities of the branch manager can be delegated to assistants. In the case of independent owners, however, there is a certain essential entrepreneurial function which by its very nature cannot be shifted. Kaldor's reason may be valid in the case of individual independent stores, but in the circumstances encountered it must be rejected as inapplicable to a chain of stores.

The envelope curve is also explained in terms of increasing and then diminishing effectiveness of managerial control as the firm's size increases. Professor Knight apparently holds this latter opinion and mentions "the varied and multiplying costs of maintaining internal stability as size increases."[44] Again, in the case of retail shoe stores, the organization of large stores is not much more complex than that of small stores. It is true that there may be some growth of specialization (e.g., the employment of porters and cashiers), but the extent to which this is possible in a shoe store is limited. Furthermore, the limited specialization which does take place is a matter of display and convenience. The increasing complexity in organization encountered in large-sized stores is intended to maintain the quality of the sales service, but this does not imply higher operating costs.

There is one characteristic of retail selling, however, which would provide a rationale for rising average cost for extremely high output. Such a circumstance would arise where selling costs are necessary to maintain the market of the individual store. With the growth of the store and the spreading-out of its market, progressively increasing promotional expenditure is necessary to draw in peripheral buyers. In the case of the stores of the shoe chain, however, there is no evidence that this explanation is relevant. The advertising of the individual stores is very small, almost the whole of such expense being incurred on behalf of the chain as a whole, the cost being prorated over the members of the chain.

These explanations for the observed behavior of costs must be rejected as inapplicable, therefore, and it is necessary to examine more carefully the nature of the statistical observations in the light of the theoretical model for retail selling to see if any more tenable reason for the shape of the empirical cost function can be found.

Let us consider the nature of the observations of a group of four stores of different sizes ranging from very small to very large. The behavior of the costs of such stores was illustrated previously in Chart 6–9, where there were assumed to be discontinuities in total cost with increases in output but not for decreases in output below each critical level. The total-cost curve for a store becomes *DTE* when its output grows beyond the level Ox_3. Similarly, a store whose output moves beyond Ox_2 possesses the cost curve *NCF*. The same process of change is illustrated when a store's output exceeds Ox_1 and its new cost curve becomes *BRG*. It is possible to look on these curves as pertaining to different stores, which for convenience will be numbered I, II, III, and IV. This is illustrated in Chart 6–12, and in addition a

Chart 6–12 Hypothetical Distribution of Total-Cost Observations for Different-Sized Stores

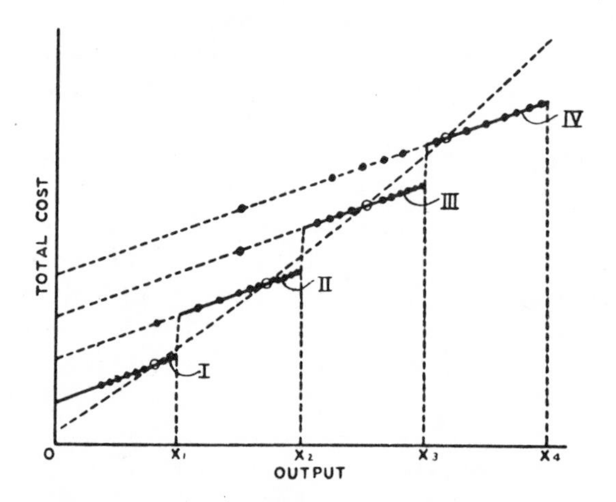

sample of hypothetical daily rates of operation is marked in on the curves. These dots represent daily sales expressed in terms of a rate per annum. The actual statistical observation of cost for any store is the average of all these daily observations. For each store the hypothetical annual observation is marked with a circle. The annual observation tends to lie comparatively closer to the dotted section of the cost curve for large than for small stores. This is because for large stores the observations can range between daily rates of operation of, say, 0–50,000, while for small stores the range is only 0–5,000. Compare, for example, the annual average of Store II and Store IV. This means that the problem of unevenness of the distribution of sales over time may lead to a comparatively more serious underutilization of capacity in large than in small stores. It is apparent from the diagram that a curve fitted to the annual averages shown by the circles will yield a cost function which accords very closely with the empirically determined functions portrayed in Charts 6–1 and 6–2.

Appendix

The Measurement of Intensity of Utilization

The behavior of the costs of an industrial plant, when input prices are independent of the plant's activity, depends on the response of costs to changes in the level of output. The analysis of this functional relation between cost and output for a given plant deals with the short-run as distinguished from the long-run analysis of the relation between costs and different-sized plants. Consequently, the magnitude of the cost observed for each of the retail shoe stores is related to the particular level at which each store is operated. For retail selling outlets the notion of capacity is a nebulous one, the difficulty being that at extreme levels of output the product changes. This means simply that the quality of the service cannot be maintained when individual salesmen are subject to great pressure. Nevertheless, it may be true that some stores are consistently busier than others and operate closer to the critical level at which service quality begins to deteriorate.

In order to be able to isolate the influence of size on cost, it would be desirable ideally to have cost observations for each store apply to strictly comparable levels of operation for each store. To select observations even roughly comparable it would be necessary to investigate empirically the course of the short-run cost function. The statistical investigation of short-run cost behavior would require a number of observations, at successive short intervals of time, showing the cost behavior of a *given* plant with variable output volume. Unfortunately, in the data available there are only two observations for each retail outlet. These are, moreover, annual observations and are, therefore, unsuitable for the derivation of short-run cost functions.

Although such a statistical investigation is not feasible, it may, nevertheless, be possible to obtain some information concerning cost behavior as intensity of utilization of fixed plant and personnel fluctuates. Since a record of hourly or daily sales is lacking, it is necessary to attempt to use the meager information afforded by annual data to find a measure of intensity and variability of utilization of capacity.

Capacity is here taken to mean not maximum capacity in some sense but rather some rate of operations which represents a normal or modal state of business or rate of selling. It is, therefore, a vague concept whose quantification is difficult, but, nevertheless, some notion of relative rates of utilization of different stores may be derived from it.

There are two separate aspects to the problem of utilization which are, however, difficult to separate. First, a store may be subject to chronic underutilization,[45] because it was originally constructed on the basis of overoptimistic forecasts of the sales potentialities of its market area, because of poor management, or because of changes in regional conditions. Second, periods of underutilization may be a result of the existence of extreme peak loads, i.e., variability in the distribution of demand through time. Indices designed to measure variable utilization may, therefore, reflect both chronic and periodic aspects of underutilization.

Indirect indices may be developed on the basis of some measure of capacity, either physical or output capacity or both, which will permit the ranking of stores on the basis of sales per unit of capacity. Several alternative measures which may reflect the physical size or output capacity were considered—(*a*) rent, (*b*) window-display expense, (*c*) salesmen's salaries, (*d*) electricity cost—each of which will be briefly discussed.

a) *Rent.* Presumably rent, which itself is intimately related to sales volume, might measure size in the sense of sales capacity. Rent, however, is unreliable as an index of capacity, because rent per unit of floor space and frontage shows wide variability depending upon the estimated potential sales of the store, its location, and the bargaining skill and foresight exercised at the time of the lease agreements. For example, some stores have high rent because of lease agreements entered into in periods when overoptimistic forecasts of sales potential were made. Furthermore, many of the rental arrangements are based upon a minimum rent *plus* a percentage of sales or of the gross margin. Since these arrangements vary from store to store, annual rent is a very unreliable indication of the output capacity of a given store.

b) *Window-display expense.* This cost element might at first glance be considered a good index of size, if there existed complete standardization of store buildings. The cost of window display is, however, unsuitable as a measure of size, first, because of random variability due to peculiarities of the structure of the store and, second, because the data concerning this item represent primarily the materials used for window displays. The amounts of these materials vary not only with the store frontage but also with the design and character of the display window and the layout of the store itself.

c) *Salesmen's salaries.* In the earlier discussion of input capacity as an objective measure of size, the conclusion was reached that for retail selling this measure lacked the necessary precision. Nevertheless, there are certain virtues in a measure of input capacity of some sort, for example, the number of salesmen in a store. There are two objections to the use of the number of salesmen employed in each store as a measure of its capacity. First, there is not enough variation in the number of the selling force from store to store. Second, such a measure makes no allowance for individual difference in capacity, which is an important consideration for our purposes. However, there is more to be said for an allied concept—the earning ability of the salesmen—which is, indeed, a more satisfactory measure of input capacity than the mere number of the selling force.

The factors that are taken into consideration in determining the individual's salary, as explained earlier, are his length of service, his appearance, his faithfulness in following instructions, but only indirectly the volume of physical sales of the store in which he is employed. If we consider the intangible aspect of the output of a store—sales service—it is fairly clear on a priori grounds that the salary of an individual is likely to be closely related to his own output capacity, i.e., his ability to serve customers efficiently and courteously.

d) Electricity cost. The expenditure of a store for electricity appears to be the most satisfactory index of physical size available. A high degree of standardization of store layout and fixtures, lighting arrangements, and wattage of bulbs for particular positions in the store makes the electricity consumption a fairly accurate measure of the cubic volume of the store's selling space and the size and character of its display windows. Furthermore, little variation exists in the public utility rates for electricity among the several regions in the metropolitan area under consideration.

On the basis of these considerations an index of intensity of utilization was constructed for each store by dividing physical shoe sales by the sum of electric-light expense and weighted salary cost exclusive of reserve salaries for that store.[46] The weighting of salary costs was necessary to avoid giving complete predominance to salary costs which were very much greater than electricity costs.

In addition to the indirect indices just discussed, a more direct measure of certain aspects of demand peakedness is found in reserve-salary expenses, which consist of remuneration to part-time salesmen employed during certain rush periods, usually evenings and Saturday afternoons. This expense may also reflect indirectly intra-daily demand fluctuations. Reserve salaries are, however, not directly comparable from store to store, since they may depend on the absolute size of the store. The deflation of reserve salaries by some aspect of store capacity or size was necessary, therefore, to permit interstore comparability. This deflation was accomplished by dividing reserve salaries by the previously mentioned measure of physical capacity—electricity cost. It was hoped that the resulting measure would throw some light on variability of utilization occurring from store to store.

In this discussion of the derivation of the indices of (1) intensity of utilization and (2) peakedness of selling activity, there has been no explicit mention of the way they could be used. Ordinarily, such measures would be used in a multiple correlation analysis to detect their influence on cost, the dependent variable. Unfortunately, however, the indices are composed largely of the elements of total cost itself. When elements of the dependent variable are used as independent variables, little confidence can be placed in the regression equations. It was, nevertheless, possible to carry out a more limited analysis by investigating the relations of the residuals about the cost–output function to the measures of utilization rate. As far as the index of utilization was concerned, some graphic analysis was carried out, but the results were disappointing; and, since no confidence could be placed in the measure derived, the attempt to take account of intensity of utilization was abandoned.

The measure of peakedness of selling activity (the ratio of reserve salaries to electric-light expenditure) was plotted against the residuals about the cost–output function in an attempt to detect the influence of peak loads on cost. Apart from the spurious correlation problem, there is a further limitation on the value of this index, because it may be true that stores subject to extreme peak loads make a practice of increasing their permanent selling force rather than relying on the services of reserve salesmen. The possibility exists, therefore, that the magnitude of reserve salaries affords a poor measure of the existence of selling peaks—an opinion which is reinforced by the inconclusive results obtained by using such a measure.

PART THREE

Cost and Plant Location

Introduction to Part Three

"Drug Chain" (Study 7) examines the geographic structure of labor costs. It was designed to reveal the topography of retail-store wage rates in the United States. The main purpose of the investigation was managerial: to help executives make better decisions about pay and perquisites of the retail employees of a nationwide drug chain.

I. Management Problem

Labor markets are local. Consequently, a corporation whose operations are nationwide must cope with geographic differences in competitive wage rates. Given the wide diversity in community wage levels, management is faced with the problem of developing a geographic pay plan that attains three marginally conflicting goals. First, the plan must be competitive as to wage rates in numerous geographically separate and diverse labor markets. Second, the plan must be equitable in the sense of having the integrity and consistency of a defensible nationwide pay pattern. Third, the plan must be economic in the sense of attaining the goals of competitiveness and equity cheaply and of being simple enough to be comprehensible and believable to employees.

II. Theories Tested

Four rival theories about the geographic structure of labor costs espoused by executives were tested by the drug chain study.

One theory is that given the geographic and technical mobility of the people hired by chain retail stores, there is a single nationwide market for this kind of labor. If mobility is directed by good information about opportunities and policed by employers' competition for competent help, then wage rates for chain store employees would, like water, seek a uniform level. If this were the case, no significant or durable geographic wage differentials, either between regions or in respect to city size, could reasonably be expected. The solution to the managerial problem that follows from this theory is a uniform nationwide pay scale.

A second theory is that there are myriads of local labor markets with wage rates in each set strictly by local supply and demand. Accordingly, wage levels have no national geographic pattern: they do not vary regionally in any systematic or identifiable way, nor do they differ predictably by size of city or degree of suburbanization. The solution this theory suggests to the salary administration of a nationwide chain is local autonomy—freedom of each store manager to set his own pay scales, with no centrally developed guidelines as to the topography of wage rates or their relation to city size.

A third theory is that wage rates differ geographically with cost of living. The solution for salary administration that follows naturally from this theory is central determination of pay scales that differ among communities in conformity with cost of living as measured by geographic differences in the consumer price index.

A fourth theory is that the market for this kind of labor is regional, i.e., that mobility plus competition at the edges of local labor markets connect them and cause a regional topographical structure and that a city size structure exists independently of these wage plateaus. The solution that this theory suggests for the managerial problem is that salaries of a nationwide chain should be geared to the geographical structure of wages by means of research on the regional topography of wage rates and their relationship to size of city.

The purpose of the study was to determine, first, whether wage rates did differ geographically and, second, if so, how? What did the findings reveal about these four rival theories? They disproved the theory of a single nationwide labor market for this kind of employee, they rejected the theory of an isolated labor market for each locality, and they failed to support the theory of conformity of wage rates to the geography of cost of living. The findings did support the theory that chain store wage rates had a significant regional structure of salary-level plateaus and also that they differed systematically by size of city.

Consequently, the results of the research made it possible to establish for all positions in all stores of a nationwide merchandising chain salaries that were equitable and internally consistent and yet were competitive with the diverse community pay levels of the scattered cities in which the company operated. Based on knowledge of geographic structure of wages, the company was able to attain reasonably well its goals of competitiveness, equity, and administrative efficiency.

III. Research Problems

The main research problems in the study were determination of (1) the isoquants that map the topography of drugstore wage rates and thus, by contour analysis, delineate wage plateaus; (2) the relationship between city size and wage rates; (3) the relationship between suburbanization and pay level; and (4) the relationship between job grade and pay rate. The study describes how each problem was tackled.

IV. Contributions to Technique

The principal contributions of the drug chain study to the technique of geographic salary administration were:

1. The concept of the relationship between city size and pay levels[1] was developed further, measured for drug merchandising salaries, and applied as a simplifying device for nationwide job-grade pricing.

2. The concept of pay-level plateaus was further developed and put to commercial use.[2]

3. A tested technique for delineating these pay-level plateaus for merchandising salaries was outlined and applied to area sampling, which streamlined nationwide job-pricing.

4. A theory of systematic relationship between salary level and skill differentials was set forth; the relationship between the minimum pay level of an area and the slope of its salary curve was measured for pay-level plateaus of merchandising salaries.

Thus the recommended structure of salaries for a nationwide drugstore chain was founded on three economic patterns which accounted for a substantial portion of the vast geographic diversity of salaries: a system of pay-level plateaus; a pattern of skill differentials (i.e., salary curve slope) that was systematically related to plateau level; and a structure of city-size differences superimposed upon these plateau salary curves. From these structural members a tested, streamlined method of job-pricing for chain store salary administration was built.

Study 7

DRUG CHAIN

I. The Problem

Successful salary administration on a regional or a nationwide scale requires comprehensive knowledge about the geographic structure of salaries.

Nationwide salary administration involves reconciling the necessity of control and internal consistency with the necessity of meeting local competition for employees. The rising spiral of wages and prices accentuates the difficulties of reconciling these two objectives and makes the problem of salary administration for a company that has a wide geographical compass extremely difficult.

Most regional and nationwide operators are fully aware of this problem. They recognize its importance and its complexity. But comparatively few have developed a solution which entirely satisfies them.

This paper summarizes a streamlined research approach to the problem which has been successfully used by a large merchandising chain.[1] The basic methods can be applied to any company that operates numerous units scattered over a broad geographical area, especially where these units are so small as to preclude the expense of a separate job price survey for each.

Methods of Establishing the Pay Scale

Several different methods of setting the geographic structure of pay scales[2] can be used by such a company. Each of the following methods has been used by nationwide operators:

1. A *uniform* salary scale throughout the nation, regardless of the location or size of the city.

Reprinted from Joel Dean, "Geographical Salary Administration," in *Plant-Wide and Geographical Salary Administration*, American Management Association, Personnel Series, no. 114 (New York, 1947), pp. 24–42, by permission of the American Management Association. The chain was Rexall Drug Stores, then owned by Liggett Drug Company. Permission to reveal the name of the company in the present volume was generously granted by Mr. Justin Dart, Chairman of Dart Industries, who was President of Liggett Drug Company at the time the study was made.

2. Complete *local autonomy*, granting to each unit manager freedom to set his salaries on the basis of his knowledge of community pay levels.
3. Pay scale differences that are proportional to differences among communities in *cost of living*.
4. Pay scale differences based solely on the *size* of the city, regardless of its location.
5. Pay scale differences that are based solely on the *location* of the community, and disregard its size.
6. Pay scale differences that are based on grouping cities without trying to examine the causes for differences.
7. Pay scale differences that are developed by analysis and combination of three separate influences: (*a*) differences in city size; (*b*) purely regional or locational influences; and (*c*) skill differentials (that is, changes in the slope of the salary schedule from jobs of low to those of high skill).

Let us look briefly at the merits and shortcomings of each of these possible methods.

Many companies have in the past paid uniform rates for the same job throughout the nation; and some large operators still cling to this method. It has the merit of simplicity and apparent equity. But it can get you into serious difficulties. It flies in the face of great and persistent regional differences in local pay rates.

In some areas pay rates will be below community rates, making it difficult to attract and hold a working force; in other areas, rates will be above community levels, resulting in unnecessary expense.

The second method—local autonomy—is the opposite of nationwide uniformity. This method was widely followed at one time and is still used by some. It may have the merit of being in tune with the trend toward decentralization of operating responsibilities. But local determination of pay scales has many deficiencies. It is unreasonable to expect that a local manager will be equipped or will have the time to perform the specialized and highly skilled task of job pricing within his community. Moreover, if effective salary control and internal consistency within the company are wanted, it is hard to get while permitting local wage autonomy. Control and consistency are almost unattainable by this method. Most large companies have found it unsatisfactory.

The third method, which makes cost of living the sole determinant of geographical differences in wage rates, has the virtue of apparent equity, in the sense that the purchasing power of wages is made the same. But there are several difficulties. Cost-of-living information is available for only about 50 cities, and these indexes are not fully satisfactory for our purposes. Moreover, cost-of-living differences and wage-rate differences among cities are not closely correlated. Other and more powerful influences govern wage differences. Consequently, the cost-of-living method results in significant departures from community pay levels. Salaries are above the community level in some areas and below it in others.

The fourth method, namely basing differences in pay scale solely on city size, has an important element of validity because the size of the city does affect market wage rates. It represents an oversimplified solution, however, because it neglects the broad regional differences existing in salary topography. Pay scales in the southeast, for example, are significantly lower for most industries than pay scales on the Pacific coast.

The fifth method places sole reliance on these regional differences in pay level, to the neglect of city-size and skill differentials. Although it is founded on one of the most important sources of geographical wage differences, it excludes other important factors. Moreover, this kind of wage plan is often based on inadequate research in salary topography. Typically ready-made political or administrative areas, such as states, Federal Reserve Regions, or company sales divisions, are used, and differentials are not founded on comprehensive job-price surveys.

Plans developed by the sixth method typically set up two or three wage groups of cities, with a uniform pay scale differential between groups. A salary curve of the same slope is used for all groups. No attempt is made to separate and analyze the various determinants of salary differentials. Control and internal consistency are obtained by this method, but whether it produces salary schedules that fit community pay scales depends upon the adequacy of research that determines into what group a given city goes. Normally this method is more appropriate when operations are confined to fairly large cities.

We see that each of the first six methods has significant deficiencies, particularly for a nationwide operation with many small units. For such a company the seventh method has greatest promise. This approach takes cognizance of each of the important influences on salary levels throughout the country. It tries to isolate each factor, measure it independently, find out the relationships between factors, and combine them eventually in a geographic structure of salary schedules for the entire nation.

Central control of the company's pay scales, if based on knowledge of geographic wage rates, meets the problems of equity, control, and conformity with diverse community wage levels. But to develop centrally controlled salary schedules that are tuned to community pay levels requires basic research.

The expense and difficulty of surveys of pay rates for all jobs in each community where company stores are located have discouraged a research approach to chain-store salary administration. Hence if this research attack is to be made useful, it will be necessary to develop reliable short cuts. Such a short cut is outlined next.

Preview of Basic Approach

Wage-making forces operate differently in different geographical areas, with the result that there are many labor markets rather than one national labor market. Each of these markets is fairly distinct but definitely related to the others. This interrelationship results in a basic pattern of geographical salary structure.

The only economically sound short cut is to dig beneath the surface of the hodgepodge of individual salary variations and find that underlying pattern of relationships.[3] This research required four steps:

1. *Determine the relationship between the city size and salary level.* The size of a city affects its salary level, and this effect appears to be the same regardless of the area in which the city is located. The influence of size is stable and pronounced. Knowledge of the effect of city size upon pay levels is an essential implement. It facilitates analysis of salary data to reveal the pure patterns of salary-level plateaus and of skill differentials; and it is useful in adjusting normal plateau salary curves for city size in order to apply them to specific unsurveyed cities.

2. *Map out major salary-level plateaus.* There is considerable evidence of the existence of broad geographical areas in which merchandising salary levels are substantially uniform for similar jobs in cities of the same size. Delineation of homogeneous areas, by means of contour analyses, is much less expensive and possibly more accurate than job-price surveys in every store locality. Salary plateaus facilitate scientific selection of a representative sample. Job prices established for sample cities can then be applied to all the other cities on the same plateau, merely by making city-size adjustments.

3. *Select a representative sample of jobs, cities, and companies and conduct a job price survey.*

4. *Develop salary curves and salary schedules that are based upon the statistical relationships between job grades and compensation.* This involves statistical analysis of the survey data (*a*) to determine salary curves for individual survey cities; (*b*) to develop the geographical pattern of skill differentials and its relation to salary-level plateaus; (*c*) to develop a composite normal salary curve for each *plateau area;* (*d*) to construct salary ranges; and (*e*) to develop the nationwide structure of salary schedules. Each of the steps comprising this research approach to salary determination is explained more fully in subsequent sections.

II. Determination of Size-of-City Patterns

Knowledge of the relationship between size of city and salary level is needed for three purposes; first, to remove the city-size influence from the crude salary data used to reveal and mark off areas of uniform salary level; second, to avoid the expensive necessity of surveying cities of each size category in each homogeneous area; and, third, to derive salary curves for unsurveyed cities of different sizes from the largest city curves developed for their respective plateaus.

A summary of the methods used to obtain size-of-city patterns will explain and illustrate this phase of the basic approach.

Three sets of store-salary data collected by the Bureau of Labor Statistics[4] during the spring and summer of 1943 were used to determine size-of-city patterns: (1) limited-price variety stores; (2) department and clothing stores; and (3) drug stores.

To develop basic salary averages, a nationwide average of actual salaries for each job in each store type was computed for each of the following four city-size categories: (1) 500,000 and over; (2) 100,000 and under 500,000; (3) 25,000 and under

100,000; and (4) under 25,000. The political city rather than the metropolitan district was used for this purpose, since the latter unit had not been worked out for smaller cities.[5]

The existence of significant city-size differentials is clearly substantiated by this study. Within the job-grade range, merchandising salaries differ substantially and systematically among the four categories of city size which were studied. The remarkable consistency of this pattern among chains of different types is illustrated by Table 7–1.

The city-size spread tends to be proportional, so that the relationship is the logarithmic type. Within the limited range of the data there is evidence that salary curves for different-sized cities have a constant percentage differential. Although there is considerable dispersion among the percentage differentials, a percentage relationship summarizes the data better than does a uniform cents-an-hour differential.[6]

Table 7–1

PERCENTAGE SUMMARY OF CITY-SIZE SALARY DIFFERENTIALS*

City-Size Category	Drug-Store Average (4 Jobs)	Variety-Store Average (4 Jobs)	Department Store Average (14 Jobs)	Weighted Average† (21 Jobs)	Salary Curve Differential
500,000 and over	100.0	100.0	100.0	100.0	100.0
100,000 and under 500,000	87.0	90.3	87.8	88.0	90.0
25,000 and under 100,000	78.8	81.9	....	79.8	80.0
Under 25,000	70.1	....	....	....	70.0

*Based on job rates from Bureau of Labor Statistics data for the spring and summer of 1943.
†Weights: drug store, 2; variety store, 1; department store, 1.

To determine whether or not there is a significant difference in the size-of-city pattern among broad geographic regions, the methods described above were used in developing city-size patterns separately for each of the following areas: (*a*) northeastern, (*b*) border states, (*c*) southern, (*d*) middle western, and (*e*) mountain and Pacific Coast. There were some differences among regions in city-size patterns, but these variations were erratic and not statistically significant. Thus, for the period studied there was no evidence of valid differences in this pattern among regions.[7]

III. Mapping Salary-Level Plateaus

Mapping broad geographic areas in which salary levels are substantially uniform facilitates sampling and simplifies the field research. In order to build a sound economic foundation for selecting the sample localities, it is necessary first to mark off major areas in which salary levels are relatively uniform. Cities that are representative of each of these areas can then be selected for the survey. And the resulting area salary curves can be applied later to other cities in that salary area, with appropriate size-of-city adjustments.

The most economical and reliable methods for marking off compensation plateaus can be summarized by the following six guiding principles:

1. A contour map of pay levels for the same job in different areas should be constructed. Such a map is highly useful in marshaling unwieldy masses of geographical salary data in an orderly, simplified form. Pay contours, like the altitude contours of a topographical map, indicate dividing-lines between areas of different salary levels.

2. A reliable pay-level contour map must be based on pay data which have broad geographical coverage. These data must also be highly comparable with respect to positions, survey periods, and collection techniques.

3. Jobs for mapping salary-level plateaus should be selected from among the lower ranges of the salary hierarchy. Work requirements are more likely to be comparable. Furthermore, the labor market for these lower-grade jobs is broader, giving greater reliability to the survey data. Finally, differences in salary curve slope (i.e., skill differentials) cause less distortion than for higher-level jobs.

4. Salary information collected for each city must be adjusted for the size-of-city differential before developing contours. The actual rate reported is the product of two distinct forces: (*a*) the broad regional pattern of salary levels and (*b*) the effect of the size of that city. Only after the city-size influence has been removed is it possible to form a realistic picture of the geographical contours.

5. Salary-level contours developed for one kind of merchandising chain (e.g., variety stores) should be the synthesis of the contours developed from the salary data of several types of merchandising establishments. This is desirable because the competition of a chain store for employees in the labor market goes beyond the bounds of its product competition. Hence, broadening the factual base in this way increases its reliability.[8]

6. In constructing a salary contour map for this purpose, only major contours that have market and administrative significance should be used. Variability is one of the most striking statistical characteristics of geographic wage data. Salary plateaus are not perfectly flat; the most that can be expected is *relative* uniformity in salary levels.

IV. Selection of Survey Samples

We have now laid the groundwork for scientific selection of the survey sample by measuring the influence of city size and mapping salary-level plateaus. Three major phases of survey sample selection which deserve special attention are (1) selection of key jobs to be priced; (2) selection of cities to be surveyed; and (3) selection of companies to participate in the survey.

Selection of Jobs to Survey

Salary curves portray the relationship between compensation and job grades. To determine job grades, positions must be described and analyzed. Their requirements are then measured in order to classify jobs into numbered grades representing relative degrees of difficulty and importance.

Jobs are ranked in this way without reference to the wages paid to persons who fill them. It is the task of job-pricing to translate this job-grade relationship into a dollar scale of market salary rates. The positions which are priced are not significant in themselves; they merely represent their job grades. Pricing of these grades is the major concern, since it establishes peg-points from which salary schedules can be developed.

An essential economic function of salary schedules is to construct an equitable bridge between jobs for which a salary rate is set by comparatively active operation of market forces.[9]

Selection of Cities for Survey

Well-defined salary plateaus can provide a sound economic basis for selecting cities to be surveyed to obtain job-grade price data for development of salary schedules.

The metropolitan district[10] is more useful for salary survey purposes than the political city. There are several reasons:

1. Metropolitan districts are continuous urban areas which are essentially homogeneous economic units. The political accident of city limits has no bearing on the problem.
2. By adopting the metropolitan district as the sampling unit, it usually is possible to pick a relatively small number of survey localities that include a high proportion of the stores in a chain.
3. The use of metropolitan districts brings into clear focus the problem of suburban differentials.[11] By considering cities as components of major metropolitan districts rather than as isolated political units, irrelevant city-size pattern is avoided. The real problem of wage differentials within the metropolitan district is thus set apart.

Cities selected for a job-grade pricing survey should meet the four following conditions:

1. They should be *representative* of each salary-level plateau as defined by the salary contour map. One or more sample cities should be selected for each area that is contiguous and uniform in level. Areas that have a particularly broad geographical sweep should usually have more than one survey city.
2. They should be *important* to the particular chain by including a large proportion of the employees and stores in the locality.
3. They should be *strategic* in being good bench-mark localities for applying size-of-city differentials.
4. They should be *reliable* by providing enough job-price quotations for a valid salary curve.[12]

Selection of Companies to Survey

Certain general principles provide useful guides in the selection of survey companies:

1. Similarity of the job is more important than similarity of products sold in selecting survey companies. Chain stores compete in the labor markets with firms in entirely different businesses.
2. Companies that have similar personnel standards will yield more relevant salary data than those that are merely similar as to products.
3. Large organizations are generally better data sources than small ones. Nationwide firms are usually best, but large regional operations with comparable personnel standards should also be included. Both types provide the economy of central collection of salary data.
4. An adequate sample of companies in a few key spots will yield more reliable results than scanty coverage in many areas.
5. It will usually be desirable to include unionized as well as nonunion establishments.

V. Development of Salary Schedules

The basic economic concepts of the geographic structure of salaries are: the delineation of salary plateau areas, the development of city-size differentials, and the selection of sample jobs, cities, and companies to survey. These concepts are all brought to focus in the collection of the survey data.

Collection of Survey Data

Good job descriptions are prerequisites to accurate job classification. They must be used as the foundation for job-pricing if the jobs surveyed are to be fully comparable.

There is frequently great diversity both among and within surveyed companies with respect to length of work-week, amount of regular compensation, and amount and kind of additional compensation. This diversity accentuates rather than reduces disparities among firms in a given locality. Frequently the best perquisites are found in firms that pay high wages. The amount and complexity of information required on job prices, job conditions and supplementary compensation in a survey of this kind necessitate personal interviews.[13] Comparableness of selected jobs can best be assured by personally comparing their job requirements.

Averaging and Weighting

Our central concern in this phase of the analysis is to price job grades. The key jobs for which salary data are obtained are significant only as representatives of job grades. Consequently, an average salary for each job grade should be computed for each store type by developing a weighted average salary for each key job within each grade. These are then combined into a single weighted average for each job grade.

Developing Salary Curves

The next step is to develop the curve of normal relationship between salaries and job grades. This involves fitting curves to the job-grade averages. To do this, visual fitting was used because it has several practical advantages over the more rigorous and orthodox method of mathematical fitting.[14]

Chart 7–1 indicates the process by which the salary curve for a typical survey city is evolved. The job grade is plotted on the horizontal scale. This ranks jobs on the basis of their relative requirements. The hourly wage is plotted on the vertical scale. Each dot represents the average hourly pay rate for jobs of that grade for all stores of the type specified. The Arabic numerals for each observation which is plotted show its weight, both in respect to the number of employees and the relative importance of that type of store. Type-of-store curves are fitted visually, with these weights taken into account. The city curve is a consolidation of the individual type-of-store curves.

Delineation of pay-level plateaus makes possible the economy of determining salary schedules throughout a broad area on the basis of a representative sample of job prices in a few survey cities.

When individual city-salary curves have been determined, it is then possible to construct a composite salary curve for each homogeneous area. The method used to develop these area salary curves may be summarized as follows:

1. Survey cities were grouped according to homogeneous areas.
2. The salary curve for each city in a homogeneous area was adjusted for appropriate city-size differentials.
3. A representative plateau curve was derived from the adjusted city curves by visual consolidation that took into account the quality and adequacy of the data. Such area curves were regarded as interim curves, since further slight adjustments had to be made as will be indicated later.

The method is illustrated by the typical results shown in Chart 7–2.

Some variation among the city curves in a given plateau area should be expected even after adjustments for city size. Salary plateaus are not perfectly flat; their salary surface is only *relatively* uniform; for considerable area imperfection exists in the labor market. Moreover, the boundaries of plateaus are not steep cliffs but gentle gradients of slow transition in salary level. In addition to this roughness of the pay terrain, considerable chance variation in the sample is to be expected. Nevertheless, the differences among cities within a plateau area appear to be significantly less than the differences among the various plateaus themselves. Moreover, in most of these areas, the range of deviation of the individual city curves from their adjusted area curve is too small to merit administrative recognition in establishing salary schedules.

Chart 7–1 Salary Curve for Typical Survey City

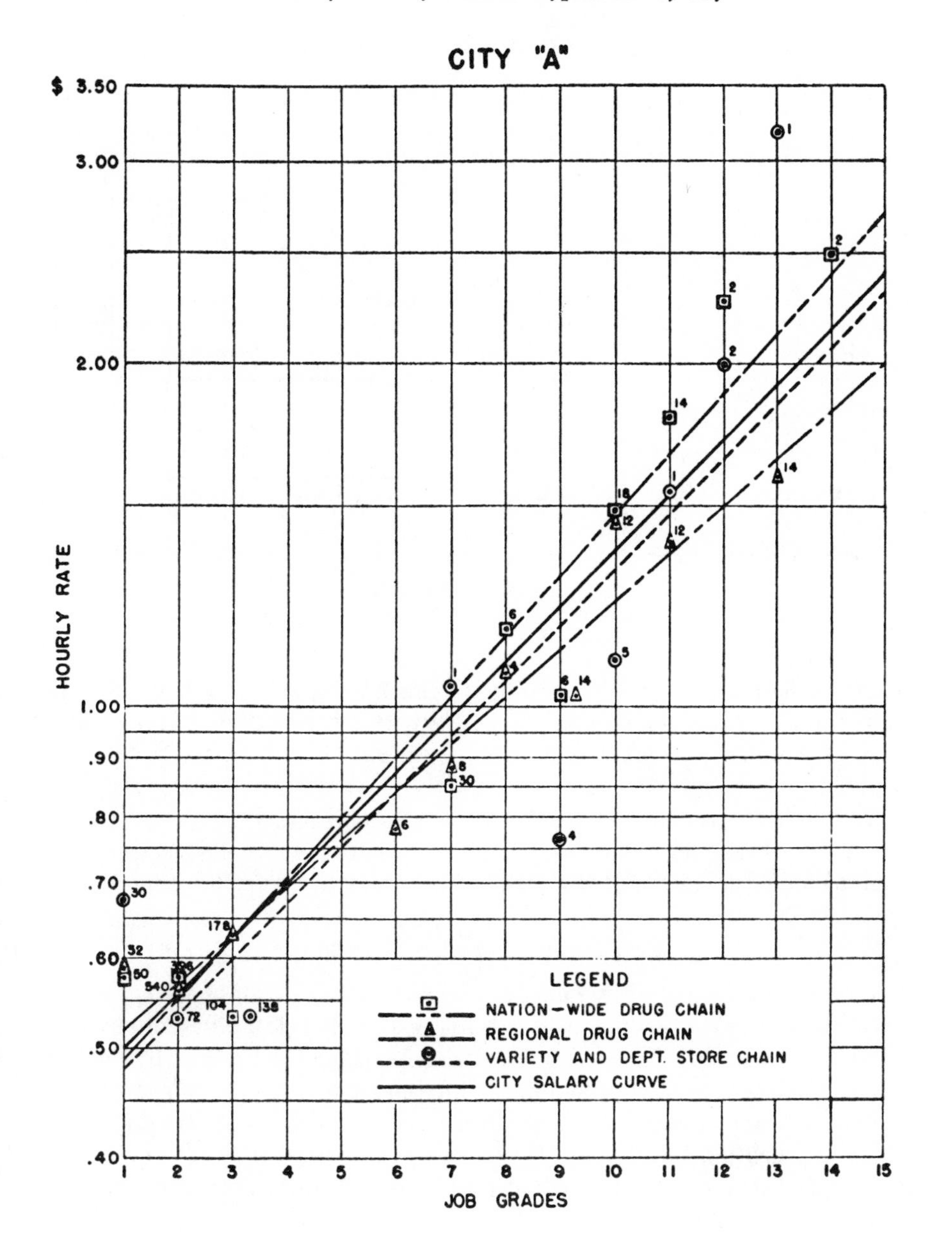

Chart 7–2 Salary Curve for Homogeneous Salary Area IV

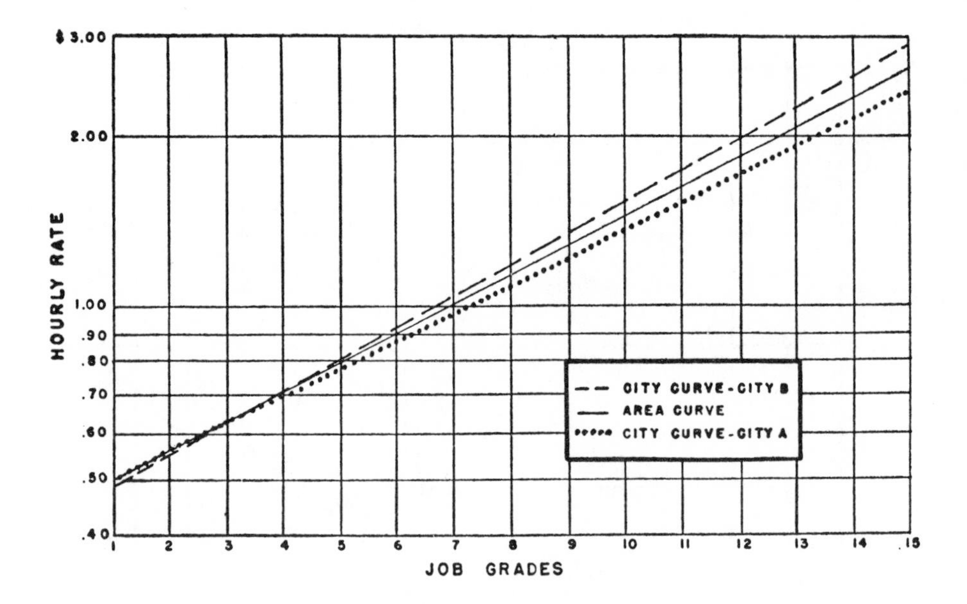

To determine plateau salary levels, the Grade I level was used because it had been used in the earlier analysis to develop tentative homogeneous areas and because more complete and reliable job-price data are available in the lower grades than for the higher ones. Table 7–2 shows the salary levels of the interim area curves at Grade I, both in cents per hour and as a percentage of the level for the base city. This table also contrasts, for one study, these final salary levels with those projected at the outset of that study. They show a comforting conformity.

Developing Slope Coefficients

Should the slope of the salary curve be the same in all localities throughout the nation? This is common practice. But is it reasonable that skill differentials should be uniform despite wide differences in salaries for jobs at the Grade I level? Our fitted salary curves showed pronounced differences in slope among areas. The Army Service Forces and other investigators have found differences in curve-slope for different localities. Is there any systematic relationship between these slopes and the Grade I salary level? This whole problem has been neglected in geographical salary administration.

To try to find answers to some of these questions, we analyzed our data further. When salary curves for individual cities were grouped by plateau areas, there were distinct, and apparently significant, differences among areas in the *slope* of their curves. Comparison of the interim area curves (obtained by consolidation of

Table 7–2

RELATIONSHIP OF SALARY-LEVEL PLATEAUS TO SALARY CURVE SLOPE

Area No.	Representative Area Cities	From Survey Data 1945		From Government Data, 1943		Selection	Salary Curve Slope	
		Cents per Hour Grade I	Per Cent Relationship (City D = 100)	Per Cent Relationship (City D = 100)	Mid-point of Range	Per Cent Relationship (City D = 100)	Actual	Theoretical
1	C, D, E, F, and G	46.5	100	100–110	105	100	26°0	27°5
2	H, I, J, and K	53.0	114	110–120	115	115	24.0	24.8
3	L, M, N, and O	44.0	95	90–100	95	95	28.5	28.5
4	A and B	49.5	106	100–110	105	105	27.5	26.3
5	P, Q, and R*	40.0	86	....	....	85	29.2	30.4
6	S	47.8	103	110–120	115	105	32.5	27.0
7	T	38.2	82	80	75	80	31.0	31.3
8	U, V, W, X, and Y	44.0	95	80– 90	85	90	28.5	28.5
9	Z and A1	49.5	106	110–110	105	105	27.0	26.3
10	B1	43.7	94	80– 90	85	90	29.5	28.7
11	C1	66.5	143	Over 130	135	140	24.0	19.0
12	D1 and E1	57.0	123	120–130	125	125	23.0	23.0
13	F1*	....	....	90–100	95	90	....	28.5
14	G1*	....	....	100–110	105	105	....	26.3
15	H1a*	....	....	90–100	95	90	....	28.5
16	I1*	....	....	100–110	105	100	....	27.5
17	J1	66.5	143	Over 130	135	140	....	19.0
18	K1	66.5	143	Over 130	135	140	....	19.0

* Cities which were *not* surveyed.

the city curves for each area) revealed an underlying pattern of relationship between salary level and curve slope. However, a few individual curves departed slightly. To develop this underlying pattern of relationship, the slope of the salary curve was plotted against the level of salaries at Grade I for each plateau as shown in Chart 7–3. The visually fitted line shows the normal relationship derived from these observations.

In order to develop a set of final area curves, the normal slope coefficient read from the relationship line of Chart 7–3, was used in conjunction with the scheme of salary levels at Grade I that had been obtained by adjusting the interim area curve levels. This systematic normal pattern of salary plateau curves is illustrated in Chart 7–4. These final area curves converge at Grade X, where geographical wage differences (which tended to grow smaller as the grade levels rose) finally disappear altogether.[15] For Grades X-XXX a single salary curve was used for all areas. The a nationwide or a region-wide market. This means that salaries become more and slope of this upper segment of the salary curve was the average of the slope of the area curves from Grades I to X; thus it was a projection of the composite of the several area curves.

In the course of collateral investigations, I have found some additional empirical

Chart 7–3 Slope Coefficient of Area Curves

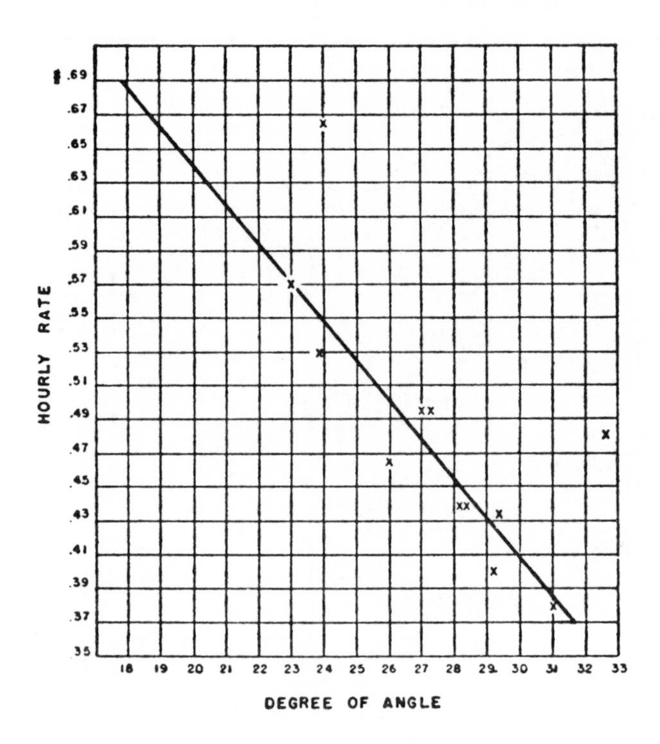

evidence and a good deal of collective wisdom leading to the expectation that as the managerial level is approached, the market for men broadens and approaches more uniform, and geographical influences tend to disappear. It will be very interesting to find out, as more studies of this type are made, whether there is some general tendency as to the point of convergence.

Salary Schedules

The next step in developing salary schedules is to make adjustments for city-size differences. Earlier we used our city-size findings to isolate pure locational influences and to make city curves within a given homogeneous plateau comparable, so that we could combine them and get one plateau curve. At this point we use our city-size findings to reverse the process, i.e., to derive from the plateau curve the specific salary curves for cities of various sizes. To do this, the final plateau curve is adjusted by the appropriate city-size differential or suburban differential.

Chart 7–4 Normal Pattern of Area Salary Curves

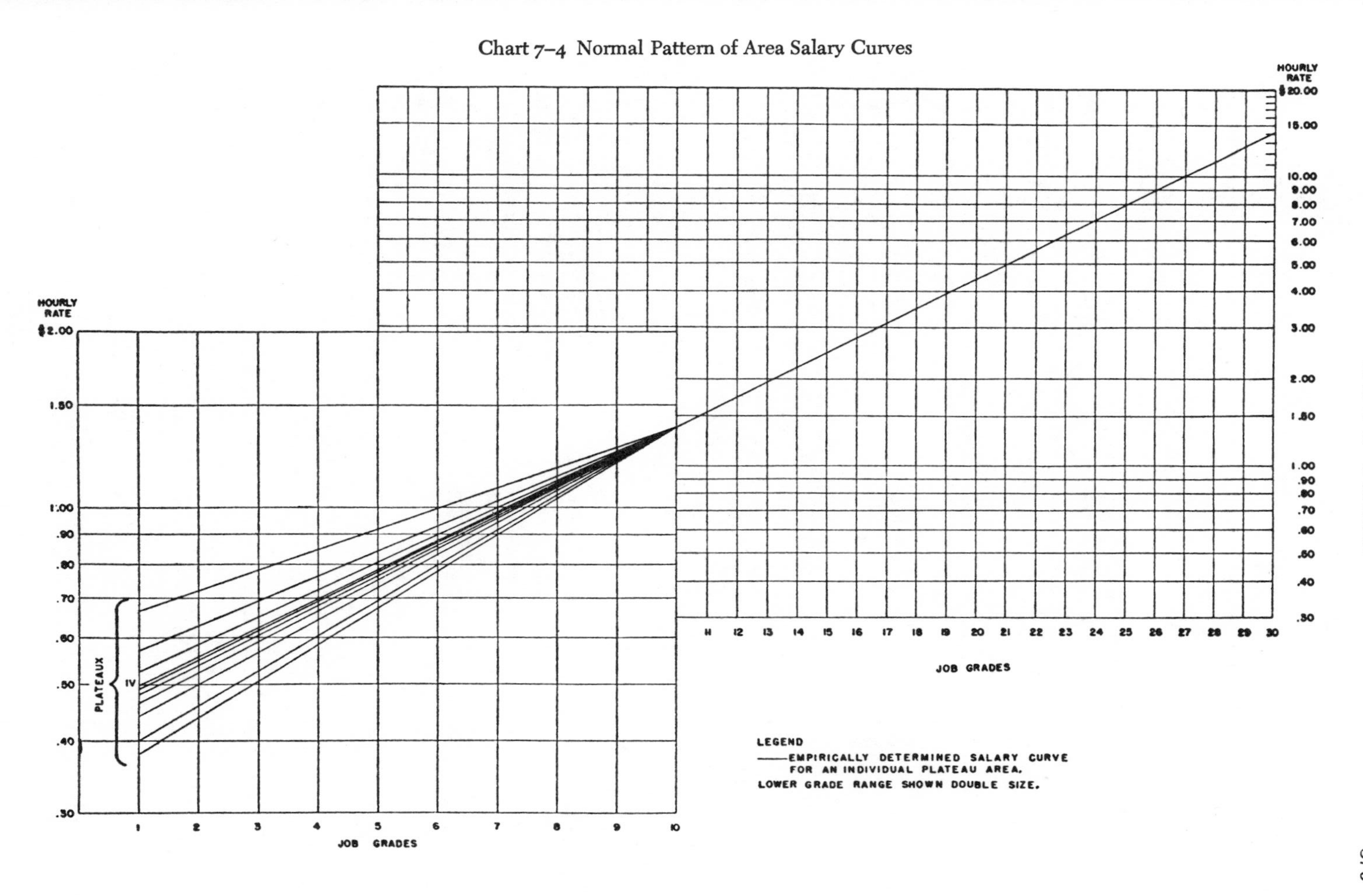

Table 7–3

RETAIL STORE SALARY SCHEDULES SHOWING MINIMUM AND MAXIMUM RATES, GRADES I-X

Job Grade	Size of City*							
	A	B	C	D	A	B	C	D
	Area I				Area II			
I	$0.41–$0.52	$0.39–$0.49	$0.37–$0.46	$0.33–$0.41	$0.47–$0.58	$0.45–$0.55	$0.42–$0.52	$0.38–$0.46
II	0.47– 0.58	0.45– 0.55	0.42– 0.52	0.38– 0.46	0.52– 0.65	0.49– 0.62	0.47– 0.59	0.42– 0.52
III	0.53– 0.66	0.50– 0.63	0.48– 0.59	0.42– 0.53	0.58– 0.73	0.55– 0.69	0.52– 0.66	0.46– 0.58
IV	0.59– 0.75	0.56– 0.71	0.53– 0.68	0.47– 0.60	0.64– 0.82	0.61– 0.78	0.58– 0.74	0.51– 0.66
V	0.67– 0.86	0.64– 0.81	0.60– 0.77	0.54– 0.68	0.71– 0.92	0.67– 0.87	0.64– 0.83	0.57– 0.74
VI	0.75– 0.97	0.71– 0.92	0.68– 0.87	0.60– 0.78	0.79– 1.03	0.75– 0.98	0.71– 0.93	0.63– 0.82
VII	0.84– 1.10	0.80– 1.05	0.76– 0.99	0.67– 0.88	0.87– 1.14	0.83– 1.08	0.78– 1.03	0.70– 0.91
VIII	0.94– 1.25	0.89– 1.19	0.85– 1.13	0.75– 1.00	0.96– 1.27	0.91– 1.21	0.86– 1.14	0.77– 1.02
IX	1.06– 1.41	1.01– 1.34	0.95– 1.27	0.85– 1.13	1.07– 1.41	1.02– 1.34	0.96– 1.27	0.86– 1.13
X	1.18– 1.59	1.12– 1.51	1.06– 1.43	0.94– 1.27	1.18– 1.59	1.12– 1.51	1.06– 1.43	0.94– 1.27
	Area III				Area IV			
I	$0.39–$0.49	$0.37–$0.47	$0.35–$0.44	$0.31–$0.39	$0.44–$0.55	$0.42–$0.52	$0.40–$0.50	$0.35–$0.44
II	0.44– 0.56	0.42– 0.53	0.40– 0.50	0.35– 0.45	0.49– 0.62	0.47– 0.59	0.44– 0.56	0.39– 0.50
III	0.50– 0.64	0.48– 0.61	0.45– 0.58	0.40– 0.51	0.55– 0.69	0.52– 0.66	0.50– 0.62	0.44– 0.55
IV	0.57– 0.73	0.54– 0.69	0.51– 0.66	0.46– 0.58	0.61– 0.77	0.58– 0.73	0.55– 0.69	0.49– 0.62
V	0.64– 0.83	0.61– 0.79	0.58– 0.75	0.51– 0.66	0.68– 0.87	0.65– 0.83	0.61– 0.78	0.54– 0.70
VI	0.72– 0.94	0.68– 0.89	0.65– 0.85	0.58– 0.75	0.76– 0.98	0.72– 0.93	0.68– 0.88	0.61– 0.78
VII	0.81– 1.07	0.77– 1.00	0.73– 0.96	0.6[illegible]– 0.86	0.85– 1.10	0.81– 1.05	0.77– 0.99	0.68– 0.88
VIII	0.92– 1.21	0.87– 1.15	0.83– 1.10	0.7[illegible]– 0.97	0.95– 1.24	0.90– 1.18	0.86– 1.12	0.76– 0.99
IX	1.04– 1.37	0.99– 1.30	0.94– 1.23	0.8[illegible]– 1.10	1.05– 1.39	1.00– 1.32	0.95– 1.25	0.84– 1.11
X	1.18– 1.59	1.12– 1.51	1.06– 1.43	0.9[illegible]– 1.27	1.18– 1.59	1.12– 1.51	1.06– 1.43	0.94– 1.27

* A = cities of 500,000 and over; B = suburbs of cities of 500,000 and over; [illegible] = cities of 100,000 and under 500,000; D = cities under 100,000.

The final area salary curves permit the establishment of the single-rate salary schedules. However, a single-rate salary schedule does not provide leeway for merit review and/or length-of-service salary adjustments. To develop schedules which do have this desired characteristic, minimum and maximum salaries are needed.

By applying rate ranges to the final plateau area salary curves (with appropriate city-size adjustments) salary schedules are produced. These schedules are tuned to community pay scales throughout the nation in any city of any size in any area.

Table 7–3 shows the kind of nationwide salary schedules that can be developed by this method. It sets forth hourly rate ranges for Grades I-X for each city size in each area. Salary schedules for Grades X-XXX are the same for all areas but differ only for city size.

General Conclusions

In closing, I offer a few general conclusions. For successful salary administration it is essential to have salary schedules fit the geographical structure of salaries. Communities differ greatly in pay levels and in skill differentials. Failure to reflect these differences is particularly serious in a period of full employment such as this.

To take account of geographical salary structure adequately, it is desirable to make a separate analysis of each of the major forces that govern salary differences among localities. These forces are: (*a*) the purely locational influence, (*b*) the influence of city size, (*c*) the influence of skill differentials and (*d*) random variations. To do this requires basic research in job-price geography. Alternative methods for nationwide job-pricing sketched at the beginning of this paper are proving increasingly unsatisfactory.

Skill differentials—that is, the slope of the salary curve—are a dimension of geographical salary administration that has been neglected. These salary gradients differ among areas and are likely to be systematically related to the salary level of each area. The geographical pattern of slopes will probably be different for different industries. The pattern discussed here applies precisely only to the retail drug industry during the war, but the methods devised would, I think, be useful for any such study. Systematic relationships between the slope of the salary curve and the basic pay level of the region are to be sought—partly because they exist and have to be reckoned with, and partly because once established, they facilitate accurate salary adjustments.

A reasonably close approximation to community scales is satisfactory for many companies. Considerable variation in pay is found within the same locality. Differences in work conditions and in job prequisites do not, as was once supposed, compensate for these differences. Frequently, instead, they accentuate them. Case studies of how workers actually select jobs show that typically they know comparatively little about wages in other firms in their locality. In fact, workers apparently have surprisingly little interest in such information as long as they have a job. When they do not have a job, instead of shopping the market as would be logical to do, they generally work through friends. Strangely enough, even in this market many consider jobs scarce because of their general apprehensiveness about unemployment. They accordingly tend to take the first job that comes along or hang on to the one they have.

This situation implies imperfections in the labor market within a given locality—immobility, ignorance. It means that internal consistency of salaries, within the company, is as important as external consistency with the local pay level. Workers are better informed and are certainly more emotional about the rates above and below them in the same plant than they are about differences between companies. This situation also means that reasonably close approximation to local wage patterns, such as the short-cut method I have outlined, yields salary schedules that are satisfactory in actual practice.

Vigilance and continual research are essential to keep a nationwide or a regional salary plan up to date and tuned to the changes in geographical pay structure. Changes are continuously taking place in the geographical structure of salaries. What are some of these changes now in the making? It is evident that there have been dramatic shifts in area wage differentials as a result of the war. These changes have persisted into the postwar period, raising some plateaus and lowering others. They are still in progress. Note particularly the sudden spurt of wage levels in Southern California as it expands its industrial and commercial activity.

Skill differentials (i.e., the slope of these salary curves) have also changed geographically during this period. There has been such a sharp increase in pay for low-skill jobs that there is some indication of a general trend toward the reduction of the gradient of salary curves. These patterns are not fixed—they are changing continuously. This calls for research on a spot-check basis, grounded upon a thorough knowledge of the geographical structure of wages, if a salary administration plan which was once keyed with community pay scales is to be kept consistently attuned to them.

PART FOUR

Cost and Real Profits

Introduction to Part Four

Profits as reported by the accountant are the primary measure of a business firm's success. In managing the firm, the executive is normally concerned with maximizing the growth of earnings per share as measured by accounted profits. This is the way the score is kept on Wall Street. Yet this goal is in partial conflict with a higher professional managerial responsibility, namely, to maximize the wealth of stockholders. Shareholder wealth means purchasing power in terms of real goods; hence it is basically determined by the corporation's real profits (and stockholders' equity in them) rather than by its book profits.

Inflation subtly undermines the efficiency of an enterprise economy by scrambling its signals and subverting the optimization decisions of managers of individual firms. Professional managers in the United States today persist in thinking about dollars as if their purchasing power were constant; they do not think in terms of real goods. Government control of inflation is episodic, unpredictable, and usually ineffectual. Inflation therefore increases risk, which dulls incentives and requires higher rewards in prices and profits. In addition, inflation distorts the semiautomatic adjustments of a competitive economy. Real tax rates of corporations and of individuals are automatically increased during inflation, which widens government's share of income. Supply of capital for corporate investment (the mainstay of technological progress and economic efficiency) is reduced because incentives for individuals to save are dulled and the power of corporations to save (by capital consumption allowances) is impaired. Inflation should force corporations to require a higher criterion of minimum profitability in pricing products and in rationing capital among rival investments because (1) book earnings overstate real earnings of corporations during inflation; (2) bond holders demand higher nominal interest to make up for buying-power erosion; and (3) equity investors demand an even higher rate of return to preserve the normal reward for their greater risk as compared with bond holders.

As a result, during a period of inflation the executive needs measurements and forecasts of real costs and constant-dollar profits, which alone are pertinent to basic economic choices that govern the long-run maximization of shareholder wealth.

The two statistical cost studies in Part Four—"Electronics Producer" (Study 8) and "Machinery Manufacturer" (Study 9)—concern the way cost and real profits should be measured when inflation causes changes in costs that invalidate the profit measurements of conventional accountancy.

I. Measurement Problem

The hypothesis tested by these two studies is that book profits reported by accountants correctly measure real profits during inflation. This hypothesis is, with rare exceptions, explicitly avowed by accountants and implicitly accepted by business executives, by investors, and by bankers, brokers and other financial institutions, who continue, even during double-digit inflation, to take seriously the corporate profits reported by conventional accounting.

To test this hypothesis, we sought to quantify the concept of real accounted profits during inflation—namely, what the books would have reported as net profits if the dollar had remained constant in buying power. The two statistical studies measure constant-dollar *accounted* profits, which is not the same as real *economic* earnings, even though Study 9 was billed (erroneously I now think) as measuring "real economic earnings." Economists' notions of earnings—what to include in costs and how to value assets—are quite different from those incorporated in conventional accounting procedures.

II. Measurement Methods

Alternative Techniques

There are several ways to determine what a firm's reported earnings would have been in the absence of inflation:

1. Adjusting the end product of the income statement—book profits—by a single index number, usually the BLS consumer price index but sometimes the GNP deflator index.
2. Directly adjusting certain components of the income statement for the effect of price changes upon major categories of cost, specifically by replacement-price depreciation, LIFO valuation of inventories, and repricing of liquid assets.
3. Adjusting each bookkeeping journal-entry for price changes by converting it to dollars of constant buying power and then classifying and combining the resulting deflated entries by conventional accounting procedures.
4. Restating in constant dollars each year the market value of all the corporation's assets taken together, as determined by the stockmarket; adjusting for outflows (dividends and capital transactions); and computing the year-to-year change in net inflation-adjusted market-value assets, which is an approximation to real *economic* earnings.
5. Restating by means of index numbers the accounted value of all the corporation's assets each year in dollars of constant purchasing power; adjusting for dividends, new capital, and other capital flows; and computing the year-to-year change in net adjusted assets, which is the year's real *accounted* profits.

The fourth method estimates the real economic earnings of a corporation by changes in the constant-dollar market value of its securities. This concept is somewhat different from the real (constant-dollar) profit, which is measured by the other four methods. This deflated-financial-market method is, I think, new. A brief description of this approach is therefore in order. Real profit is estimated by measuring year-to-year change in the market value of the stock. To the economist, a corporation's profit during a year is the net increase in the present value of the stream of future shareholder benefits (dividends, including a terminal payout). This is a measure of the corporation's earnings if, as is common, the corporation has no long-term debt. If the corporation has long-term debt, then it is a measure of the (inaccessible) gain to the stockholders. For clarity, we assume the absence of long-term debt.[1]

The price of the stock is a tangible approximation to this ethereal concept of shareholder wealth. Profit estimated by the change in the stock market's valuation of the company is measured in dollars of fluctuating purchasing power. If wealth, thus measured, had risen 10% during a year when the cost of living of shareholders had gone up 10%, then the real earnings of the shareholders would have been zero. To find real earnings in constant dollars, the aggregate stock price at the beginning and at the end of the year must be expressed in dollars of the same purchasing power. For this purpose, a stockholders' cost of living index is needed.

Stock prices, adjusted for consumers' cost of living, are sufficient for measuring real corporate income in the long run, provided that markets are adequately responsive. When the value of a corporation to its shareholders is measured by market price of the stock, we do not need to rectify the valuation of individual corporate assets as we must if, instead, shareholder wealth is measured by the book value of the corporation's net worth. If we calculate earnings by book value, we must restate asset accounts in constant dollars by means of a battery of different price indexes, each appropriate to a particular category of assets. This individualized deflation is not needed when we use stock price as our measure because, given enough time and responsiveness, the stock market should do the job for us. It should do this for four reasons.

In the first place, faced with the prospect of significant inflation, investors should eventually realize that the profits reported by conventional accounts overstate the corporation's real profits, not only during a period of inflation but for many years after the price level has ceased to rise. This overstatement differs among companies. For many it is serious: real earnings are much smaller than book earnings. Second, investors should know that the power to pay any dividends, even variable-dollar dividends, is determined by the corporation's real earnings, not by its profits as measured by conventional accountancy. Third, investors will recognize that what matters to them is not the nominal dollar value of dividends and capital gains, but instead their purchasing power. Dollar dividends must be larger during inflation in order for their purchasing power to stay the same.

Fourth, investors possess a marvelous mechanism for expressing and enforcing the foregoing expectations. This mechanism is the market rate of capitalization of future dividends (i.e., the market cost of equity capital). It is essentially the same mechanism that investors have already used to express their expectations as to future inflation in the market capitalization rate of long-term bonds. To compensate for the future rate of inflation that investors anticipate, the rate of return (interest rate) which they require is raised enough to allow for inflation while still leaving an adequate real return. For example, if the expected inflation rate is 5% and the required real rate for that level of risk is 4%, the market capitalization rate gets pushed up to 9%.

The same mechanism can, and I think will, be employed by investors in common stocks to compensate for the inroads of anticipated inflation. It will compensate by raising the market rate of equity capitalization high enough to leave a total return in real-goods terms (dividends plus capital appreciation, in constant dollars) that will be competitive with the bonds. The total return on stocks will have to be high enough to include a premium for the higher risks of common stock compared with high-grade bonds, in addition to a similar premium to compensate for anticipated inflation.

The market capitalization mechanism is already in use. Investors apply it to bonds and they can be expected eventually to use it also for stocks, as they come to recognize that inflation causes reported profits to overstate a corporation's real earnings and its power to pay dividends in real-goods terms. The resulting change in the capitalization rate for stocks, as for bonds, will be determined by investors' expectations as to the rate of inflation in the future rather than by actual future inflation.

This market capitalization mechanism has the capability of reflecting the impact of inflation upon the corporation's future real dividends (and shareholders' capital gains). Hence, the change in the price of the stock, while directly measuring the change in shareholder wealth and thus the earnings of the corporation, will indirectly measure inflation's erosion of future real earnings. The book value computation of real earnings, we have seen, required restatement of the value of individual corporate assets by means of index numbers that reflect buying power of corporate dollars spent on that category of asset. This direct detailed rectification is performed indirectly by the securities market through change in investors' expectations of earnings and dividends, and change in the rate at which the market price capitalizes them.

As a consequence, when we use security prices for measuring a corporation's real earnings, the only price level correction we need to make explicitly is to restate the aggregate stock price at the beginning and the end of the year in dollars of the same consumer buying power. For this purpose, the BLS consumer price index is what we are stuck with. It is not quite suitable since it measures the cost of living of middle income urban workers who, as a group, are not important investors.

Method Used

In both the electronics producer (IBM) and machinery manufacturer studies I used the fifth method—repricing of all asset accounts—to estimate real accounted profits. This concept—buying-power net profits that accountants would have reported in the absence of inflation—is only approximated by the statistical estimates, since the analysis makes no attempt to redo the bookkeeping in constant dollars. Rather, it only doctors the asset-account results of jumbled-dollar bookkeeping. The doctoring is done with price indexes that imperfectly reproduce, by means of migrating averages, the diverse behavior of the individual prices they purport to represent.

How close this estimate comes to the concept of buying-power profits depends on how faithfully the index numbers represent actual price behavior. Faithfulness is determined partly by index-number technology but mostly by the uniformity of behavior of the prices summarized by the price index. Thus the quality of the indexes governs the precision of the real-profits estimate.

To reprice with maximum precision the wide spectrum of products purchased by a corporation like IBM would require a host of price indexes specially constructed at great expense. To try to do the job with a single ready-made index would sacrifice precision. Finding the best trade-off of cost against precision is part of the art of commercial measurement of real profits. The optimum compromise is different for each company. It depends on the budget and on the precision required. In the IBM study I used eleven price indexes, each specially constructed for the job.

III. Findings

As shown in the electronics producer study, IBM's real profits for the 16-year period 1939–54 were estimated as 77.4% of its reported profits ($273.8 million vs. $353.9 million) and 75.6% of its deflated book earnings. A fivefold growth of earnings during this period was reported ($10.4 million in 1939 to $53.7 million in 1954), whereas real earnings merely doubled ($17.9 million in 1939 to $34.8 million in 1954). IBM is famous for its niggardly payout of dividends, yet these, plus long-term interest in real terms, averaged 67% of real earnings for the 16-year period. IBM's average rate of return on net investment for the same period, measured by conventional accountancy, was 50% higher than its real rate of return (12.4% vs. 8.2%). The main source of overstatement of real earnings by reported earnings was understatement of depreciation, which accounted for 62% of the disparity over the entire 16-year period; 23% was caused by overstatement of inventory growth and 8% by decline in the buying power of net liquid assets.

Findings for the machinery manufacturer were similar. Reported profits fell from about $23 million in 1937 to $12 million in 1950, whereas real profits in constant dollars fell from about $21 million in 1937 to zero in 1950. During the period

1946–50 dividends were lower than reported profits, resulting in substantial plowback per books. Real plowback, however, was negative: in every year except one during the analysis period 1937–50, this company dipped into its real capital by paying out in dividends more than its real profits. Net investment, instead of growing (as stated by the books) from about $160 million in 1937 to $175 in 1950, declined in constant dollars from about $190 million in 1936 to $140 million in 1950. The real return on net real investment was substantially lower than the reported return and it dropped drastically during the period from about 12% in 1937 to zero in 1950.

IV. Usefulness of Findings

Real-profits measurements and forecasts have important practical uses for managerial decisions on products, promotion, pricing, dividends, and capital rationing and sourcing. Some of these uses are described for IBM in Study 9.

In addition, real accounted earnings can be of value for public policy. Real-earnings analysis shows that the real rate of taxation of corporate income is, during inflation, much higher for most corporations than the purported rate. Hence measurements of real earnings can add to the pressure on government to curb inflation by revealing this important hidden aspect of its devastating effect. Real earnings analysis also reveals inflation's impairment of real capital formation by showing the large disparity between real corporate savings and those reported by conventional accounts.

V. Criticisms and Limitations

The concept and methods used for measuring real accounted profits in these studies have been much criticized. Six of these criticisms are examined below.

1. Real-profit measurements are unnecessary because inflation is ephemeral: the inflation distortion will reverse and correct itself as prices drop back again.

This criticism is without merit. A return to preinflation price levels is fantastically improbable. Once the tether to gold is cut, democratic governments of industrialized nations do not seem to have the knowledge or the willingness to prevent inflation, to say nothing of reversing it. A return of prices to earlier levels would be politically intolerable. It would bring vast unemployment and much hardship to those who have gone in debt at recent price and income levels to purchase houses or businesses. Those who lose from inflation—including pensioners, bondholders, and consumers generally—have little political clout. Those who gain from inflation—including the politicians and bureaucrats who denounce it most roundly—have great political influence. Politicians gain insidiously because the real rate of taxation of both corporations and individuals rises inexorably during inflation, but this rise is invisible so that it is hard to man the barricades against it.

2. Adjustment of corporate profits for inflation is not needed because all prices go up at about the same rate, leaving profit margins unchanged.

This contention, although widely believed, is incorrect. Selling prices do not usually move up in conformity with prices or with correctly measured costs during inflation. This is partly because of government pricing restraints and partly because delusions about real costs and profits result from conventional accounting and cause delays in raising prices. Even if the same percentage margin in variable buying-power dollars were reported by the accounts, and the total dollars of reported profit rose with inflation, these profits would usually be inadequate for maintaining real accounted profits at preinflation levels as a return on real investment.

Orthodox accounting vigilantly tries to keep price-level revaluations from getting into the profit and loss account, either by ignoring them or by treating them as surplus adjustments. But during inflation such revaluations do find their way into the accounts—through the back door of turnover of assets. On the books these inflation "profits," which are the result of one-shot revaluations of assets, are treated as ordinary income and are indistinguishable from regular income during the life of the assets and before the time of replacement at higher price levels. They therefore help to maintain profit margins in variable dollars. Moreover, because taxable profits vastly overstate real profits during inflation, a 50% corporate income tax may result in the corporation giving to the government all the corporation's real earnings and possibly more. Thus, maintaining the variable-dollar book profit margin on sales, even though dollar sales are growing as a result of inflation, will not suffice to maintain real earnings per share or real return on investment. The situation is shown in Chart Four–1 from the testimony of Secretary of the Treasury William E. Simon before the House Ways and Means Committee, January 22, 1975.

3. No inflation adjustment is needed for large corporations because they are immune to the effects of inflation and to government efforts to control it.

According to this criticism, the top thousand U.S. corporations comprise a separate segment of the economy, which Professor John Kenneth Galbraith labels the "planning system." These top corporations are, he says, immune to inflation because they have the power to set prices without regard to market demand and to influence government activities so as to create for themselves an inflation shelter. It is only the "market system"—which is comprised of the other twelve million U.S. firms—that must bear the brunt of both inflation and price controls.[2]

This argument assumes that great size is tantamount to monopoly power. This is simply not so. The top thousand corporations do not have enough market power to raise prices at will so as to pass along the rising costs of inflation. Nor can they avoid the effects of inflation (or the strictures of government monetary, fiscal, and price-ceiling policy aimed at controlling it) by foisting off products that consumers don't want or that are not up to competitive standards. During inflation large as

Chart Four–1 Undistributed Profits of Nonfinancial Corporations as a Percent of GNP, 1946–1974

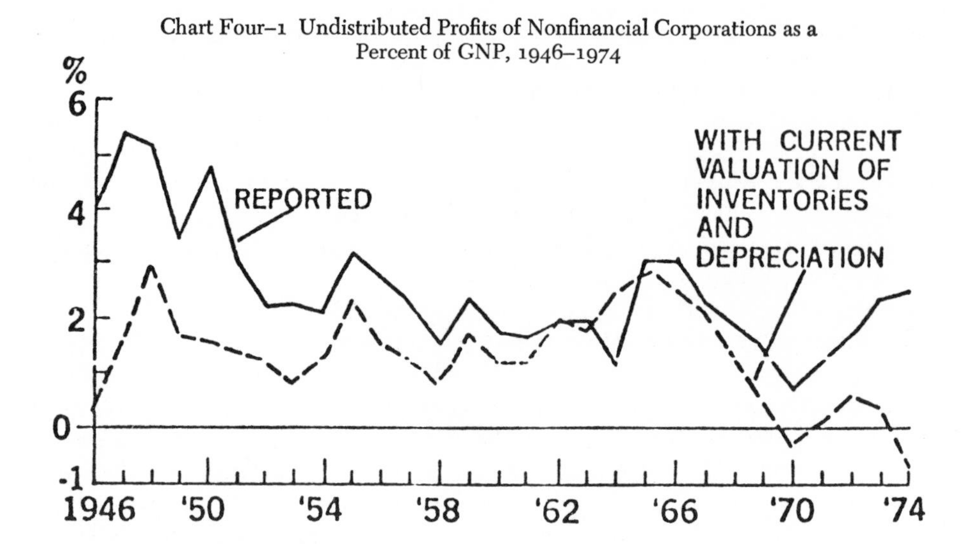

well as small companies face a sharp compression of real profit margins, squeezed by cowed or deluded competitors and by zealous government price controllers who mistake book profits for real profits. Moreover, because taxable profits overstate real profits during inflation, all corporations are drained of cash earnings by real rates of federal and state income taxes that are, as we have seen, much higher than the stated rates. Thus inflation causes the same shortfall of after-tax real earnings from reported earnings for a big corporation in the top thousand as for a small company.

4. Real-profits analysis fails to take into account advancing technology, which offsets part or all of the impairment of earnings caused by inflation.

This criticism asserts that technical progress is ignored in real-profits measurements because the repricing of asset accounts assumes that machines will be replaced in kind; but they will instead be replaced by more modern machines: repricing asset accounts by price indexes incorrectly assumes like-for-like replacement.

The benefits of advancing technology are not, it seems to me, ignored in these estimates of real profits, since these benefits have been embodied in the assets and in the resulting reported earnings. Replacement of equipment because of obsolescence occurs continuously throughout the analysis period. The fact that a machine was still there shows that it had not yet become obsolete—for that use in that plant. In the year in which it was repriced, that machine clearly was not obsolete and hence would not have been replaced by an advanced model. All machines that did become outmoded in that year would have been replaced. Only nonobsolete machines would have been repriced, i.e., valued at depreciated book value, adjusted for changes in purchasing power. The ideal way to revalue productive assets that are displaceable by technological advances is to deflate their used-machine market value as established in active markets. Lacking such markets, the revaluation method used here is probably the best practical approximation to that ideal.

5. Real-earnings measurements should be made by directly adjusting the profit and loss statement rather than by computing the annual increment in the price-adjusted book value of the corporation's net assets.

This is certainly one way to do it. Whether it is preferable is doubtful. To directly purge the corporation's income statement of the distortions caused by inflation is tricky. As noted earlier, accountants try to keep price-level revaluations from getting into the income account. During inflation, however, they do come in, through the process of turnover of assets (e.g., of inventory). These delusive gains from a rise in the price level are then counted as ordinary profits, even though they are caused by nonrecurring revaluations. And they cannot, during the life of the assets, be distinguished in the books from recurring income.

The basic method of measuring real profits, i.e., adjusting the asset accounts for inflation and then (after making allowances for capital transactions) finding the year-to-year change in their constant-dollar value is not in essence different from the accountant's method of getting net book profits. Basically, the accountant computes profits as the yearly change in the company's net investments, measured in dollars of fluctuating buying power. His income statement is only a supporting schedule which shows some of the transactions that were made to achieve the annual change in net investment. Within the limits of the precision of the price indexes, the asset-adjustment method measures the income that would have been reported by conventional accounting if transactions had been recorded in dollars of constant purchasing power, instead of dollars of varying power.

6. If the corporation were to report to stockholders its real profits per share, which are significantly lower than its book earnings, the price of the stock would be driven down.

This criticism has great political force, but little intellectual merit. It is voiced frequently by executives because they fear that public reporting of a corporation's real profits during inflation will lower not only measured profit per share but also the market's capitalization of earnings (i.e., reduce the stock's price-earnings multiple). Publication of real profits will undermine the projectability of any past growth gradient of book earnings per share and raise doubts about the future growth of real profits per share.

Therefore, according to these critics, measurement of real accounted profits should not be initiated by an individual company. Instead it must be compelled by the auditing profession or by the SEC, so that all companies will be on the same footing as to market capitalization rate. Otherwise Gresham's law will operate: bad accounting will drive out the good.

This objection underrates the intelligence of investors. The drastic drop in stock prices during 1973 and 1974 was caused primarily by investors' intuitive recognition (1) that the corporate profits that were reported and certified by CPAs were simply not believable and (2) that the rate of capitalization of real profits per share and of dividends would have to become competitive with the inflation-anticipating yield on high-grade bonds. Ultimately the economic reality will win out. Conceivably, corporations that take the lead in restoring their credibility will gain more than they lose in the market's rate of capitalization of stockholder benefits.

To have real-profit reporting required by the SEC or by CPAs would be very nice. However, enlightened executives and sophisticated investors cannot wait for this millennium. Executives (and investors, too) are already using a variety of home-made inflation adjusters in an effort to make the correct economic decisions despite bad bookkeeping.

The statistical estimation of real profits exemplified by the studies in Part Four does have some significant limitations. These are discussed below.

1. The inflation adjustment of *accounted* earnings is sometimes inadequate because differences between the book values of a corporation's assets and their economic value are so great that no amount of repricing of balance-sheet assets to correct for inflation will accurately reflect the market value of the corporation and hence reveal what is happening to its real *economic* earnings.

This is a serious limitation. These studies do not measure economic earnings. They measure real profits only in the accounted sense of what the books would have reported had the dollar's buying power stayed the same. They do not incorporate economic concepts of earnings and of valuation and they fail to reflect other changes (e.g., in the market rate of capitalization) that affect the value of the corporation's stock. In this sense they measure only real *accounted* earnings, not real *economic* earnings. However, the goal of our analysis was a limited and modest one: to correct accounted profits for inflation alone, not for any other omissions or mismeasurements.

For some corporations, the disparity between real accounted profits and real economic earnings is great. The book value of IBM, both at the beginning and at the end of the analysis period, greatly understated the economic worth of its assets as measured by the market value that the securities market placed on the company as a whole. Year-to-year change in this market value, adjusted for inflation, is, as mentioned earlier, one way to measure real *economic* earnings. These were quite different from real accounted profits estimated in this study.

2. Stockholders gain from corporate debt during inflation and this gain should be added to the corporation's real earnings.

This is true, but the potential gain is limited in three ways. First, the stockholder can gain from inflation only if the rate of inflation turns out to be greater than the premium that was embodied in the market rate of interest on the corporation's bonds to allow for anticipated inflation. In other words, the stockholder reaps benefits from this gain (which is caused by the lessened economic burden of paying interest and of repaying principal in shrunken dollars) only to the degree that inflation is faster than expected, so that the interest rate is not high enough to compensate.

Second, in order for the stockholder to gain from inflation, the corporation must be a debtor on balance—after offsetting its creditor activities, both formal and implicit (such as long-term leasing equipment at fixed rentals). Because long-term fixed-rental leases have the economic effect of lending money to customers, they (together with other lending) counterbalance borrowings in determining whether the corporation is a *net* debtor.

For the machinery manufacturer, lending activities more than offset the company's direct borrowing. Long-term leases that were equivalent to loans to customers were greater than the corporation's direct debt, making it a net creditor. Consequently, the impairment of stockholders' earnings by inflation was in this case not overstated but instead understated, because of the omission of an additional loss caused by indirect loans to customers in the form of long-term leases at fixed rates.

For the electronics producer, the situation is clouded by uncertainty as to whether its leases are de facto long-lived and fixed-price and therefore are equivalent to long-term loans. IBM has no long-term fixed-rental lease *contracts*. All leases can be canceled in ninety days and the company is then legally free to raise the rent. However, IBM has never raised the lease charges on an installed machine. Hence rentals have been fixed de facto (if not de jure) in the face of inflation. Although the leases are not long term by contract, they tend to be so in fact. IBM computers, once embedded in the customer's information system, stay there for long periods, usually until they are replaced by a more advanced piece of IBM equipment. It may be argued cogently that in practice these computer leases have the economic traits of long-term, fixed-rental contracts and are therefore equivalent to term loans at fixed interest. If so, they should be netted against the corporation's direct debt to calculate its net debtor status for purposes of determining whether there is a stockholders' gain from net corporate debt.

Third, the gain from corporate debt during inflation is relatively inaccessible to stockholders. As is true of some other economic gains of the corporation, shareholders cannot cash this benefit except to the extent (a) that the stock market values the potential gain of stockholders correctly and/or (b) that it results in bigger dividends.

My studies first measure the real accounted profit of a corporation as an economic entity, without regard to how it is apportioned between stockholders and bond holders. Inflation can give shareholders an economic gain at the expense of bond holders as a result of the corporation being in debt. This debtor-gain should be added to that portion of the corporation's real profits attributable to stockholders. Given the foregoing limitations, however, measurement of this gain is difficult and forecasting it is inevitably inexact. To estimate and forecast the resulting real earnings of stockholders from the real profits of their corporation usually requires a separate study. In the case of IBM, such a study would be particularly complex because of the debatable impact of computer leases upon stockholders' gains from corporate debt.

3. The value of computers and other equipment leased to customers at fixed rental rates does not rise with inflation because their rental earnings cannot be increased; therefore, this equipment, which constitutes a high proportion of the assets of each of the two companies studied, should not be repriced in the same way as assets that are used in production.

This is a serious limitation. It was more relentless for "Machinery Manufacturer" (whose leases ran contractually for ten years at fixed rentals) than for IBM, who was *legally* free to raise rents on 90-day notice. Fundamentally, this limitation arises from deficiencies of the basic method of valuation employed by accountancy. Value is measured by the cost of the asset, not by its earnings value, either as a production tool in an IBM factory or as a rental-earning asset in a customer's office. Thus the trouble is not with our methods of removing the effects of inflation from the values of book assets that are stated in dollars of differing purchasing power. Instead, the difficulty is with the concepts themselves, i.e., the original accounting valuations: cost-value rather than earnings-value. To remove this limitation would require a change in the concept of valuation embodied in the accounts. This is beyond the scope of a statistical real profits analysis, which only attempts to measure what profits the accounts would have reported had the dollar remained constant in purchasing power. This limitation goes beyond such measurements by raising more basic issues of what value to deflate.

Original cost (less accumulated book depreciation) is an especially poor measure of value for rent-earning equipment during inflation. A lease that is long (whether by contract or de facto) and that has rentals that are fixed (whether by contract or by IBM policy) prevents inflation-geared hikes in the price charged for the services of the equipment. Such a lease is the economic equivalent of a term loan to the customer at an interest rate which does not embody a premium for anticipated inflation. Hence it causes the lessor (e.g., IBM) an economic loss, invisible on the books but nevertheless real. Reflecting the resulting real loss, the present value of the future rental income will shrink during inflation because the rate of capitalization rises as market interest rates rise to incorporate a premium for anticipated inflation. Offsetting this to some degree, the market value of equipment at the end of the lease is likely to rise with inflation unless the equipment is obsolete. This is apt to be the case with IBM equipment, the lease-life of which is determined, not by contract, but by obsolescence. Even when the lease-life is contractual and the term is long—e.g., ten years in the case of the machinery manufacturer—the offset is small because the present value of this distant disposal revenue is discounted down by high inflation interest rates.[3]

4. The reported profits of a corporation like IBM understate its economic profits, inflation aside, because of the conservatism of accounting conventions; this understatement offsets overstatements caused by inflation.

This understatement is quite different from, not related to, and unlikely to offset the inflation-overstatement of accounted profits. "Conservatism" can and often does cause accounted profit to understate economic profit in a period of unchanging price level as a result of the following practices: (1) expensing of research, product development, and product launching investments; (2) expensing of investments in the training and developing of salesmen and executives as if no lasting benefits resulted; (3) immediate writing off of outlays for market development, customer education, and institutional promotion, even though these constitute, in part, true investment; (4) refusal to bookkeep appreciation in the earning value or the market value of an asset until such appreciation is "realized" by its sale.

The amount of this overstatement is not related to the rate of inflation; it would be the same if there was no inflation. The magnitude of the overstatement is, therefore, unlikely to be the same as the spread between real and reported profit.

Inflation causes accounted earnings to overstate real profit of the same content, i.e., measured by the same accounting rules, the only difference being change in the purchasing power of the dollar. Hence keeping the dollar yardstick constant can correct only partially the inadequacies of book profit.

Economic profit is more comprehensive than conventional accounted profit. Economic profit of a corporation for a year is the difference between (a) the present worth at the beginning of the year of the future stream of its real economic earnings and (b) this present worth at the end of the year. This concept is sometimes approximated by the year-to-year change in the stockmarket value of the company. Usually the approximation is poor because the market capitalization rate changes by amounts that exceed the correctly appraised change in risk. Year-to-year change in book value (book profit) is not distorted by shifts in capitalization rate, but other valuation errors and omissions make book profits inherently inferior intellectually.

5. Real profit measured in these studies still overstates true real profit because the book value of land, unlike other assets, was not restated to reflect the effects of inflation.

This criticism is correct. Failure to restate in constant dollars the book value of land is a limitation on the accuracy of these studies and does result in overstatement of true constant-dollar accounted profits by the measured real profit. Clearly the market value of land had gone up. Hence continuing to value it at original cost for the purpose of measuring real profit was an error. To have avoided this mistake, however, would have required creation of a specially tailored index of the price behavior of many scattered parcels of land. The price of each parcel, dominantly determined by its locational peculiarities, was not reported each year. Hence market values would have had to have been estimated by real estate specialists in each locality. To have repriced land correctly was therefore out of the question as a practical matter. To have repriced it *approximately* by using some nationwide average of land prices was a possibility that we rejected. The error of inapplicability of such an index might have been as great as the error of foregoing revaluation altogether.

Study 8

ELECTRONICS PRODUCER

I. Summary

A. Purpose

The purpose of this study is to measure real profits of the International Business Machines Corporation for the years 1939–54 by eliminating the effects of the spectacular inflation in prices and wages that made reported net earnings overstate IBM's real earnings in every year.

Inflation distorts orthodox reports of earnings in three ways. First, during and after a rise in prices, current depreciation charges are less than adequate to meet current replacement costs. Second, as prices go up, the purchasing power of a dollar of working capital goes down so that more dollars of working capital are needed just to keep the company operating on the same scale as before. Third, inventories purchased at lower prices during prior periods are charged to expense during the current year as the goods are used up. This cost understates the true cost of replacing inventories, thus revaluation and physical replacement are falsely reported as profits.

B. Practical Uses of Real Earnings

Estimates of IBM's real economic earnings have important practical uses for (1) dividend policy, (2) pricing policy, and (3) public policy.

1. DIVIDEND POLICY

Measurement of real profits can provide management with estimates of the amount of dividends that can reasonably be paid without impairing IBM's real assets—that is, without paying out dividends that were not earned in terms of real goods. Also, real earnings estimates can explain to stockholders the reasons for what may seem to them excessive retention of earnings.

This study, made for the International Business Machines Corporation in 1955 and originally titled "IBM's Real Economic Earnings, 1939–1954," is published here for the first time with the generous permission of Mr. F. T. Cary, Chairman of the Board of Directors of IBM.

2. PRICING POLICY

Rental and sale prices of equipment set on the basis of accounted costs may be inadequate to provide for the long-run maintenance of real investment. Within the limits imposed by competition, the knowledge of IBM's real rate of return may encourage management to set rentals and prices at levels that will produce an adequate return on real investment. Insofar as costs and earnings have a bearing on setting prices and rentals, they should be real costs and real earnings, not book costs and earnings measured conventionally in a hodgepodge of dollars that differ in purchasing power.

3. PUBLIC POLICY

For public policy concerning the market power of a corporation, real earnings rather than book earnings are pertinent. The reasonableness of the prices of a corporation that has market power could be tested in terms of the minimum earnings required (1) to survive as a viable supplier, (2) to obtain the capital needed to serve a fast-growing demand for its products, and (3) to do so without inequity to its present suppliers of capital. The performance of each of these functions requires real earnings. For survival, for the internal investments needed to take care of burgeoning customer needs, and for fairness to suppliers of the needed capital as indicated by a competitive level of return, buying power alone matters. The minimums must be measured in terms of real earnings and real rate of return on investment rather than book earnings and book return. Consequently, objective tests of the reasonableness of earnings and prices and of the existence or absence of monopoly profits should, particularly during a period of drastic change in price levels, be in real terms instead of in the conventional book profits of accountancy.

C. Methods

There are a number of ways to remove price-level distortions in reported income. The simplest and best method, in our judgment, is to measure real profits by restating the annual balance-sheet values in dollars of constant purchasing power, using a series of special-purpose index numbers to do the job. The year-to-year changes in net investment, after adjusting for dividends and new security flotations and/or security retirements during the year, represent asset growth due to the year's real earnings. The same approach, when applied to obtain book earnings, yields a profits figure identical in concept to that reported in the company's financial statements.

The distinction between current-dollar wages and real wages has long been recognized. A similar distinction may be made between current-dollar reported earnings of a corporation and its real earnings, although the method for determining real earnings of a corporation is more complicated than the single index number deflation used to translate current-dollar wages into real wages.

Both reported earnings and real earnings can be stated either in current dollars or in dollars of constant purchasing power. Either way, during inflation real earnings will be lower than book earnings stated in comparable-value dollars. Thus the difference between book earnings and real earnings is not just a matter of the unit value in which they are expressed. Rather it reflects the difference between conventional original-cost bookkeeping and accounts that state all costs and revenues in terms of a single price level.

Behavior of real profits is seen more easily when dollars of the same purchasing power are used for all years. In this report, IBM's annual real earnings for the period 1939–54 are stated in terms of a single constant price level: the average level for the years 1947–49, a widely used base period for price indexes. In order to make valid comparisons, book earnings for each year are also restated in terms of 1947–49 price levels.

The merits of traditional accounting procedures are not at issue here. Conventional bookkeeping is basic to the measurement of corporate earnings. Concepts of conventional book earnings have been integrated into the legal and tax structures of the economy. The measurement of real earnings in this analysis conforms to the rules of orthodox accountancy insofar as they govern the distribution of current expenditures to the balance sheet or to the income statement. Similarly, depreciation rates and account classifications in the conventional accounts have been carried over intact to the real accounts. The only orthodox accounting postulate that is called into question by the analysis is the assumed constancy in the value of a dollar.

D. Findings

Net investment is defined as the total investment commitment of long-term investors—that is, total assets less current liabilities. Earnings are stated on a comparable basis, as the earnings available for payment to long-term investors, both bond holders and stockholders—that is, after-tax earnings available for payment of dividends and interest on long-term debt. Rate of return is simply the ratio of earnings to net investment, both defined in this way.

IBM's capital investment has been provided partly by IBM's stockholders and partly by long-term creditors. Funds from these two sources have no separate identity once the investment commitment is made; a realistic measure of company investment cuts across the distinctions of ownership claims and concentrates on companywide investment. Accordingly, the real earnings measured are those of IBM as an economic entity, before payment of interest to bond holders or of dividends to stockholders.

1. EARNINGS, BOOK VS. REAL

IBM's real earnings have been substantially lower than reported earnings for the period 1939–54. Real profits during these years, stated in terms of dollars of

1947–49 purchasing power, totaled only $273.8 million, as compared with reported earnings (also in terms of 1947–49 dollars) of $361.8 million.

Chart 8–1 compares the course of IBM's real earnings, current-dollar reported earnings, and reported earnings as restated in 1947–49 dollars by means of a single index of the purchasing power of an IBM dollar. Book earnings since 1946 have been about four times their prewar level. When deflated like a man's wages, i.e., by a single index of the purchasing power of the IBM dollar, these reported earnings in the postwar period have been less than double prewar earnings. When earnings were measured in real terms by methods that reflect the full impact of inflation upon IBM's costs, these real profits increased very little since the prewar years, although they moved up sharply in 1954. The gap between deflated book earnings and real earnings grew steadily until 1948 and has been about $9 million (1947–49) since then, except in 1949, when there was a slight decline in some price levels.

2. DIVIDEND AND INTEREST PAYOUT, BOOK VS. REAL

IBM's dividends and long-term interest payments totaled $174 million for the 16-year period 1939–54, or 49% of available book earnings. In 1954 these payments reached $22.8 million, four times their 1939 level; yet they represented less than twice as much purchasing power as they did in the prewar period.

Real earnings available for dividend and interest payments can be considered as the amounts that could be paid out by the corporation without impairing the real value of the assets underlying investors' equity in the company. Between 1939 and 1954 real payout in constant-dollar terms averaged 67% of real earnings, compared with 50% of reported earnings for the period. In the postwar years the real payout averaged 62%, reaching a high of 80% in 1951. Chart 8–2 contrasts annual dividend and interest payments with real earnings available for such payments. Values are stated in dollars of average 1947–49 purchasing power.

3. NET INVESTMENT, BOOK VS. REAL

Chart 8–3 compares three measurements of IBM's total net investment (total assets less short-term liabilities) for the period 1939–54. The dashed line indicates the amount shown on the books, which represents a combination of cash and near-cash assets and liabilities measuring purchasing power directly in current prices, inventories of goods valued near current prices, and fixed assets and long-term investments valued at many different price levels of past years. The total amount grew about sixfold from the end of 1938 to the end of 1954.

The thin solid line in the chart shows book net investment deflated to dollars of 1947–49 value. Like current-dollar investment, this series still includes some effects of a jumble of different price levels. Only the total has been deflated: the price differentials that underlie the total remain. This series had a twofold increase during the period 1939–54.

The thick solid line in Chart 8–3 shows real net investment in constant dollars,

Chart 8–1 Earnings: Book vs. Real

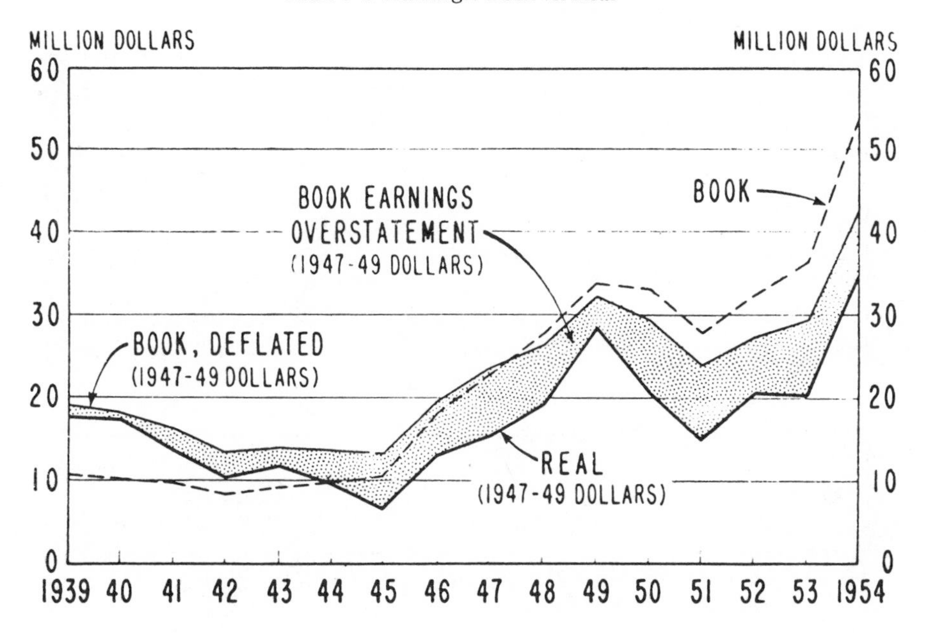

YEAR	REPORTED BOOK EARNINGS	DEFLATED BOOK EARNINGS *	REAL EARNINGS *
	(in millions of dollars)		
1939	10.4	19.1	17.9
1940	10.0	18.2	17.5
1941	9.7	16.3	13.9
1942	8.4	13.4	10.1
1943	9.1	14.0	11.6
1944	9.6	13.5	9.6
1945	10.5 ≁	13.2 ≁	6.9 ≁
1946	18.1	19.8	12.9
1947	22.9	23.3	15.2
1948	27.7	26.4	19.1
1949	33.8	32.1	28.4
1950	33.2	29.7	20.4
1951	28.0	24.0	15.1
1952	32.4	27.1	20.3
1953	36.4	29.5	20.1
1954	53.7	42.6	34.8

Both book and real earnings represent earnings available for dividends and interest on long-term debt.

* Stated in dollars of 1947-49 average purchasing power.

≁ Reflects accelerated amortization of emergency war facilities.

Chart 8–2 Real Earnings vs. Dividends and Interest Paid
(Dollars of 1947–1949 Value)

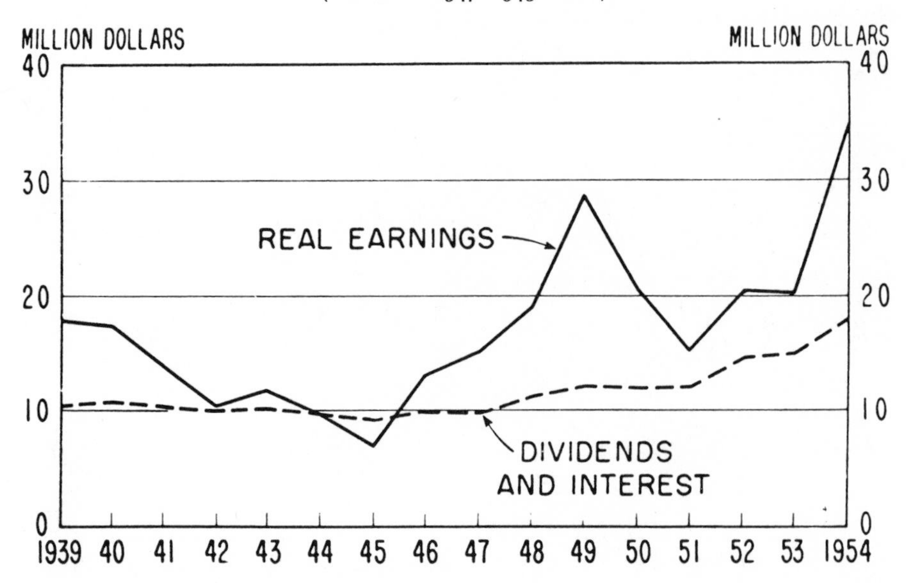

YEAR	DIVIDENDS AND INTEREST	DIVIDENDS & INTEREST (1947-49 DOLLARS)	REAL EARNINGS (1947-49 DOLLARS)
	(in millions of dollars)		
1939	5.6	10.2	17.9
1940	5.9	10.6	17.5
1941	6.0	10.1	13.9
1942	6.3	10.0	10.1
1943	6.6	10.1	11.6
1944	6.9	9.7	9.6
1945	7.2	9.1	6.9
1946	9.1	9.9	12.9
1947	9.7	9.9	15.2
1948	11.7	11.2	19.1
1949	12.8	12.2	28.4
1950	13.3	11.9	20.4
1951	14.1	12.1	15.1
1952	17.4	14.6	20.3
1953	18.5	14.9	20.1
1954	22.8	18.0	34.8

Chart 8–3 Net Investment: Book vs. Real

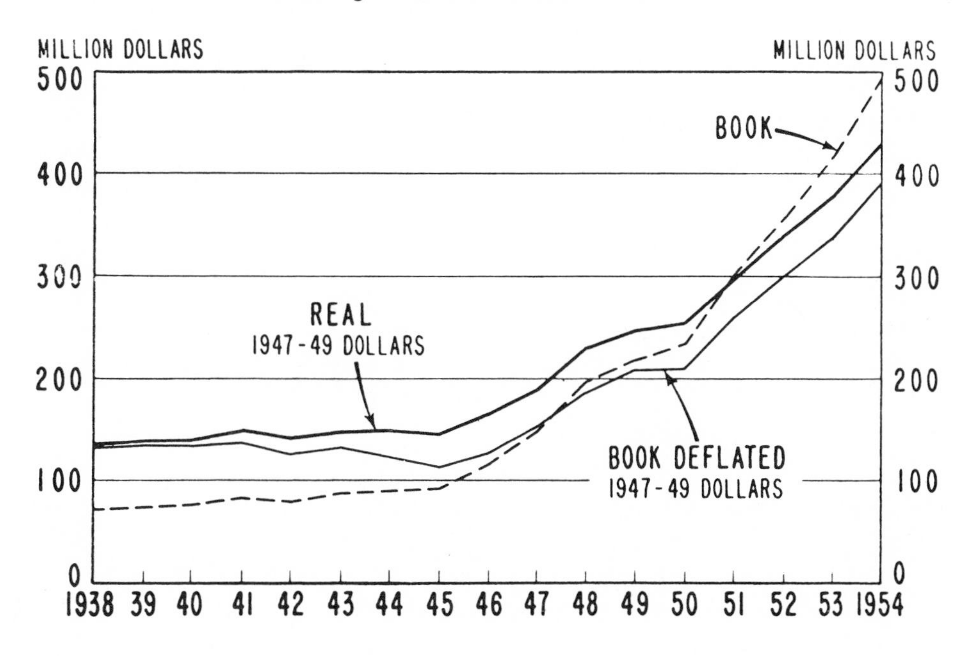

YEAR	BOOK INVESTMENT	DEFLATED BOOK INVESTMENT	REAL INVESTMENT
	(in millions of dollars)		
1938	71.0	130.9	132.4
1939	72.5	132.8	134.1
1940	72.9	132.8	134.1
1941	80.1	137.7	147.1
1942	79.0	125.5	139.0
1943	85.1	130.5	·145.9
1944	88.6	124.8	147.0
1945	90.6	113.6	143.1
1946	114.7	125.6	162.6
1947	147.1	149.5	187.4
1948	196.9	187.9	227.6
1949	217.8	207.5	243.8
1950	236.7	211.9	251.4
1951	303.2	259.8	299.5
1952	358.1	299.2	338.6
1953	416.1	337.2	376.1
1954	494.2	391.3	430.2

Net investment is the sum of net worth and long-term debt. Deflated book investment and real investment are stated in dollars of 1947-49 average purchasing power.

where every component of every asset account is stated in terms of 1947–49 price levels. Real net investment grows less rapidly than current-dollar book net investment because it shows the growth in purchasing power in the current-asset accounts, rather than dollar growth. It grows more rapidly than constant-dollar book net investment because it values all fixed assets at the same price levels as current assets, rather than in prices of many earlier years. Over the period it increased about two-and-a-half times (in contrast to the sixfold increase of book investment).

4. RATE OF RETURN, BOOK VS. REAL

In order to test the reasonableness of earnings, it is necessary to relate them to a measure of the capital employed to produce the earnings. Chart 8–4 shows IBM's annual book and real rate of return on net investment for the years 1939–54. Over the entire period, book rate of return was about 50% higher than real rate of return.

5. DEPRECIATION, BOOK VS. REAL

Chart 8–5 presents details on the most important source of earnings overstatement for IBM during this period, underdepreciation of fixed assets. It is a comparison of original-cost depreciation as recorded on the books with replacement-cost depreciation as computed for the real-earnings measurement. Both series are stated in constant dollars of 1947–49 value. If, instead, depreciation is stated in current dollars, the two series grow much more rapidly, and the gap between them increases over the period at a correspondingly greater rate than that shown in Chart 8–5.

For the entire 16-year period, underdepreciation (shown on the chart by the spread betwen the two lines) accounted for approximately 62% of total earnings overstatement. Of this total overstatement, underdepreciation of rental machinery accounted for 48% and plant and equipment were responsible for 14%. In 1951–54 underdepreciation amounted to about 67% of the total earnings overstatement, with rental machinery underdepreciation making up 19% and plant and equipment depreciation 48%.

Underdepreciation of rental machinery grew until 1951 but remained relatively stable after that, averaging about $4 million in 1947–49 dollars (or about $5 million in 1954 value). Plant and equipment underdepreciation, on the other hand, has grown steadily since 1943 and is still increasing; it is currently (1954) about $1.7 million in 1947–49 dollars (or $2.1 million in 1954 dollars).

The findings for the entire period 1939–54 and for the nine postwar years 1946–54 are summarized in Table 8–1.

II. Disparity Between Book Earnings and Real Earnings

As noted earlier, for a business enterprise with invested capital, the measurement of real earnings poses an entirely different problem from the measurement

Chart 8–4 Rate-of-Return: Book vs. Real

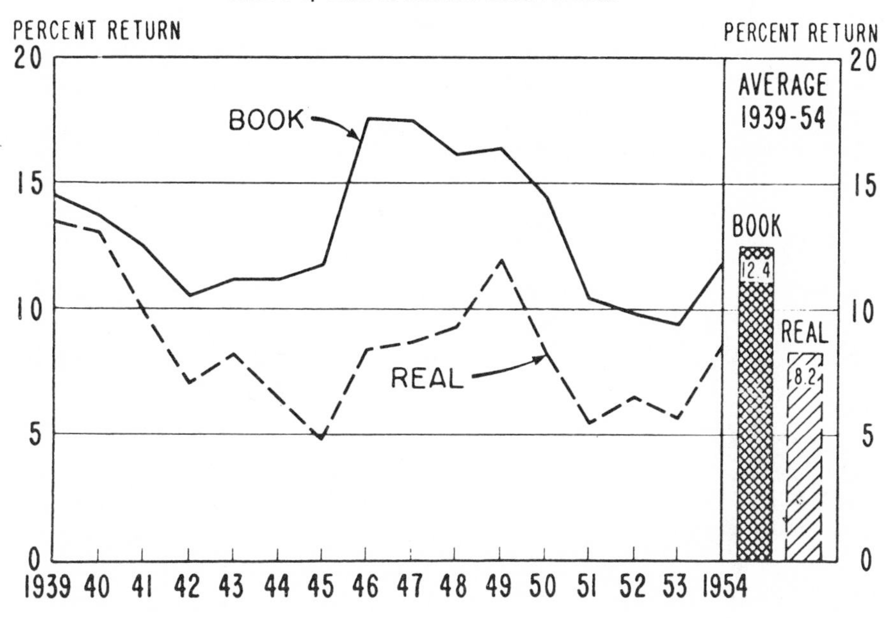

YEAR	BOOK RATE OF RETURN	REAL RATE OF RETURN
	(in millions of dollars)	
1939	14.5	13.5
1940	13.7	13.1
1941	12.5	9.9
1942	10.5	7.1
1943	11.1	8.2
1944	11.1	6.5
1945	11.7	4.8
1946	17.6	8.4
1947	17.5	8.7
1948	16.1	9.2
1949	16.3	12.0
1950	14.6	8.3
1951	10.4	5.5
1952	9.8	6.4
1953	9.4	5.6
1954	11.8	8.6

Chart 8–5 Depreciation: Book vs. Real (1947–1949 Dollars)

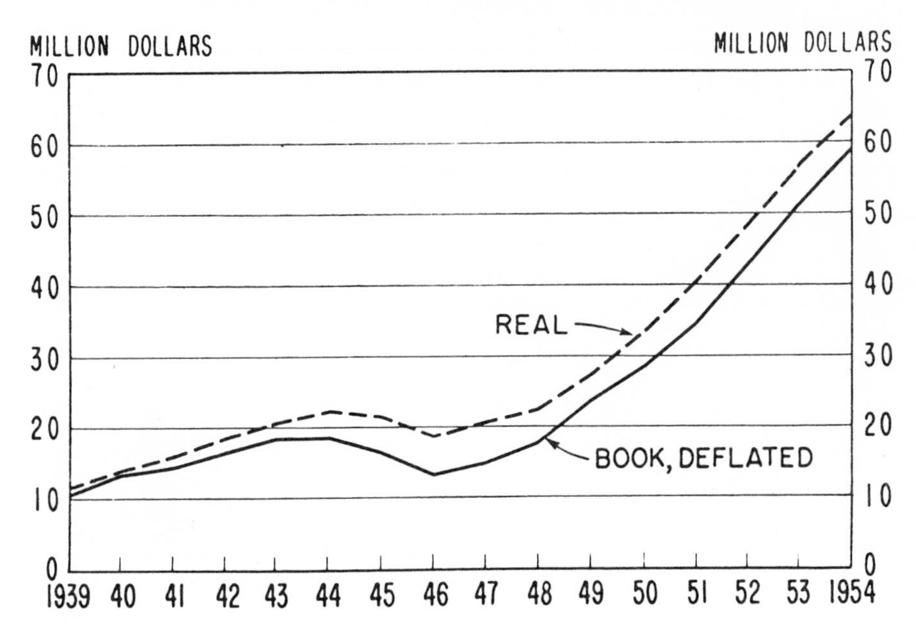

YEAR	DEFLATED BOOK DEPRECIATION	REAL DEPRECIATION
	(in millions of 1947-49 dollars)	
1939	10.66	11.22
1940	13.22	13.66
1941	14.37	15.91
1942	16.25	18.49
1943	18.08	20.34
1944	18.22	22.04
1945	16.39	21.43
1946	13.05	18.62
1947	14.56	20.50
1948	17.56	22.47
1949	23.39	27.34
1950	28.64	33.66
1951	34.71	40.71
1952	42.87	48.22
1953	51.09	56.52
1954	59.06	63.95

Table 8–1

SUMMARY OF IBM'S BOOK EARNINGS VS. REAL EARNINGS
1939–1954 AND 1946–1954

Period	Average Earnings (Millions)	Average Investment (Millions)	Rate of Return
1939–1954			
Book	$22.1	$177.7	12.4%
Book, deflated*	22.6	183.6	12.4
Real*	17.1	209.9	8.2
Inflation distortion	$ 5.5	−$ 26.0	4.2%
1946–1954			
Book	$31.8	$253.7	12.5%
Book, deflated*	28.3	225.7	12.5
Real*	20.7	263.7	7.8
Inflation distortion	$ 7.6	−$ 38.0	4.7%

*Stated in dollars of 1947–49 average purchasing power.

of real wages of labor. Unlike labor income, business income is not a simple dollar inflow that can be measured by adding up the dollars. Thus, before any determination can be made of whether to state earnings on a current-dollar or constant-dollar basis, business income must first be expressed in a way that reflects a uniform valuation basis for the individual asset accounts underlying the income estimate.

A. *Nature of Business Income*

For a firm operating with invested capital, business income in a given year is the net result of many receipt and payment transactions over that year and over preceding years as well. Some of the receipts are income, but other receipts are the result of asset liquidation or outside funds for new investment. Some of the payments are for operating costs, but other payments represent additions to assets or retirement of investment. From all these transactions net income for the year emerges as the growth in net assets plus payments of dividends and other reductions of investment minus receipts that increase liabilities or net worth accounts.

1. MISREPRESENTATIONS OF ORIGINAL-COST VALUATION

The crux of the income-measurement problem lies in estimating the net growth in assets. (No measure of assets is more than an estimate since assets can never be valued absolutely except upon liquidation of the firm.) Certain problems here are well recognized in established accounting procedures: valuation of inventories and receivables, measurement of the consumption of fixed assets over the years, etc. But with few exceptions—notably bad-debt reserves and use of market values of inventories when they are lower than book values—accountants have valued assets on the basis of their original cost, which they consider the most objective and conservative basis for measuring income.

But original-cost valuation, while factual, is no more "correct" a basis for income measurement than are several other valuation methods, for example replacement price or disposal value, each of which produces its own concept of income. Indeed, in a period of spectacular inflation such as has occurred since 1940, income computed on an original-cost basis can be dangerously incorrect for some decisions. It overstates seriously the amounts that a firm can pay out as dividends without dipping into its real operating assets and leaving it poorer at year-end than at the beginning of the year.

This overstatement of earnings during inflation comes about in the asset valuation process. It takes several forms:

a) *Liquid assets.* In original-cost valuation there is an overstatement of the growth of liquid assets (or an understatement of their decrease), since each dollar of such assets at the end of the year can buy fewer goods and services than it could at the beginning of the year. This is an unrecorded capital loss, felt by any firm that has to have some liquidity during inflation.

b) *Inventories.* Original-cost measurement overstates the net growth in inventories. This is because old, low-price charges to the account are written off during the year and are replaced by new, high-price charges. This overstatement is not a capital loss (to the extent that inventories maintain their market value relative to other goods). Instead it is the recording on the books of dollar growth that does not represent growth in the real value of inventories.

c) *Depreciation.* Original cost measurement undervalues the depreciation of fixed assets. In a period of continually rising replacement costs, any system of original-cost depreciation will be inadequate as a measure of the predepreciation cash earnings that must be retained by the firm in order to maintain the real value of its assets. Similarly, retirements charged against income at original-cost book value are far less than their book value in current prices. These are in economic reality neither capital gains nor losses. Instead they are inadequate measurements of operating costs for the year that happen to impinge on net asset values.

2. NATURE OF THE REAL-EARNINGS COMPUTATION

These misrepresentations in original-cost accounts can be corrected substantially by using price indexes to state all assets on hand at the beginning and end of the year in dollars of the same value and then remeasuring the change in net assets during the year. This is the procedure used in the real-earnings computation.

Two qualifications on the precision of this restatement of IBM's earnings should be noted. First, the scope of asset revaluation in the real-earnings computation is severely restricted. It represents no more than a restatement of the many price levels reflected in book accounts into a single, homogeneous price level. It does not venture into such areas as *ex post* revaluation of accounts to reflect expensed investments, unforeseen obsolescence, capital gains and losses, and other speculative revaluations, such as an appraiser or an investor would make, based on future earning power and some capitalization rate. Real profits thus do not purport to measure a concept of income different from book profits.

Second, price indexes are at best approximations, based on the representativeness of a sample of price data that is never completely appropriate for a company's operations. It is indeed not possible to define the index that is absolutely correct for a specific problem. We are faced, however, with the alternatives of (1) abandoning the indexes and starting anew from the appraiser's point of view or (2) relying on the original-cost data, some of which are stated in terms of prices less than half of current levels. Given this situation, the price indexes represent a simple and fairly efficient way to build on conventional accounts and at the same time derive information from them that is not apparent in them as they stand.

B. Sources of Earnings Overstatement, 1939–54

Original-cost accounting, as we have seen, overstates income in three ways during inflation: (1) it fails to reflect certain real capital losses; (2) it reflects nonexistent capital gains; and (3) it fails to measure current costs adequately. These various sources of earnings misstatement are separately measurable in the real-earnings computation, and an examination of them for the period 1939–54 can give some insight into future trends of IBM's real earnings relative to book earnings.

1. WORKING CAPITAL

The overstatements of growth in liquid assets and inventories can be grouped together, since both depend on the size of the account and the amount of change in prices during the year. Such overstatements are offset by a similar effect on current liabilities, where the net growth is also overstated. When these are combined and netted, the result may be called working-capital inflation.

Working-capital inflation is a combination of (a) losses in purchasing power of net cash and near-cash assets and (b) overstatements of the growth of physical inventories. It is the most volatile element of earnings overstatement, since its principal determinant is the annual change in prices, which can vary widely from one year to the next. This type of overstatement for IBM was about $3.2 million in 1948, after an 8% price rise, but was slightly negative in the following year, when prices related to IBM's working capital faltered in their upward course.

Over the entire period 1939–54 working-capital inflation accounted for about 31% of the total overstatement of earnings, of which 8% was in net quick assets and 23% was in inventory (including rental machine parts inventory). Between 1951 and 1954 working capital accounts produced 27% of the total.

It should be noted that for most manufacturing companies working-capital effects are the dominant source of income overstatement, generally accounting for about 75% of the total for this period. They do not dominate in IBM accounts because of the importance of IBM rental machinery assets.

2. PLANT AND EQUIPMENT

Total effects over the period arising from original-cost accounting for plant, manufacturing equipment, and furniture and fixtures amounted to 17.2% of the total difference between book and real earnings. Of this 13.9% was underdepreciation related to replacement price levels and 3.3% was the difference between (1) net book value of retirements charged against income and (2) their net value stated in current price levels.

Underdepreciation is a far more stable element of earnings overstatement than is working-capital inflation, since it depends not on one year's price movements but rather on the cumulation of price increases over the period for which the present fixed assets have been on the books. Thus for fixed assets acquired by IBM in 1938 the shortfall of the book depreciation charge in 1955 is determined by the entire increase in fixed-asset prices since 1938. Moreover, this shortfall will continue on the books and will continue to grow with cost increases, as long as depreciation is charged against the 1938 assets. Underdepreciation is therefore a persistent phenomenon that can be eliminated only by years of price stability or substantial and permanent declines of price levels. Neither of these conditions is likely to occur in the economic setting of our times.

Underdepreciation of IBM's plant and equipment has grown steadily through all years since 1939 and is currently (1955) running at about $2 million per year in 1955-value dollars. (The rise over the period is illustrated in Chart 8–5.)

3. RENTAL MACHINES

The accounting for rental machines gives rise to the same type of underdepreciation and understatement of retirement as in the accounting of IBM's production plant and equipment. However, because of the high depreciation rates used in the rental machine account, net underdepreciated book values are closer to current costs in rental machines than in plant and equipment, and depreciation charges are correspondingly closer to replacement-cost charges. Thus in 1950 book depreciation charges on rental machines were 87% of current-cost charges, whereas book charges on plant and equipment were 75% of current-cost charges. Moreover, the shorter economic-life estimates (i.e., higher depreciation rate) on rental machines means that if prices were to stabilize today the underdepreciation of rental equipment could be substantially eliminated in about five years, considerably less than would be needed for manufacturing facilities.

Nevertheless, earnings overstatement attributable to underdepreciation of rental machines comprised about half of IBM's total earnings overstatement for the years 1939–54. This is because rental machinery is the major asset account on IBM's books. Book depreciation charges for rental equipment since 1952 have been several times higher than plant and equipment depreciation charges. For this reason underdepreciation of IBM's rental machinery has been almost four times as great as for its production plant.

Underdepreciation of rental machinery was at its greatest in 1947, when it reached about $5 million, measured in 1947–49 dollars. Since then it has settled down to about $4 million, indicating a rough balance between write-off of older machines on the one hand and continuing increases in manufacturing costs on the other.The $4 million in 1947–49 dollars, it should be noted, is the equivalent of about $5 million in 1955 dollars.

The understatement of net value of retirements has been a relatively minor item in the rental machine account, contributing about 4% of total earnings understatement during 1939–54.

The findings on the sources of earnings overstatement for IBM are summarized in Table 8–2, which shows the relative importance of each source for the period as a whole and for recent years.

Table 8–2

SOURCES OF DISPARITY BETWEEN
IBM'S BOOK EARNINGS AND REAL EARNINGS
1939–1954 AND 1951–1954

	Percentage of Total Disparity	
Sources of Earnings Overstatement	1939–1954	1951–1954
Underdepreciation:		
Plant and equipment	13.9	18.5
Rental machinery	47.9	48.3
Understatement of retirements:		
Plant and equipment	3.3	3.4
Rental machinery	3.7	2.9
Decline in value of net liquid assets	8.2	9.7
Overstatement of inventory growth	23.0	17.2
Total	100.0	100.0

C. Forecast of Real Earnings, 1955–60

1. SUMMARY

IBM's reported earnings will seriously overstate its real earnings for many years to come, even if inflation stops tomorrow. To attempt to forecast a corporation's real earnings is perilous and unprecedented. The prediction is bound to be wrong. Nevertheless, the future alone matters for management. Therefore forecasts of real earnings are needed for decisions on rental rates, equivalent sale prices, investment projects, dividends, and plowback.

The first step in such a forecast is to estimate the disparity between book and real earnings available for interest and dividends. The findings of this analysis are summarized in Table 8–3.

Table 8–3

FORECAST OF OVERSTATEMENT OF IBM'S REAL EARNINGS BY ITS REPORTED EARNINGS, 1955–1960

(Earnings Available for Interest and Dividends)*

	1954	1955	1956	1957	1958	1959	1960
Total earnings overstatement	$9.7	$5.9	$5.0	$3.6	$2.7	$2.2	$2.0
Source of overstatement:							
Underdepreciation:							
Plant and equipment	$2.1	$2.1	$2.0	$1.9	$1.8	$1.7	$1.6
Rental machinery	4.1	3.2	2.4	1.2	.4	—	—
Understatement of retirements	.6	.6	.6	.5	.5	.5	.4
Loss in purchasing power of liquid assets	.8	—	—	—	—	—	—
Overstatement of inventory growth	1.5	—	—	—	—	—	—
Other	.6	—	—	—	—	—	—

*This analysis assumes as a benchmark for future forecasts that prices will stabilize at their 1954 levels, which is unlikely. So these are estimates of the *minimum* overstatement of IBM's real earnings.

To be most useful to management in the future, the forecast should be made in stages. The present analysis covers only the first stage, which is based upon two simplifying but unrealistic assumptions. One assumption is that the nature and scope of facilities additions, retirements, and depreciation will, in the future, be similar on average to those of the recent past. This assumption will, of course, prove incorrect. Nevertheless, adjustments can be made for departures from this assumed base to the extent that they can be more clearly foreseen by management in the future. The other assumption is that prices will stabilize at their 1954 levels. If they do, the disparity in current dollar terms between real earnings and book earnings will come from only two sources, underdepreciation and understatement of the value of retired assets.

The assumption of price stability will undoubtedly prove wrong. Wage rates and metals prices are more likely to rise than to fall. Rising prices will increase the disparity between book and real earnings. Nevertheless, as a first approximation, this unlikely assumption provides a sturdy benchmark for estimates of the *minimum* overstatement of IBM's real earnings by reported earnings.

The first step in forecasting IBM's real profits was to predict its book profits. This was done in two stages. The first was to compute the annual rate of growth of IBM's adjusted earnings per share, which was 9%, and mechanically project it forward. The second stage was to appraise and modify this mechanical forecast. Because the profits predicted by extrapolation of this rate of growth seemed to us unattainably high, the forecasted trend of adjusted earnings per share was arbitrarily reduced to 6%.

The second step was to build on this forecast of book profits by forecasting the probable future disparity between IBM's book profits and its real profits, based on the assumptions just mentioned. The forecasts of IBM's real profits thus obtained are summarized in Table 8–4.

Table 8–4

FORECAST OF IBM'S REAL EARNINGS 1955–1960
ASSUMING NO FURTHER INFLATION

(Earnings Available for Interest and Dividends)

	1954	1955	1956	1957	1958	1959	1960
Book net profit[ac]	$46.5	$52.3	$54.9	$57.6	$60.2	$62.8	$65.4
Interest	7.2	8.3	10.2	12.1	12.1	12.1	12.1
Total book earnings[b]	$53.7	$60.6	$65.1	$69.7	$72.3	$74.9	$77.5
Overstatement of real earnings[d] by book earnings	9.7	5.9	5.0	3.6	2.7	2.2	2.0
Real earnings[bc]	$44.0	$54.7	$60.1	$66.1	$69.6	$72.7	$75.5

[a] After tax but before interest, earnings available for dividends only.
[b] After tax earnings available for interest and dividends.
[c] Stated in constant dollars of 1955 purchasing power.
[d] Assumes that prices will stabilize at their 1954 levels.

2. REAL EARNINGS FOR 1955

If prices do in fact remain stable during 1955, reported earnings will probably overstate real economic earnings (in 1955 dollars) by approximately $5.9 million (Table 8–3). This means that if book earnings (after taxes but before payment of interest and dividends) total $60.6 million, real earnings will be approximately $54.7 million.

Of this $5.9-million overstatement, about $2.1 million represents underdepreciation of plant and equipment accounts and $3.2 million is underdepreciation of the rental machinery inventory. The remaining $60,000 would be accounted for by the understatement of the real value of net retirements of fixed assets.

The 1955 book earnings forecast was based on a comparison of the reported earnings for the first quarter of 1955 with those for the first quarter of 1954. Reported earnings for 1954 totaled $46,536,625, or 4.59 times the $10,134,000 earnings reported for the first quarter of 1954. In 1955, first quarter earnings totaled $11,402,000; multiplying this figure by 4.59 yields a reported profit projection of $52.3 million for 1955. Interest payments of $8.3 million were added to this figure to produce the estimated $60.6 million book earnings available for interest and dividends.

The postulate of price stability will not prove correct. If 1955 price levels average two percentage points higher than 1954 prices, book earnings will overstate real earnings (in 1955 dollars) by $8.5 million. Real earnings would therefore be only

$52.1 million. About $2.4 million of the earnings disparity would stem from underdepreciation of production facilities, $4.1 million would come from underdepreciation of rental machinery, and $1.0 million would result from the loss of purchasing power of net liquid assets. Real inventory values would decline by $400,000 and the net value of retired property would be understated on the books by approximately $600,000.

3. REAL EARNINGS PROJECTIONS, 1956–60

Real earnings (after tax) available for interest and dividends, stated in dollars of 1955 purchasing power, are expected to grow from $44 million in 1954 to $75.5 million in 1960, even in the unlikely event of no further inflation, as seen in Table 8–4. This forecast is based on the ultraconservative projection of a 6% rate of growth in book net profits, which is a third lower than the 9% rate averaged in our analysis period. This benchmark forecast is based on the analysis summarized in Table 8–3, which indicates that if prices cease to rise after 1954 the overstatement of real earnings by book earnings, almost $10 million in 1954, will fade away to only $2 million in 1960 as the stored-up consequences of past inflation tail off.

The future course of real earnings will depend on the underlying earning power of IBM and on the extent and timing of future inflation. It is much more likely the prices will go up than that they will remain constant. But were they to stay the same, almost all of the earnings overstatement would arise from underdepreciation of fixed assets; little of the earnings disparity would be due to realization of stored-up past losses in the purchasing power of working capital or declines in the real value of inventory.

Even if prices remain constant however, underdepreciation of fixed assets will be substantial in the next five years. The forecasts of real earnings for the years 1956–60 shown in Table 8–4 are based on this assumption.

Appendix A

Computation of Real Earnings

The real-income data summarized in this report were derived directly from IBM balance sheets rather than from income statements. This appendix describes the computation procedures used in deflating balance-sheet items and capital transactions in order to compute IBM's real earnings. The various price indexes used in the computation are described separately in Appendix B.

Like book earnings, real earnings are equal to (1) the growth in assets during the year, less (2) the growth in liabilities, less (3) new capital funds raised, and plus (4) earnings paid out in dividends and interest. The essential difference in the computation of book and real earnings arises from the way that growth of assets is measured. Book earnings measure asset changes from dollars of one value to dollars of another, whereas real earnings measure asset changes in dollars of constant value. Since the difference in treatment lies in valuation of asset changes, it is not necessary or convenient to develop a separate income statement for real earnings, since this would be only a summary of certain information bearing on the ways in which the asset changes occurred.

In brief, real earnings measurement consists of (1) restating the values of year-end assets and liabilities for a number of years in terms of dollars of constant purchasing power by the application of appropriate price indexes; (2) measuring annual changes in constant-dollar net investment; (3) adding dividends, interest on long-term debt, and disbursements for debt retirement, all stated in constant dollars; and (4) subtracting new long-term funds raised during each year.

A. *Basic Data*

Some definitions are needed at this point. The *basis of consolidation* for both the book and real earnings accounts in this study is the parent company. *Net investment,* or net assets, refers to the parent company's total assets less current liabilities; thus it includes, as a capital account, long-term debt and special reserves as well as net worth. *Net income,* correspondingly, refers to earnings before interest charges on long-term debt but after corporate income taxes, since these are current liabilities. It differs from consolidated net income after taxes as reported to the SEC mainly in that it excludes the retained earnings of subsidiaries. (For 1942 it also reflects a renegotiation settlement charged on the books against surplus in 1943.)

The financial data for this study were assembled by the staff of IBM's treasurer. Sources of the data for computation of special price indexes are described in Appendix B.

B. *Account Groupings*

Different types of assets require different kinds of price index and different treatments in the deflation process. The first step in the deflation procedure was therefore to group all assets (and current liabilities) into homogeneous categories. The groupings were as follows: (1) working capital (including parts inventory), (2) land, (3) plant and equipment, (4) rental machines, (5) investments, (6) patents and goodwill, and (7) capital transactions.

1. WORKING CAPITAL

Working capital at year-end, defined as current assets and deferred assets net of current liabilities, represents a combination of (1) cash and near cash, offset by liabilities soon to be paid with the cash, and (2) inventories valued at prices prevailing near the end of the year. The book value of working capital is therefore an accurate estimate of its current-dollar value. For example, at the end of 1953 working capital on the books was virtually equal to working capital stated in end-of-1953 prices. To state working capital in 1947–49 prices, therefore, it is necessary only to divide an appropriate index of end-of-1953 prices into book values. This procedure was followed for each year from 1938 to 1954.

Rental machine parts inventory at year-end was deflated to 1947–49 dollars by a similar process, but by a different index reflecting parts prices.

2. LAND

Land was not restated in constant dollars, but instead left at its book value. Because of the important influence of location on the market value of land, it was not possible to develop any meaningful index of changes in the dollar value of land. As a resort of desperation, therefore, the book value of land was treated as representative of its real value and no adjustment was made for the effects of inflation on land values.

3. PLANT AND EQUIPMENT

The deflation of fixed-asset accounts is a very different operation from working-capital deflation. Whereas working capital is accurately stated on the books in terms of current dollars, the fixed-asset accounts, having a slow turnover, reflect a number of different acquisition-cost price levels running back many years. In order to sort out the mixture of price levels, an incremental system was applied to plant and equipment deflation.

First, the end-of-1938 gross balances and depreciation reserves in the plant and equipment accounts were restated in 1947–49 dollars. This was done by means of a special price index reflecting the mixture of past prices underlying the 1938 accounts rather than 1938 prices themselves.

Next, plant account transactions during 1939 were restated in terms of 1947–49 dollars. Additions to the gross accounts were deflated by indexes of 1939 plant and equipment prices; depreciation charges were deflated by indexes reflecting the mixture of prices underlying them; and retirements from gross accounts and reserves were deflated by a set of indexes appropriate to them. Constant-dollar gross balances for the end of 1939 were then computed as 1938 deflated balances, plus deflated additions to the accounts, less deflated retirements from the accounts. The 1939 deflated depreciation reserves were similarly computed by adding deflated 1939 depreciation charges to deflated reserves as of the end of 1938 and deducting deflated retirements from the reserves.

The operation just described for 1939 was then carried out for each of the years through 1954. The operation was performed separately for each of four groups of assets: (1) buildings, (2) machinery, (3) office fixtures and furniture, and (4) emergency war facilities (buildings and machinery separately). Separate sets of price indexes were used for each group. Thus for each year, the deflated net fixed asset balances reflected in real net investment is the sum of the deflated gross account balances less the deflated depreciation reserves as of the end of the year.

The plant and equipment accounts were adjusted frequently during the years covered to reflect reclassification of assets from one account to another. These adjustments were recognized explicitly in the deflation, and each was restated by an index that appeared to be most appropriate to the particular adjustment. No single procedure was applicable to all adjustments.

4. RENTAL MACHINERY

The deflation procedure for rental machinery is similar in principle to the deflation of plant and equipment in that it is to a large extent an incremental computation. This account, too, was adjusted frequently to reflect reclassifications. The details of the procedure, however, differed in many respects. With data supplied by the company on various types of age distribution in the rental machinery account, we were able to apply more refined techniques in deflation of rental machinery than was possible in the deflation of plant and equipment.

For example, data were supplied for 1938 and four other years up to 1954 as to the gross value of machines on hand that had been acquired in the preceding year, the year before that, and so on for as many years back as were necessary to account for the entire gross value of the account. Similar age distribution data were available for depreciation reserves. A combination of the gross value and reserves data yielded an age distribution for the net value of the rental machine account in each of the five years. Thus, the net value in 1953 of machines acquired in 1950 could be deflated to 1947–49 value with the appropriate cost index for 1950; 1951 acquisitions could be deflated by the 1951 index; and so on for each year represented in the 1953 net account. The result was a benchmark estimate of the deflated net value of the 1953 account that reflected specifically the age composition of machines on hand at that time.

Such benchmark net values were computed from the company data for 1938, 1945, 1948, 1951, and 1953. (In the plant and equipment accounts, in comparison, for lack of data on age distributions, benchmark values could be computed only for 1938, using a national index of prices underlying the plant accounts.) Net values of the account for other years were computed incrementally—that is, by starting at a benchmark level, adding the deflated value of new machines acquired, and deducting the deflated values of depreciation charges and net retirements. The deflated net values calculated in this manner were then adjusted so that they would be consistent with the deflated net values computed for the later benchmark years from age distributions.

Age-distribution data were also used in the computation of price indexes for depreciation and net retirements in the incremental method. These indexes are described in Appendix B.

5. INVESTMENTS

The investments account was treated in the deflation as an original-cost account, i.e., as representing the capital funds actually paid to subsidiaries (or the value of goods transferred to them). IBM's long-term investments in outside companies, which were included in this group, were treated in the same way. The account was therefore deflated by an incremental process similar to that used on fixed-asset accounts. That is, the end-of-1938 level of the account was restated in 1947–49 dollar value, and succeeding year-end levels were determined by adding deflated increments to the account and deducting deflated decrements. The net changes were deflated by a general index of the value of cash in the company.

For this operation, increases in the account were considered to reflect expenditures from liquid assets, either directly or indirectly; decreases were taken to reflect receipts on subsidiaries' trade payables to the parent, liquidation of IBM's paid-in equity, or (possibly) increases in trade debt owed by the parent to subsidiaries. This treatment is a good approximation to reality except (*a*) when net changes in investments are transfers of old fixed assets at original-cost values and (*b*) when there are writeups and writedowns of the value of investments in subsidiaries. However, both of these types of investment transactions are offset in other accounts in the real-income computation. Transfers of fixed assets are offset by adjustments to the plant and equipment and rental machinery accounts of the parent; a small error is introduced in that in the investments account the transfer is deflated by an index of cash values, whereas in the plant accounts it is deflated by an index of price levels underlying plant accounts. Capital adjustments, such as writedowns of investment in foreign subsidiaries during the war, are offset by adjustments to the capital transactions group of accounts, which are discussed below. Since capital transactions are deflated by the same index as net changes in investments, these adjustments and their offsets are equal both on the books and when deflated.

Decreases in the investments account representing liquidation of subsidiaries reflect the return of only the paid-in capital invested by the parent. Earned surplus of such subsidiaries enters the real-earnings computation as an adjustment to the capital transactions account.

To summarize the deflation procedure: (*a*) the level of the investments account as of the end of 1938 was deflated to 1947–49 dollar value; (*b*) succeeding annual net changes in the account were deflated by a general index of cash value; and (*c*) the deflated net changes were added (or deducted if negative) from the deflated level of the account for the preceding year to get the current year's net deflated value. The starting level of the account, for the end of 1938, was deflated by the

1938 level of the index used to deflate increments to the account. This index was used on the assumption that assets represented in the account were valued at prices near 1938 price levels; specific information was not available.

6. PATENTS AND GOODWILL

Patents and goodwill, although different in asset nature from investments, were deflated by the same procedure and with the same index. The objective here, however, was to deflate amortization of the account so as to have equal effects on deflated book earnings and real earnings, thus neutralizing the account as a source of disparity between the two earnings estimates. This was accomplished by deflating net changes in the account by the cash-value index, which was the index used in computing deflated book earnings and also the index used in deflating net changes in the investments account.

7. CAPITAL TRANSACTIONS

The accounts described up to this point are those that make up net investment. The year-to-year change in the aggregate of these deflated asset accounts is the crude measurement of real profit. To refine it requires adjustment for capital transactions. These constitute another group of transactions, not included in the balance-sheet accounts, that also enters the real-profit computation. This group comprises payments of interest on long-term debt, dividends, debt retirements, and other capital distributions from assets, and receipts from new funds raised in the capital markets.

The capital transactions group also includes two types of adjustments that are needed in computing real income from changes in net investment: (*a*) offsets to capital adjustments that are reflected in net investment but not in earnings, and (*b*) items that enter the earnings computation but that are not reflected in net investment. An example of the first type is a writedown of the value of foreign investments during the war. The second type is illustrated by provisions for after-war adjustment charged against income and credited to a surplus reserve. Such reserves are included in net investment; hence charges against income to increase them are not reflected in changes in net investment. These transactions were all deflated by the cash-value index, which is a general index of the prices paid by IBM in current operations. This index was used because current operations are the principal application of new capital funds and the principal alternative use of funds paid out in dividends and debt retirement.

The cash-value index is also the index applied to investments, patents, and goodwill. Moreover, it is the index applied to book earnings in computing deflated book earnings. From the nature of the deflation procedure applied to these accounts, it follows that capital transactions and net changes in investments and patents and goodwill are reflected in real earnings by the same amounts as in deflated book earnings. They therefore are not sources of disparity between the two earnings concepts.

This can be explained in another way: capital transactions are cash flows during the year, and net changes in investments are assumed also to be net cash flows or their equivalent. As cash flows, these items are correctly valued on the books in terms of current-year dollars, and when they are deflated for the real-earnings computation they are also correctly stated in 1947–49 dollars. They enter into the computation of real earnings in exactly the way they enter a deflated-book earnings computation, which represents a straightforward deflation of all balance-sheet changes and capital transactions by the same index. Thus they do not give rise to differences between real earnings and deflated book earnings. (Surplus adjustments included in the capital transactions group and changes in patents and goodwill were both deflated expressly in a way which would not affect the comparison between book and real earnings.)

C. The Real-Earnings Computation

Once all account groups have been deflated by the procedures just described, the actual computation of real earnings is simple. Real net investment is compiled for each year by adding up the deflated net-asset accounts. The preceding year's real net investment is subtracted to get a net change for the year. Deflated dividends and interest on long-term debt are added and deflated net new capital funds raised during the year and surplus adjustments are subtracted. The result is real profits, stated in 1947–49 dollars.

Appendix B

Construction of Price Indexes

This appendix describes the construction of the price indexes used in the IBM real-earnings computation. Indexes are discussed in the order in which the account groups to which they apply were presented in Appendix A. All indexes are of the Laspeyres (fixed-weight) type, in which the quantity weights are 1947–49 average dollar amounts, as far as possible, and all are computed so as to have 1947–49 average values equal to 100. The weighting is explained as follows.

A. *Working Capital*

The working capital account was deflated by a single index that can best be described as a combination of five price indexes computed for five types of working capital: (1) net quick assets, (2) card stock, (3) raw materials and supplies, (4) goods in process, and (5) finished inventory. The five indexes were weighted in the combination in proportion to the average relative importance of the five types of assets in the years 1947–49.

1. NET QUICK ASSETS

Net quick assets consist of total current assets and deferred assets less inventories and current liabilities. It is a meaure of the company's liquidity—its ability to pay for labor services and purchased goods and materials.

The buying power of the dollars of quick assets depends on what they are spent for. These dollars are mostly spent for labor and for purchases of goods and materials. Therefore, the index used to deflate this asset group has two components, prices of labor and prices paid for purchases, each weighted in proportion to the relative amounts of expenditures for it in the base period 1947–49. To construct such an index it is necessary first to compute indexes for the labor and purchases components separately.

a) *Labor-cost index.* This index is a measure of the companywide average total cost of labor per man-hour. For the years 1939 and 1947–53 the index was computed from data supplied by the company on total net payrolls, total employee benefit costs, and total hours worked. The employee benefits covered a broad group of costs, ranging from vacation allowances and employment taxes to awards, contests, and social activities. The ratio of total net payroll and employee benefits to total hours worked measures average hourly labor costs for each of these years. For the years 1940–46 basic data were not available in a form covering all company employees; thus index values for this period had to be estimated on the basis of the only accessible data: average hourly rates paid by IBM to hourly plant employees. The 1947–49 average of hourly labor costs was divided into each year's hourly labor costs to compute a labor cost index with 1947–49 average equal to 100.

b) *Purchases index.* The index measuring prices of materials, supplies, and purchased parts was computed as follows. From data provided by the company, an analysis was made of the relative importance of various types of commodities in IBM purchases. On the basis of the analysis, an appropriate group of price series covering commodities other than card stock was selected from wholesale price statistics published by the U.S. Bureau of Labor Statistics (BLS); to this group was added a time series on card stock prices, obtained from IBM files. The price data were then given weights reflecting the data on types of commodities purchased, and these were combined into a single index of purchase prices.

The commodity composition of IBM purchases was determined from data for

in the purchases index computation, and the other ten, none of which accounted for more than 3% of the total, were discarded. The following six groups were used:
1947 and 1953 supplied by the Poughkeepsie and Endicott plants. The data show total purchases for each year classified according to the commodity code presently in use at these plants. To each IBM commodity class (excluding card stock, which was treated separately) we assigned the most appropriate commodity code in the new BLS wholesale price index, which was introduced by the bureau in 1952. Sixteen BLS commodity groups were needed to cover all IBM purchases other than card stock, but purchases in some of the groups were very small, and six BLS groups covered about 90% of total noncard purchases. These six were used in the purchases index computation, and the other ten, none of which accounted for more than 3% of the total, were discarded. The following six groups were used:

BLS Commodity Code (New Index)	Commodity	Weight in Total-Purchases Index
11–70	Electrical products	29%
10–40	Nonelectrical metal products	16
10–25	Nonferrous metals	12
10–14	Steel	7
6–73	Plastics	5
9–31	Paper products	4
	Total	73%

Card stock accounts for the remaining 27% of total purchases.

These commodity-group components of the new BLS wholesale price index go back only to 1947. Thus it was necessary to use components of the old BLS wholesale price index (with base year 1926 equal to 100) for years before 1947. Components of the old index were selected to correspond as closely as possible to the six groups in the new index. BLS material relating component series of the two indexes was used in the selection. For the electrical and nonelectrical metal-products groups in the new series, we had to use an old BLS series on "manufacturers' prices of general and auxiliary machinery" (with August 1939 equal to 100), which is a special index outside the coverage of the old wholesale price index itself.

Purchases of tabulating-card paper are given 27% of the total weights in the purchases index. Price data on card stock purchases were supplied by the company. They consist of the prices per pound charged by Hollingsworth and Whitney to IBM from 1938 to 1954, dated as to the specific period of months in which each price was in effect. Prices for both manila and colored paper were available, but the two price series had the same movement, so only the manila paper series was used.

c) Combined index for net quick assets. At this point we had a labor-cost index based on IBM data and an index for total purchases taken from BLS price series and card prices with weighting by IBM purchases data for 1947 and 1953. The next step was to compute weights for combining the labor and purchases indexes into a single net-quick-assets index. For this we used company data on total cash disbursements for the years 1947–49, divided between payables and payroll (including employee benefits). Certain irrelevant disbursements were deducted from the total and from payables. These deductions included renegotiation, interest, dividends, taxes other than employment, contributions, royalties, and purchases of securities. Of the remaining total, labor costs accounted for 59% and purchases for 41%. These were the weights used in computing the price index for net quick assets.

2. OTHER COMPONENTS OF THE WORKING-CAPITAL INDEX

The remaining components of the working-capital deflation index are four types of inventory: card stock, raw materials and supplies, work in process, and finished inventory. The price behavior of card stock and of raw materials and supplies was used to deflate other elements of working capital (as well as quick assets) because that is how these working capital dollars were spent. These materials (plus labor) are embodied in the four kinds of inventory.

The index for card stock, as noted earlier, was based on data supplied by the company. The index for raw materials and supplies was computed from the six BLS price series used in the total-purchases index described in the preceding section. The same relative weights for the six series were used.

The index for finished-goods inventory combines a factory payroll index and the raw materials index, with weights for the two components determined from company data on cost of production in the years 1947–49. The factory payroll index was computed in the same manner as the total labor-cost index. The only differences were that the data on average hourly labor costs covered plants only, rather than the company as a whole, and included all pay and benefits, rather than pay to hourly workers alone.

Computation of the relative weights for the labor and purchases components of the finished-goods index was a complex procedure. It required making allowances in the 1947–49 cost-of-production data for labor cost elements that were hidden in nonlabor accounts and for labor and nonlabor elements of the manufacturing costs in card production. The final weights arrived at for finished goods were: labor 56% and purchases 44%.

The index for work-in-process inventory is also a combination of the index for factory payroll and the raw-materials index. In this index the labor component received less weight than it was given in the finished-goods index. The weights in this case were: labor 39% and purchases 61%.

3. COMBINING THE COMPONENTS

The last step in deriving the working-capital deflation index was to combine the five component indexes. This was done by (1) computing each of the five working-capital component groups as a percentage of total working capital in the years 1947–49; (2) multiplying the percentages so determined by the five respective index values for all years 1939–54; and (3) summing the five products for each year to obtain the working-capital deflation index. This index is shown in Chart 8–6.

4. RENTAL MACHINERY PARTS INVENTORY

The stock of parts for machines was deflated by the same procedure as working capital, but it was restated separately, using the index for finished-goods inventories, which is a component of the working-capital index.

B. Plant and Equipment

Four types of plant and equipment were repriced separately: buildings and building equipment, plant machinery, office fixtures and furniture, and emergency war facilities.

Each of these asset types requires three price index series for deflation of the account: one for additions to the account, another for depreciation charges, and a third for net retirements. This makes a total of twelve indexes needed for the deflation of plant and equipment. However, the buildings and machinery components of war facilities were deflated by the price data used for nonwar buildings and machinery, and no retirements index was computed for this group (because the group was written off the books in 1945, before retirements were significant). There were therefore ten rather than twelve indexes to derive.

One other type of index required for the computation was a measure of 1939 asset levels. This index expresses the ratio, as of the beginning of 1939, between the value of plant and equipment as stated on the books and their value expressed in 1947–49 dollars. Two price series were used here, one for buildings and the other for machinery.

The following description of plant and equipment indexes is arranged by type of index rather than by type of asset.

1. ASSET LEVELS FOR 1939

The opening 1939 balances in fixed-asset accounts were deflated to 1947–49 dollars on the basis of data compiled by the National Bureau of Economic Research (NBER), a private nonprofit research organization. The data were indexes, for construction and for "other producers' durable goods" separately, of prices underlying business depreciation charges. While not directly applicable to the deflation of gross assets and depreciation reserves, these indexes were the most appropriate information available on the question at hand. The values of the indexes for 1919–35 have been published by the NBER.[1] Values for later years are available directly from the NBER.

These indexes were computed on a base of 1929 equal to 100. For the real-earnings study, the 1938 levels of the two indexes were converted to a 1947–49 base

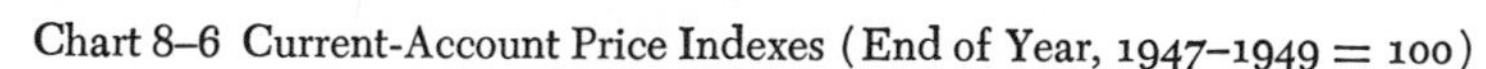

Chart 8–6 Current-Account Price Indexes (End of Year, 1947–1949 = 100)

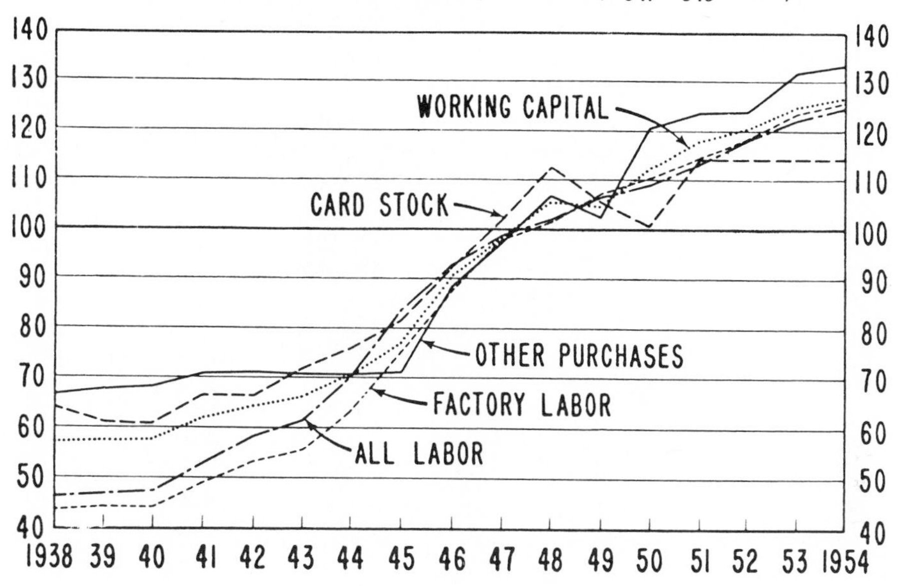

YEAR	CARD STOCK INDEX	OTHER PURCHASES INDEX	ALL LABOR INDEX	FACTORY LABOR INDEX	WORKING CAPITAL INDEX
1938	64.0	66.2	46.4	43.7	57.0
1939	61.2	67.6	46.8	44.1	57.4
1940	60.7	68.0	47.2	44.1	57.7
1941	66.1	70.6	52.9	49.0	61.9
1942	66.1	70.8	58.5	53.8	64.3
1943	71.6	70.7	61.3	56.0	66.2
1944	76.0	70.6	70.3	63.6	70.5
1945	81.7	71.2	83.6	75.1	77.1
1946	92.5	88.4	92.7	88.2	90.3
1947	101.8	97.2	98.4	97.9	98.3
1948	112.7	106.9	102.2	102.0	105.8
1949	105.9	102.4	106.1	106.3	104.5
1950	101.0	120.4	109.4	110.4	112.9
1951	114.6	123.6	113.6	114.9	118.1
1952	114.6	123.9	118.6	118.6	120.2
1953	114.6	131.6	122.2	123.1	125.1
1954	114.6	133.2	125.0	126.0	127.0

by indexes of 1929 current costs of construction and equipment respectively. They were then applied to the gross values and depreciation reserves on IBM's books as of the end of 1938 to establish the initial values of fixed assets in 1939, in 1947–49 dollars. The construction index was applied to the buildings accounts, and the index of producers' durables was applied to the machinery and office equipment accounts. According to these indexes, the net book value of the buildings account in 1938 was 78% of its net value in 1938 dollars; the other accounts were about 5% higher than their net value in 1938 dollars.

2. ADDITIONS TO PLANT AND EQUIPMENT

Additions to the buildings account were deflated by an index of construction costs of commercial and factory buildings, published by E. H. Boeckh and Associates, Washington, D.C. Plant equipment additions were deflated by an index of equipment costs in metal-working industries, published by the engineering firm of Marshall and Stevens, Chicago, Illinois.

Additions to the office furniture and fixtures account were deflated by an index constructed from components of the BLS wholesale price index. The components taken from the new BLS index are commercial furniture (BLS code 12–2) and office and store equipment (code 11–5–3); the two components were combined with weights derived from a tabulation of the composition of the IBM office equipment account as of 1953. The first year covered by the new BLS index data is 1947, and for earlier years office furniture components of the old wholesale price index were used. Additions to the office equipment account were small before 1947.

The indexes used to deflate additions to fixed asset accounts are shown in Chart 8–7.

Additions to the rental machinery account were deflated by an index which combined a factory payroll index and the raw materials index. Weights were based on cost of production in the base years 1947–49, after making the same adjustments described above for the index of finished goods inventory. The index used to deflate additions to rental machinery is shown in Chart 8–7.

3. DEPRECIATION CHARGES

The indexes for deflating depreciation charges were computed in the course of the plant and equipment deflation process. The index for a given year is the ratio of that year's book value of gross assets to the year's deflated, constant-dollar value of gross assets. For example, the 1948 depreciation index for the buildings account is 1948 gross value of buildings on the books divided by 1948 gross value of buildings in 1947–49 dollars. The gross constant-dollar value of the account in 1948 is the result, to that point in the computation, of the cumulative process of adding deflated plant additions and deducting gross deflated value of retirements to the deflated initial balance in the account as of the beginning of 1939.[2]

Chart 8–7 Capital Equipment Price Indexes (Annual Averages, 1947–1949 = 100)

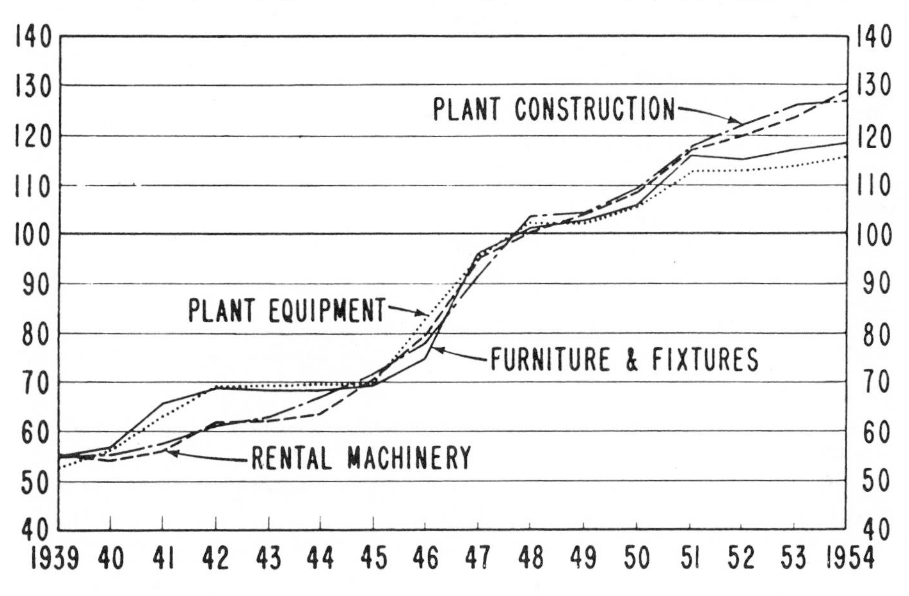

YEAR	PLANT CONSTRUCTION INDEX	PLANT EQUIPMENT INDEX	FURNITURE & FIXTURES INDEX	RENTAL MACHINERY INDEX
1939	54.3	52.8	54.5	54.8
1940	55.2	56.0	56.4	54.1
1941	57.9	63.3	65.5	56.1
1942	60.8	69.2	68.6	61.1
1943	63.2	69.5	68.4	61.9
1944	67.5	69.7	68.4	63.3
1945	71.6	69.9	69.7	70.4
1946	78.1	82.9	75.0	79.8
1947	91.7	95.6	96.0	95.4
1948	103.6	102.2	101.0	100.4
1949	104.8	102.2	102.5	104.3
1950	109.5	105.4	105.7	108.5
1951	117.9	112.5	115.9	117.3
1952	121.9	112.8	115.1	120.0
1953	126.3	114.3	117.2	123.9
1954	127.1	116.6	118.8	129.0

Because the depreciation price index is computed in this way, it is necessary to deflate emergency war facilities separately. In war years new buildings that are being amortized at 20% depreciation per year carry far more weight in depreciation charges than in the gross value of the buildings account on the books. Book depreciation charges, therefore, reflect high wartime construction costs to a greater extent than does the gross-value account, and the ratio of book to deflated gross value of buildings is an inappropriate index for deflating book depreciation charges. This potential distortion is avoided by deflating war facilities separately. The indexes for deflating depreciation charges are shown in Chart 8–8.

4. RETIREMENTS

Since fixed assets are carried on the books at original cost, the appropriate index for deflating a retirement is a measure of prices in the year in which the retired asset was acquired. No information was available on the ages of retired assets other than rental machinery, however. For non-rental assets it was necessary to estimate acquisition dates roughly from the relation of (*a*) the amounts retired from depreciation reserves to (*b*) the amounts retired from gross-asset accounts. For example, when a reserve retirement was almost as large as a gross retirement, this was taken as an indication that the retired facilities were old and that an index based on the more distance past should be used. Development of retirement deflation indexes was based largely on judgment; no formal technique was used.

For rental machinery, in contrast, information on various types of age distribution made it possible to employ the more precise methods described in Appendix A.

C. *Other Account Groups*

In the remaining groups—investments, patents and goodwill, and capital transactions (including payments of interest and dividends)—only surplus adjustments, amortization of patents, and goodwill and net cash transactions were deflated. The treatment of surplus adjustments and amortization of patents is discussed in Appendix A. The value of net cash transactions is measured in the real-earnings analysis by the alternative uses to which the cash involved could be put in the company.[3] The major alternative uses of funds are the operating expenditures that are made with liquid assets. The constant-dollar value of cash capital transactions is therefore measured by deflating them with the price index for net quick assets. This index is a component of the working-capital index, described in an earlier section.

As noted in Appendix A, the cash-value index was applied to the deflation of book earnings as well as capital transactions. It was therefore also used to deflate changes in patents and goodwill and surplus adjustments, since the objective was to deflate these two types of transactions in a way that would not produce any differences between deflated book earnings and real earnings.

Chart 8–8 Price Indexes Underlying Depreciation Charges
(Annual Averages, 1947–1949 = 100)

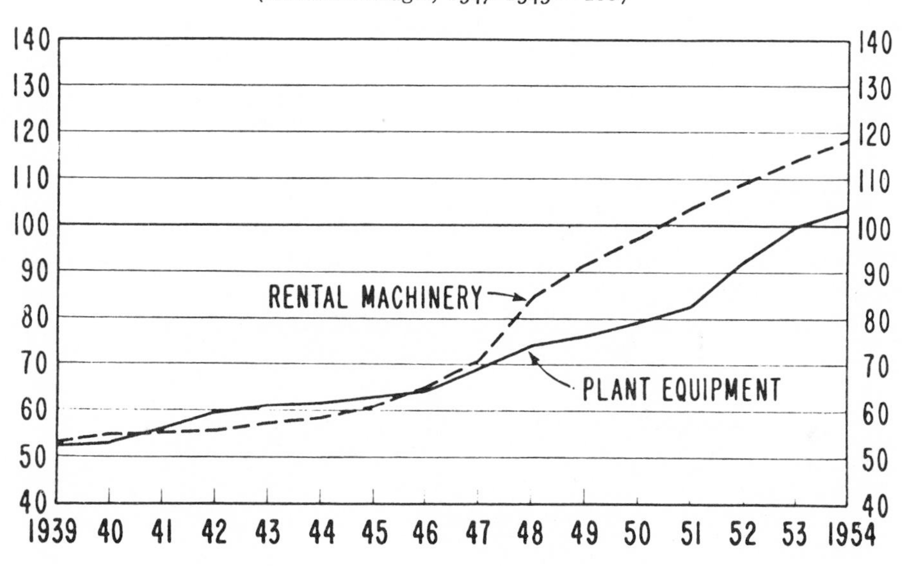

YEAR	PLANT & EQUIPMENT DEPRECIATION INDEX	RENTAL MACHINERY DEPRECIATION INDEX
1939	52.6	52.9
1940	53.6	54.1
1941	55.2	54.9
1942	59.6	55.8
1943	60.8	57.5
1944	61.3	58.4
1945	62.7	60.9
1946	64.2	64.9
1947	69.2	70.7
1948	74.0	84.2
1949	76.6	91.4
1950	79.4	97.4
1951	82.8	103.9
1952	92.3	109.3
1953	100.0	114.1
1954	103.5	118.8

Appendix C

Forecasting Real Earnings, 1955–60

This appendix describes in more detail the analysis and results of a benchmark forecast of IBM's real earnings for the years 1955 through 1960. The prediction is based upon two simplifying but unrealistic assumptions: first, that the changes in the composition of assets in the future will be like those during the analysis period, and second, that the inflation will stop. These assumptions serve the purpose of laying down a benchmark estimate of the future disparity between real earnings available for interest and dividends and those that will be reported. The forecast is confined to the overstatement that will occur as a result of the inflation that has occurred in the past. Building on this base, additional allowances for inflation distortion can be made by IBM on the basis of the price increases that at future dates are experienced or forecasted.

In this forecast of real earnings we used the price indexes whose character and construction are described in Appendix B. The computation procedures are fundamentally similar to those used in the basic study and described in Appendix A. By these methods we forecasted the spread between book earnings and real earnings separately for each individual source of disparity year by year, six years into the future. The results of this analysis are summarized in Table 8–3. For simplicity, the forecast assumes that future additions, retirements, and depreciation of IBM's facilities will, on average, be similar to those of the past. It assumes also that the inflation has miraculously come to an end and that prices will remain stable at their 1954 levels. Both these assumptions will prove incorrect, but the resulting simplified model supplies the baseline for making the adjustments in the forecast required by departures from these two assumptions. As the future unfolds and as forecasts of future rates of inflation become firmer, this simplified forecast of the overstatement of real profits by book profits can be expanded and refined into a dynamic prediction of detailed disparities that can be continuously up-dated.

This predicted spread between book and real profits can then be added to a forecast of IBM's book profits that can be similarly up-dated. To lay the foundation the simplified forecast of the overstatement of real earnings by book earnings was converted into a prediction of real earnings by building on a forecast of IBM's reported earnings made in an earlier study of IBM's cost of capital. The results are shown in Table 8–4. Like our estimate of real earnings, the concept of real profits that is forecasted consists of earnings of the whole economic entity, without regard to IBM's capital structure, i.e., profits after taxes available for interest and dividends. A brief description of the two analyses follows.

A. Overstatement of Real Earnings by Book Earnings

As can be seen from Table 8–3, underdepreciation of rental machinery is a major source of overstatement of real earnings by reported earnings. This underdepreciation was at its greatest ($6.2 million in 1955 purchasing power) in 1947. From 1948 through 1953 it remained relatively constant at about $5 million (1955 value). In 1954, with a slackening of the upward surge of prices, underdepreciation of leased machines fell to $4.1 million. With substantial price stability, this would drop again to about $3.2 million in 1955, as noted in the preceding section, and would disappear entirely by 1960. The explanation for this disappearance is to be found in the high depreciation rates and use of the item method of depreciation for rental machinery, which means that book depreciation can catch up with real depreciation very quickly once price levels stabilize. Underdepreciation of rental machinery would average about $800,000 a year for the five years 1956–60, again assuming stable prices over the period.

Underdepreciation of plant and equipment, on the other hand, has been slowly increasing over the entire period since 1939 and is currently, as noted above, about $2.1 million per year in 1955 dollars. The prospects for reducing this discrepancy are less bright than for rental machinery because of the relatively low depreciation rates and turnover of plant and equipment. Even with price levels holding steady over the entire replacement period, to eliminate the underdepreciation from this source by 1960 would require the rapid replacement of most of the company's present plant and equipment. In lieu of such an extraordinary experience, this underdepreciation is likely to decline only slowly. Thus, on the average, for 1955–60, underdepreciation of plant and equipment is likely to run about $1.8 million if prices remain steady.

Understatement of the real value of retirements is practically negligible. As long as plant and equipment are carried at such low depreciation rates, however (2% to 2½% for buildings, 6% to 10% for other facilities), facilities retirement will occasion some value write-off. In round numbers this is likely to average a half million dollars a year during the period 1956–60.

B. Forecast of Book Earnings

In order to convert the disparity analysis into projections of real earnings, forecasts of reported earnings were needed. In our study of IBM's cost-of-capital we predicted earnings per share of common stock, assuming a constant number of shares outstanding (3,278,777 shares). To do so we first computed the past rate of growth in earnings, 9%, and projected it forward. The resultant earnings appeared to us so unattainable that we arbitrarily scaled down the forecasted rate of growth to 6%. This produced the following net earnings forecasts:

Year	Per Share	Total
1954	$15.20	$49,800,000
1955	16.20	53,100,000
1956	17.00	55,700,000
1957	17.80	58,400,000
1958	18.60	61,000,000
1959	19.40	63,600,000
1960	20.20	66,200,000

A number of factors, including the change in the basis for reporting the income of foreign subsidiaries, combined to cause 1954 results to fall short of our earlier forecast. An income of $52.3 million was projected for 1955 on the basis of up-to-date information, as described in the preceding section. Since this is $800,000 lower than the $53.1 million forecasted earlier for 1955, the projections of reported net income for each year, 1956 through 1960, were adjusted downward by this amount. The adjusted figures are shown in Table 8–4.

Real profits are earnings available for dividends and interest, but after taxes. The book earnings estimates were placed on a comparable basis by adding estimated interest payments. For this purpose it was assumed that the final $100 million installment of 3¾% debt would be taken up during 1956 and that the 12-year debentures maturing in 1958 would be refunded at no change in interest rate. These calculations produced the estimates of total income shown in Table 8–4.

Study 9

MACHINERY MANUFACTURER

Manufacturers' real earnings are seriously overstated by conventional financial reports in a period of inflation such as we have gone through and still face. This overstatement by conventional accounting reports creates an illusion of higher company earnings than actually exist, and thus may distort top management's decisions on vital policies. The accounting profession has become increasingly aware of this problem. Its leaders are actively seeking solutions.

The purpose of this paper is to illustrate by means of a case study (1) the impact of inflation in distorting the financial results of a manufacturer of equipment; (2) the kinds of management decisions for which the resulting disparity between book earnings and real economic earnings is important; (3) the meaning of real economic earnings; and (4) principles and methods of measuring real earnings by one tested method.

I. Financial Results Distort Economic Reality

Conventional accounting seriously overstates a manufacturing company's earnings during an inflationary period when prices rise drastically. We all recognize that in a period of inflation our personal cost of living goes up. It is obvious that a company's cost of operations also goes up, that the cost of replacing worn equipment goes up, and that the cost of expanding capacity goes up. Yet accounting procedures generally fail to take adequate account of these increases in the corporate cost of living. Orthodox accounting vigilantly keeps price-level revaluations from getting into the profit and loss account by ignoring them or by treating them as surplus adjustments. But when revaluations find their way into the accounts indirectly, by the process of turnover of assets during inflation, they *do* get into the earnings account. These "inflation profits" that result from such revaluations are treated as ordinary income and cannot in the books be distinguished from regular income during the life of the asset, and before the time of its replacement at higher price levels. Hence accounting profit overstates real earnings.

When plant and equipment are important, as in machinery manufacturing, con-

Reprinted from Joel Dean, "Measurement of Real Economic Earnings of a Machinery Manufacturer," *Accounting Review*, vol. 29, no. 2 (April 1954), pp. 255–66, by permission of *Accounting Review*.

ventional accounting analysis becomes an unreliable tool for managerial policy. "Inflation profits" make book earnings seriously overstate real economic earnings. Moreover, these illusory profits will continue to exaggerate real profits long after the price levels have become stabilized. It took twenty years after World War I before prices underlying depreciation charges caught up with current replacement costs. For sound managerial use of business accounts it is desirable that top management know how big this inflation distortion has been.

II. Managerial Uses of Real Earnings

Among the more important practical applications of estimates of a company's real economic earnings to top management policies are: (1) dividend policy; (2) capital budgeting; (3) pricing policies; (4) appropriations for advertising; and (5) government negotiations.

Dividend Policy

Perhaps the most important single use of the real earnings study lies in its implications with regard to dividend policy. Cash dividend payments may be far less than reported book earnings and still be in excess of the maximum consistent with the maintenance of the real capital of the corporation. This was the situation of the firm studied here, which we shall refer to as "Machinery, Inc." The data are examined in a later section. Maintaining a record of real earnings may serve a twofold purpose in this respect: (1) as a guide to the determination of dividend policy; and (2) as a device for explaining to stockholders the reasons for what may seem to them to be excessive retention of earnings.

Many companies in such a period pay out dividends that are not earned in a real economic sense. This is legal, and quite inadvertent. It is the function of real earnings analysis to provide management with estimates that will prevent this kind of impairment of the company's real assets through ignorance.

Capital Budgeting

Concepts and measurements of real earnings outlined here have great practical usefulness in plough-back decisions and in estimates of the profitability of future investment. They also have a vital bearing on the choice of sources of outside capital—notably, on borrowing versus equity capital, and lease-back versus conventional financing. Knowledge of the firm's real rate of return, coupled with data on the cost of various types of investment capital, can be of real assistance in considering expansion or change in capital structure.

Pricing Policies

A more informed approach to price policy may also stem from an analysis of the real earnings series. Prices set on the basis of book costs will be inadequate to provide for the long-run maintenance of real investment. Provided that competitive conditions permit, a knowledge of the firm's real rate of return may encourage the firm to set prices which are adequate to produce a normal return on the real investment. Only those firms which have recognized this, either explicitly by analysis of their real earnings or implicitly by intuition, have set prices in the postwar period which have been designed to maintain their prewar rates of return in real terms.

Insofar as costs and profits and return on investment have a bearing on setting prices, these should be real costs and real investment, rather than those measured conventionally in a hodge-podge of dollars of different purchasing power.

Appropriations for Advertising

One way to decide on the size of the advertising budget is for the company to spend all it can afford. This all-you-can-afford method has been used with signal success by some new enterprises, and it has been an important factor in deciding on advertising appropriations for established enterprises. The essence of this criterion is what the company has earned in the past. And the earnings that are relevant for this purpose are after-tax real profits.

Negotiations with Governmental Agencies

The company's profits are a pivotal consideration in negotiations with agencies such as OPS, and in antitrust, military price negotiation, public utility rate regulation, and in wage negotiations. In each arena of negotiation, concern about the reasonableness of a corporation's earnings is vital, so when there is discretionary latitude and open-mindedness about the basic economic purpose of government control, measurements of real earnings can be helpful to both parties.

Summary

In a period of rising prices, reported earnings overstate the real increments to the value of the firm's assets stemming from operations. Book earnings can be deflated and real earnings estimated by applying principles of economic and statistical analysis. Knowledge of real earnings and real rate of return can have great practical value, not only for management decisions on various operating and financial policies, but also in the education of stockholders, governmental agencies and the general public.

III. The Meaning of Real Economic Earnings

Business income ought to be measured in terms of the real goods that can be purchased with it. *No business may be said to have earned a profit if at the end of the accounting period the value of its total assets (plus the current purchasing power equivalent of dividends paid) is the same as the value of its total assets at the beginning of the period.* Strictly speaking, it is the earning power of the assets that must be maintained before economic earnings may be said to exist. However, earning power is essentially a speculation about future demand, technology, competition, and management ability, and thus involves a number of intangible considerations that rest entirely upon judgment. In practice the only feasible way to measure maintenance of capital value appears to be in terms of physical capital—inventories, equipment, and working capital in dollars of constant purchasing power.

Real income may be defined, roughly, as the increase in the value of the firm's assets during the year plus dividends paid during that year. Reported dollar statements of income do not reflect these economic concepts accurately in periods of rapidly changing prices since accounting principles are based on the postulate of a fixed purchasing power of the dollar over time.

Deflating Book Earnings

Making adjustments for price level changes in analysing personal incomes has gained wide acceptance since the war. Wages, salaries, and prices of personal services have all been raised to help their recipients keep up with increases in the cost of living. This is a sensible recognition that the purchasing power of a dollar of income is lower than it used to be and that more dollars are necessary for the individual to maintain the same command over goods and services that he had prior to the rise in prices.

Business firms are faced with much the same problem, but income is a more complex concept than personal income and the effects of inflation upon it are correspondingly more difficult to measure. Personal income normally takes the form of a simple inflow of money, typically for wages or salary. Business income, on the other hand, is a residual from the flow through the accounts of the costs of depreciable and non-depreciable assets acquired at different price levels and being used up at different rates. Since these assets have been acquired at different times, their accounting records, which are almost always historical costs, reflect the price fluctuations for many preceding years. Hence the book value of assets is in terms of dollars of different purchasing power, and the residual accounting income reflects a hodgepodge of big and little purchasing-power dollars.

Consequently, in measuring real income it is not enough merely to deflate the accounting income itself, as is done for personal income; instead it is necessary to restate the quantities of individual assets in terms of dollars of the same purchasing power size and recalculate the income.

Current Dollars vs. Constant Dollars

The problem of "unscrambling" the jumbled dollars in order to determine the real income of any one year should be clearly distinguished from the problem of comparing income from one year to the next. An estimate of real earnings can be stated in either current dollars (e.g., 1939 dollars for 1939 earnings and 1949 dollars for 1949 earnings) or in some constant dollar (e.g., average purchasing power for the period 1936–1939). In either case homegenous dollars are being used for the earnings estimate for any one year. It is for comparisons among years that the two measures differ, and in order to see long-term trends in the real position of the company, earnings for the entire analysis period must be expressed in constant dollars. This has been done in the chart series for "Machinery, Inc."

Furthermore, it is not correct to compare reported earnings in current dollars with real earnings in constant dollars. The effect of using jumbled dollars in conventional accounting is illustrated correctly by comparing the deflated reported earnings with deflated real earnings. Reported earnings for "Machinery, Inc." have therefore been shown in Chart 9–1 in 1936–1939 dollars as well as in current dollars. For example, deflated book earnings for 1948 represent the 1936–1939 purchasing power of the dollars reported for this company in its 1948 financial statements.

It should be understood that none of the above reasoning is meant to imply that existing accounting methods are either "wrong" or "right." Accounting records of historical transactions are indispensable in making any kind of an income analysis. However, the measurements of income which flow directly from the unmodified historical accounting records cannot be as appropriate and accurate for purposes of economic appraisal of income as are income statements in which values have been deflated to homogeneous dollars.

Measuring Income

The concept of real economic earnings mentioned above, dividends paid plus net growth in real capital during a year, is equivalent to the accountant's concept of book earnings. The accountant reports earnings as a total and then segregates this total into dividends paid and growth of the earned surplus account. In book earnings, as well as in real earnings analysis, the growth of surplus represents the net change in assets during the year. In real earnings analysis, however, the change in assets is stated in physical terms. In book earnings estimates, assets changes represent only the dollar transactions that have taken place in the past. The significant difference between the two lies in the ways in which the changes in the values of the assets are estimated.

IV. Principles and Methods of Measuring Real Earnings

The purpose of this section is to outline how economic, statistical, and accounting analysis is applied in measuring real income. Principles are illustrated by showing how they were applied to one company, "Machinery, Inc." Methods and real earnings results for this company are presented in a series of charts.

The keystone of the method lies in the determination of the real value of the firm's assets at the beginning and ending of each accounting period. This is done by

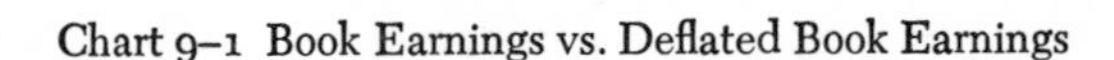

Chart 9–1 Book Earnings vs. Deflated Book Earnings

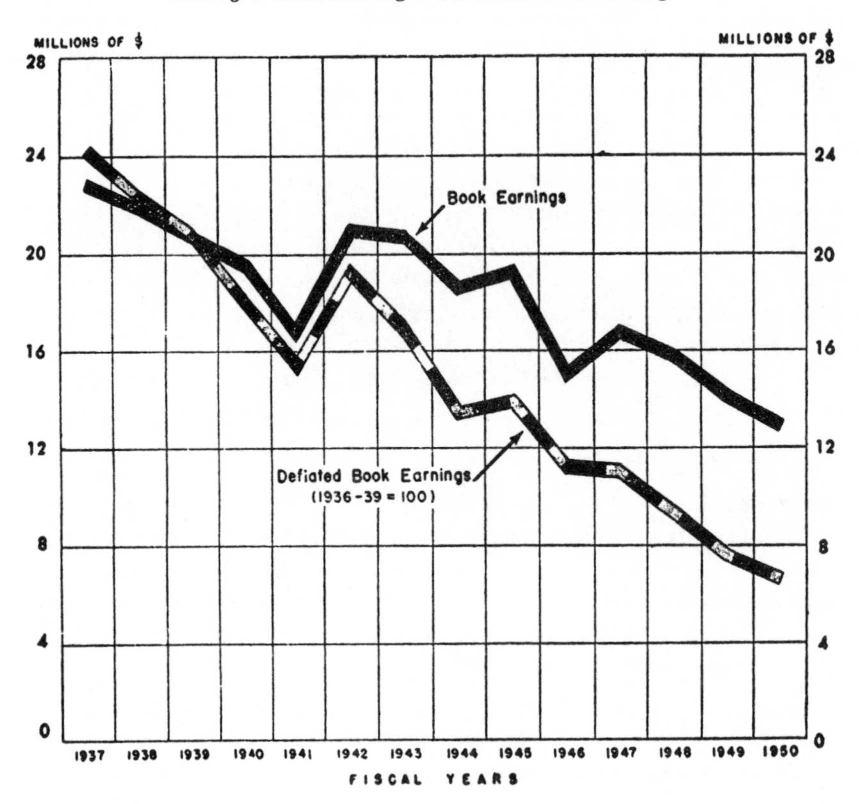

restating the book values of assets in dollars of constant purchasing power, e.g., buildings acquired 40 years ago and inventories acquired last month are all restated at what they would cost during a single base period (in this study, 1936–1939). Depreciation allowances are then recalculated in constant dollars on the basis of these restated values.

In estimating the real value of assets, different methods are used for different kinds of property, and it is convenient to discuss each asset group separately.

Liquid Assets

When the price level rises, the purchasing power of cash and other liquid resources declines correspondingly. Thus, merely holding on to cash during an inflationary rise results in real losses through the shrinkage in the purchasing power of these liquid assets, which become inadequate at the higher price level to handle a given level of operations. To the extent that these liquid assets lose purchasing power this kind of productive facility is reduced in capacity and in economic value.

Liquid assets are a reservoir of current purchasing power and can be valued in terms of the things which they buy, primarily labor and materials. Cash and cash-equivalent items that will normally be turned into cash in the near future (i.e., current assets less inventory), are used to pay current liabilities and to pay wage bills and buy raw materials and supplies. By deflating liquid assets net of current liabilities, it is possible to eliminate the need of indices for such troublesome items as tax liabilities. Liquid assets net of current liabilities are then deflated by a specially weighted index of wage rates and material costs.

Inventory

Orthodox accounting vigilantly keeps revaluation of fixed assets from getting into the profit and loss accounts by not recording them until they are realized in cash by sales. Inventories, however, turn over much faster than fixed assets, and inventory revaluations become an important element of the income statement in years of substantial change in the price level. These revaluation profits are treated as ordinary income on the books and are not distinguished from other incomes, even though they reflect not changes in the real value of the assets but reverse changes in the value of the dollar. When inventory accounts are restated in dollars of constant size, only change in the real volume of inventory is shown and illusory inventory profits are removed from the income statement.

It might be pointed out that the use of LIFO (last-in, first-out) inventory valuation reduces the amount of revaluation profit which is treated as ordinary income on the books but does not necessarily eliminate it. The use of LIFO accounting for inventories requires the use of a slightly different, more complicated method of deflation to arrive at real values, however, inasmuch as book values of inventories under this system are often purely nominal.

Assuming that inventories turn over about once a year, the book values would reflect an average of the year's costs, under FIFO (first-in, first-out) accounting, and would in that event be deflated to constant dollars by an annual average price index. Inventories may be grouped into relatively homogeneous categories and deflated according to the average age of the items in each category. Different index numbers ought to be used to deflate inventory at different stages in the production process. These indices assign progressively more weight to labor as the product moves from the raw material stage through the in-process stage to finished inventory.

Plant and Equipment

Plant and equipment are a stock of physical goods whose balance sheet values reflect the price levels of many different years in the past. The first step in deflating these accounts is to express the opening net value of these assets in dollars of constant purchasing power. This is done in order to make the initial year's values comparable to those of later years. It is necessary to develop a special-purpose index of the prices that underlie the company's depreciation charges and net book values of fixed assets. Constructing this index is one of the more intricate problems of real earnings analysis.

The next step is to deflate *additions* to plant and equipment. New construction should be deflated by the current value of a suitable index of the cost of factory building construction. New purchases of equipment should be deflated by an appropriate index of equipment costs.

The third problem is to deflate *retirements* of plant and equipment by the price levels at their probable dates of acquisition. For this purpose special retirement indexes for different categories of assets must be worked out. These indexes reflect prices for many years past. They also should reflect mortality rates of each asset category. The figures that are obtained by the application of these special indexes in these three steps express the gross value of all plant and equipment, every year, in a uniform size dollar.

To obtain net plant and equipment in real terms, real depreciation allowances must be computed. These deflated depreciation charges measure roughly the consumption of physical "usefulness" of the facilities through time, whereas conventional depreciation is merely an allocation of a historical cost to subsequent production.

The choice of the appropriate index numbers is vital to the success of a real earnings study. An important feature of wide swings in the price level has always been the consequent disruption of the price structure—that is, of the relationships among prices. Some prices react to inflationary pressure much more rapidly than others, and although the former structure may eventually be re-established at a new price level, the transition period (such as that covered by the charts in this study) will at any point in time see some costs much higher than others relative to their former stable levels. Some idea of this disparity of price movement is shown by the price indices plotted on Chart 9–2.

There are available in this country a great many price indices, published and unpublished, prepared by government agencies and by private sources, relating to general and special commodity groups at the several levels of production. In studying income in real terms it is important to choose from this host of indices the ones that are truly appropriate for the particular firm. Often none are; hence it is necessary to construct special price indicators whose movements reflect most accurately the changes in prices of the particular types of assets to which they are to be applied.

There are two general types of construction cost indices, one of which relates to the cost of one or a few specific structures and the other to a tailored mixture of company wage rates and market prices of materials. Equipment costs are notoriously difficult to gauge because the range of products is vast and innovations are rapid. Intimate technical knowledge of the company and considerable familiarity with index number theory and sources is required to select or create the kind of indices that will meet a particular company's needs for restating its equipment accounts.

Chart 9–2 Published Price Indexes (1936–1939 = 100)

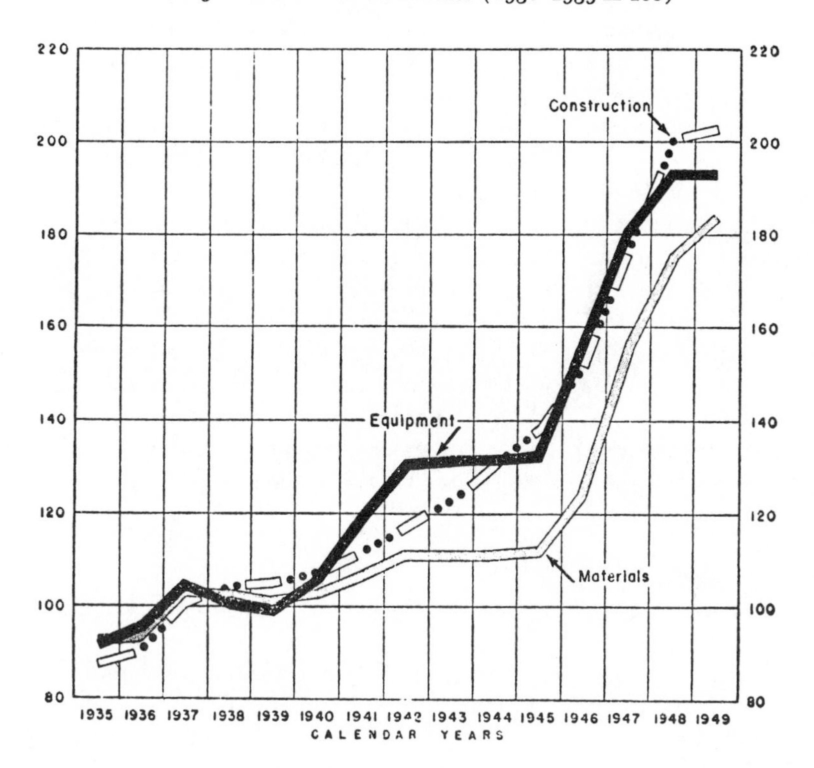

Other Assets

"Other assets" may be taken to represent an accumulation of cash transactions. After estimating the total value as of a base date, net changes year by year can be deflated by a special index of current cash purchasing power used for liquid assets, since this index reflects the value of cash involved in alternative uses. If this category is very large or if the nature of the assets warrants, a different treatment will be required.

Real Earnings in Constant Dollars

By restating each category of assets in terms of dollars of constant purchasing power for each year during the period studied, it is possible to obtain an adequate measure of the real change in the company's aggregate assets from year to year during the period. To get a measure of real economic earnings annually it is necessary to make adjustments for dividends and capital transactions so that the year-ending and year-beginning real asset values will show the results before dividends paid and will not show the effects on the assets of securities sold or retired during the year. This is done by adding dividends plus security retirements minus security sales during the year (deflated by a special index of the real value of cash to the company) to the year-ending real asset value for each year. The annual differences measure the company's real earnings.

This method of getting a company's income from the year-to-year change in its net assets is not essentially different from the accountants' method of getting income. Accounting measures of income are essentially estimates of the yearly change in the company's net investment, measured in scrambled dollars. The income statement is a supporting schedule showing some of the transactions that were made to achieve the annual change in net investment. In measuring real income, we determine the real change, rather than the dollar change, in the company's net investment during each year. We have disregarded the historical accidents of what the price level happened to be at the time the various assets were acquired, and determined what income would have been reported by conventional accounting if transactions had been recorded in dollars of constant purchasing power instead of dollars of vastly varying purchasing power, as was the case.

V. Findings

Next we shall examine briefly the findings for our illustrative company. The top curve in Chart 9–3 shows the book earnings of "Machinery, Inc." Reported earnings fell jaggedly from about $23.0 million in 1937 to a little more than $12 million in 1950. The gray curve in the middle represents the results of a simple-minded attempt to get the corporation's real earnings, that is, by taking the book earnings and dividing by a single index number as we do to estimate the real earnings of a person.

"Machinery, Inc.'s" real earnings measured by the method outlined in the preceding sections, are charted as the broken line in Chart 9–3. These real economic earnings declined during the period 1937 to 1950 from around $21.0 million a year to an amount not significantly different from zero in the last two years shown.

The significance of this economic measurement is highlighted by a comparison of real earnings with dividends paid. The period from 1946 to 1950 was one in which conventional accountancy reported that this company and most companies like it earned big money. It was also a period in which dividends were niggardly as compared with these gigantic reported earnings, i.e., of big plow-backs, per books.

The black line in Chart 9–4 shows what "Machinery, Inc." actually paid out in dividends during this period. To keep the comparison accurate, these dividends have been restated in terms of prewar dollars and this is shown by the broken line in the middle. The black and white line at the bottom shows the company's real economic earnings stated in dollars of the same purchasing power as the deflated dividends. In every year except one "Machinery, Inc." dipped into its real capital by paying out in dividends more than it took in in real earnings. The peril

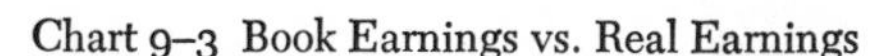
Chart 9–3 Book Earnings vs. Real Earnings

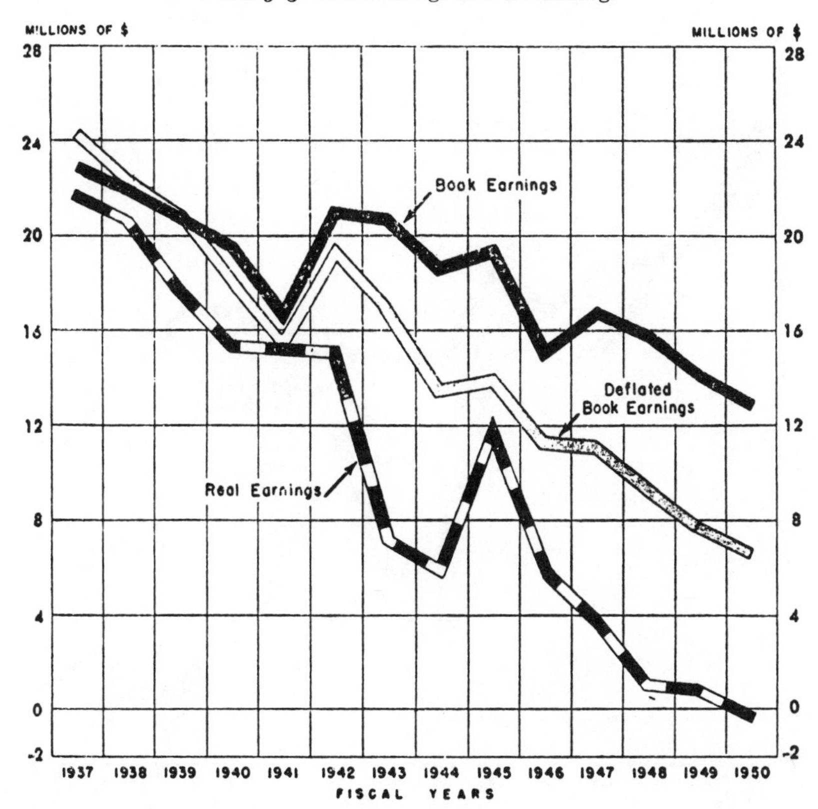

of making managerial decisions on dividend policy and capital expenditure plowback policy without this kind of information is clear.

Rate of return estimates should be computed in terms that have real economic meaning for policy decisions. In order to place the real earnings data in relevant ratio form it is necessary to compute the company's real net investment for each year. The result of this computation is illustrated for "Machinery, Inc." in Chart 9–5.

Net investment is total assets minus current liabilities. We excluded from asset values Treasury stock and we included marketable securities at market value rather than at cost, since the purpose here is not conservative accounting but a portrayal of the company's probable real value.

Instead of growing as the books state, the company's net investment in real goods terms declined continuously from 1936 to 1950.

Findings on real return on net real investment are compared with book return in Chart 9–6. This chart shows that the real rate of return has been much lower and

Chart 9–4 Dividends vs. Real Earnings

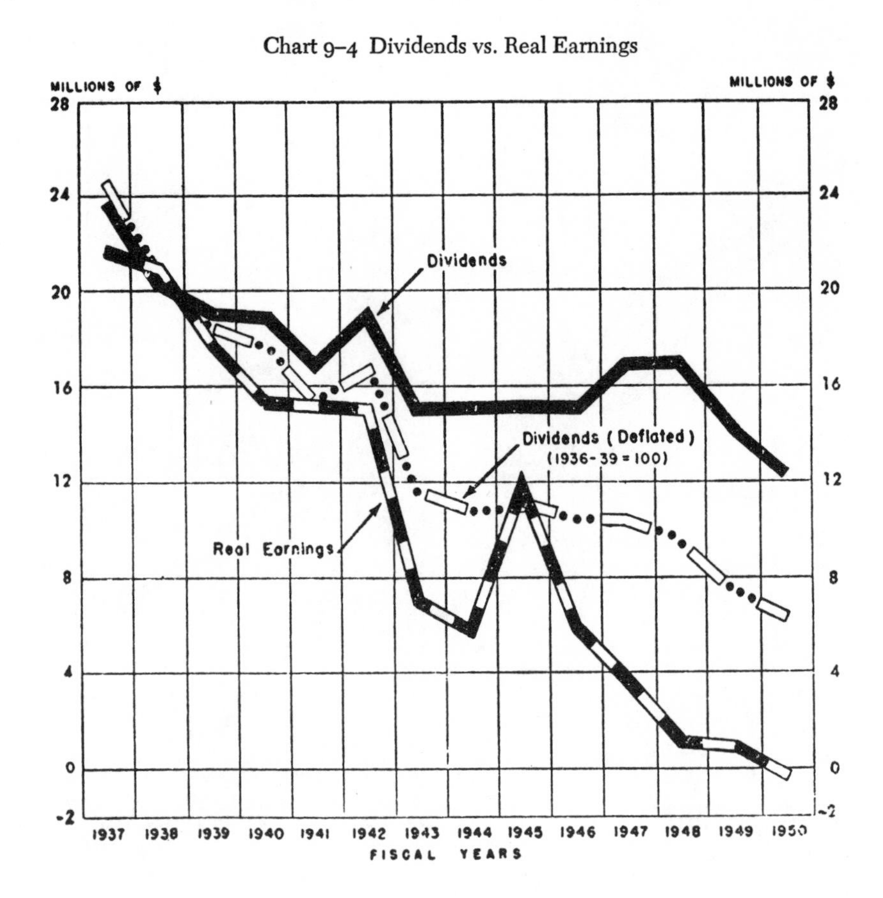

has fallen more rapidly than the rate of return revealed by the conventional accounts.

Disparity Between Real and Book Earnings

There are three ways in which conventional measures of income are distorted by price fluctuations. First, the accounts of any one year are a jumble of past and present dollars of different value that cannot validly be added together to show a statement of the company's real economic situation and performance. Second, the earnings and assets of one year are not comparable to those of another year if the two price levels are different. Third, since prices of some commodities rise or fall more than prices of others, there are gains and losses in the values of particular kinds of assets that are not reflected in conventional accounts. In order to measure real earnings over a long period these three kinds of disparity have to be eliminated.

An analysis of the disparity between real earnings and the reported earnings of "Machinery, Inc." shows the relative importance of these several sources of distor-

Chart 9–5 Net Investment: Book vs. Real

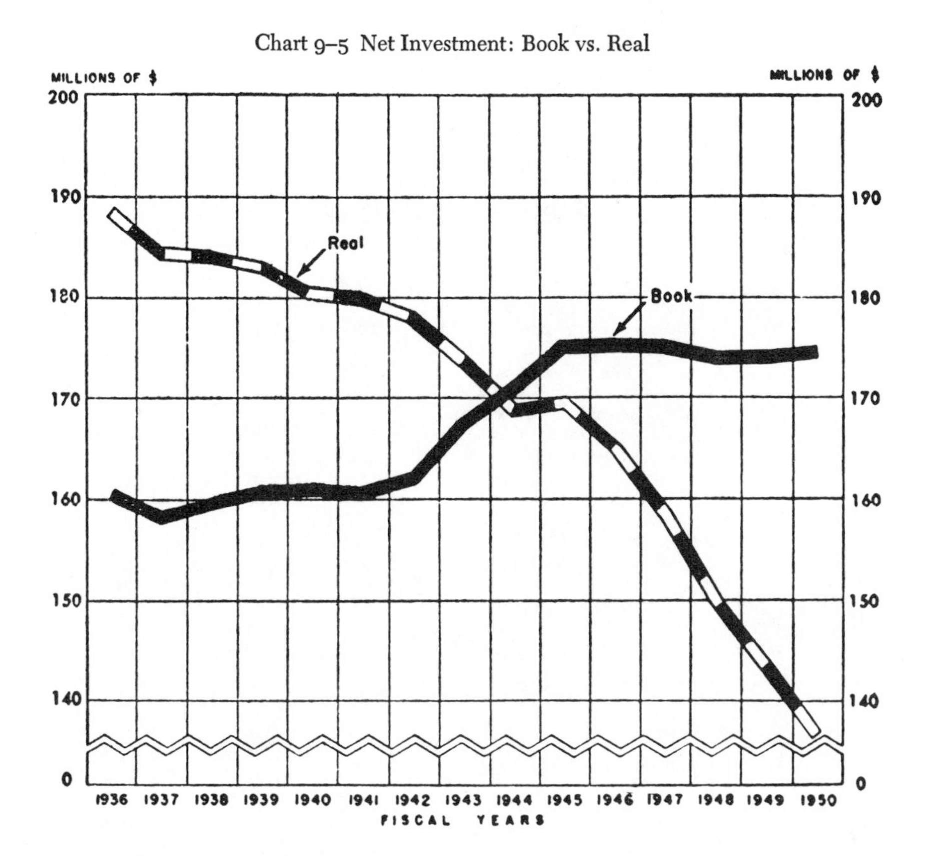

tion. The explanation of the difference between reported and real income varies from year to year, since the disparity is the net result of many changes in assets and price levels.

A convenient method of analysis is the conventional table of sources and uses of funds. By comparing the book statements of sources and uses with the real fund flow, the causes of the disparity in earnings can be made quite clear. The following table presents the analysis for "Machinery, Inc." for 1949. In this table book data are as recorded and real data are stated in terms of prices of the most recent year. The choice of price level for the analysis is arbitrary; the important thing is to have book and real figures stated in the same size dollar. The disparity figures in the lower half of the table are simply the net differences between real and reported data in the first part of the table.

VI. Summary

It is the responsibility of management to know what is happening to the investment with which it is entrusted. What the real economic investment is, what real

Table 9–1

MACHINERY, INC.

Sources of Disparity Between Real and Reported Income
Fiscal Year Ended December 31, 1949

A. SOURCES AND USES OF FUNDS

	(Thousands of Dollars) (a) Reported	(b) Real[1]
Sources:		
1. Retained income	566	(12,328)
2. Depreciation		
Plant	1,686	2,922
Machinery	6,780	14,250
3. Decrease in cash account	756	1,666
4. Inventory reduction	4,008	6,780
5. Investment liquidation	460	460
Total sources	14,256	13,750
Uses:		
6. Machinery additions	12,296	11,836
7. Plant additions	1,874	1,828
8. Retirement of capital	86	86
Total uses	14,256	13,750

B. ANALYSIS OF INCOME DISPARITY[2]

	Amount	*Proportion —%*
2. Underdepreciation		
Plant	1,236	9.5
Machinery	7,470	58.1
Understatement of cash decline	910	7.1
4. Understatement of inventory decline	2,722	21.6
5. Overstatement of growth in fixed assets	506	3.7
Total overstatement of earnings	12,894	100%

[1] 1951 dollars.
[2] Column (b) minus column (a) for individual rows.

earnings on it are, and what real rate of return it is producing for stockholders are matters management should know. When prices are rising, accounting reports on these three matters are not likely to be correct. Conventional reports can be seriously in error and thus produce an illusion, an inflation puff, that can lead management astray.

Measurement of a company's real economic earnings has great practical usefulness. And it can be done. It cannot be done simply, as in the case of personal income, by dividing one single index number into the reported accounting earnings.

Chart 9–6 Return on Net Investment: Book Return on Book Investment vs. Real Return on Real Investment

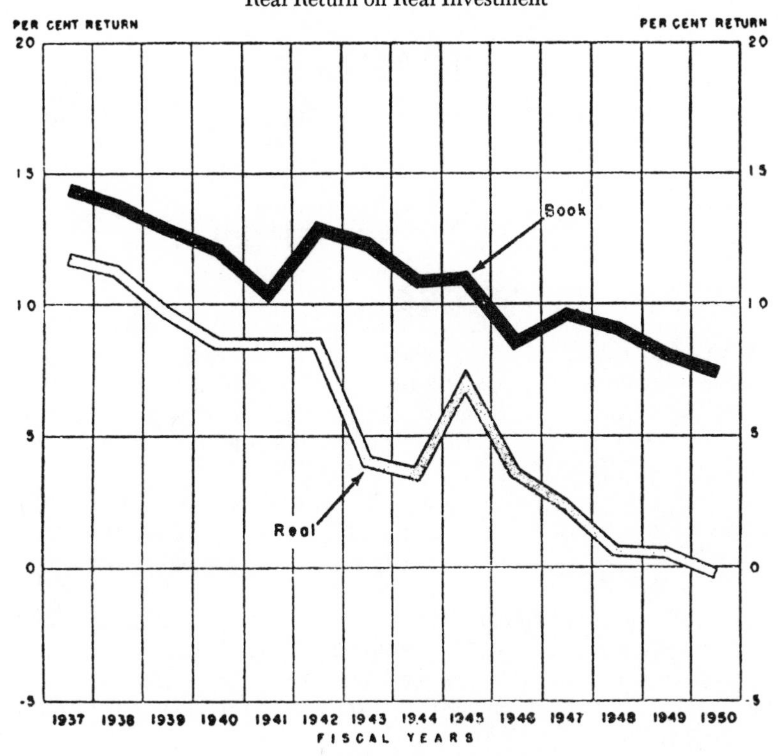

Instead it must take the form of restating each asset or each group of assets in terms of dollars of constant purchasing power. This can be done with tolerable accuracy and applied with economic and statistical sophistication.

Notes

Part One. **Cost and Output Rate**

Introduction

1. The law of diminishing returns has been stated as follows by George Stigler: "If the quantity of one productive service is increased by equal increments, the quantities of the other productive services remaining fixed, the resulting increments of product will decrease after a certain point." George J. Stigler, *The Theory of Price* (New York: Macmillan Co., 1947), p. 109.

2. Again, George Stigler puts it neatly: "The units of the variable productive service are homogeneous. The presence of diminishing returns is not due to the employment of less and less efficient men, for example, but because men of equal ability are being employed less efficiently." Ibid., p. 117.

3. Complete adjustment of the productive services of *all* input factors so as to obtain the least-cost mix for a rate of output (and for each hypothetical rate of output) produces the idealized condition known as the long-run cost curve. The label is, however, misleading. Total adaptation is achieved by investments (tangible and intangible), not by the passage of time.

4. C. Karl Menger, "The Laws of Return: A Study in Meta-Economics," in Oskar Morgenstern, ed., *Economic Activity Analysis* (New York: John Wiley & Sons, 1954). This is an English translation of two articles published by Professor Menger in the *Zeitschrift für National Ökonomie* in 1936. As Professor J. Johnston has pointed out: "The net result of Menger's work is to demote the law of diminishing product increments from the position of being necessarily true and to place it alongside other promising hypotheses for empirical testing." John Johnston, *Statistical Cost Analysis* (New York: McGraw-Hill, 1960), p. 9

5. An alternative theory of short-run production cost behavior that is built upon the results of testing the conventional hypothesis will be proposed later in this section.

6. For advertising and other persuasion outlays whose significant effects on sales are immediate, the pre-persuasion incremental profit is an adequate measure of worth. But for persuasion outlays whose impact is delayed and cumulative, purchasing customers by promotion is really an investment, like purchasing annuities. The value of the customer is the present worth of the stream of future profits he will produce, which is mainly determined by his loyalty-life expectancy. Investment in persuasion ought to be forced to compete for funds with alternative ways of investment, on the basis of profitability. The present worth of the stream of profits which persuasion outlays can yield should be compared with the present worth of the stream of profits from rival corporate investment opportunities and with the company's cost of capital.

7. It is also an efficient method when the operation in question is susceptible to complete theoretical analysis. An engineering cost function with a high degree of accuracy was developed for an airline on the basis of aerodynamic theory. A. R. Ferguson, "Empirical Determination of a Multidimensional Marginal Cost Function," *Econometrica*, vol. 18, no. 3 (July 1950), pp. 217–235.

8. These conditions were approximated by grocery stores of the same chain, similar in size but located in areas that differ in economic activity. See *Organization and Competition in Food Retailing*, National Commission for Food Marketing, Technical Study no. 7, p. 141.

9. The present analysis is from the standpoint of top management of a multiplant firm faced with the problem of determining which plants can best be studied by statistical analysis (as opposed to other methods of determining the cost function). From the standpoint of the manager of a single plant, the problem is whether his plant qualifies for a statistical approach. The substance is the same from either viewpoint.

10. A simple but adequate way to adjust future short-run operating costs for changes caused by modernization investments is to presume that the minimum required rate of return is attained in the form of operating cost savings and therefore to reduce the forecast of future costs by a corresponding amount. This method is inadequate to the extent that the investment return differs from cost of capital or takes the form of greater salability because products are of higher quality or greater consumer acceptability. Also, the investment-return adjustment is too low if the modernization is forced by competitors' advances so that the return is adequate only by comparison with some disastrous alternative, e.g., plant shutdown.

11. They were met by a plant whose costs were analyzed statistically in this way by Johnston (*Statistical Cost Analysis*, pp. 87ff.).

12. In the statistical analysis of average cost, selection of the most suitable specification for the average cost function is more difficult, and slight errors in the choice of the function produce magnified errors in the derived marginal cost function. Moreover, since average cost is a quotient of two variables, each of which is subject to error, the statistical distribution of the quotient may be less likely to conform to the assumptions upon which multiple regression analysis is based.

13. Whether or not it is correct to assume that the firm's rate of output does not affect the prices it pays for input factors, such an assumption is the only practical approach to the determination of the firm's cost–output functions. Changes in factor prices were removed without regard to their cause. A rise in the marginal manufacturing cost curve that is caused by a cyclical rise in prices of inputs correlated with the firm's output is excluded by hypothesis.

14. This procedure is substantiated by the experience of Ehrke and Schneider in their statistical analysis of cost in a cement mill. Their correction for price changes in the factors was first undertaken by using the *Groszhandels Preisindex*. Finding this unsatisfactory they constructed a special index for the prices of labor, limestone, clay, coal, and coke. See Kurt Ehrke, *Die Ubererzeugung in der Zementindustrie* (Jena: Gustav Fischer, 1933).

15. In the hosiery study, reported production was adjusted for a recording lead based upon the length of the production cycle.

16. Use-depreciation is the *net* loss in market value caused by use of productive assets,

i.e., loss in excess of that caused by deterioration or obsolescence and not prevented by repairs. Use-depreciation will be zero if the loss in value occasioned by obsolescence or by physical deterioration due to the passage of time is not made greater by more intense use. For example, a die for stamping automobile body parts will be obsolete in one year by a planned change in design. It will have zero use-depreciation if no conceivable rate of production could diminish its efficiency or hasten its scrapping. Since the loss in value attributable to use may be reduced or completely prevented by maintenance, use-depreciation represents only the loss in value not restored or avoided by maintenance. To the extent that productive assets are fully maintained in the sense that no residual loss in value results from use, use-depreciation, as we have defined it, may be neglected in estimating marginal cost.

17. In a study of U.S. Steel Corporation costs by Theodore Yntema, the effects of technological progress over the 17-year analysis period were eliminated in four steps: (1) annual observations of input-price-corrected total cost were plotted against output; (2) a least-squares straight line was fitted to the cost–output observations; (3) cost deviations of the observations from the line were plotted against time; (4) a straight line was fitted to measure the relation of these cost deviations in time. This line was used to adjust cost observations to a 1938 level of technology. See U.S. Steel Corporation, *Temporary National Economic Committee Papers,* vol. 1 (New York, 1940), p. 253.

18. Several of these criticisms were ably appraised by Professor J. Johnston in a paper read at the 1957 meetings of the American Economic Association, later published as J. Johnston, "Statistical Cost Function: A Reappraisal," *Review of Economics and Statistics,* vol. 40 (1958), pp. 339–50, and expanded in chapter 6 of his book, *Statistical Cost Analysis.* I have drawn upon Professor Johnston's appraisal in this discussion.

19. The cost-output relationships are valid within the institutional pattern of the number of shifts. Most modern plants operate a single day shift. An extra full shift, half shift, or bottleneck-factor shift, might cause quite different cost behavior at the extreme of the output range.

20. Economic theory is of little help, mainly because of its dedication to the fictional firm that produces a single, homogeneous product. Little attention is given to the practical problems of the multiproduct, multimarket firm, despite the fact that all modern firms produce many products.

21. Hans Staehle, "The Measurement of Statistical Cost Functions: An Appraisal of Some Recent Contributions," *American Economic Review,* vol. 32 (1942), p. 270.

22. J. R. Hicks, *Value and Capital* (New York: Oxford University Press, 1939).

23. Caleb A. Smith, "The Cost–Output Relation of the U.S. Steel Corporation," *Review of Economics and Statistics,* November, 1942.

24. Staehle, pp. 321–33.

25. J. Johnston examines this possible bias in detail and concludes that factor price correction is unlikely to cause a linearity bias, that in fact "in the case postulated by orthodox theory a likely result is an *increase* in the curvilinearity displayed by the cost–output data." *Statistical Cost Analysis,* pp. 171–76.

26. Higher marginal cost caused by lower quality of inputs is explicitly excluded from the diminishing-returns hypothesis we are testing. But cyclical deterioration does occur and is likely to be positively correlated with the plant's output. However, decision rules in some modern plants check a rise in marginal manufacturing costs from this source. A

rule of no second or third shifts and no overtime is not uncommon in U.S. firms. A propensity for equipment of uniform modernity removes or restricts the required condition of laddered inferiority of standby equipment.

27. There isn't just one short run: there are many. Hence the cost observations statistically are not on a single curve but on several curves representing different degrees of incompleteness in the adjustment of the various costs to the rate of output. Complete adjustment is the limit defined as the long-run cost, but incomplete adjustment covers a broad spectrum. Moreover, the cost observation for any one month may be an average of positions on several short-run curves that may mask any one of them.

28. This non-efficient timing is designed to stabilize reported profits by mitigating cyclical fluctuations and to minimize borrowing strain by spending when the kitty is full, and perhaps to minimize after-tax costs by meshing big maintenance outlays with big taxable profits.

29. Richard Ruggles, "The Concept of Linear Total Cost–Output Regression," *American Economic Review,* vol. 31 (1941), pp. 332–35.

30. Stigler, p. 167.

31. Staehle, p. 274.

32. Professor Johnston attacked this criticism on its own theoretical ground, concluding that, on the assumptions made, "the power of accounting observations to detect curvilinearity in the cost function is greater than the power of comparable observations on unit periods, even though the effect of the merging of unit periods is to show a curvature less than that of the 'true' short-run cost function." He added that, "again, even in the case of random outputs, it is still quite possible for the accounting data to have greater power of detecting curvature than do the unit data." *Statistical Cost Analysis,* pp. 181–82.

33. In the late 1930s Sune Carlson attempted to remedy this defect of accepted theory. He expanded the hypothesis by removing the simplifying restraints of homogeneous single-product output and also of a single variable input factor. See Sune Carlson, *A Study of the Pure Theory of Production,* Stockholm Economic Studies no. 9 (London: P. S. King & Staples, 1939). The resulting hypothesis provided for several variable inputs and for multiple products. Nevertheless, the assumption of diminishing marginal increments was retained. The factor mix and the product mix could be varied only subject to assumptions that force marginal cost to rise. Diminishing incremental rate of substitution of one variable input factor for another and increasing marginal rate of substitution of one product for another were assumed. The cost consequence of expanding the model is, as Sir John Hicks pointed out, "that the marginal cost (in money terms) of producing a particular product must rise when output increases, even if the supplies of all factors . . . are treated as variable." *Value and Capital,* p. 87.

34. Exactly what constitutes the "fixed factor" is not specified by theory, partly for greater generality and partly because it differs from decision to decision. Usually it is thought of as the factory building and the machinery. It should also include human capital: the core of executives and the nucleus of skilled, experienced workers who have been accumulated by continuous intangible investment. The bundle does have, in the short run, some outer limit of capacity (stretchable though this appears in any period of emergency). But even thus viewed, the fixed factor is segmentable.

35. In the theoretical analysis the assumption is that the machine units are of uniform efficiency. In practice, however, the machines and their labor complements may differ in

efficiency. If so and if the machines are brought into operation in order of decreasing efficiency when output is increased, a progressive cost curve would result. This source of variation may be unimportant in a particular plant where union rules prevent management from taking advantage of the hierarchy of machine efficiency.

36. Premium pay rates for overtime or for second- and third-shift work can, of course, be quantitatively important and should be taken into account both in optimizing manufacturing operations and forecasting costs for other short-run decisions. Nevertheless, for the narrow purpose of measuring the relation of cost to output rate, these and similar changes in the price of labor and other input factors should be ruled out and examined separately because they are not necessarily and unavoidably related functionally to the rate of output of the plant.

37. Changes in the input mix, even when materials look highly substitutable and when their relative prices have changed enough to make price-motivated substitution appear imperative, occur only after substantial investment in research, testing, and learning. For example, the substitution of man-made rubber for artificial rubber in making Ace combs and billiard balls occurred long after relative prices had signaled for synthetics. Why? Because the switch required a big investment in research, product testing, and market testing, and in learning the changed procedures.

Study 1. Furniture Factory

1. A more debatable use for flexible cost standards is in evaluating inventories. For this purpose they share with fixed standard cost the advantages of timeliness (since they do not await the closings of the books) and of economy (since detailed records for particular lots of product are not required—at least, not for inventory purposes). At the same time flexible cost standards exhibit some of the advantages of historic cost in that they approximate more nearly the actual costs peculiar to the operating conditions of the period.

2. This advantage holds *in toto* only when all the conditions which were allowed for in adjusting the standard fall outside the executive's realm of control and when the cost adjustment to these altered conditions is likewise not subject to his control. But important practical advantages result even when this ideal situation is not realized.

3. The significance of marginal cost and marginal revenue for short-period output adjustments has been pointed out by many economists. A precise exposition is found in Joan Robinson, *The Economics of Imperfect Competition* (London, 1933). A briefer discussion is found in Jacob Viner, "Cost," *Encyclopedia of Social Sciences.* The long-period as well as the short-period marginal behavior is discussed by R. F. Harrod in "Doctrines of Imperfect Competition," *Quarterly Journal of Economics,* May, 1934, pp. 442–70.

4. It is the assumption of orthodox accounting that costs, once they have been reduced to dollars and cents, are homogeneous in behavior and co-ordinate in importance. Accordingly, a figure representing imputed rent upon an owned machine (i.e., overhead) is often considered as having the same managerial significance as a cash outlay for direct materials. Some departures from orthodox average cost acounting have, nevertheless, found their way into business practice in the form of compromises between average cost and outright marginal cost accounting. Examples of this groping for marginal estimates

are seen in the concept of idle burden and in the use of standard burden rates. Both these practices bring cost estimates nearer to marginal figures when the plant is operating at less than the normal rate. Some recognition of the usefulness of such departures from orthodoxy is also being found in the business journals; cf. M. R. Benedict, "Opportunity Cost Basis of the Substitution Method of Farm Management," *Journal of Farm Economics,* XIV, 548.

5. See chap. ii, "Survey of Previous Studies of Cost Behavior."

6. Long-period cost behavior was also analyzed by the author. By using procedures similar to those developed in the short-run study, long-run average and marginal cost curves both for combined cost and for several components were obtained from cost data of a chain finance company. These findings are summarized in Appendix E of Joel Dean, "A Statistical Examination of the Behavior of Average and Marginal Costs," Ph.D. dissertation, University of Chicago, 1936. [For Appendix E, see "Finance Chain" (Study 5).]

7. By fitting curves to total cost data by freehand methods marginal cost estimates which are almost as accurate are obtained in much less time. No objective test of curvilinearity is provided, however, by this method.

8. Curves for most of these cost items are shown in this abstract. Curves depicting the behavior of the remaining cost items can be found in the author's dissertation, chaps. vi and viii.

9. See chap. v, "Reliability of Findings," for a more detailed discussion of these limitations.

10. Marginal cost will remain constant as output increases if average cost is made up of a constant term plus a hyperbolic function—in other words, if variable costs remain constant per unit and fixed costs per unit decline; see Jacob Viner, "Cost and Supply Curves," *Zeitschrift für Nationalökonomie,* III, No. 1 (September 25, 1931), 1–48.

11. This statement is not incompatible with the estimate, for the firm studied, that marginal cost is constant with respect to output. A constant variable cost per unit together with a hyperbolic fixed cost function will yield constant marginal cost yet permit variability in average and total cost.

12. Cost increments can also be computed for other independent factors, such as the size of lot, number of new styles, etc. But the increment associated with output is most important and most familiar and is, therefore, the one usually meant by the term "marginal cost."

13. C. B. Williams, "Treatment of Costs during Periods of Varying Volumes of Production," *Journal of Accounting,* XXXII (1921), 329–33.

14. *Budgetary Control* (New York: Ronald Press Co., 1923), p. 193.

15. "A Technique for the Chief Executive," *Bulletin of the Taylor Society,* VII (1922), 47–68.

16. The essential principle is thus stated by C. E. Knoeppel, "Must Fixed Expense Be Fixed?" *Factory and Industrial Management,* September, 1931, p. 323: "While the sales income line begins at zero and crosses the 100% capacity line at a given point, the total cost line does not begin at zero, but at a point where fixed costs end. Somewhere up the capacity scale costs and sales lines cross; profits lie above this cross-over point; losses below."

17. "Three Budget Compasses Give SKF Industry Smooth Sailing," *Automotive Industries,* April 30, 1932, p. 211.

18. "Budgeting Simplified by Separating Fixed and Fluctuating Costs," *American Accountant,* XVI (1931), 40–45.

19. Ibid., pp. 47–68. He suggests here a formula for interpolating such a cost in the intermediate range: Cost = (Expense at minimum expected level of output) + (Proportional cost as a percentage of sales volume).

20. "Accounting for Fixed Expenditure," *Factory and Industrial Management,* LXXVI (1928), 914–16.

21. "Smoothing the Road to Predetermined Costs," ibid., LXXIX (1930), 1069–70.

22. "Cost Accounting and Budget Making," *Bulletin of the Taylor Society,* XVI (1931), 102.

23. *The Economics of Overhead Cost* (Chicago: University of Chicago Press, 1923), p. 180.

24. "The Economic Characteristics of Manufacturing Industries," *Mechanical Engineering,* November, 1932, pp. 830–55.

25. "Actual Examples of the Advantageous Use of Operating Budgets," *Yearbook of the National Association of Cost Accountants* (1923), pp. 140–44.

26. Ibid., p. 140.

27. Clark, p. 185.

28. Smith, pp. 40–43.

29. Bates, pp. 143–44.

30. "Overhead in Economics and Accounting," *Bulletin of National Association of Cost Accountants,* VII (1926), 587.

31. E. L. Grant, "Increment Costs and Sunk Costs," *Mechanical Engineering,* October, 1934, p. 586.

32. "Variability of Overhead Costs," *Journal of Accounting,* XLIX (1930), 202–206.

33. "Taking the Guess out of Overhead Costs," *Industrial Management,* LXV (1923), 324–26.

34. Floyd F. Hovey, e.g., in his "Present Cost Facts Graphically" (*Factory and Industrial Management,* LXXXI [1931], 599–601) recognizes the complexity of the causes for cost fluctuation.

35. This marginal cost interpretation, however, was not given to the results by Professor Crum; see his "Statistical Allocation of Joint Costs of Railroad Freight and Passenger Service," *Journal of the American Statistical Association,* XXI (1926), 18.

36. "Control of Operating Expenses," *Mechanical Engineering,* LIV (1932), 711–17.

37. The prices of input factors frequently constitute an important distorting influence upon cost behavior. Effective wage rates are sometimes positively correlated with output, however, and hence are one factor which would cause marginal cost to rise.

38. Unpublished computations of marginal cost for a finance company.

39. "A Study of Factors Influencing the Cost of the Small Loan Business," a master's thesis written under the direction of Theodore O. Yntema, University of Chicago, 1934.

40. Previous statistical studies of cost behavior have not, so far as the author can learn, carried the analysis to the constituents of combined cost. See chap. ii for a discussion of these studies.

41. See chap. iv of the author's thesis, "A Statistical Examination of the Behavior of Average and Marginal Costs," for an elaboration of this statement.

42. See J. M. Clark, *The Economics of Overhead Cost,* p. 50.

43. Input measures are strictly suitable for determining marginal costs only when the input item is proportional to output. They have, nevertheless, been employed as a rough index of the level of output when the variety of finished product is so great as to prevent a common denominator.

44. In a study of joint costs in railroad operation W. L. Crum used only two independent variables and made no modifications of data. This was defended on the ground that disturbing variables would tend to be distributed sufficiently evenly so as to have negligible effects ("Statistical Allocation of Joint Costs of Railroad Freight and Passenger Service," *Journal of the American Statistical Association*, XXI [1926], p. 22).

45. It should be emphasized that the attempt to remove from the data the effect of all price fluctuations is based upon the assumption that these price fluctuations are not a function of the output of the particular firm. The indications are that this assumption is not entirely correct. It appears reasonable to suppose that there is some positive correlation between the rate of output and the effective prices paid for the input factors (labor and material). If the output rate of this concern is positively correlated with the output of the furniture industry and if the output of the industry is positively correlated with the prices of input factors, then the output of the concern is related to the prices of cost components even though competition were approximately atomistic. If the market is monopolistic, that is, if this firm represents an appreciable proportion of the demand for a certain type of factor, then its price may fluctuate more directly and markedly in response to changes in the output of this firm. For example, executives are of the opinion that savings from quantity discounts on certain materials are more than counterbalanced by premiums demanded for rush delivery required when output is great. Similarly the effective price paid for certain types of labor may fluctuate roughly with output, because this firm's requirements constitute the most important demand for this labor in the town in which the company is located. Even though wage rates are not very flexible, the quality of worker obtained at a given rate does appear to vary with output; this results in variation in the price per unit of work. To remove these price changes which are caused by output fluctuations results in an erroneous estimate of the effect of output upon cost.

46. For a more complete description of methods used and a discussion of alternative procedures see the Appendix of the author's dissertation previously cited.

47. A brief summary of the methodology is found in this chapter. The reader who is interested in a more complete description and evalution of the methods employed will find such a discussion in the author's dissertation, "A Statistical Examination of the Behavior of Average and Marginal Costs," Appendix B.

48. For a discussion of this contribution see chap. viii, "Practical Application of Findings."

49. Marginal cost can be assigned more definitely, but not necessarily more accurately. This allocation, of course, only arbitrarily cuts the Gordian knot of a problem which strictly is soluble only when it is possible to vary the proportions of the cost elements entering into the product and to determine the incremental cost of alternative combinations. All allocations of joint cost short of this rather theoretical solution are arbitrary. For a more complete discussion see the author's dissertation, chap. x.

50. Costs are reduced to averages per dollar of old warehouse value of output. The starred items are discussed briefly in this abstract and more fully in the author's dissertation, chap. vii and Appendix B.

51. This ratio is not a pure index of the quality of materials since standard cost differs from old warehouse value in other respects.

52. Margins of combined cost for size of production order (X_3), number of new styles (X_4), change in output (X_5), and labor turnover (X_7) were computed in addition to the more familiar output margin (in this study called marginal cost). In investigating the components of combined cost, however, only the output margin was studied.

53. Chap. viii and Appendix B.

54. If the output of this concern varied with fluctuations in general business, then the prices of those input items which responded sensitively to business conditions would tend to be positively correlated with output. In that event the undeflated money cost curves would be related indirectly to the quantity of the firm's product which was demanded, but not to the shape or the position of these demand schedules themselves.

55. Professor Viner suggests that the workers selected for retention are likely to be those having high output during good periods. These men, because of their greater psychological instability, may be the very ones whose capacity to produce is most curtailed by the psychological effects of bad times.

56. Mordecai Ezekiel, *Methods of Correlation Analysis* (New York: John Wiley & Sons, Inc., 1930), p. 169.

57. H. R. Tolley, "Economic Data from the Sampling Point of View," *Journal of the American Statistical Association,* XXIV (1929), 70 (suppl.).

58. The difficulty of finding a random sample for economic data led Yule to develop the "stratified sample" and Bowley to work out a "purposive sample." The small sample of accounting periods which was used for the subsidiary correlation analyses is a sort of cross between these two types of samples. It attempts to cover the range of all independent variables, but was selected with certain definite ends in view. It has already been noted that this selection, with respect to the values of the dependent variables, does not tend to alter the regression curves, but does somewhat change the correlation coefficient.

59. Some serial correlation was noted in the corrected data, since successive accounting periods tended to have deviations of the same sign from fitted curves. This common characteristic of time series data makes error formulas tend to overstate the probability of recurrence of observed relationships in other samples taken from the same universe.

60. Charts 1–2 and 1–9 clearly indicate that two of the independent factors (X_3 and X_5) have approximately normal distribution, that two others (X_2 and X_8) have a flat-topped, fairly symmetrical distribution, and that still two others (X_4 and X_7) have extremely irregular distribution. The distribution of most cost items (dependent variables) seems to be skewed toward low costs.

61. The neoclassical theorists talk about other cost factors in addition to output, but concentrate their attention upon the effects of output and dispose of the other variables by the assumption that "all other things remain equal." The present analysis recognizes that, in fact, these other factors do not remain the same and that it is important to measure the actual effects upon cost of their variation.

62. A parallel analysis of several of the components of combined cost is summarized in

Appendix A of this monograph. For a more complete statement of these results see the author's dissertation, "A Statistical Examination of the Behavior of Average and Marginal Costs," chap. vii and Appendix C.

63. Indexes of part correlation and of part determination were computed for direct labor cost. The experiment was not extended to other cost items because this type of result appeared to be irrelevant to the major objective of the study. These experimental computations are found in the author's dissertation, Appendix B.

64. See Jacob Viner, "Cost and Supply Curves," *Zeitschrift für Nationalökonomie,* III, No. 1, 27.

65. No attempt has been made in this summary to explore exhaustively the causes for the cost behavior which was found to be associated with output. Cost reduction consequent upon higher output may be due to steadier rate of flow of production through the plant, finer subdivision of task, or better morale of the force, as well as to the spreading of fixed costs.

66. Strangely enough, however, this curve actually rises for a brief interval in its initial stages. The explanation for this illogical initial rise is difficult to find. Apparently it is caused in part by incomplete removal of irrelevant cost influences from the corrected data (the curve for uncorrected cost rose throughout its course) and in part by too flexible curve fitting.

67. Moreover, because these grudgingly shrinking costs lag behind the rapidly contracting volume, it is probable that a large output contraction would cause a higher cost than a small contraction. Hence it is clear that the curve should have high values for large decreases and lower values for small decreases. Whether the curve should continue downward is less certain. There are, however, two considerations—the limited divisibility of most cost items (as compared with the divisibility of output) and the tendency to overload a productive factor during a brief period of expansion—which make it appear reasonable that the curve of combined cost should continue to decline throughout the range of increases.

68. The amount and direction of change in output from that of the previous period represent only part of the pertinent characteristics of this change. The output level from which the increase took place, the number and amount of successive increases or decreases preceding this one, the length of time for which the new level could be expected to be maintained, and their combinations and permutations should also be recognized in a complete allowance for the pattern of change in output.

69. Error formulas based upon a simple sample theory are not strictly applicable to a selected sample from a time series, especially when these observations appear to be intercorrelated chronologically. Additional error, also not allowed for in these formulas, was added when it was found necessary to drop a few extreme observations from the charts; see chap. v for evaluation of the applicability of these error formulas.

70. Ezekiel, *Methods of Correlation Analysis,* p. 271.

71. Ibid., p. 178.

72. The peculiarities of cost behavior of a particular firm are not a very sound foundation for generalizations regarding the validity of a method. It must be recognized that such correction is necessary only when the distortion removed would affect results to a quantitatively significant degree.

73. By marginal cost is meant the increment in cost which is consequent upon a unit increment in output. In this study that unit is one dollar of old warehouse value. The curve of marginal cost is defined as the locus of this increment in cost over a given range of output. Cost increments associated with other operating conditions are also of interest to management. They are discussed at the end of the chapter.

74. The procedure actually followed for corrected combined cost totals differed from that for data in the form of averages primarily in the step at which the conversion from averages to totals took place. With average cost data the transformation came after a function had been fitted to the observations; with total cost data, on the other hand, the conversion was affected individually for each accounting period before fitting a curve.

75. This parabola was found to have the following equation:

$$X_T = 5.46 + 1.7759X_2 - .116X_2^2.$$

When differentiated, this equation yielded as the marginal cost function the following declining straight line:

$$X_M = 1.7759 - .0232X_2.$$

76. Jacob Viner makes the suggestion that neoclassical theory may not have taken enough factors into consideration. The quasi-physical relationships considered in the law of diminishing returns may be more than offset by organizational and psychological influences which work counter to them.

77. There is some question whether marginal production costs are more likely to be constant or declining under monopolistic competition than under perfect competition. Much depends upon the conditions of supply of the input factors. The point at issue here is the conditions for individual firm equilibrium. Although equilibrium for the economy as a whole is never realized and even though there may not even be a tendency toward such equilibrium, equilibrium for the individual firm does tend to be approached if the enterprise is rationally managed.

78. This procedure consists of: (1) visual fitting of a curve to the average cost observations; (2) tabular transformation of this curve into a total cost series; and (3) arithmetic differentiation of this function into a marginal cost series of first differences which is smoothed, by means of moving averages, to a marginal cost curve.

79. As noted above, a hyperbolic function plus a constant would have paralleled the total cost function which gave reliable results.

80. The least-squares-derivative method has, however, some important superiorities over other techniques of deriving marginal cost from average cost series. In the first place, its least-squares method of curve fitting yields an average cost curve which is definite and comparatively free from personal bias. In the second place, the marginal and the total cost curves are related to average cost in a precise manner which assures the internal consistency of the findings. And, in the third place, the results of the cost analysis are expressed in a form which, while concise, is yet sufficiently general to determine the curve precisely for all values of output. However, to offset these advantages of objectivity, precision, and conciseness we find certain defects: (1) the curve may fail to represent actual cost behavior because of poor choice of function or inability to abstract from nontypical cost behavior; (2) the procedure is laborious and somewhat difficult to understand; and (3) the results are difficult to interpret to the group who will use them.

81. The omission of cost items whose total is unchanged by changes in output in no way influences marginal cost, for the derivative of a constant is zero. Some administrative and selling items are of this type.

82. Quite apart from changes in the prices of input factors (for which allowance can easily be made), the character of the cost relationship itself may be different for various phases of the business cycle. The marginal cost function obtained here is based upon the assumption that the combination of components found at each level of output is the least cost or optimum combination for that output. Cyclical forces may cause fluctuation in the degree to which this assumption holds true.

83. One previously discussed measure of statistical significance is the standard error of departures of a net regression curve from the curve value at the mean of output (Table 1–3). Another measure of reliability is the standard error of estimate about the functions fitted to total cost residuals. For the straight line this constant was found to be $4,243 (O.W.V.) and for the second-degree parabola, $3,763 (O.W.V.). However, the width of these bands exaggerates the reliability of these functions since the departures of observations from these curves are actually serially correlated. A more pertinent measure of the reliability of this marginal cost estimate would be the standard error of the slope of the total cost net regression for output. Unfortunately such standard errors cannot yet be determined for graphic correlations.

84. W. L. Crum, in his statistical study of railroad costs, indicated his belief that the distortion of his cost data was distributed in such a manner as to have little effect upon the relationship under investigation ("Statistical Allocation of Joint Costs of Railroad Freight and Passenger Service," *Journal of the American Statistical Association*, XXI, 22).

85. These data had been adjusted only by the omission of three unimportant accounts (teaming, traveling, and interplant prorations). These omissions were necessary in order to make the findings comparable since these accounts had also been dropped from the refined cost data.

86. It was suggested in chap. iii that the prices of input factors may be positively correlated with output. If this correlation is quantitatively significant and if this type of price variation has been removed by our correction procedure (which is doubtful), then corrected marginal cost would be more likely to decline than would uncorrected. The fact that the reverse was found to be true indicates that this type of price variation was not important but that the price and lag corrections, taken together, were actually negatively correlated with output.

87. The possibility of an abridged procedure which would effect enormous savings is suggested by the improvement in accuracy and reliability resulting from this latter partial correction. Under certain conditions even this correction could be dispensed with and rough approximations to marginal cost could be made by visually fitting a function to the raw total cost data. This procedure could be used, however, only during periods of relative price stability when no important output lag existed and when factors other than rate of production did not cause important cost distortion.

88. The finance company whose data are used for the analysis of long-period cost behavior attempted to measure and control this quality margin by grading loans according to quality (unpublished study by T. O. Yntema).

89. It should be pointed out that these increments need not be computed with great

precision for, as a rule, the estimates with which they are compared are subject to such a wide margin of error that meticulous refinement of these cost margins is not worth while.

90. Statistical exploration of this type is useful in calling attention to the limitations of the simplified assumptions of the neo-classical price theory.

91. The level at which the other factors are held constant may also have some effect upon the magnitude of the increment for the varied factor. Such information is not provided by this study since the independent variables were, in effect, held constant at their means.

92. Even allowance for these other margins of cost is occasionally insufficient since a particular order may require special additional expenses (such as retooling) over and above those included in the typical increments of cost.

93. Since the distribution of residuals was approximately linear and since insufficient a priori grounds could be found for a curvilinear regression, a straight line was thought to be sufficiently accurate for the purpose.

94. This margin should be negative since it represents the cost increment of making the same output in larger rather than in smaller lots. The proportionately variable costs would have no marginal cost since they would be constant, whereas the fixed costs of the production order would decline, since there would be fewer setups.

95. See also chap. viii and Appendix C of the author's dissertation, "A Statistical Examination of the Behavior of Average and Marginal Costs," for a more detailed discussion of the regression curves of the constituents of combined cost.

96. Livingston advocates the use of the gross regression line of a linear correlation analysis as a standard of expense (*Control of Operating Expenses,* p. 716).

97. These minor uses have been discussed in chap. vi.

98. For a brief description of the break-even chart see Rautenstrauch, *The Economic Characteristics of Manufacturing Industries,* p. 223. Note that allowance is made for only one operating condition, namely, output.

99. See Jacob Viner, "Cost," *Encyclopedia of Social Sciences,* p. 467. Frequently considerations other than maximization of immediate profit influence price policy.

100. If the market is not perfectly sectionalized, i.e., if sales to chain stores involve loss of sales through other channels, then the lowest acceptable price will be higher than marginal cost since the marginal revenue, which must not be lower than marginal cost, will then be less than the price.

101. In computing marginal cost the other operating conditions were, in effect, held constant at their respective means. The level at which they are held constant may have some effect upon the marginal cost function. This effect was not investigated in this study.

102. This average method assumes, first, that the old warehouse value of the article under discussion is a correct measure of output and, second, that this particular article corresponds to the average of the products manufactured with respect to the behavior and the relative importance of its cost constituents. Only under such conditions could the computed average ratio of marginal cost to old warehouse value be validly applied to any particular article of furniture. See Appendix B for discussion of these assumptions.

103. In the first place, it should be remembered that these components of marginal cost are not partial derivatives of the total cost function; they represent, instead, the increment of cost of the particular component which is associated with a unit increase in output with other variable cost items increasing as they typically do. In the second

place, it is assumed that old warehouse value measures output accurately and that standard average cost and its components are satisfactory indicators of the proportions of the cost constituents. Third, we assume that the ratio of marginal cost under any particular set of operating conditions to average cost under standard or normal conditions is the same for all articles manufactured. Fourth, it is supposed that for any given size and combination of output the proportions of cost constituents are fixed within rather narrow limits. These assumptions are discussed more fully in Appendix B.

104. The need for extrapolating any cost estimate used in making policy decisions should be obvious. Estimates of past marginal cost are of managerial interest solely as a basis for predicting future marginal cost, for it is this extrapolated cost which is balanced against expected marginal revenue.

105. It should be pointed out that the average cost curves of components therefore do not precisely represent what the economist calls the "technical coefficients of production," for these curves do not show the cost of a particular input factor per unit of output when all other input components have been held constant. These curves may, nevertheless, give some rough indication of the behavior of these technical coefficients.

106. Direct labor cost, though labeled "direct," is not a purely variable cost but contains elements of fixed cost. Moreover, labor tends to be more highly specialized and better organized as the plant approaches the output levels for which it was designed. Output had apparently not reached (in the range of observations included) the point at which the operation of the law of diminishing returns begins to cause cost increases. Psychological and organizational factors may more than offset this latter tendency over the practical range of output.

107. When the constant total of this portion of maintenance cost is spread over a larger output, the cost per unit declines.

108. This practice results in incorrect allocation of cost between periods. Part of this cost should properly have been charged to past periods when the machines were actually worn out. Part should be charged to future periods when the working force which is held together in this manner will begin to function again with smaller initial recruiting cost than otherwise would be the case.

109. The meaning of correlation indexes obtained from cost data in the form of averages per unit of output is not parellel to, or exactly comparable with, these secured from data in the form of totals.

110. "A Statistical Examination of the Behavior of Average and Marginal Costs," chap. vii and Appendix C.

111. In this particular study the check could not be used since not all the elements of combined cost were studied.

112. To the total cost data of any particular component a function (usually a straight line) was fitted freehand, then converted, by the method of selected points, into an equation which was differentiated to obtain marginal cost. Data in the form of averages were studied in a similar fashion. To a curve, drawn freehand, an equation was fitted by the method of selected points. This equation was transformed to a total cost equation and differentiated to obtain the marginal cost function.

113. The symbols used in these equations have the following meanings:

X_2, output in terms of thousands of dollars of old warehouse value

X_T, total cost in thousands of deflated dollars

X_A, average deflated cost per unit of output, i.e., per dollar of old warehouse value

X_M, marginal cost (unit cost, one deflated dollar) of an increase in output of one dollar of old warehouse value.

114. For discussion of the average cost behavior of this cost item see the author's dissertation, chap. vii and Appendix C.

115. In the first place, data in the form of averages rather than in the preferred form of totals were used to determine this curve; second, the scatter of observations about the average cost curve was so great that its departures from the horizontal appeared to have limited statistical significance beyond $30,000 output (Table 1–8); third, the fit of the average cost equation to this graphically determined curve was not good at large outputs. Consequently the increase in marginal cost estimates for these higher production levels is probably incorrect.

116. The sample was small and was widely scattered in a distribution which seemed to be linear. The standard error of this total cost curve (Table 1–9) indicated that its convolutions were not statistically significant.

117. There are three serious limitations upon the reliability of this graphic average curve: (1) imperfect correlation of the data for price changes; (2) the limited success of the multiple correlation analysis; and (3) the small number of widely scattered observations upon which the curve was based. These limitations are more fully discussed in chap. vii and Appendix C of the author's dissertation.

118. Confidence in this curve of marginal cost is shaken, first, by the fact that data in the form of averages were used; second, by the small size of the sample and the wide scatter of the observations; and, third, by the fact that only a small part of the distorting effect of other variables was removed by the multiple correlation analysis (see Appendix A).

119. To be controllable variance must be capable of reduction and means of reducing it must fall within the jurisdictional domain of the executive held responsible for the variance.

120. The extent to which this statistically assignable variance is actually noncontrollable will be discussed in a subsequent section of this paper.

121. It is the contention of the author that cost variation in excess of that which can be accounted for by this statistical investigation is more likely to be attributable to the effectiveness of executives than that which has been measured statistically. This contention, which is almost of the order of an assumption, will be discussed in the third section of this appendix.

122. As a common denominator for measuring the diverse output of this firm the deflated standard cost of the various products was used. By removing price fluctuations through deflation a strictly physical common multiple was devised for summing the various items of finished product. The total deflated standard cost value of output was expressed as a percentage of total deflated standard cost value at rated capacity to obtain a measure of output.

123. It is recognized that number of styles is a crude index of style change. An attempt to weight this index for the varying amount of production disturbance caused by each new style was not, however, successful.

124. The method of graphic multiple correlation analysis is sufficiently well-known

so that it need not be described in detail in this paper. Those unfamiliar with this simple and highly useful technique should consult Ezekiel, *Methods of Correlation Analysis,* chap. 16, and L. H. Bean, "A Simplified Method of Graphic Curvilinear Correlation," *Journal of the American Statistical Association,* vol. 24, pp. 386–397, December, 1929.

125. It is significant to note that actual costs do not typically go down to the level of standard costs until the firm is operating at 90% or more of rated capacity. This may mean that the entire level of standard cost is too low; that is to say, that the firm records unattainable goals rather than estimates of likely cost. This constant error does not affect the shape or the slope of the regression curve, but only to intercept (general level). It is, however, isolated by the curves as an element of variance for which the production management should not be held accountable.

126. Variance due to price changes was presumably removed by adjusting the standard cost of each item in response to changes in the prices of its labor and material ingredients. The relatively small proportion of variance which was accounted for by the correlation analysis may indicate that the effect of price fluctuations was only incompletely removed.

127. Variance attributed to output, size of lot, and number of new styles is, of course, due in part to managerial incompetence in adjusting costs to these altered operating conditions. But since this cost sluggishness was typically associated with these operating factors in the past, it may, therefore, for control purposes be regarded as primarily a function of the operating condition. Moreover, it should be remembered that, for the most part, this statistically allocated variance is actually noncontrollable. Not only are the operating conditions themselves beyond the control of the production management, but the short-period adjustment of cost to the changed conditions probably is largely beyond control.

Study 2. Hosiery Mill

1. It should be made quite clear at the outset that the costs referred to throughout are short-run costs, i.e., costs incurred by a firm whose physical characteristics are given in the form of fixed plant facilities, etc.

2. The level at which the prices of input services are stabilized can, of course, be different, in which case one can observe the influence of changes in operating conditions for different sets of prices. This procedure is illustrated in chap. v on the "Adjustment of Cost Functions for Practical Use."

3. It cannot be too strongly emphasized that the short-run cost function which is sought here will not be descriptive of actual cost behavior. It will in general be true that near-capacity rates of output of any one firm will occur in the rising phase of the business cycle when other firms will also be expanding operations. While an individual firm may be in itself sufficiently small to have no influence on factor prices, increases in its output will be highly correlated with a rise in the output of the industry, which will generally exert a significant influence on prices of factors. Consequently, as a firm increases its rate of output to a marked extent, this will be accompanied by a rise in factor prices with consequently rising marginal cost. The analysis of cost behavior under these conditions is, of course, a significant problem; but the treatment here is an attempt to approximate a static model where prices are assumed to remain unchanged no matter what the rate of output of the individual firm.

4. The total cost of operation of an enterprise for a specified period will be the sum of the products of the amounts of productive services employed and their prices. Since the prices are considered to be independent of the actions of the enterprise, i.e., constants from the firm's point of view, the total cost is a function of the amounts of the various productive services employed. However, because these services are themselves functions of output, total cost can be shown as a function of output alone.

5. In the theoretical analysis the asumption is that the machine units are of uniform efficiency. In practice, however, the machines and their labor complements may differ in efficiency. If so and if the machines are brought into operation in order of decreasing efficiency when output is increased, a progressive cost curve would result. This source of variation is inadmissible in the theoretical treatment and is furthermore practically unimportant in the particular plant studied because union rules prevent management from taking advantage of the hierarchy of machine efficiency.

6. It is conceivable that the proportions in which the services of the agents of production are used may be either constant or variable for each of the kinds of segmentation, but it is likely that unit segmentation and time segmentation will be accompanied by constant proportions of factors over the relevant range, whereas speed segmentation may involve variability of factor proportions.

7. E.g., this is the situation envisaged in the Knightian product curve (Frank Knight, *Risk, Uncertainty, and Profit* [Boston, 1921 (London, 1933)], p. 100).

8. "Combined" cost is here used synonymously with the sum of fixed and variable cost or aggregate cost and is the sum of the component costs: overhead, productive labor cost and nonproductive labor cost, a classification which is adopted in order to secure conformity with the accounting terminology of the hosiery mill. Combined cost does not necessarily include all costs, however, some of which may be omitted for a number of reasons discussed below. The individual constituents of the components of cost are referred to as elements of cost throughout the paper.

9. The choice of a cubic function to represent the pattern of cost behavior in this instance is based on its convenience and simplicity. While it is the simplest functional form descriptive of the behavior postulated, it is, of course, not the only admissible form.

10. Essentially, the procedure in the graphic method of multiple correlation analysis is to determine the change in the dependent variable attributable to variations in one of the independent variables in a range where the other independent variable is relatively constant, i.e., where it exerts little influence. The values of the dependent and independent variables are then plotted on co-ordinate axes and a regression line is drawn. This is repeated for different ranges of constancy of the independent variable whose influence is to be isolated. From the set of estimated regression lines thus obtained one general regression line is drawn in. The deviations about this regression line are then plotted against the remaining independent variable, and a second regression line is drawn to show the new relation. The deviations about the second regression line can then be replotted about the first in order to improve the first approximation. This process of successive approximation may be repeated ad libitum until a satisfactorily close fit is obtained. The regression lines obtained will be net regressions showing the net effect of one independent variable when the influence of the other has been allowed for. For a complete description of the technique see Mordecai Ezekiel, *Methods of Correlation Analysis* (New York: John Wiley & Sons, Inc., 1930), chap. xvi. See also a series of articles by Bean, Ezekiel, Malenbaum,

and Black on "The Short-Cut Graphic Method of Multiple Correlation," in the *Quarterly Journal of Economics,* Vol. LIV (February, 1940). An illustrative case of use of graphic methods in cost analysis is found in Joel Dean, *Statistical Determination of Costs, with Special Reference to Marginal Costs* (Chicago: University of Chicago Press, 1936), pp. 24–107 [see "Furniture Factory" (Study 1)].

11. For a brief rigorous treatment of the technique see Henry Schultz, *The Theory and Measurement of Demand* (Chicago: University of Chicago Press, 1938), Appen. C.

12. A more complete discussion of these independent variables is offered in chap. viii, "The Selection and Measurement of Independent Variables."

13. The "simple" regression referred to should be carefully distinguished from the "net" or "partial" regressions referred to earlier in the description of the statistical method.

14. Furthermore, statistical examination of the relation of cost and output first differences (an approximation to marginal cost) and of the relation of average cost to output supported the hypothesis of linearity of the total cost function. Despite the support given the linear total cost specification by the analysis of the production techniques and by the distribution of total cost observations and the behavior of average cost and the approximation to marginal cost, a cubic function was also specified and fitted by least-squares regression analysis. The results are shown in Chart 2–12. The higher-order function did not appear to fit the data significantly better than the linear function. This problem is discussed in more detail and the results of the analyses are presented in Section B of chap. iv.

15. This is also shown by comparing the simple regression equation for combined cost, ${}_cX_1 = 2935.59 + 1.998X_2$, and the partial regression equation, ${}_cX_1 = 762.54 + 2.068X_2$.

16. The mathematics of this transformation may be illustrated by the following equations for monthly combined cost, where ${}_cX_1$ is total combined cost, and X_2 is output. The partial regression equation for combined cost in the form of monthly totals was found to be

$${}_cX_1 = 762.54 + 2.068X_2.$$

By dividing this equation through by X_2, the following equation for the combined cost per dozen was obtained:

$$\frac{{}_cX_1}{X_2} = 2.068 + \frac{762.54}{X_2}.$$

By differentiating the total cost function with respect to X_2, the output, the resulting first derivative gives the marginal cost function as

$$\frac{d\,{}_cX_1}{dX_2} = 2.068.$$

17. It should be remembered that these costs do not include raw-material costs.

18. To be more precise, the marginal cost derived from the simple regression function is \$1.998, while the estimate of marginal cost obtained from the partial regression function is \$2.068.

19. See chap. v, Sec. B, for a more complete discussion.

20. Obviously, silk cost must be added to the marginal operating cost estimated in this study to determine the absolute level of over-all marginal cost. This procedure is illustrated in chap. v, where the practical applications of the marginal cost estimates are discussed.

21. This figure of 2.2 per cent includes the following elements of cost: suppers and overtime; electric, sundry, and sanitary supplies and expense; coal; stationery and print-

ing; line-elimination labor and parts; traveling expenses, oil, learners (all operations), style changes, machine cleaners, starting up machines, building repairs, and replacement labor. The costs omitted but not accounted for in the above percentage are not properly production costs and are small in any case. They are samples, freight and express, machine conversion, and chauffeurs and helpers.

22. This was the case for one small cost element, the heat, light, and power account for four winter months.

23. Since marginal cost is given by the slope of the total cost function, the shape of the total cost function can be determined immediately when the marginal cost function is known.

24. For the geometric derivation of marginal and average cost curves see H. von Stackelberg, *Grundlagen einer reinen Kostentheorie* (Vienna: Julius Springer, 1932), pp. 25–27.

25. The determination of the marginal cost approximation is shown in the following table:

Output x	Cost y	First Difference $y_i - y_i - 1$	Average First Difference $y_i - y_i - 1/x_i - x_i - 1$
x_1	y_1	$y_2 - y_1$	$y_2 - y_1/x_2 - x_1$
x_2	y_2	$y_3 - y_2$	$y_3 - y_2/x_3 - x_2$
x_3	y_3	$y_4 - y_3$	$y_4 - y_3/x_4 - x_3$
x_4	y_4	....	....

If the output series is plotted against the corresponding average first difference, an approximation to marginal cost at various levels of output can be obtained.

26. It is true that an increase in the number of constants in the regression equation will improve the fit in the sense that the sum of the squared residuals about the regression line will be diminished. This process can, of course, be carried to the point where no further improvement is possible. However, the question is whether or not the improvement in the fit obtained by adding more constants can be considered significant.

27. For an example of the use of several applications of the analysis of variance in testing the form of a total cost function and for a discussion of the underlying assumptions see Joel Dean, *Statistically Determined Cost Functions of a Leather Fabricator* (New York: National Bureau of Economic Research, 1941), Sec. 7. [Reprinted in the present volume as "Belt Shop" (Study 3)].

28. The regression equations on which these estimates are based are shown in Table 2–3. Corresponding similarity was found for estimates of average cost and marginal cost.

29. The fixed factors embodied in a given machine unit are not fixed in the most inclusive sense of the term, since they may be withdrawn or introduced into the production process at will. However, costs such as depreciation, interest, taxes, and insurance on the machinery are not allocated to products on a machine-center basis but are recorded as constant in total regardless of how many machine units are used. Variable overhead costs such as power, oil, and maintenance, on the other hand, are reported as incurred (as nearly as accounting practice permits). Thus the method of recording costs obviates discontinuities in the cost function that might appear as a result of other methods of recording fixed costs. If variations in proportions of factors used in a machine unit occur and cause curvi-

linear total cost functions for the units, their effect on mill costs is small and is averaged out in the statistical process of fitting a function to the observations.

30. If the correction can be made for the major components of cost rather than for their constituent elements, the procedure of reflation is considerably simplified. If this simplification introduces considerable error, a procedure similar to that suggested here for components can be followed for each constituent element.

31. By a multiple correlation analysis of the relation of the general firm overhead to the output of this hosiery mill and the output of other plants used as independent variables, it is possible to obtain some knowledge of the influence of the firm's output on the general overhead. The partial regression of general overhead and output of this particular mill would indicate the contribution of firm overhead to this mill's marginal production cost. Rough graphic analysis indicated that this regression was not statistically significant, so that apparently no allowance is needed for this omission.

32. Because of the relative homogeneity of the output of this mill, style and specification variety were found to have no significant influence on cost behavior. Hence, no allowance was made for them in the index of output, and the index was essentially a simple count of whole production *plus* weighted transfer production.

33. Whereas some years ago the threads were twisted only enough to hold them together, about three or four turns to the inch being necessary, more recently it has been found that more twisting can be used to give increased sheerness and dullness, as well as greater elasticity, until today stockings are knit from yarn with as many as seventy turns to the inch.

34. The reason for making legs on one machine and feet on another is that, in order to secure an exact fit and especially in order to have the right amount of fabric across the instep, the foot must be put on at right angles to the leg.

35. The reason for the existence of the two types of joining is that the seam, though very fine, forms a light ridge which would be objectionable at the two points looped, since the pressure of the foot is greatest at these points. While it might be desirable to loop the entire stocking, this is not feasible from an economical manufacturing standpoint. The looping operation is too slow and costly, as each stitch, in order to be looped, must first be put on points—similar to the transfer points—of the footer.

36. For a discussion of the significance of three-thread carrier attachments see below, Sec. B, subsec. 2, on the characteristics of the plant selected.

37. Although one knitting mill was chosen, a cursory survey of two of the other mills was undertaken after completion of the study of the plant selected. The results of these additional surveys are not presented.

38. In addition to the eighty-one eighteen-section knitting machines there were four sixteen-section knitters, but the latter were not in operation during the period selected for analysis.

39. Technical changes were confined to the installation of three-thread carrier attachments and half-speed shock absorbers, which affect cost negligibly. No new machinery was installed, and machines were not speeded up.

40. There were originally fifty-four observations, some of which were dropped subsequently because of special considerations that made them noncomparable with the other months.

41. It was explained earlier that raw-material costs were not included in the analysis

for several reasons. First, expenditures for silk were available only in quarterly form and could not be accurately allocated to constituent months. Furthermore, this omission was not considered to affect the usefulness of the results, since the technical coefficient of production for raw material was almost certainly constant and accurately known by the executives of the hosiery company.

42. Price rectification specifically assumes that substitution among input factors in the short run did not take place as a result of changes in their relative prices. Examination of the mill's production technique indicated the first assumption to be correct for the time period studied. The procedure assumes, moreover, that changes in the plant's rate of output had no influence upon prices paid for input factors. The mill's individual effects upon the factor markets also appeared to be negligible.

43. It is likely also that there exist certain deficiencies in the accounting technique which give rise to errors and inadequacies in the accounting data. There is little to be done to allow for this, however.

44. The hosiery labor market in which this plant hired was so large and so well organized that the changes in demand for workers resulting from changes in the output of this one plant could have had no significant effect upon wage rates. Hence the assumption of the independence of wage scales and plant output rate, implied in this rectification, appears to be justified.

45. A list of these accounts has been given in the earlier section of this chapter.

46. These ratios, which are shown in Table 2–14, gave support to the assumption that a large part of the heat, light, electricity, and power account was made up of expenses that were a function of the season of the year and not related directly to variation in output volume.

47. This procedure is possible only if the additional independent variables which are introduced are not functions of any measure of output which may be selected as the principal independent variable.

48. The influence of rate of output will be different for the different dependent variables, i.e., combined cost and its three components. Cost elements may, in fact, be arrayed from those that increase more than proportionately to output, at one extreme, to those that are constant, irrespective of output, at the other.

49. This analysis revealed that the average standard direct cost was $2.23 per dozen, with a standard deviation of $0.084 and a coefficient of variation of 3.8 per cent.

50. The complications which are introduced when it is desired to estimate the marginal cost of particular styles have been treated in chap. v.

51. Seconds may be detected either at the examining stage in the knitting mill or in the finishing mill. In the former case they may be either perfectly mended and considered as firsts, imperfectly mended and classified as seconds, or re-wound. When knitting damages are discovered in the company's finishing plant, they are charged to the knitting mill from which the stockings came.

In addition to the revenue loss of producing seconds instead of first-quality stockings, seconds may have slightly higher production costs since they frequently incur additional mending costs (the percentage of seconds among mended hosiery is higher than the percentage of seconds in all hosiery). On the other hand, other labor costs were lower for seconds, since piece rates were not paid for the operation that caused the imperfection.

52. Moreover, basic changes in other conditions that affect the number of seconds

did not appear to have occurred during the period of analysis. The proportion of seconds appeared to be related to the style and the sheerness of the stocking, the type of equipment on which it was made, the quality of silk, and the efficiency of the workmen, as well as the quality standard of the customers. None of these factors except the above-mentioned customer standard changed much, and there was no indication in the data or in the manufacturing processes that the proportion of seconds was a function of the output rate.

53. Had the importance of the item justified it, it would have been convenient to introduce the output of seconds as an independent variable, either by separating output into two components, firsts and seconds, or by combining the two and using the proportion of seconds as an additional variable.

54. Backwinding itself involves certain reclamation expenses, which were not included in the cost records of this mill and could not be accurately reallocated to it without much labor, since backwinding was not carried on at this mill, the product being transferred to another mill for this operation.

55. A study of the period from January to June, 1939, by weeks, revealed, moreover, that the proportion of backwinds to total output was very small.

56. Moreover, this is true even if the defect is not detected until the stocking has passed through subsequent operations to substantial completion. After the substitution of a "replacement" for a defective stocking had been made, further operations were recorded in the productive labor account, which also included all operations of the defective stocking except the one causing the defect. Payments for production of stock to be used for replacement were recorded in a nonproductive labor account. The productive labor account, therefore, contained only one payment for each operation on the finished stocking, unless the defective stocking passed through operations subsequent to the point at which the error was made. In this case the productive labor account included double payments for all operations between the defective one and the one at which the defect was discovered. For this reason the nonproductive (replacement) account does not provide a perfect measure of the waste resulting from errors in production, even if this waste had been confined to labor cost.

57. During the period when the replacement stock was being built up, costs were incurred for production which are not recorded. On the other hand, if replacement stock was being used up in connection with cessation of production of any particular style, then production was recorded on which costs were incurred in an earlier period.

58. Replacements were too important to omit altogether, however, and were retained for use as an independent variable as shown in the following section.

59. To obviate the possibility of spurious correlation this small account was excluded from cost.

60. The problem that arises from changes in style should be distinguished from that of style variation discussed above. Style variation is relevant to the question of measuring output, while in style change the concern is with the frequency of changes in style in so far as it is related to cost.

61. The distinction between differences in specifications and specification changes is analogous to that between style variations and style changes which was mentioned above.

62. To determine the importance of the effect of these technological changes upon

cost behavior during the period of analysis, it was necessary to carry out a historical study of the machinery in the plant to find the exact date at which each piece of equipment was replaced or modified in any way.

63. Three-thread carrier attachments were installed on twenty-nine leggers and seven footers late in 1937. Forty-six leggers and nineteen footers had this equipment, and ten leggers and three footers were without it throughout the period covered by this study.

64. See chap. vi, Sec. A, on the "Description of Hosiery-Mill Operations" for a more complete discussion.

65. Had the changes been found to be significant, several alternative solutions would have presented themselves. Since the improvements in the equipment all occurred at one time, two methods are available:

1. The unit of output in the earlier period could have been adjusted in order to make it comparable with the increased output resulting from increased productive capacity of the machine in a later period, e.g., if speeding up the machine increases the output per time unit by 10 per cent, then the production reported in the period before the speedup would be augmented by 10 per cent.

2. The relation of cost to output could be separately determined for the periods before and after the technical changes. The two functions might then have been spliced by modifying the recorded cost of the period before the adjustment in taking account of the relation between the two cost functions. Or each could be treated as a distinct function.

66. Since the equipment is approximately homogeneous, all leggers being eighteen-section machines and all footers twenty-section and since union regulations restrict the management's freedom to take advantage of whatever differences may exist in the efficiency of the equipment, multiple-shift operation would not be expected to affect cost through difference in machine efficiency.

67. Quarterly data do not reflect a different degree of "partial adaptation" from that observed in the monthly data. That is to say, the short-run cost functions determined do not pertain to different short-runs, i.e., periods in which the quantities of the fixed factors change. This is true even if adjustment time is correlated with degree of "partial adaptation," for monthly observations do not reflect a different adjustment time than quarterly observations, merely a different observation unit.

68. The output range for monthly data is approximately 4,000 dozen to 48,000 dozen pairs; when quarterly data are reduced to rough equivalence by dividing by three, the corresponding range for quarterly data is somewhat narrower, approximately 6,000 dozen to 42,000 dozen pairs.

69. The omission of raw material cost (silk) was discussed earlier in chap. iv, Sec. A.

70. It should be apparent that the inclusion of an account in the dependent variable as well as employing it as an independent variable would destroy the meaning of the relation.

71. Since replacement cost was used as an independent variable, it is more fully discussed in chap. viii on the "Selection and Measurement of Independent Variables."

72. The role of style change in this mill and the reasons for its omission from the output index are discussed in chap. viii on the "Selection and Measurement of Independent Variables."

73. Although the propriety of omitting some of these items may be questioned, it is not likely that leaving them out had any material effect on the results, since their total

value for the entire period of analysis was only about 1.5 per cent of total corrected cost. For the period April, 1935, to June, 1938 (excluding November and December, 1937, and January, 1938), the omitted accounts—suppers and overtime; electric, sanitary, and sundry supplies; coal; stationery and printing; line-elimination parts and labor; oil; and traveling expense—amounted to only 1.4 per cent of the total corrected cost.

Study 3. Belt Shop

1. Cf. Alfred Marshall, "The sums of money that have to be paid for . . . [the] efforts and sacrifices will be called either its *money cost of production,* or, for shortness, its *expenses of production;* they are the prices which have to be paid in order to call forth an adequate supply of the efforts and waitings that are required for making it [a commodity] . . ." *Principles of Economics* (8th ed.; London: Macmillan, 1920), p. 339.

2. Cf. F. Y. Edgeworth, "The relations between cost of production and quantity produced present such a variety of aspects as almost to defy the subtlety of speech, even when rendered precise by mathematical conceptions." "On Some Theories Due to Professor Pigou," *Papers Relating to Political Economy* (London: Macmillan, 1925), II, 429–30.

3. Throughout the subsequent discussion combined cost will be used in preference to total cost to refer to the aggregate of the components: overhead and direct cost. In the theoretical discussion combined cost includes all costs incurred, while in the empirical analysis it is less inclusive because of the omission of some cost elements. Total cost will instead refer to the form in which cost is stated—accumulated expenditure for the four-week accounting period—as contrasted with average cost per unit and with marginal cost. Thus combined cost may take several forms: total combined cost, average combined cost, and marginal combined cost.

4. The central position accorded marginal cost follows from a simple principle of maximization. Upon the assumption that the entrepreneur is desirous only to maximize the difference between total revenue and total combined cost, the requisite output is that at which marginal cost and marginal revenue are equal. For the sake of simplicity, the possibility of rivals' reactions are not taken into account.

5. There are at least two basic means of segmentation: (1) the use of a series of nearly identical production units, such as machines, so that the rate of output can be altered by varying the number of machines operated; (2) the introduction of flexibility into the time the plant is operated by varying (*a*) the number of days in a work week, (*b*) the length of working day, (*c*) the number of shifts in order to vary the rate of output.

6. Professor Viner has pointed out that if the overhead cost of the segments that are intermittently used were allocated to the portion of output they make possible, such an allocation would influence the marginal cost estimates that could be derived from accounting records. The level of marginal cost would be higher, but if each segment were of equal efficiency and its overhead cost were distributed evenly over the block of output to which it contributes, average first differences of cost (incremental cost) would still be constant. If, on the other hand, its cost were allocated to the first unit that required its introduction into the process of production, the marginal cost curve would be discontinuous.

7. Changes in factor prices resulting from changes in the firm's own output are, of course, not excluded from the discussion of cost theory in general although they were

omitted from the models outlined above. The firm did not appear to be in a monopsonistic position, i.e., so influential in the factor markets that the prices paid for materials and labor were affected by the quantities of these factors purchased.

8. The effects of variation in the composition of the output were tested by introducing as independent variables in a multiple correlation analysis various aspects or dimensions of the product, such as width, thickness, and weight per square foot. This procedure is described in Section vi.

9. This procedure is substantiated by the experience of Ehrke and Schneider in their statistical analysis of cost in a cement mill. Their correction for price changes in the factors was first undertaken by using the *Groszhandels Preisindex*. Finding this unsatisfactory they constructed a special index for the prices of labor, limestone, clay, coal and coke. See Kurt Ehrke, *Die Ubererzeugung in der Zementindustrie* (Jena: Gustav Fischer, 1933), §2, Die Verbesserung der Daten, in the statistical part.

10. During the observation period machinery repairs, supplies, and cement together constitute 4.22 per cent of combined cost.

11. A method of obtaining a corrected series by means of moving averages was suggested but not used, because the fluctuations followed a pattern that would lead to considerable distortion of the values in parts of the series.

12. Roy W. Jastram has pointed out to us that this rectification procedure may conceal to some extent any tendency to increasing marginal repair cost. We used a constant ratio for distributing repair cost between the current month and the third month preceding, although the ratio may be expected to vary with the ratio that current output bears to earlier output. The nature of repair expenditure makes it impossible to allocate it accurately. If there is a tendency to concentrate repairs in slack periods repair cost for high rates of output would be understated. The approximation used in this study is admittedly somewhat arbitrary, but seems better than no reallocation. Repair cost, however, constituted so small a percentage of combined cost (0.23) that significant alteration of the form of the marginal cost function by an elaboration of the method seems unlikely.

13. This admittedly crude rectification was necessitated by the lack of records of supplies charged to production or of monthly inventories of supplies. The resulting error is unlikely to affect the findings greatly since purchases of supplies averaged only 0.61 per cent of combined cost for the period of analysis.

14. The degree of possible error is indicated by the fact that cement cost is only 3.62 per cent of direct cost.

15. There seemed to be no possibility that cutting room work was a function of the output rate of the leather belt shop. Consequently, this treatment of leather cost did not distort the findings concerning the relation of cost to output.

16. The second procedure would be equivalent to ignoring the stabilized items in the correlation analysis, which is justifiable only if no relation exists between them and the independent variables. Since there was no proof of this independence, the charges were included at a uniform figure in the original computations.

17. Constant weights were considered satisfactory on a priori grounds as well as on the basis of empirical evidence which supported the belief that the physical proportions were kept fairly constant by the technical requirements of production. In addition, the relative importance of the cost elements thus rectified did not seem to justify construction of a more refined index.

18. Since the firm operates its own curriery, it might be thought that, if higher currying cost and consequent higher leather cost arose from increases in the level of operations in the curriery, and if the rate of operations of the curriery were determined by the activity of the belt shop, such variations in leather cost should not be eliminated from the analysis. In the present instance, a large proportion of the leather used did not come from the firm's own curriery. Furthermore, because of storage facilities, the activity of the curriery was not highly correlated with that of the belt shop. Even if it had been considered desirable to include leather cost variations, however, our findings would remain unaffected, since an analysis of curriery cost showed marginal cost to be constant.

19. Use-depreciation may be defined as the loss in value of productive assets not offset by maintenance that is in excess of time-depreciation. Use-depreciation may be zero or negative as well as positive. It will be zero if the loss in value occasioned by physical deterioration due to the passage of time is not increased by more intense use. For example, an automobile body die, which will be rendered obsolete in one year by a planned change in design, may have zero use-depreciation if no conceivable rate of production could diminish its efficiency or hasten its scrapping. This will be true even if physical deterioration results from use, provided its effectiveness and scrap value is unaffected. Use-depreciation will be negative if the loss in value is greater when the equipment is idle than when it is in use. Since the loss in value attributable to use may be reduced or completely balanced by maintenance, use-depreciation represents only the loss in value not restored or avoided by maintenance. To the extent that productive assets are fully maintained in the sense that no residual loss in value results from use, use-depreciation, as we have defined it, may be neglected in estimating marginal cost.

20. Such adjustments amount to a rough, arbitrary assignment of use-depreciation which is unsuited to our purposes.

21. The minimum coverage was so high and the production cycle so short and uniform that changes in inventory arising from changes in the rate of output did not affect the amount of insurance carried.

22. Since electricity is produced by the company as a joint product with needed heat and steam, the amount of electricity used may not be closely related to changes in output. Allocations to the various plants are based upon engineering estimates which take account not only of the number of lighting units and the rated power consumption of each machine, but also of the utilization of power plant by-products.

23. The sample was too small for cross tabulation on a multiple basis in order to reflect the influence of various operating conditions; moreover, well defined measures of any existing relations could not be determined. Confluence analysis did not seem necessary, for reasons discussed later; nor were the factors sufficiently numerous or inter-correlated to justify factor analysis.

24. The precise meaning of the partial correlation coefficient, such as $\gamma 12.3$, should be pointed out. It measures the closeness of the relation between combined cost (X_1) and output (X_2) after allowing for the effects of average weight (X_3). It shows the correlation between cost and output (as measured by the type of function used) excluding the portion due to the co-variation of cost with weight and of output with weight (as measured by the form of the relation used). Thus it measures the correlation between cost and output which is incremental to any correlation between each and average weight (so far as the functions used define their actual relations). This procedure for allowing for

the influence of another variable is frequently referred to by a verbal short-cut, by saying that the influence of the other variable is "held constant." In this paper, at the suggestion of W. L. Crum, this possibly misleading phrase has been replaced by "allowing for the effects of."

25. Although the graphic method is useful for preliminary analyses, its reliability is so difficult to assess that it cannot safely be applied in precise analyses. A function fitted by the graphic method may reduce the degrees of freedom but not allow adequately for the reduction in the estimate of the degree of interrelation of the variables. See Wilfred Malenbaum and J. D. Black, "The Use of the Short-Cut Graphic Method of Multiple Correlation," *Quarterly Journal of Economics,* LII (Nov. 1937); "The Short-Cut Graphic Method of Multiple Correlation," a discussion by L. H. Bean, Mordecai Ezekiel, J. D. Black, and Wilfred Malenbaum, ibid., LIV (Feb. 1940), 318–64.

26. To determine the true relation, all variables may be assumed subject to error, and some such technique as confluence analysis used.

27. Several problems are encountered in the statistical analysis of average cost. First, selection of the most suitable specification for the average cost function is more difficult. Second, slight errors in the choice of the function produce magnified errors in the derived marginal cost function. Third, since average cost is a quotient of two variables, each of which is subject to error, the statistical distribution of the quotient may be less likely to conform to the assumptions upon which multiple regression analysis is based.

28. This involves a special use of the term marginal cost. It is used exclusively hereafter to mean the addition to total cost caused by a unit increment in output equal to one square foot of single-ply equivalent finished belting. An analogous marginal cost could be found for an increment in average weight, average width, or one of the other independent variables, provided that these variables can vary independently of one another.

29. The coefficients of multiple correlation and multiple determination test this criterion objectively under some circumstances. If the observations are derived from time series, however, there may be an element of serial correlation in the successive observations that accounts for part of the high degree of correlation attributed to the influence of the independent variables. A major defect of many economic time series is positive serial correlation due to lack of independence of successive observations and to the effect of common cyclical influences. In this study, the coefficient of serial correlation of the residuals of the multiple regression equation was found to be −0.311. Since in this instance the coefficient is negative and small, serial correlation did not impair the usefulness of the multiple regression coefficients as estimates of the degree of dependence of cost on the independent variable. John H. Smith has suggested that the explanation for the negative serial correlation may lie in compensatory errors of allocation between adjacent accounting periods.

30. Weight was defective as a measure of output since it did not reflect certain manufacturing operations performed only upon the surface of the leather. It was, therefore, rejected, even though a measure computed in pounds would have yielded results comparable with price data and with cost computations for other departments and companies.

Dollar value of output was subject to irrelevant fluctuations arising from changes in the prices of finished belting and from variations in the proportions of different qualities of output. Variability in quality was not relevant to cost because the manufacturing processes were approximately identical for all qualities of belting, and the effects of differences in the quality of raw material on cost had already been eliminated by the use of a uniform leather price.

In order to use standard cost as a measure of output it would have been necessary to remove the effect of changes in annual manufacturing cost and material prices. Standard cost is fixed at the beginning of each year on the basis of the average material, labor, and overhead cost of the preceding year. Its raw material cost component reflects differences in quality, but does not vary during the year with changes in hide prices. The annual variation in the differentials between the actual cost of various types of product presented such difficulties in the construction of an accurate deflation device, however, that the use of standard cost as an output measure was abandoned.

31. The conversion was performed simply by muliplying double-ply output by 2, and triple-ply output by 3. Although labor and cement cost seemed more than proportionately greater for double- and triple-ply belting, other costs were thought to be less than proportionate, so that the errors tended to compensate. The accuracy of this conversion was tested by introducing the percentage of single-ply belting in total output as an independent variable. No significant net correlation was found to exist between this percentage and cost.

32. Because the evidence supporting a linear hypothesis may not be conclusive, a third degree function was also fitted to the observations of total combined cost. Discussion of the problem of specification is postponed to Section 7 where the results of the fitting of a cubic function are shown.

33. The partial regression equation for total combined cost (${}_tX_c$) and output (X_2) (allowing for the influence of average weight of belting) is:

$${}_tX_c = 2.974 + 0.770\ X_2$$

The equation for average cost, derived from the above equation, becomes:

$${}_aX_c = 0.770 + \frac{2.974}{X_2}$$

Marginal cost is estimated from the first derivative of the total combined cost function with respect to output. In this instance, marginal combined cost (${}_mX_c$) is constant at \$0.77. The reliability of this estimate of marginal cost is indicated by the standard error of the coefficient of net regression, \$0.0063. However, this error formula is posited upon conditions of sampling and of distribution that are not entirely met by time series data of the type encountered.

34. The information in the chart and in the regression equations is also presented in Table 3–2 as a schedule showing the cost associated with different levels of output.

35. The net relation of cost to average weight was established more precisely in the partial regression equation derived by least squares regression analysis:

$${}_tX_c = 7.394 + 70.18\ X_3$$

36. For forecasting purposes, all estimates obtained from such a regression equation would have to be modified to take into account changes in wage rates, prices of materials, and tax rates. This type of adjustment is discussed below.

37. The meaning of these regression curves can be illustrated by computing the total, average, and marginal overhead cost at different levels of output. The magnitudes of

these costs for a hypothetical set of output rates are shown in Table 3–4. In addition, the total overhead cost associated with various percentage changes in output from the preceding month can easily be calculated (Table 3–5).

38. The partial regression equation of overhead cost ($_tX_o$) on output (X_2) is:

$$_tX_o = 2.109 + 0.001\ X_2$$

The partial regression equation for magnitude and direction of change in output (X_4) is:

$$_tX_o = 3.108 - 0.004\ X_4$$

The multiple regression equation for the overhead cost of the belt shop, when both X_2 and X_4 are variable, is:

$$_tX_o = 2.158 + 0.001\ X_2 - 0.005\ X_4$$

39. The equation relating cost to output (X_3 allowed for) is:

$$_tX_d = 0.663 + 0.760\ X_2$$

When the effect of output measured by surface area is allowed for, the independent relation of cost to average weight is:

$$_tX_d = 5.06 + 69.32\ X_3$$

The combined effect of output and weight upon total direct cost is expressed by the multiple regression equation:

$$_tX_d = -61.639 + 0.760\ X_2 + 69.324\ X_3$$

The standard error of estimate for this equation is approximately $985, the coefficient of multiple correlation, 0.999, and the coefficient of multiple determination, 0.997.

40. Incremental cost is determined indirectly from the accounting records and is obtained by dividing the difference in total cost for two adjacent months by the corresponding difference in ouput. Although an approximation to marginal cost, incremental cost is to be carefully distinguished from it.

41. The calculations necessary for the analysis of variance were made by Phyllis van Dyk. John H. Smith made helpful suggestions in their interpretation.

42. This procedure is not completely satisfactory since it assumes the correctness of the linear regression of cost on average weight, which was determined by least-squares fitting. The linearity of this function was strongly indicated by the graphic correlation analysis but it was not subjected to a more objective test.

43. Since small changes in the shape of the total cost function cause relatively large fluctuations in marginal cost, the distribution of incremental cost determined independently constitutes an additional test of the linearity of the total cost function.

44. In computing these first differences, cost was corrected for errors in accounting allocation, for rate changes, and time lags, but was not adjusted to remove the estimated effect of average weight. Since one objective of this analysis was to test an alternative short-cut method of estimating marginal cost, it would have been inconsistent to use a correction that presupposed a multiple correlation analysis of total cost.

45. The standard error of the mean was $.055, the standard deviation $.342, and the coefficient of variation, 44.6 per cent. In the calculation of these estimates one observation was omitted because it showed a change in output of only .4 per cent of the mean output, which made the associated incremental cost unreliable. All other changes were greater than 1.25 per cent.

46. The ratio of the difference between marginal cost and average incremental cost to the standard error of the difference was .0438.

47. The value of z is 0.1930 when $n_1 = 9$ and $n_2 = 30$, to be compared with the 5 per cent point, 0.3925.

48. The numerical values of the regression coefficients and the standard errors for the multiple regression equation are presented in Table 3–10.

49. Demand for leather belting has declined since about 1923 in a secular movement transcending the business cycle, the peak of 1929 being lower than that of 1923, and the peak of 1937 much lower than that of 1929.

50. The cost of storage, insurance, and interest on inventories of finished goods were not charged to current production since they are caused by uneven distribution of demand and are more properly attributable to cost of selling.

51. Variation in the quality of labor inputs as a result of longer hours and night shifts lie outside the compass of static cost functions, although the effects of such influences are difficult to remove in empirical investigations.

52. Apart from these conventional rigidities, evidence of technical rigidities was to be found in the behavior of one component of cost, overhead cost. A significant relation of overhead cost to magnitude and direction of change in output from the preceding month was observed, after the influence of output itself had been removed. It might have been expected that if output were unchanged for two successive periods, overhead cost would be less, because of the longer time allowed for adjustment, than if a given output followed either lower or higher outputs. As a matter of fact, increases in output were accompanied by lower overhead cost, while decreases in output were more costly (see Chart 3–7). One possible interpretation is that increases in output can be attained by temporarily overloading the staff, which results in a low cost per unit, while, when output is reduced it is difficult to adjust the personnel immediately. There are many indications that overhead cost reductions in the face of output contraction (especially when personnel changes are involved) occur less easily than cost increases when output is increased. The lack of evidence of rigidities in the behavior of combined cost can be attributed only to the relatively small importance of overhead cost in total combined cost.

53. Although such conventional rigidities cannot be classed as dynamic forces, they may cause even greater deviations from the static model. For purposes of prediction, however, empirical cost curves possess some advantage over cost curves predicated upon the existence of a minimum cost complex of factors at every output because of their closer approximation to reality.

Study 4. Department Store

1. See H. Smith, "Advertising Costs and Equilibrium," *Review of Economic Studies*, Vol. II (1934–35), and R. M. Shone, "Selling Costs," in the same volume. On polyperiodic production see Sune Carlson, "A Contribution to the Pure Theory of Production" (Ph.D. dissertation, University of Chicago, 1936), chap. vi.

2. Shone has used the term "allocated costs" for costs of production because they "vary directly with output, and their redistribution, increase, or decrease for any given output is excluded by hypothesis. On the other hand, the unallocated expenditures (selling costs) may be incurred in any form, increased or decreased, in order to alter the price and the output of the product defined by the allocated costs" (p. 225).

3. Particularly for goods that are purchased infrequently, the advertising, displays, demonstrations, and advice of salespersons required before a vague desire is transformed into a sale of goods may be viewed as creating a utility for which consumers are willing to pay. Even newspaper advertising may be considered as contributing to the store's sales service since it acquaints the consumer with prices and qualities of goods available, and since its "tone" may augment the prestige satisfaction derived from the merchandise.

4. Cf. Frank H. Knight: "In so far as they [changes in wants] result from a deliberate expenditure of resources, they become as all other economic operations. . . . In fact, as we have previously observed, the advertising, puffing or salesmanship necessary to create a demand for a commodity is causally indistinguishable from a utility inherent in the commodity itself" (*Risk, Uncertainty and Profit* [Boston, 1921; reprint, London, 1933], p. 339).

5. Considerable stability in the store's standards with respect to sales personnel, fixtures, delivery, returns, and internal displays would be expected in view of the importance of prestige in this particular store's merchandising policy. Prestige is an important aspect of product differentiation in the oligopolistic department-store market in which this firm competes.

6. For a further discussion of the index of output see Sec. II, Part C.

7. Two papers concerning empirical studies of other department stores should be mentionel in this connection: "Cost Functions in Merchandising," delivered by Miss Phyllis van Dyk at the 1940 research conference of the Cowles Commission in Colorado Springs, Colorado, and "Cost Functions in the Department Store," delivered by Roswell H. Whitman at the 1940 annual meetings of the American Economic Association (Philadelphia).

8. The indirect advertising costs allocated to departments on the basis of dollar sales include direct mail, compositions and cuts, art work, interior decoration, car advertising, Foster Bureau, fashion shows, magazines, addressograph, window-dressing and sign-writing, salaries of advertising office.

9. Sales volume is probably not proportional to the number of selling days, since the occurrence of holidays often results merely in a time redistribution within a relatively short period and not a loss of sales. Hence, even the cost items most closely related to volume do not vary proportionately with the length of the month.

10. It is important to distinguish between changes arising from career advancement of the individual (i.e., improved selling effectiveness) and those associated with changes in factor prices.

11. Rectification of other kinds of salaries was regarded as undesirable for several reasons. Variation in the salary of buyers and assistant buyers in the departments in question over the years studied was small. In two departments the same buyer, at almost the same salary, was employed during practically the entire period. In the other department the salary changes which accompanied the change of buyers were not great. The changes which did occur were supposedly a reflection of increases in the "effectiveness" of the buyer rather than changes in the labor market. However, the buyers' labor market is so imperfect and the measurement of "effectiveness" of merchandising executives so ineffective that little solace can be derived from this possibility. Nevertheless, the absence of objective bases for rectification of these costs made it seem more questionable to try to correct them than to leave them as recorded.

12. The index employed was a weighted arithmetic average of price relatives based on January, 1931, rates, with total linage purchased in the base year as the weighting factor. Although these weights are not constant throughout the five-year period for each newspaper, the composite result differed negligibly from an index weighted with the five-year average annual linage.

13. Cost residuals from these net relations were first plotted as a time series to ascertain whether there was any trend in the residuals or any perceptible shift in the seasonal pattern during the period of study, or whether the magnitude of the residual deviation was systematically different for different years. This analysis indicated that no significant trend, shifting seasonal, or variation in absolute magnitudes characterized the years studied.

14. The index was constructed by taking the arithmetic mean of the absolute deviations for each month. This figure was then added to or subtracted from the original observations. The use of an additive, rather than a multiplicative, index was the logical outgrowth of employing an additive process in establishing the relationship of cost to number of transactions and average value of transaction. Moreover, the preliminary analysis indicated that the magnitude of the residual deviation was not proportional to the total cost, which further confirmed the choice of an absolute rather than a relative method of rectification.

15. E.g., lower standards of service during the Christmas rush and during the January clearance sales may partially explain the low costs of these two months in the coat department. Differences in the character of the merchandise are also seasonally regular and would be expected to affect cost. The low costs of May and June in the coat department, for instance, are attributed by executives to sales of summer wraps which require relatively little sales service.

16. The effect of this correction as well as of most rectification is to increase the correlation between costs and sales volume.

17. Cost deviations from the net regression curves of number of transactions and average value of transaction were plotted against dollar volume of sales in order to determine whether this factor was significantly correlated with the residual variation in cost not accounted for by the net regression curves. The application of this test indicated that the inclusion of dollar volume of sales as an additional independent variable in the multiple-correlation analysis was unnecessary. The influence of number of transactions and average value of transaction upon cost accounted for all the cost variation associated with dollar volume of sales, leaving no net correlation between the cost residuals and the latter.

18. As has been noted, differences in the character of transaction were held to a minimum by selecting departments whose products were relatively homogeneous and whose sales service was similar for different types of articles. A large part of the remaining variation was probably removed by the use of a monthly time interval and by correction for seasonal pattern. Nevertheless, it is likely that important residual differences remained, part of which are reflected in average value of transaction.

19. Two other influences were given consideration: rate of stock turnover and the proportion of inexperienced workers employed. Inquiry indicated, however, that the effect of inventory turnover was likely to be confined to cost and profit items not included in the analysis, eg., interest, storage space, and mark-downs.

The proportion of inexperienced salespeople employed did not appear to exert a serious influence on cost for several reasons: (1) the low rate of labor turnover in these departments; (2) the store's policy of developing a reserve of experienced extras and a versatile flying squadron of salespeople trained to work in several departments; and (3) the policy of starting inexperienced girls at low salaries and raising salaries only as their sales increase.

No suitable index of inexperienced personnel could be found. Average length of service is an inexact measure because of marked individual differences in learning speed and because it does not reflect a new clerk's previous training in selling the same merchandise at other stores. Selling cost per salesperson was equally unsuitable, since there appeared to be no correlation between length of service and the ratio of salary to sales.

20. To a certain extent, successful adjustment depends upon ability to forecast sales accurately. Nevertheless, the problem is larger than this, for it also relates to the limited divisibility of production factors; e.g., the hour-to-hour fluctuations in business cannot be completely met by hourly adjustments in the selling force.

21. Although the pattern of hour-to-hour variation had not been accurately determined, it was thought by executives to be approximately the same from month to month. The day-to-day variation within the week had been subjected to an analysis which showed that, although the fluctuations were marked for the departments under study, the pattern did not appear to differ significantly from month to month. Week-to-week fluctuation showed marked differences among months; but this type of variability could be more easily coped with because regularity of the seasonal pattern made it more exactly predictable and because a week is a long enough period to permit tolerably accurate adjustment of personnel needs.

22. This method was based on the assumption that relative variation within the week was more significant than absolute variation. To have measured deviations from the expected pattern of fluctuation might have been preferable but was considered too laborious.

23. Whether unseasonable weather results in actual loss of sales or merely in postponement is of no consequence here, for concern is only with the extent to which it may cause errors in planning personnel needs.

24. Average variation of temperature from the normal for past years is not satisfactory, for it does not spot the timing and number of unseasonable days; while buyers' records are incomplete and are likely to rationalize unfavorable showings.

25. By studying daily sales records over the five-year period, in conjunction with average noontime temperature, it was found that, on the average, sales fell off when the noontime temperature was 85° or higher in summer and 15° or lower in winter. The number of selling days with temperature above or below these critical points was therefore determined for each month of the period under study.

The effects of precipitation were studied by comparing daily sales with records of rain and snow during store hours. Snow appeared to have little or no ill effect upon sales, whereas rain was frequently associated with subnormal sales. The number of days in which it rained during store hours in each month was therefore included in the index of bad weather. Duplications between the two criteria of bad weather were then eliminated, leaving a net count of the number of selling days with disagreeable weather in each month. This involved equal weighting of days regarded as uncomfortable from each of these conditions, but no a priori or empirical evidence was available to indicate other weights.

26. The hierarchy was based upon opinion of executives plus closeness of simple correlation as determined by inspection of scatter diagrams. Space limitations prevent introduction of these and the multiple-correlation charts in this presentation.

27. The order of introduction may have a significant effect upon results, hence the care with which the variables were arrayed (cf. W. Malenbaum and J. D. Black, "The Use of the Short-Cut Graphic Method of Multiple Correlation," *Quarterly Journal of Economics,* LII [1937], 66–112).

28. "Average value of transaction" will hereafter be referred to as "average gross sale" in deference to department-store terminology.

29. Seasonal variation, on the other hand, did not exhaust its influence on cost by working through number of transactions and dollar value of sale. Rather it apparently exercised an independent additional influence since the cost residuals from the net regression of X_2 and X_4 had a subdued but well-defined seasonal pattern. Instead of including the seasonal factor as a third independent variable, however, the residuals were used to rectify the original cost data. This procedure is discussed under "Data Rectification."

30. Both the number of selling days per month and disagreeable weather, for example, were clearly correlated with volume of sales, although multiple-correlation analysis revealed no additional independent effect upon cost. Likewise, the original data revealed a clear cyclical pattern in volume of transactions and in average dollar value of transaction. Yet there was no significant residual cyclical variation in corrected cost over and above that attributable to X_2 and X_4. Both number of transactions (X_2) and average gross sale (X_4) had a strong seasonal pattern. It was, in fact, this high intercorrelation which precluded correction of the original data for the full seasonal influence. The residual seasonal pattern which served as a basis for the correction index was not, however, correlated with X_2 or X_4.

31. Cf. Joel Dean, *Statistical Determination of Costs, with Special Reference to Marginal Cost* (Chicago: University of Chicago Press, 1936).

32. A strong gross relation between dollar volume of sales (X_3) and both seasonally corrected and uncorrected cost was evidenced for each department. The relation of total cost appeared to be convex downward for coats and shoes (falling marginal cost), and linear for hosiery (constant marginal cost). Since dollar volume is more commonly used in store management than is physical quantity, a least-squares analysis of its relation to cost was made which confirmed these graphic results. Simple correlation was indicated because deviations from the gross regressions did not have a significant relation to average gross sale or to other factors.

33. The following subscripts and superscripts are used in these equations: T = total cost, A = average cost, M = marginal cost; C = coat department; H = hosiery department; S = shoe department; X_2 = index of number of transactions; X_4 = index of average gross sale.

34. Graphic correlation analysis paralleled the least-squares findings by revealing the same type of net relation, although the uneven and scattered distribution of the observations from which the relations were obtained somewhat restricts the reliance to be placed in these functions.

35. "When a field is fertilized or a pig fed, the profit per dose decreases with the quantity of doses, and it is time to stop when profit and dose have the same value. Similar phenomena of fatigue appear when we feed a market, which may also be considered as a kind of living being. . . ." (Ivar Jantzen, "Increasing Returns in Industrial Production," *Nordisk tidsskrift for teknisk økonomi,* March–June, 1939, p. 6).

36. It was not possible to isolate the promotional or selling constituent in the various accounting categories of cost. Nevertheless, in the coat and hosiery departments the behavior of newspaper advertising costs (which approximate pure promotional costs) indicated the existence of a proportional relation to output. In a graphic correlation analysis the functions appeared to be linear and to pass through the origin, although the scatter was wide, particularly for high outputs. An ill-defined curvilinear function was found for the shoe department. It seems reasonable, moreover, to expect substantial constancy of store standards over time, which would further buttress the assumption of approximate constancy of promotional components of cost. The results of this partial analysis are not conclusive, however, because of the failure to stabilize demand and other impinging conditions.

37. The corrected data employed are intended, however, to represent a fairly uniform quantity of sales service. Day-to-day and hour-to-hour variation in sales service was probably averaged out by taking monthly totals rather than daily or weekly totals, and the effect of seasonal variation in sales service was at least partially removed by means of the index of seasonal variation.

38. Since the cost components entering into sales service are not independent of marginal revenue, it is therefore possible that declining marginal cost indicates that the department has passed the output which will maximize revenue over a period of time extending more or less indefinitely into the future.

39. Cf. M. F. W. Joseph, "A Discontinuous Cost Curve and the Tendency to Increasing Returns," *Economic Journal,* XLIII (September, 1933), 390–98.

40. See a recent article by Mr. George Stigler, entitled "Production and Distribution in the Short-Run" (*Journal of Political Economy,* XLVII [1939], 305–28), for a discussion of the concepts of adaptability and flexibility. Mr. Stigler has pointed out that technical flexibility may be deliberately provided by building a plant in a way which will maximize profits over time through efficient operation over the expected range of outputs, even if this involves the sacrifice of the optimum technical adjustment for any one output.

Part Two. **Cost and Plant Size**

Introduction

1. For the firm, there is agreement on an early and often substantial decline in the long-run average unit-cost curve followed by a size range in which costs remain approximately constant. There is disagreement, however, as to whether ultimately a rising phase occurs when problems of coordination cause cost increases that more than offset waning economies of scale. Professor Sargant Florence dissents as to the existence of net diseconomies of size, largely on empirical grounds: "There is little to prove the universality and inevitability of any such law of increasing costs in the long run when manufacturers have time to get new equipment in order to meet enlarged orders or anticipated orders,

and have time to reorganize and delegate responsibilities." P. Sargant Florence, *The Logic of British and American Industry* (London: Routledge & Kegan Paul, 1953), p. 63.

2. A. A. Walters, "Production and Cost Functions: An Econometric Study," *Econometrica,* vol. 31, no. 1–2 (1963), pp. 2–65. A valuable and distinguished roundup and appraisal of plant-size savings. Not only do plant-size savings have an important impact upon the growth and profitability of individual enterprises, they also account for most of the scale economies of the industry and of the national economy.

3. *Business Concentration and Public Policy,* National Bureau of Economic Research (Princeton, N.J.: Princeton University Press, 1955), pp. 230–38.

4. Presented and evaluated in J. Johnston, *Statistical Cost Analysis* (New York: McGraw-Hill, 1960), pp. 189–92.

5. Ibid.

6. This was done in a study of a grocery chain. See *Organization and Competition in Food Retailing,* National Commission on Food Marketing, Technical Study no. 7 (Washington, D.C.: U.S. Government Printing Office, 1966), p. 141.

7. For example, the addition of roadside and shopping-center shoe stores of changed format by the Thom McAn chain in the 1950s made obsolete the shoe chain findings of 1937 and 1938.

Study 5. Finance Chain

1. Costs excluded were head office supervision, federal taxes, interest, and advertising. They were omitted because accounting allocations of these centrally-incurred costs were essentially arbitrary. Consequently their inclusion could have distorted the relationship between cost and size of loan office.

2. These measures were computed from the results of the graphic multiple-regression analysis by methods described in Mordecai Ezekiel, *Methods of Correlation Analysis* (New York: John Wiley & Sons, 1930), pp. 384–90.

3. In graphic multiple-regression analysis, the curve of net relationship of cost to size of loan office was obtained by successive approximations. The first cost–size curve only imperfectly removes the contaminating effects of the other cost determinants, e.g., delinquency (X_3) and city size (X_4). The second approximation is more successful in isolating the effect upon cost of each of these other independent variables (in the form of its net regression) and thus removing this cost influence by holding it constant at its mean. The third-approximation net regression of cost on size presumably is purged still more completely of distortions caused by the other four independent variables.

4. Standard errors of deviation from the curve indicate that the departure of the curve from the horizontal is most significant in this sector.

Study 6. Shoe Chain

1. As far as is known, the first use of this method of analysis was in Joel Dean, "A Statistical Examination of the Behavior of Average and Marginal Costs," an unpublished doctoral dissertation of the University of Chicago (1936). In this case the branch units of a financial enterprise were studied in an effort to determine the most efficient size.

2. A firm is a section of the economy operating under a single control, so that in reality

the firm is the parent shoe corporation which controls the individual retail stores. In this case, however, the existence of the external control over the stores is abstracted from, and each retail store is treated as a firm in its own right. It so happens, moreover, that business and technical units—i.e., plants—coincide so that firms and plants can be identified.

3. These two years were selected both because of their recency and because they occur in markedly different phases of the business cycle.

4. Actually, decisions regarding price policy are made by the central office of the shoe corporation, which, nevertheless, does not prevent us from classifying the individual stores in this way.

5. E.g., in Chamberlain's treatment of the individual firm under monopolistic competition, output is established at some level less than the "economic" capacity.

6. The use of value measures, while it has the advantage of comparing heterogeneous physical units, raises complex problems of valuations which need not be considered here.

7. The superiority of an input measure arises when the degree of input homogeneity is greater than the degree of output homogeneity, e.g., in oil-refining it is simpler to use the input of crude oil rather than the complex of end-products.

8. The lack of data concerning the physical volume of sales of hosiery, rubbers, and sundries constituted an important gap in the data, which prevented examination of the effect on cost behavior of varying the proportion of total *physical* sales composed of these miscellaneous items.

9. Since the styles and sizes stocked were adapted to the peculiarities of the individual store's clientele within the limits of the standard styles carried by the chain, it appeared unlikely that important heterogeneity in the sales-service units resulted from the relatively minor store-to-store variation in assortment of styles.

10. The possibility of employing the joint production analysis to take account of heterogeneity in the quality of the sales service because of selling peaks and troughs is discussed in the theoretical section (chap. iii).

11. Intensity of store utilization is apparently related to sales service. This relation is discussed in the Appendix on the "Measurement of Intensity of Utilization."

12. Although men's shoes and children's shoes are weighted equally and may, therefore, misrepresent the actual selling conditions, there is reason to believe that the sale of men's shoes induces greater sales of allied products than does the sale of children's shoes. It may be, therefore, that the association of the sales of men's shoes with allied sales, which probably require a disproportionate amount of service, balances the larger sales-service output involved in the sales of children's shoes. If this is true, it offers a further justification for the allocation of equal weights.

13. The inclusion of window display under building expense rather than under selling expense may be questioned. It is classed as it is mainly because of its similarity to interior displays. On the other hand, almost the whole of a store is devoted to displaying and advertising, so that if the window display were classified as advertising so should much of the other building expense. Its inclusion in one or the other is, however, not an important matter.

14. It is defective as a measure of this operating condition, since reserve salaries reflect only the iregularities of demand for which allowance can be made by the employment of part-time salesmen. Temporary selling help can be used, for example, during the peak

selling period of Saturday afternoons and evenings. The magnitude of reserve salaries does not, however, reflect the volume fluctuations during or among week days.

15. When costs such as advertising are allocated to the retail stores on the basis of sales volume, the rationale of the cost–output relation may be distorted. Cost elements allocated in this way vary in direct proportion to sales and, if important enough, tend to induce an undue linearity in the total-cost function. In the case of advertising, however, the cost is sufficiently small to have only a negligible influence. It amounts to 2.06 per cent of the total expenses of all the stores in the sample.

16. E.g., where total cost is represented as a function of output, i.e., $K = K(x)$, the elasticity of total cost is

$$\frac{dK}{dx} \cdot \frac{x}{K} = \frac{d(\log K)}{d(\log x)} .$$

Since in the charts the abscissae and ordinates are the logarithms of x and K, the elasticity of total cost at any point is given by the slope of the regression line or, alternatively, by the first derivative, d_tX_1/d_cX_2, derived from the regression function.

17. Where s is average cost, the elasticity of average cost is

$$\frac{x}{s} \cdot \frac{ds}{dx} .$$

But,

$$\frac{x}{s} \cdot \frac{ds}{dx} = \frac{x^2}{K} \cdot \frac{d\frac{K}{x}}{dx} = \frac{x^2}{K} \cdot \frac{x \cdot \frac{dK}{dx} - K}{x^2}$$

$$= \frac{x}{K} \cdot \frac{dK}{dx} - 1 .$$

18. It is possible to determine the optimum level of output mathematically from the regression functions themselves. The equations are, for 1937 and 1938,

$$_tX_1 = 9.029 - 3.128\ _cX_2 + 0.458\ _cX_2^2,$$

$$_tX_1 = 7.617 - 2.460\ _cX_2 + 0.380\ _cX_2^2.$$

The output at which the elasticity of total cost is unity is given by the solution of the equation

$$\frac{d_tX_1}{d_cX_2} = 1 .$$

The optimum output expressed in pairs of shoes is thus found to be, for 1937, 32,000 pairs and, for 1938, 38,864 pairs.

19. A second-degree regression curve was also fitted but was rejected. In view of the extreme dispersion of the observations, it was decided that the simpler form of regression function was to be preferred.

20. The percentage composition of total cost (less administrative and general indirect expense) made up by each of the components for the two years is as follows: Selling expense: 1937—46.92 per cent, 1938—48.76 per cent; handling expense: 1937—5.62 per cent, 1938—5.27 per cent; building expense: 1937—47.46 per cent, 1938—45.96 per cent.

21. A. L. Bowley, *Mathematical Groundwork of Economics* (Oxford: Clarendon Press, 1924), p. 35.

22. There have, of course, been numbers of empirical studies of the relation of cost and size, but usually they are concerned with heterogeneous establishments. Such investigations were particularly common during World War I in the attempts to determine "bulk-line" costs as a basis for price-fixing. These studies are, however, more intimately related to the statistical determination of the Marshallian "particular-expenses" curve than to the cost functions of the individual firm.

23. K. E. Boulding, "Professor Knight's Capital Theory: A Note in Reply," *Quarterly Journal of Economics*, May, 1936, p. 524.

24. F. H. Knight, "Capitalistic Production, Time and the Rate of Return," *Economic Essays in Honour of Gustav Cassel* (London: George Allen & Unwin, 1933), p. 332.

25. Cf. Ragnar Frisch, "The Principle of Substitution," *Nordisk tidsskrift for teknisk økonomi*, September, 1935, p. 13.

26. Even though these services may be intended to mislead the prospective purchaser, as long as he does not perceive the deception no account need be taken of this.

27. For an excellent discussion of the problem of demand fluctuation in connection with heavy industries see the article by I. M. Biss, "Overhead Costs, Time Problems, and Prices," in *Essays in Political Economy in Honour of E. J. Urwick* (Toronto: University of Toronto Press, 1938).

28. George Stigler, "Production and Distribution in the Short Run," *Journal of Political Economy*, June, 1939, p. 316..

29. An alternative statement of the condition is that the producer will maximize output for any given total expenditure.

30. E.g., in digging and excavation at least two methods are available, the one involving men and shovels; the other, men, power, and steam shovels.

31. On the grounds of convenience and customary usage we shall continue to speak in this theoretical section as though the production of a commodity involves the whole production process from the inputs of raw material to the appearance of the physically finished product, although it must be remembered that in the actual study under consideration attention is restricted to a small part of the total process.

32. It should be noted that when the minimum-cost combination of productive resources, which is the best possible quantitative combination for a given output, has been achieved, this does not imply that this is the best possible resource complex of all, since the given output of the firm may not be "optimum," i.e., the scale of output chosen may not be that which affords the lowest average cost of production.

33. It must again be stressed that the aim of this study is the determination of the functional relation between output and cost, not the historical relation which may have existed over the course of time.

34. Erich Schneider, *Theorie der Produktion* (Vienna: Julius Springer, 1934), p. 28. Schneider's terms corresponding to total adaptation and partial adaptation are *totale Anpassung* and *partielle Anpassung*.

35. Whether such distinctions have any meaning depends upon how broadly the variables in the production function are grouped. One could conceive of a broad functional grouping which would effectively rule out changes in the production function—a process which would just as effectively rob the concept of the production function of any heuristic value.

36. Alfred Marshall, *Principles of Economics* (8th ed.; London: Macmillan, 1923), p.

377. Cf., further, Hans Neisser, ". . . . the term ['short run'] usually includes . . . . the premise that equipment for productive purposes is to be utilized in a given, immobilized shape, changes in the quality and quantity of equipment representing a long-run affair" ("The Economics of the Short-Run," *Social Research,* February, 1940, p. 36).

37. The derivation of the cost-output function is as follows. For a given output x, where there are n factors, the minimizaton of cost prescribes the condition that the marginal productivities of the n factors are proportional to their prices. The solution of these n equations for a given x yields the minimum-cost combination of the factors. Where x is variable, the solution of the equations gives the amounts of the factors as functions of the output. Since we have the total cost K given by the relation

$$K = \sum_{i=1}^{n} \pi_i v_i ,$$

where π_i is the price of the ith factor v_i and

$$v_i = v_i(x) \qquad (i = 1, 2, \ldots . n) ,$$

we have

$$K = \sum_{i=1}^{n} \pi_i v_i(x) ,$$

so that with π_i given, $K = K(x)$.

38. Cf. this discussion and what follows in Schneider, esp. "Das Harmoniegesetz," pp. 51–57.

39. An example of the situation visualized is given by H. von Stackelberg. He speaks of a bookkeeper, who, while able to make 1,000 entries per day, is faced with the necessity of making 1,001. This means an additional bookkeeper must be hired to make the single entry (*Grundlagen einer reinen Kostentheorie* [Vienna: Julius Springer, 1932], p. 12).

40. Although it is true that the assumption of a specific capacity for salesmen is not justified, salesmen must, nevertheless, be employed in such numbers that service standards in rush periods do not deteriorate to such an extent as to jeopardize the store's good-will or customer relations. This sets an upper limit on the amount of sales which each individual salesman can make.

41. Besides the fact that there is a certain range within which variability of the capacity of the salesmen exists, there is some flexibility in the number of salesmen, made possible by the employment of reserve selling help for temporary periods. Reserve salesmen, however, are not employed in sufficiently large numbers to solve the peak-load problem. Therefore, while this is an additional complication, the principles of behavior enunciated are not changed.

42. It should be recalled that the statistically determined cost functions are in logarithmic terms. There is, however, no essential difference in the behavior of these cost functions if they are transformed into arithmetic terms. Consequently, in the following discussion no account is taken of this small problem.

43. N. Kaldor, "The Equilibrium of the Firm," *Economic Journal,* March, 1934, p. 67.

44. See F. H. Knight, "Cost of Production and Price over Long and Short Periods," in *The Ethics of Competition* (New York and London: Harper & Bros., 1935), p. 210.

45. Underutilization which is a result of immaturity has been excluded by the nature of the sample selected.

46. The weighting method was the following: The amount of salary costs for each store used in the index was calculated by subtracting $2,000 from the expenditure of each store on regular salesmen's salaries.

Part Three. **Cost and Plant Location**

Introduction

1. In the early stages of the drug chain study, Robert Myers of the Bureau of Labor Statistics called my attention to this relationship and its possible usefulness here. The bureau had studied the effect of city size on salaries for variety-store jobs. It provided data for measuring this relationship for other merchandising salaries.

2. This concept was implicit in the excellent pioneering work on wage contours done by the Army Service Forces under the direction of Dean C. C. Balderston. (C. Canby Balderston, "Wartime Lessons in Wage Administration," *Advanced Management,* January–March 1945.)

Study 7. Drug Chain

1. This paper presupposes that all jobs—from porter to president—have been analyzed, measured, and classified into a series of numbered job grades. Such a process is a necessary prelude to the basic approach.

2. By "pay scale" we mean the series of basic wage rates that apply to the whole ladder of jobs which have been ranked according to differences in job content, i.e., responsibility, skill, difficulty, etc. The problem here is not to determine the job ladder but to find out what pay scale should be applied to this ladder in different localities.

3. The short cut developed in the analysis has not been fully validated by pricing all jobs in all localities and comparing results with those of the short cut. Such a procedure, though obviously desirable, was quite impractical because of its tremendous cost.

4. I am indebted to Mr. Robert Myers and his staff of the Division of Wage Analysis of the Bureau of Labor Statistics for making these data available and for assistance in their analysis and interpretation.

5. In developing these differentials, satellites that were in metropolitan districts or near large cities were excluded. In applying the results, small satellite cities can be treated as suburbs.

It has been suggested that jobs of the same title will probably have different difficulty and content in large cities than in small. This is possible; but the job descriptions sought to prevent it. Moreover, most of the jobs were simple and highly standardized.

6. Although there are not sufficient data regarding the upper reaches of the salary curve, the simplification of a uniform differential appears sound. Its effective use is limited to the range ending at Job Grade X, because of other aspects of the approach explained later. Hence, it did not appear to be worth the cost of collecting data specifically for the purpose of verifying this conclusion for higher-level jobs. Such jobs are more difficult to compare, and therefore, large masses of data would have to be collected in order to develop averages that would be really significant statistically.

7. It should be emphasized that this conclusion was based on fragmentary data and that it applies only to the industry and the time period studied. But even if subsequent development should cause differences among regions in this city-size pattern, our basic

method would not be invalidated. It would need only to be elaborated by separate adjustment for size of city in regions that differed significantly in this respect.

8. In principle it would be desirable to include other industries that compete for labor of the same type. In practice the difficulty of finding key jobs of known comparability makes inclusions of nonmerchandising establishments difficult.

9. Appropriate representative positions, accordingly, should meet these three tests:

1. They should represent as complete a coverage of the entire salary scale as possible. The positions should also be appropriate bench marks from which salaries for other positions are computable.

2. They should be selected from jobs for which the forces of supply and demand operate effectively in the labor market. There should be a large number of persons throughout industry employed in that job, with no significant restraint on new people wishing to undertake the job.

3. They should be relatively easy to describe accurately and to verify for what people really do. And they should be comparable as to requirements and job content among companies for whom market-price data are obtained.

10. In marking out the boundaries of the metropolitan district, the Census Bureau used two standards: population density and contiguity. To be included, a suburb must have a specific number of persons per square mile and must also be contiguous with the political city or with another suburb that meets both standards. Thus, a metropolitan district is an area that is essentially urban and constitutes a continuous economic unit.

11. The special problem caused by suburbs appears to be significant only in the case of the largest cities. The compensation level in the suburbs of cities of over a million population is about 5 per cent lower than that in the central city. The methods of arriving at this suburban differential, of applying it to rectify homogeneous area data, and of adjusting findings in the final salary schedules are all very similar to those already discussed for city-size differentials. Hence, they are not described in detail here. In cities of less than a million population the pay-level differential between suburb and city appears from the available data to be too small for administrative significance.

12. Cities for which suitable salary data are already collected and those for which such information will be available in future on a continuing basis are likely candidates; but this is not a necessary or a sufficient condition for selection.

13. Detailed information about all kinds of supplementary compensation was obtained, with the aim of developing a coherent system of salary supplements that would also be tuned to market facts. Hence stripping the base pay of these "fringe" perquisites did not ignore or neglect them. It merely set them aside for separate treatment.

14. It offers greater flexibility in the exercise of judgment as to the quality and relative importance of the data. And it provides greater flexibility in the shape of the curve, since the cost and complexity of mathematical curve-fitting usually confines the analysis, for practical reasons, to straight lines.

15. This does not imply that there was no dispersion beyond Grade X, but rather that it was not systematically related to geography for the higher jobs. This convergence tendency may be peculiar to the industry studied, and the job level where geographical impacts weaken is likely to differ. We are not propounding a universal law of Grade X convergence, nor is the validity of this approach in any way dependent upon the existence of such a tendency.

Part Four. **Cost and Real Profits**

Introduction

1. Existence of net long-term corporate debt merely means that the market's valuation of the company is the sum of the price of the stock and the market value of the bonds (plus other net debt). The year-to-year increment in this aggregate value—after allowance for capital transactions—is a market measurement of the corporation's yearly earnings in dollars of fluctuating buying power. Deflating them with the consumer price index gives an estimate of constant-dollar profits.

2. John Kenneth Galbraith, *Economics and the Public Purpose* (New York: Houghton Mifflin Co., 1973), pp. 44ff.

3. If there were an active market for used computers of all kinds, the decline in market price would be an appropriate measure of capital consumption for IBM because buying and selling on this market would be an alternative both for IBM and for its customers, becoming a determinant of their actions on the leasing front. The price in such a market would, in principle, be determined by contending forces.

For IBM equipment, the critical determinant of the market value of a used computer is technological advance, as embodied in new equipment that obsoletes the existent computer by superiorities of capacity, speed, cost, etc., which make it cheaper per unit of service to the customer. Although obsolescence-life is calibrated by time, it is a function not of time but of future unborn technological advances. Other forces on the demand side would be (1) the decline in the present worth at a given rate of interest of the stream of future rental receipts of the remaining lease life or economic life of the machine; (2) the rise in the rate of capitalization caused by inflation, which would cause a further decline in present worth of the given stream; (3) the shrinking buying power of this falling present worth. On the supply side three forces would generally work in the opposite direction: (1) reluctance of customers to part with machines with bargain rentals compared with the prices of new equipment; (2) higher costs of creating new machines of that design; and (3) higher costs and a higher birth-rate criterion for commercializing equipment of advanced design which would obsolete that machine.

Clearly these forces would differ in power and in length of lags. The beauty of an active market, such as we are envisaging here, is that it would measure impersonally the product of these forces and lags and would be the reality confronting both buyer and seller.

Study 8. Electronics Producer

1. Solomon Fabricant, *Capital Consumption and Adjustment* (New York: National Bureau of Economic Research, 1938), p. 183.

2. Deflating depreciation charges in this manner is algebraically equivalent to applying the book *rate* of depreciation to the deflated gross-value account. This can be seen easily if it is recalled that deflation of a quantity by a price index consists of dividing the index into the quantity, or multiplying the quantity by the reciprocal of the index.

3. If these transactions were deflated from the viewpoint of investors in the company, the value of the amounts might be measured by the uses to which investors could put the funds. The real-earnings computation, however, is made from the viewpoint of the company.

Index